BIODIVERSITY ENRICHMENT IN A DIVERSE WORLD

Edited by **Gbolagade Akeem Lameed**

INTECHOPEN.COM

Biodiversity Enrichment in a Diverse orld
http://dx.doi.org/10.5772/3088
Edited by Gbolagade Akeem Lameed

CBS, Edition 2016

Contributors

Jessica P. Karia, A. G. Lameed, Jenyo-Oni Adetola, Rosario Tejera, María Victoria Núñez, Ana Hernando, Javier Velázquez, Ana Pérez-Palomino, Benjamin L. Allen, Peter J.S. Fleming, Matt Hayward, Lee R. Allen, Richard M. Engeman, Guy Ballard, Luke K-P. Leung, Analía Alvarez, Flavia del Valle Loto, Yoshitaka Oishi, Thiago Metzker, Tereza C. Spósito, Britaldo S. Filho, Jorge A. Ahumada, Queila S. Garcia, I. A. Orellana, E. Raffaele, Hector Duarte Tagles, Alvaro J. Idrovo, Suzane M. Fank-de-Carvalho, Sônia N. Báo, Maria Salete Marchioretto, Stefanos Kalogirou, Ernesto Azzurro, Michel Bariche, Andrzej Kędziora, Krzysztof Kujawa, Hanna Gołdyn, Jerzy Karg, Zdzisław Bernacki, Anna Kujawa, Stanisław Bałazy, Maria Oleszczuk, Mariusz Rybacki, Ewa Arczyńska-Chudy, Cezary Tkaczuk, Rafał Łęcki, Maria Szyszkiewicz-Golis, Piotr Pińskwar, Dariusz Sobczyk, Joanna Andrusiak, Gabriella Buffa, Edy Fantinato, Leonardo Pizzo, Jing Tan, Tian Yan, Wang Shang-Wu, Jie Feng, Heimo Mikkola, Guilherme Fernando Gomes Destro, Tatiana Lucena Pimentel, Raquel Monti Sabaini, Roberto Cabral Borges, Raquel Barreto, Ylo Joann Systra, Mohd Nazip Suratman, Marcos del Pozo-Baños, Jaime R. Ticay-Rivas, Jousé Cabrera-Falcón, Jorge Arroyo, Carlos M. Travieso-González, Luis Sánchez-Chavez, Santiago T. Pérez, Jesús B. Alonso, Melvín Ramírez-Bogantes

Published by InTech

Janeza Trdine 9, 51000 Rijeka, Croatia

Publishing Process Manager Oliver Kurelic

Technical Editor InTech DTP team

Cover InTech Design team

Additional hard copies can be obtained from orders@intechopen.com

Biodiversity Enrichment in a Diverse World, Edited by Gbolagade Akeem Lameed
p. cm.
ISBN 978-953-51-0718-7

INTECH

open science | open minds

Contents

Preface

This book - ***Biodiversity Enrichment in a Diverse World*** - considered biodiversity (plants, animals, fungi, and microbes) from three different angles: genetics, species, and ecosystems. The relationships between them are complex and it looks at these aspects from different angles and also various interventions at different levels. The scientific approach of the book demonstrates that the three levels are closely interconnected and action is therefore needed to conserve and protect the systems if the benefits provided to human life will continue to be available. However, conservation of the biological diversity is essentially an umbrella term for traditional species, relationship to human health, ecosystem conservation and the need to manage the human use of the species and ecosystems in a sustainable way.

Gbolagade Akeem Lameed
University of Ibadan,
Nigeria

Section 1

Ecological Section

Chapter 1

Floral and Avifaunal Diversity of Thol Lake Wildlife (Bird) Sanctuary of Gujarat State, India

Jessica P. Karia

Additional information is available at the end of the chapter

http://dx.doi.org/10.5772/50073

1. Introduction

Wetlands are the ecotonal or transitional zones between terrestrial and aquatic ecosystems where the water table is usually at or near the surface of the land, which is covered by the shallow water (Mitsch & Gosselink, 1986). Due to these characteristics, wetlands provide opportunities for adaptations to different plant and animal species with high diversity of life-forms. Thus wetlands are among the most biologically diverse and productive ecosystems on earth. Wetlands can further be classified by one or more of the following attributes: (a) at least periodically, the land supports hydrophytes, (b) the substrate is predominantly undrained hydric soil, and (c) the substrate is saturated with water or covered by shallow water at some time during the growing season each year.

As per the convention on Wetlands of International importance (RAMSAR) (1971) – Article 1.1: wetlands are "Areas of marsh, fen, and peat land or water whether natural or artificial, permanent or temporary with water, that is static or flowing, fresh, brackish or salt including areas of marine water the depth of which does not exceed 6 meters." Also according to Article 2.1: "[Wetlands] may incorporate riparian and coastal zones adjacent to the wetlands, and islands or bodies of marine water deeper than six meters at low tide lying within the wetlands".

The values of the World's wetlands are increasingly receiving due attention as they contribute to a healthy environment in many ways. They help to retain water during dry periods, thus keeping the water-table high and relatively stable. During periods of flooding, they act to reduce flood levels and to trap suspended solids and nutrients directly flowing into the lakes. The removal of such wetland ecosystems because of urbanization or other factors typically causes lake water quality to worsen. In addition, wetlands are important feeding, breeding, and drinking area for wildlife and provide a stopping place and refuge for waterfowl. As with any natural habitat, wetlands are important in supporting species diversity and have a complex and important food web. The recent millennium assessment of ecosystems puts freshwater biodiversity as the most threatened of all types of biodiversity.

The interaction of man with wetlands during the last few decades has been of concern largely due to the rapid population growth, accompanied by intensified industrial, commercial and residential development. Thereby leading to pollution of wetlands by domestic, industrial sewage, and agricultural run-offs as fertilizers, insecticides and feedlot wastes. The fact that wetland values are overlooked has resulted in threat to the source of these benefits. Apart from the above the absence of reliable and updated information and data on extent of wetlands, their conservation values and socioeconomic importance has greatly hampered development of policy, legislation and administrative interventions by the state.

Fortunately in the recent years, the wetlands have received a good deal of attention. It really started with the conference held in Ramsar in Iran in 1971 where the first listing of wetlands of international importance was made and the contracting parties agreed to take necessary steps to safeguard these wetlands for posterity. India, as one of the original signatories, has made impressive efforts in initiating work for conservation and management of wetlands.

2. Indian scenario

India by its unique geographical position and with its annual rainfall of over 130 cm, and its varied terrain, and climate ranging from the cold arid of Ladakh to the warm arid of Rajasthan, with a cost line of over 7500 km, with its major river systems and lofty mountain ranges has, no wonder, a wealth of wetlands.

In addition to the various types of natural wetlands, a large number of man-made wetlands also contribute to the faunal and floral diversity. These man-made wetlands, which have resulted from the needs of irrigation, water supply, electricity, fisheries and flood control, are substantial in numbers. The various reservoirs, shallow ponds and numerous tanks support wetland biodiversity and add to the countries wetland wealth.

It is estimated that freshwater wetlands alone support 20 per cent of the known range of biodiversity in India (Deepa & Ramachandra, 1999). Wetlands in India occupy 58.2 million hectares, including area under wet paddy cultivation (Directory of Indian Wetlands).

Of about 35 Protected Areas (PAs) of India, which have been specifically notified for bird conservation, seven are in Gujarat (Grimmett et al. 1998). The State also falls within the Indus flyway a route that extents along the Indus valley from Pakistan to northwest India. This flyway is highly used by birds migrating from their breeding grounds in the Palearctic realm (Grimmett et al. 1998). The World Conservation Union (IUCN), International Wetland and Waterfowl Research Bureau (IWRB) and Birdlife International have rated this passage as the fourth major bird migration flyway in the world (Grimmett et al. 1998).

Gujarat is the State where the wetlands cover 27.1 lakhs hectares, a sizable area out of the total geographical area of the State. Of the total wetland area, inland wetlands cover 7.7%

and coastal wetland covers 92.3%. In coastal wetlands maximum area is under tidal flats/mud flats and the main contribution is from Great and Little Rann of Kachchh i.e. 1,930,581 ha. Analysis of the natural and man-made categories of wetlands indicate that, of the coastal wetlands, only 1.83% (mainly salt pans) is man-made, while in case of inland wetlands man-made wetlands account for 76.39% area.

Thol is one such man-made inland wetland situated in Mehsana district which is one of the top food grain producing districts in Gujarat (Anno. 1975). This marks the presence of well developed irrigation system consisting of wells and irrigation tanks. Thol water body is irrigation tank originally constructed in 1912 by the Gayekwadi State Rulers, built to prevent erosion and flooding and to store rainwater for irrigation purpose (Vaghela, 1993). Initially the area was declared as "Game Reserve" vides Government notification dated 29th May 1986 by Forest and Environment Department. Later on, due to its popularity amongst the bird fraternity, the area was notified as Bird Sanctuary through the notification GVN-53-88-WLP-1386-162-V.2 dated 18th November, 1988 under Section 18 of Wildlife (Protection) Act, 1972 (Anno. 2001).

Thol lake Wildlife Sanctuary which is now known as Thol Bird Sanctuary (TBS), as a part of conservation and management of Thol wetlands the biodiversity was studied to implement the Action Plan of Thol lake wildlife (Bird) sanctuary. This information will be comprehensive for preparing the management plan of the Sanctuary.

3. Study area

Thol Bird Sanctuary is situated in Mehsana district of Gujarat state, India between 23° 15′ to 23° 30′ N latitudes and 72° 30′ to 72° 45′ E longitude. It is a shallow water reservoir situated 25 km northwest of Ahmedabad and most popular birding place near Ahmedabad from Nal Sarovar Bird Sanctuary which is about 50 km away. Geographically, Thol Wildlife Sanctuary falls in the Kadi taluka of Mehsana district, North Gujarat region. Kadi taluka is head quarter of the district which is just 22 km away from the Sanctuary (Figure 1).

3.1. Salient features

Thol water body occupies a total area of 699 ha (6.99 sq.km.) and its periphery is 5.62 km long. Thol wetland catchment area is spread within six villages i.e. Thol, Jethlaj, Adhana, Vayana, Chandanpura, Jhaloda, which spreads 55.95 sq.km. It has well-developed canal based irrigation system. There are four head regulators at the water body to control the flow of water. The canals and their distributaries / sub-distributaries are about 19.97 km long.

The catchment area of the water body which covers 320 sq.km is located to its north and north-east so the spread is from Kadi taluka of Mehsana district and Kalol taluka of Gandhinagar district. These areas have seven small or big industrial areas they are, Karoli, Saij, Wamaj, Kalol, Chhatral, Indrad and Rajpur (Information from INDEXTb, Industrial Extension Bureau, Gandhinagar). Water finds its way through a number of canals draining into the feeder canal located on the north to northeastern sides of the water body. Water is received through Eastern canal, Saij-Hajipur canal, Irana-Indrad-Wamaj canal, Hajipur-Piyaj canal, Eastern feeder at Saghan drain, and Jaspur canal at Thol water body.

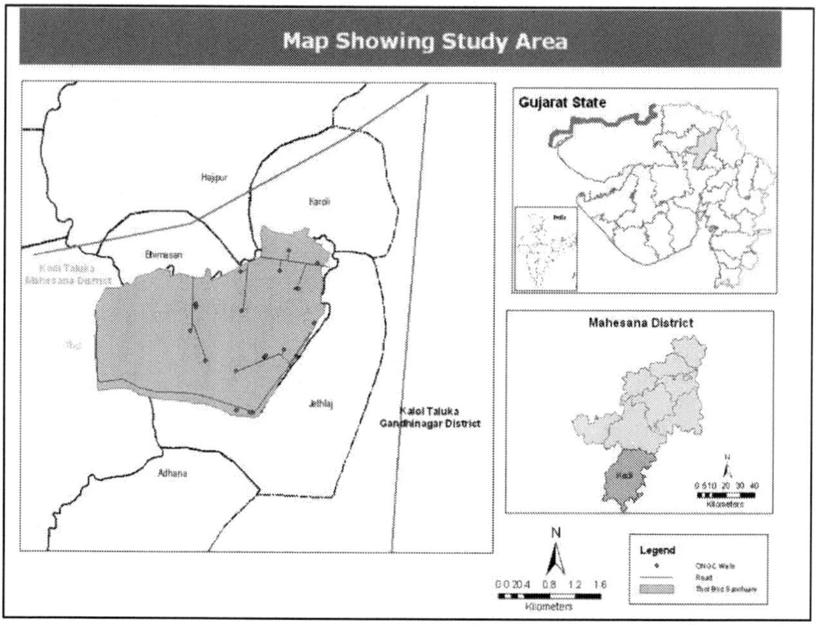

Figure 1.

In addition to the feeder canal, the water body receives run-off water directly from the catchment area. Before the feeder canal reaches the manmade wetland, there is a diversion, which is known as waste weir and is employed to control the volume of water in the water body. If the level of water reaches beyond 9 ft., the water is diverted to waste weir. Waste weir drains into a canal, which runs along the eastern boundary of the Thol pond/tank to reach Nalsarovar Bird Sanctuary located southwest of Thol Bird Sanctuary. Thol and Nalsarovar Bird Sanctuary are thus connected with each other.

There are no villages and settlements inside the sanctuary. Majority of the population is engaged in farming either as landholders or labourers. Also there are oil wells belonging to the public sector company Oil and Natural Gas Commission (ONGC) within the sanctuary area. There are total 21 number of wells among which 13 are functional. Polymer injection wells are 3 in number and Chase water wells are 5 in number. The total oil production from Thol area wells is 102 tpd.

3.2. Geology

Geologically, it is a part of the alluvial plain of recent age. The soil is clayey to sandy clay. There are no hard rock outcrops in and around the sanctuary.

3.3. Climate

Thol area experiences three distinct seasons namely winter (November to February), summer (April to May) and monsoon (June to September). Months of October and March mark the transition period from monsoon to winter and winter to summer respectively. The pond receives rainfall from July to September through the southwest monsoon. Old records for Mehsana district in general (Anno., 1975), as well as rainfall data of previous years at TS indicate that the rainfall is highly erratic and ranging from 189 to 786 mm.

4. Methodology

4.1. Land use / Cover studies

The methodology employed for preparation of Land use and land cover map included:

- Data collection
- Interpretation of satellite data
- Ground truth study
- Final map preparation

4.1.1. Data collection

- Downloading of Satellite imagery using the licensed software, Google Earth Pro having high resolution (<1.0m) data.
- Topographical maps as base map.
- Quick reconnaissance survey of the study area to get a feel of the entire ground area which can aid in the preliminary interpretation of the data.

4.1.2. Interpretation of satellite data

The downloaded satellite imagery was imported to Arc GIS 9.3 software and georeferencing of the imagery was done by registering it to the SOI maps through identification of common points between the map and the image. Considering the basic elements of interpretation such as tone, size, shape, texture, pattern, location, association, shadow, aspect and resolution, along with ground truth and ancillary information collected during the preliminary reconnaissance survey the interpretation was accomplished.

4.1.3. Ground truth study

A detailed ground truth was carried out to check the discrepancy of the interpreted data. It comprises of data collection of ground features along with the respective geographical position in terms of latitudes and longitudes.

4.1.4. Final map preparation

The interpreted file was then projected with Universal Transverse Mercator, which is universally followed projection system. The proportional presence of different land uses and land cover in terms of statistical percentages was derived for the study area. Appropriate legends were used to represent the various categories of land use and land cover, and were then written on the prepared land use and land cover maps. Based on interpreted map floral and faunal sampling site was selected so that the entire area will be covered.

4.2. Vegetation cover

The phyto-sociological studies were carried out using quadrant method with in terrestrial vegetation covered region. Quadrate plots were laid in triplicate at each selected locations. Density, frequency, abundance and dominance and their relative values were calculated along with IVI values (Ambasht, 1990). The basal area was calculated by formula using diameter at breast height (Ravindranath & Premnath, 1997). Secondary analyses like different indices were calculated using this primary data (Odum, 1983).

The lower side of embankment had species diversity within this area the phytosociological studies were done. The grass cover region along the sanctuary boundary and on the beyts was surveyed and the herbs growing in this region was enlisted. The enlisting of the aquatic floral species like floating, emergent and submerged species had also been done.

4.3. Avifaunal studies

Avifaunal diversity studies sampling location was decided based on the water level and distribution as seen from interpreted satellite data. Observations were done by conducting field visits at regular intervals. Field works were conducted during winter season by visiting the place thrice in a season mainly from 0600 hr to 1200 hrs in the morning. The observations were recorded using field binocular (Pentax 10x50) and identified on basis of standard field guides like Grimmett *et al.* 1998, Salim Ali, 2002. This was done for both waterfowls and surrounding terrestrial birds. The bird diversity was classified according to its Order & Family, and their migratory statuses were noted.

4.4. Correlation between bird diversity & macrophytes

The relationship of the availability of bird diversity and macrophytes growing in the area was studied using statistical correlation method. The number of bird diversity distributed between the six sampling location and the available macrophytes diversity was documented.

5. Results

5.1. Land use / Cover studies

Visual interpretation of satellite data categorized area into five classes, they are shallow and deep water covered area, among terrestrial area it had been classified as vegetation cover, scrub land and agriculture land (Figure 2). The major portion of the sanctuary geographical area is covered by scrub land i.e. 36 per cent followed by 27 per cent of agricultural land. The category wise percentage area is as given in Table 1.

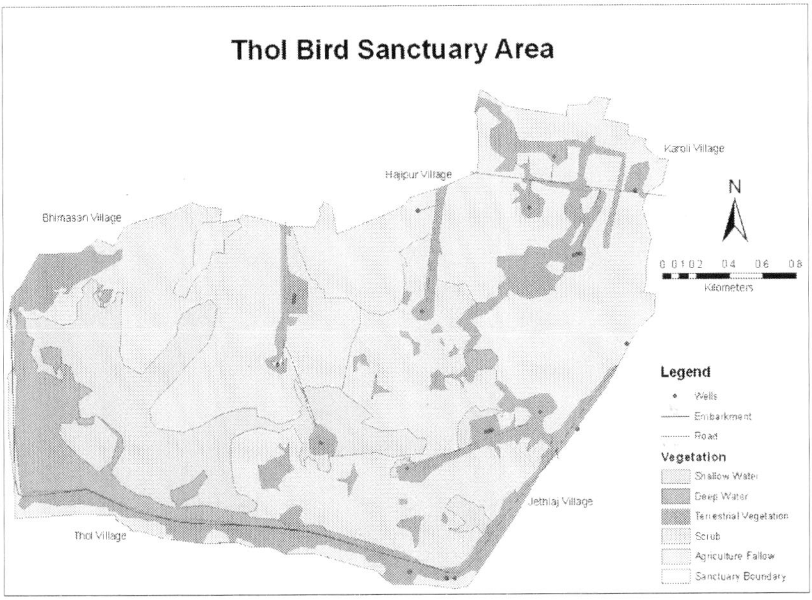

Figure 2.

Sr. No.	Class	Area (%)
1	Shallow water	15.08
2	Deep water	6.36
3	Vegetation Cover	15.69
4	Scrub	36.06
5	Agriculture	26.81

Table 1. Land use/cover Area Statistics of TBS

Each category is specific on its own as described below.

a. Shallow water: It could be delineated based on the light tone on the satellite data. Shallow water region was having less than 1 foot, as correlated on the ground. This was on the western side of the sanctuary boundary.
b. Deep water: It had dark coloured tone and smooth texture on the satellite data by which it was delineated as deep water. This was water filled region having more than 2 feet, as correlated on the ground. Its was on the western region or the corner of the sanctuary area.
c. Vegetation Cover: This was only 15 per cent of the area. This category will be described in details in following section.
d. Scrub: Most of the sanctuary area was covered by scrub, it is described as area coverage with less than 10 % of the canopy density (FSI, 2011) i.e. with scattered tree species and undergrowth dominated area.

e. Agriculture: This was delineated based on the square patterns as seen on the data. The major crop grown in the region was Wheat, Juwar, and Bajra with water source as canal, bore or rainfed.

5.2. Vegetation cover

According to the vegetation map prepared using the satellite image of the Thol bird sanctuary area. There are three main patches of terrestrial vegetation first towards the Bhimasan village, on the north-east region along the water inlet to the Thol water body and lower side of the embankment. The vegetation towards the Bhimasan village was of monoculture type i.e. the plantation of *Acacia nilotica,* done by forest department. On the north east region along the water flow there was dominance of *Ipomoea fistula* and *Acacia nilotica* (Baval) vegetation. Lower side of the embankment had comparatively more species diversity where the phyto-sociological studies where done. Apart from this there were some patches of terrestrial vegetation which was again dominated by the planted species *Acacia nilotica* (Baval).

Phytosociological studies in the mixed vegetation type on the southern side of the sanctuary area showed presence of few species only. The highest abundance was of the *Acacia nilotica* (Baval) plantlet and its tree species had the highest IVI which showed that there was good regeneration of this species (Table 2). The understorey vegetation in this region was very less.

Sr. No.	Species	Abundance	IVI
1	*Acacia nilotica*	9.5	187.50
2	*Acacia planifrons*	2	22.80
3	*Zizyphus mauritiana*	2	19.67
4	*Azadirachta indica*	2	21.02

Table 2. Plant Species Status in Mixed Vegetation

The index calculated from the field data showed the dominance index to be greater than 0.5 indicating that one or two species contribute very highly in the community. Also from the high evenness index it could be judged that there is even distribution of the species (Table 3).

Index	Value
Simpson Dominance Index	0.67
Shannon-Wiener Diversity Index	1.03
Evenness Index	1.12
Species Richness Index	0.83

Table 3. Vegetation indices estimated from Mixed Vegetation

Along with this the enlisting of the species diversity within the sanctuary boundary was done, which showed the presence of 88 plant species (Annexure 1, see Appendix). It includes herbs, shrubs, grasses and hydrophytes species. There were in all 12 floating, emergent and submerged hydrophytes species.

5.3. Avifaunal diversity

Bird diversity recorded from Thol during present study were 144 in number out of it 76 are waterfowl rest are terrestrial birds (Annexure 2, see Appendix). The enlisted birds within the sanctuary area had 9 no. of rare and endangered species according to the Red Data Book (Table 4).

Within waterfowl there are members from 21 families and the family Anatidae had the most members i.e. 15, followed by Ardeidae having 11 members rest of the family have less then 10 members. This indicates that the ducks and geese are the dominating species, followed by herons, egrets and bitterns. Anatidae family members are mostly resident migratory or migratory species, only Comb duck is resident species. Resident migratory species are five i.e. Mallard, spot billed duck, bar headed goose, white eyed pochard and ruddy shelduck.

Terrestrial birds also had members from 21 numbers of families, and the Accipitridae family had highest number i.e. 10 members, followed by Cornidae having 9 members. These shows the dominating diversity within terrestrial area surrounding the water body are shikara, kite, eagle, vulture, buzzard, osprey and besra.

The statistics of residential status of species indicates that within aquatic birds diversity the highest number of species are resident-migratory i.e. 40 % while there are 33 % of resident species and 27 % of the migratory species. While within terrestrial birds species highest is of resident species having as high as 76 %, next is resident-migratory species of 21 % and just 3 % of migratory species.

Sr.No.	Scientific name	Common name	Migratory Status	Threatened Birds
1	*Anhinga melanogaster*	Oriental Darter	RM	NT
2	Mycteria leucocephala	Painted Stork	R	NT
3	*Phoenicopterus minor*	Lesser Flamingo	RM	NT
4	*Aythya nyroca*	White-eyed Pochard	RM	NT
5	*Threskiornis melanocephalus*	Oriental White Ibis	R	NT
6	*Grus antigone*	Sarus Crane	R	V
7	*Aquila heliaca*	Imperial Eagle	RM	V
8	*Aquila clanga*	Greater Spotted Eagle	RM	V
9	*Pelecanus philippensis*	Spotbilled Pelican	RM	V
NT - Near Threatened, V – Vulnerable, R – Residant, M – Migratory, RM- Resident-Migratory.				

Table 4. List of Threatened Birds in Thol Bird Sanctuary

The habitat requirement of the waterfowl inhabiting in Thol were studied. The details are as given in the Annexure 3. The foremost requirement is that of the open water both deep and shallow waters, in terms of percentage 47 % of birds which includes all the members of the dominating family Anatidae this habitat was used. Birds inhabiting in the muddy habitats are 22 % this includes some heron & egret, plovers, godwit, greenshank and sandpiper. Thereafter 16 % of the birds require emergent vegetation habitat, followed by 7 % shoreland, 6% agriculture and fallow land and 3 % of wooded area. Among the agriculture & fallow land habitat the dominating species is Sarus crane which comes under the vulnerable status. Among this there are overlapping of the use of habitat as per the birds resting, roosting and foraging habits.

5.4. Correlation between bird diversity & macrophytes

It has been observed that there was lot of variation in the floral and faunal diversity within selected six locations (Table 5). The sampling locations were selected based on the difference in the availability of water and the congregation of birds found in the region and the accessibility of the region. The sampling location P6 had presence of more floral diversity and location P1 had more of bird diversity.

Sample No.	Location	Longitude	Latitude	Remarks	Macrophyte Diversity	Bird Diversity
P1	Towards Bhimasan village	72°23′44.6″E	23°08′34.7″N	Shallow water nearly half foot	4	27
P2	Check Post side	72°23′35.0″E	23°08′25.8″N	Deep water around 2 feet	2	17
P3	Middle region	72°23′55.2″E	23°08′26.0″N	Shallow water	2	15
P4	Camp site	72°24′09.0″E	23°08′09.7″N	Muddy Area, no disturbance	2	16
P5	Towards Jetlaj village	72°24′50.3″E	23°07′52.6″N	Emergent Vegetation in pockets	4	11
P6	Towards ONGC well No. 30	72°24'41.0"E	23°08'13.6"N	Small ponds of water, gets flooded during monsoons, less biotic disturbance	6	23

Table 5. Floral and Faunal Diversity

6. Discussion

6.1. Land use / Cover studies

Land use and land cover classification is an essential prerequisite for any management operation as it is a direct indicator of the ecology of the area. This is particularly important to identify what kind of habitats in relation to the water level is formed and to find out habitat preferences of various species of waterfowl. Habitat is the natural home of any living form, may be an animal or a plant. Mc Farland (1980) suggested that birds respond to a summation of many factors and habitat selection thus, has some variability within a species. According to him a. characteristics of the terrain, b. nesting, feeding and drinking sites, c. food availability, d. other animals, are important factors influencing habitat selection. Therefore, identification of various habitat types are important factors influencing habitat selection.

The statistics reveals that there is more availability of shallow water habitat which in due course of time will be muddy region once water gets dried out. This study is very dynamic, thereby changes the types of birds visiting the wetlands. Also the availability water depends on the rainfall and the irrigation system. So this study needs to be conducted along with the bird census on regular basis.

6.2. Vegetation analyses

Macrophytes occurring in the area clearly indicate habitats and condition prevailing in the area of its occurrence. The habitat in the study area is mostly muddy and also it is saline, as indicated by sediment analysis (GEC, 2009).

Submerged rooted aquatic vegetation in Thol water body was of *Vallisneria spiralis* found near the Bhimasan village site and *Hydrilla verticillata* on the south eastern side near the ONGC well no. 30. The *Najas graminea* was found to be grown in abundant with *Hydrilla verticillata* and the *Potamogeton sp.* in the waters of Narmada Canal reaching TBS. Also on the sedges of the Narmada canal water there was growth of emergent hydrophyte *Typha* sp., and floating *Paspalum sp.* This indicates the presence of nutrients content in the Narmada canal water. While in sanctuary area waters such abundant growth of submerged hydrophytes was not seen. The bed of Thol water body was covered with the grass *Cynodon dactylon* (Darba) and the free-floating hydrophyte *Ipomea aquatic* (Nala ni Vel) on the south-western corner of the sanctuary. *Cynodon dactylon* (Darba) shows salt tolerance capacity and at the same time these are nutritious and palatable species. Rooted floating weeds *Nelumbo lutea* (Kamal) was seen to cover the small portion of the water body. On the check post side of the pond the growth of grass on the sedge was seen that of *Eragrostis sp.* Thus the reeds and sedges provide resting, rooting or nesting habitat for many species apart from providing an excellent cover, too many birds which take shelter in such habitats. In the middle portion of the sanctuary area from northern side, which is less disturbed site had the presence of free floating *Lemna* (Kaye) sp. on the edge the Amarantheceae member herb *Alternanthera sessilis.*

On the small bets, there appear mostly abundant *Ipomoea fistulosa* and in others bets *Acacia nilotica* (Desi baval) tree. On the other side of the waterbody i.e. on south eastern side near the ONGC well no. 30, there along with *Acacia nilotica, Parkinsonia accuminata* was found. These trees are extensively used by egrets, black ibis, crows, doves etc. for roosting. They are also used for nesting by crows, doves etc.

It was observed that on the south eastern side near the ONGC well no. 30 there appears reed meadow sedge, a seral stage where due to siltation, sedges, grasses grow abundantly. These include *Cyperus* sp. which play an important role of air circulation in the lake, as they are hollow and possess aerenchymous tissues. They help in gaseous exchanges of carbon dioxide and oxygen, which are thus made available to the submerged species. Also there was growth of *Polygonum sp.* on the sedges in this area. Emergent *Scirpus* sp. was also found in this region on the sedges, it has advantage to inhibit soil erosion and provide habitat for other wildlife. The plant rhizomes have medicinal value. So in this area we find diversity of macrophytes indicating the quality of water in Thol wetlands.

Major part of the Thol wetland sanctuary area is covered with scrub area. In this area there was sparse distribution of *Acacia nilotica* tree and the mesquite *Prosopis juliflora* (Gando baval) most of which are of shrubby appearance, seldom attaining a height of more that 5 meters. The ground is covered with grass *Cynodon dactylon* (Darba) and few herb species. The *Xanthium strumarium* (Gadariyu) an abnoxius weed appears at places along the shore and on some bets, which is indicative of excessive grazing in the area. This can be confirmed by the field survey in the area. The scrub area had more growth of herbs like the *Grangea maderaspatan* (zinki mundi), *Coldenia procumbens* (basario okharad), and *Glinus lotoides* (mitho okharad).

Among the tree species growing on the boundary of the Sanctuary were *Azadirachta indica* (Limdo), *Cordia myxa* (Gunda) and *Ailanthus excelsa* (Maharukh), etc. Apart from this on the southern boundary of the sanctuary area the natural vegetation grows where phytosociological parameters were studied. Micro level vegetational studies carried out aided to bring out sharp differences in the vegetation of these areas.

Each of the species within the community has a large measure of its structural and functional individualism and has more or less different ecological amplitude and modality (Singh and Joshi, 1979). This requires the understanding of the phytosociological status of each species within a community. Importance Value Index is a measure of plant status which brings out the overall role of a plant in a community (Ambasth, 1990). The study of phyto-sociology along with floristic composition proves useful in comparison of species from season to season and year to year (Singh, 1976). The study of vegetation its spatial distribution and analyses, and on field study indicates that the anthropogenic pressure had resulted in decrease in the undergrowth of the area. This would increases the possibility of the environmental stress i.e. soil erosion. This area shows the dominance of *Acacia nilotica* (Desi baval) with highest IVI of 187.5 and the dominance index. With the changing environmental conditions, the vegetation may reflect changes in structure, density and composition as observed by Gaur, (1982). The high evenness index shows the even distribution of vegetation in the community. It could be found out from survey that there is decrease in the undergrowth since it gets subjected to more anthropogenic pressure.

6.3. Avifaunal diversity

Bird communities are often referred as an ideal indicator to monitor the ecological condition of any wetlands as they impact on all the trophic levels of an aquatic ecosystem. On the other hand aquatic ecosystems have significant impact on migratory birds. Birds carry out, diverse ranges of ecological functions among vertebrates. As consumers, they help regulate populations of smaller animals they prey upon, disperse plant seeds, and pollinate flowering plants. As prey items, birds and bird eggs are consumed by a variety of larger predators.

Birds also benefit humans by providing important ecosystem services such as regulating services by scavenging carcasses and waste, by controlling population of invertebrates and vertebrate pests, by pollinating and dispersing the seeds of plants; and supporting services by cycling nutrients (Croll *et.al.*, 2005) and by contributing to soil formation (Post, 1998).

There are two birds which has been identified as flagship specis for Thol wetlands, being fresh water ecosystem, they are Sarus Crane (*Grus antigone*) and Osprey (*Pandion haliaetus*) since they represent the present ecosystem which is in need of conservation. They are distinctive in order to engender support and acknowledgement from the public.

- Sarus Crane (*Grus antigone*)

Sarus Crane is a large crane that is a resident breeding bird with *disjunct* populations that are found in parts of the Indian Subcontinent, Southeast Asia and Australia. Having height up to 1.8m, it is tallest of the flying birds; they are conspicuous and iconic species of open marshlands. As a species, the Sarus crane is classified as vulnerable this means that the global population has declined by about a third since 1980, and is expected to continue to do so until the late 2010. Estimates of the global population suggest that the population in 2000 was at best about 10% and at the worst just 2.5% of the numbers that existed in 1850 (BirdLife International, 2001). Unlike many cranes which make long migrations, the Sarus Crane does not; they may however make short-distance dispersal movements in response to rain or dry weather conditions. They tend to be more gregarious in the non-breeding season.

- Osprey (*Pandion haliaetus*)

Ospreys are sometimes known as the sea hawk, it is a large raptor, reaching 60 centimeters (24 in) in length with a 1.8 meter (6 ft) wingspan, is a resident-migratory species. They are widespread during winters in Indian Union, Bangladesh; Pakistan; Sri Lanka; Myanmar. Ospreys are diurnal, fish eating hawk, they flies up and down over the water scanning the surface for any fish coming up within striking depth.

Thol waterbody and surrounding area is most suitable habitat for Sarus, it can be appreciated from records that large number of Sarus congregations were seen. It has presence of over 50 birds feeding in the farmlands neighboring Thol, as late as 1998; the Sarus has remained the integral part of the avifauna of this territory (Singh & Tatu, 2000).

This shows that type of habitat is very important for wetland dependent species. Different species have different set of adaptations due to which they require certain types of habitats only. In case there is habitat loss in breeding areas it may directly result in loss of birds. Also the habitat is species specific and birds differ according to the habitat availability. Thus, the foremost requirement is identification of habitats in relation to various species of waterfowl.

TBS have variety of habitat which attracts many birds to the area. It was observed that dominating family Anatidae is having members like ducks and geese using open water habitat both deep and shallow. Thus the high usage of open water habitat explains why number of birds decrease with changes in the water spread and its level. Vijayan (1991) also reported preference of open water habitat over other categories by waterfowl at Keoladeo National Park. Anatidae group could truly be regarded as an indicator of the quality of habitat. As they depend on TBS for foraging, resting as well as roasting. Almost 60 per cent of Anatid members present are migratory and some species like the Whistling duck and Spot billed duck are potential breeders at TBS (GEER, 2002). Wetland could also be acting as a staging and dispersal area for the migrant ducks, which first arrive there and later spread to other smaller water bodies.

The migratory birds which come to TBS are coming mostly from northern and central Asia, Siberia and Europe or locally from Himalayas so their path is mostly north, north-east or north-west direction of TBS.

A total of 144 birds' species including 76 waterfowl and 68 terrestrial birds had been recorded at TBS during the study. The species diversity of waterfowl is similar to as recorded by Patel and Dharaiya, 2008 as 77 species. Species diversity was compared with other wetlands falling in semi-arid region like Wild Ass Sanctuary (Little Rann of Kachchh) and Nal Sarovar Bird Sanctuary. At Wild Ass Sanctuary (a seasonal fresh cum saline water protected wetland) Singh *et al.* (1999) had recorded 100 species of waterfowl (including wagtails and oriental pratincole) belonging to 18 families (as per old nomenclature). At Nal Sarovar Bird Sanctuary, Singh (1998) recorded about 113 waterfowl species. While Patel and Dharaiya in 2008 recorded 50 species of waterfowl. If we consider the area coverage Wild Ass Sanctuary is spread within 4953 sq. km., Nal Sarovar Bird Sanctuary spread within 120.82 sq. km. and TBS within 6.99 sq. km., if area is considered species diversity of TBS can be regarded remarkable.

The earlier study reveals that Nal Sarovar Bird Sanctuary which is just 50 km away, have high vegetation and faunal diversity compared to TBS (Patel, et al. 2006) due to different physical and hydrological configuration of largest natural fresh water reservoir. TBS and Nal Sarovar Bird Sanctuary are the valuable wetlands for migratory bird species. Moreover it can be also said that the Thol lake is more favored by the wetland obligatory birds. Since the study reveals that comparatively Nal Sarovar sanctuary had high disturbance score which indicates less healthy wetland for bird integrity than that of the TBS (Patel and Dharaiya, 2008). It was observed that towards the southwest direction of TBS is Nal Sarovar Bird Sanctuary, which is known to be one of the richest food crop (mainly paddy) growing areas, so there is continuous movement of birds between TBS and agricultural areas. This makes the birds visiting nearby village tanks and water bodies, which needs to be surveyed.

Thol has privilege of sustaining nine near threatened and vulnerable species. As reported by Chase et.al. (2000), presence of individual species may serve as indicator of the overall species composition of birds, but it may say less about the species richness, so the focus should be given to a diverse suite of the range of species representative for conservation purpose. The efforts should go in the line to conserve the threatened and lower risk species so that the population should not come down and they become extinct in near future. As per the red data guidelines they should be conserved when their populations are still healthy, before they become genetically impoverished and their populations gets fragmented. Out of nine vulnerable and near threatened species six are resident migratory species, and rests are resident species. Two species like eastern imperial eagle and greater spotted eagle are terrestrial birds and they are birds of prey.

6.4. Correlation between bird diversity & macrophytes

The enlisting of the bird diversity and availability of macrophytes in the region was subjected to statistical correlation which shows that there is positive correlation with 86 per cent of variance is related.

6.5. Avifaunal population trend in thol bird sanctuary

Avifaunal density trend was studied from the year 2000 to 2008. The year 2000 data was from the GEER foundation report 2002 while 2004 data was of Forest department; this was the first census of Thol bird sanctuary. Remaining two census data were taken of the year 2006 and 2008 conducted by Forest department. The trend changes in the population density of the birds found in the Thol bird sanctuary is as given in the table 6.

Sr. No.	Group of Species	2000#	2004*	2006*	2008*
1	Grebes	0	2	40	3
2	Pelicans	120	4	321	750
3	Cormorants & Darters	0	830	942	482
4	Herons & Egrets	1	479	485	210
5	Storks	4	83	236	95
6	Ibises & Spoonbills	19	768	183	5099
7	Flamingos	2	0	273	205
8	Geese & Ducks	419	1753	5599	7671
9	Cranes	525	380	664	1651
10	Reas, Crakes, Gallinules & Coots	0	21	943	552
11	Jacanas	0	0	0	0
12	Shorebirds & Waders	188	13839	8140	8120
13	Gulls, Terns & Skimmers	1	199	143	234
14	Kingfishers	0	10	15	25
15	Wagtails & Pipits	2	0	0	53
16	Eagles & Harriers	0	4	7	15
17	Total	1281	18372	17991	25165
# GEER, (2002), * Forest Department Census					

Table 6. Comparative Account of Birds Population (2000 to 2008)

It has been observed that over the years, there is increase in the population of the Pelicans, with sudden decrease in the year 2004. This growth can be attributed to the availability of the food; Pelicans mainly depend on the fish for food. It can be concluded that the forest department initiative of releasing fresh water fishes to the wetland was fruitful. Thus it could attract the migratory species to Thol wetlands.

The group Ibises & Spoonbill also shows the increase from just 19 numbers in year 2000 to 5,099 in the year 2008. The increase in population change of nil in year 2000 to 4,876 in year 2008 of Glossy ibis i.e. shore birds. The Gloosy Ibis requires the muddy habitat and they depend mainly on benthos for food. Thus over the years there is improvement of the food and availability of muddy habitat had increased Glossy ibis population. While, there is decrease in population of Eurasian spoonbill from 661 in 2004 to 187 in 2008. So it could be inferred that as there is habitat changes in the wetland ecosystem bird population changes. Reason could be that there is shift from water availability to muddy habitat availability.

Geese and Ducks group which had maximum diversity in Thol wetlands also shows the increasing trend from 2000 to 2008. This could is due to the increase in population of the migratory species Common teal to 4,769 in 2008. They are dependent on benthos as well as vegetation matter for food and require shallow water habitat.

Whereas the group Shorebirds & Waders shows the decreasing trend from highest of 13,839 in 2004 to 8,120 in 2008. This is largely because of decrease in migratory species Ruff from 13,345 (2004) to 5455 (2008), which is being compensated by the increase in population of Black tailed Godwit from 4 (2000) to 2,156 (2008). Ruffs are sporting birds they take larger quantities of weed seeds (Ali, 2002). Due to regulated supply of water for irrigation and developmental activities there is decrease in the agricultural fields and the availability of food for the species so there is negative change in the Ruff population. This year, Ruff species are not even noted, since as per the regulations due to construction work going on, the water supply was restricted causing the negligible population availability. This information was obtained from the forest officials of Thol wetlands.

Thus from the above discussion it can be concluded that due to adopted management practices there was overall increase in bird population of 1,281 (2000) to 25,165 (2008). But, definitely there was an overall change in the habitat causing the birds population to change accordingly. If we correlate the population of birds with the rainfall of the region it also had the increasing trend from 232 mm in 2000 to 786 mm in 2008 (Table 7). Rainfall in the year 2006 was slightly more as compared with 2008, but there was decrease in total bird population. This is probably due to favorable conditions prevailing in other wetlands also of the State during that period.

Looking at the avifaunal diversity it can be concluded that the Thol is the valuable wetlands for migratory bird species and it is more favored by the wetland obligatory birds because at Thol there is less human disturbance.

Sr. No.	Year	Rainfall (mm)
1	2008 - 2009	473
2	2007 - 2008	786
3	2006 - 2007	659
4	2005 - 2006	855
5	2004 - 2005	582
6	2003 - 2004	662
7	2002 - 2003	203
8	2001 - 2002	500
9	2000 - 2001	189
10	1999 - 2000	232

Table 7. Decadal Change In Rainfall Data of Thol.

7. Conclusion

TBS is important wetland of the western region as variety of migratory birds visit this wetland during winters. The study had identified the potential of TBS as an internationally important wetland due to species richness and home for nine near threatened and vulnerable species including endangered Sarus Crane, having pre-breeding congregations and nesting grounds.

It has been observed that though TBS is facing less human disturbance in comparison to Nal Sarovar Bird Sanctuary, there are certain threats if not controlled may increase. The foremost being the location of ONGC oil well within the sanctuary boundary and catchment area. It should be monitored regularly to check for oil spills or leaks as oil spills could be a threat for birds. Also the major portion of the sanctuary area is covered by agricultural region which is given to local people for cultivation at a meager rate. This activity causes disturbance to the birds. The withdrawal of water for irrigation which is through supply canals in command area and lift irrigation causes pressure to the wetland ecosystem.

Another major pressure on the Thol Bird Sanctuary is due to livestock population. Livestock of five peripheral villages as well as those belonging to the pastoral people from Kachchh and Saurashtra visit this area for grazing in scrub lands and for drinking water. The grazing pressure was confirmed by the field visit and the type of species growing in the region. The livestock includes goats, sheep, cows, buffaloes and camel which causes disturbance to birds. The forest department should manage TBS taking into consideration the mentioned threats.

Thus the present study has shown the importance of carrying out such a study on regular basis so as to monitor the changes of dynamic ecosystem due to concomitant changes in water regime at TBS. The study had a limited scope owing to its short span and was conceived only to document bird diversity. It is being suggested to carry out movement and dispersal pattern of migratory waterfowl. This can be extended to the neighboring villages' tank and water bodies which would enhance our knowledge about these winged visitors.

Author details

Jessica P. Karia
Enviro-GIS, Makarpura, Vadodara,Gujarat, India

Appendix

Sr. No.	Scientific Name	Vernacular Name
	Family: Mimosaceae	
1	*Acacia auriculiformis*	Pardeshi baval
2	*Acacia tomentosa*	Aniyar
3	*Acacia nilotica* ssp. *indica*	Baval
4	*Acacia farnesiana*	Talbaval
5	*Acacia planifrons*	
6	*Albizzia lebeck*	Siras
7	*Prosopis cineraria*	Khijado
8	*Prosopis glandulosa*	Gandobaval
9	*Prosopis juliflora*	Gandobaval
10	*Pithecellobium dulce*	Goras ambli
	Family: Malvaceae	
11	*Abutilon indica*	Kanski
	Family: Simaroubaceae	
12	*Ailanthus excelsa*	Maharukh
	Family: Amaranthaceae	
13	*Alternanthera sessilis*	
	Family: Meliaceae	
14	*Azadirachta indica*	Limdo
15	*Melia azadirach*	Bakan limdo
	Family: Ceasalpiniaceae	
16	*Bauhinia racemosa*	Asotri
17	*Cassia auriculata*	Aval
18	*Cassia fistula*	Garmalo
19	*Cassia italica*	Mindhi Aval
20	*Cassia occidentalis*	Kasundri
21	*Cassia siamea*	Kassod
22	*Cassia javanica renigera*	Pink cassia
23	*Cassia tora*	Kuvandio
24	*Delonix regia*	Gulmohar
25	*Tamarindus indica*	Ambli
	Family: Balanitaceae	
26	*Balanites aegyptiaca*	Ingorio
	Family: Papilionaceae	
27	*Butea monosperma*	Khakhro

Sr. No.	Scientific Name	Vernacular Name
28	*Crotalaria burhia*	Kharshan
29	*Derris indica*	Karanj
30	*Indigofera oblongifolia*	Zil, Ziladi
	Family: Poaceae	
31	*Cynodon dactylon*	Darba
32	*Cenchrus ciliaris*	Shukli
33	*Dicanthium annulatum*	Jinjavo
34	*Eragrostis sp.*	-
35	*Dactyolactelum aegypticum*	
	Family: Cyperaceae	
36	*Scirpus sp.*	
37	*Cyperus sp.*	-
	Family: Capparaceae	
38	*Capparis decidua*	Kerdo
39	*Capparis sepiara*	Kanther
	Family: Asclepiadaceae	
40	*Calotropis procera*	Nano akado
41	*Calotropis gigantea*	Akdo
	Family: Verbenaceae	
42	*Clerodendron multiflorum*	Arani
	Family: Boraginaceae	
43	*Coldenia procumbens*	Basario Okharad
44	*Heliotropium indicum*	Hathisundho
45	*Cordia myxa*	Gunda
	Family: Menispermaceae	
46	*Cocculus hirsutus*	Patalagarudi
	Family: Solanaceae	
47	*Datura metal*	Dholo Dhanturo
48	*Solanum xanthocarpum*	-
	Family: Euphorbiaceae	
49	*Euphorbia hirta*	Dudheli
50	*Euphorbia nivulia*	Thor
51	*Euphorbia obiculata*	-
52	*Phyllanthus reticulata*	Kamboi
53	*Ricinus communis*	Castor
	Family: Moraceae	
54	*Ficus benghalensis*	Vad
55	*Ficus religiosa*	Peepal
	Family: Asteraceae	
56	*Grangea maderaspatana*	Zinki Mundi
57	*Launaea procumbens*	Moti Bhonpatri
58	*Xanthium strumarium*	Gadariyu

Sr. No.	Scientific Name	Vernacular Name
	Family: Molluginaceae	
59	*Glinus lotoides*	Mitho Okharad
	Family: Ulmaceae	
60	*Holoptelea integrifolia*	Kanjo
	Family: Sterculiaceae	
61	*Helicteris isora*	Maradiya
	Family: Hydrocharitaceae	
62	*Hydrilla verticilliata*	-
63	*Vallisneria natans*	-
	Family:Convolvulaceae	
64	*Ipomoea fistulosa*	Naffatiyu
65	*Ipomoea aquatica*	Nada ni Vel
	Family:Lemnaceae	
66	*Lemna sp.*	Kaye
	Family: Sapotaceae	
67	*Madhuca indica*	Mahudo
	Family: Anacardiaceae	
68	*Mangifera indica*	Keri
	Family: Celastraceae	
69	*Maytenus emarginata*	Vicklo
	Family: Moringaceae	
70	*Moringa oliefera*	Saragavo
	Family: Najadaceae	
71	*Najas graminea*	-
	Family: Nelumbonaceae	
72	*Nelumbo lutea*	Kamal
	Family: Rubiaceae	
73	*Oldenlandia sp.*	-
	Family: Fabaceae	
74	*Parkinsonia floridum*	
	Family: Potamogetonaceae	
75	*Potamogeton sp.*	
	Family: Annonaceae	
76	*Polyalthia longifolia*	Asopalav
	Family: Polygonaceae	
77	*Polygonum plebeium*	-
78	*Polygonum glabrum*	-
	Family: Salvadoraceae	
79	*Salvadora persica*	Piludi
80	*Salvadora olioedis*	Pilu

Sr. No.	Scientific Name	Vernacular Name
	Family: Verbenaceae	
81	*Tectona grandis*	Sag
	Family: Zygophyllaceae	
82	*Tribulus terrestris*	Bethu Gokhru
	Family: Typhaceae	
83	*Typha angustata*	
	Family: Rhamnaceae	
84	*Zizyphus mauritiana*	Bordi
85	*Zizyphus nummularia*	Chani Bor

Annexure 1. List of Vegetation (Aquatic & Terrestrial) Recorded in Thol Bird Sanctuary

Sr. No.	Scientific name	Common name	Migratory Status	Food Habits
AQUATIC BIRDS				
	Order: Anseriformes			
	Family: Anatidae			
1	*Anas acuta*	Northern Pintail	M	Aquatic plants,grains, insects, tadpoles etc.
2	*Anas clypeata*	Northern Shoveler	M	Water insects, snails, planktons, fish spawn.
3	*Anas crecca*	Common Teal	M	Chiefly vegetable matter, insects, crustaceans etc
4	*Anas penelope*	Eurasian Wigeon	M	Largerly vegetarian
5	*Anas platyrhynchos*	Mallard	RM	Largerly vegetarian
6	*Anas poecilorhyncha*	Spot billed Duck	RM	Chiefly vegetable matter
7	*Anas querquedula*	Garganey	M	Largerly vegetarian
8	*Anas strepera*	Gadwall	M	Largerly vegetarian
9	*Anser anser*	Greylag Goose	M	Vegeterian, winter crops, grass, aquatic weeds
10	*Anser indicus*	Bar-headed Goose	RM	Chiefly green shoots of winter crops - wheat/gram
11	*Aythya ferina*	Common Pochard	M	Vegetable matter, insects, molluscs, small fish etc
12	*Aythya nyroca*	White-eyed Pochard	RM	Vegetable matter, insects, molluscs, small fish etc
13	*Sarkidiornis melanotos*	Comb Duck	R	Grain, shoots vegetable matter
14	*Tadorna ferruginea*	Ruddy Shelduck	RM	Vegetable matter, insects, molluscs, small fish etc
15	*Tadorna tadorna*	Common Shelduck	M	Ominivorous, molluscs, algae, seeds etc.

Sr. No.	Scientific name	Common name	Migratory Status	Food Habits
	Family: Dendrocygnidae			
16	*Dendrocygna javanica*	Lesser Whistling Duck	R	Largerly vegetarian - shoots and grain.
	Family: Anhingidae			
17	*Anhinga melanogaster*	Oriental Darter	RM	Fish
	Family: Ardeidae			
18	*Ardea cinerea*	Grey Heron	RM	
19	*Ardea purpurea*	Purple Heron	RM	Fish, Frogs, snakes etc.
20	*Ardeola grayii*	Indian Pond Heron	R	Frogs, fish, crabs and insects
21	*Bubulcus ibis*	Cattle Egret	R	Chiefly grasshoppers, blue bottle flies,lizards, fish etc
22	*Casmerodius albus*	Great Egret	RM	Fish, Frogs,etc.
23	*Egretta garzetta*	Little Egret	R	Insects, fish, frogs etc.
24	*Egretta gularis*	Western Reef Egret	RM	Mainly crustaceans, molluscs and fish
25	*Ixobrychus minutus*	Little Bittern	RM	Fish, molluscs etc.
26	*Ixobrychus sinensis*	Yellow Bittern	RM	Fish, frogs, molluscs etc.
27	*Mesophoyx intermedia*	Intermediate Egret	RM	Fish, frogs etc.
28	*Nycticorax nycticorax*	Black crowned Night Heron	R	Crabs, fish, frogs, aquatic insects, etc.
	Family: Charadriidae			
29	*Charadrius alexandrinus*	Kentish Plover	RM	Insects and crustacea
30	*Charadrius dubius*	Little Ringed Plover	RM	Insects, sand-hoppers, tiny crabs, etc.
31	*Vanellus indicus*	Red wattled Lapwing	R	Insects, grubs, molluscs, etc.
32	*Vanellus leucurus*	White tailed Lapwing	M	Aquatic insects and other vertebrates
33	*Vanellus malabaricus*	Yellow wattled Lapwing	R	Insects, grubs, molluscs, etc.
34	*Calidris minuta*	Little Stint	M	Tiny insects, crustaceans and molluscs.
	Family: Recurvirostridae			
35	*Himantopus himantopus*	Black winged Stilt	RM	Worms, molluscs, aquatic insects, etc.
36	*Recurvirostra avosetta*	Pied Avocet	RM	Worms, aquatic insects and small crustacea, etc.

Sr. No.	Scientific name	Common name	Migratory Status	Food Habits
	Family: Ciconiidae			
37	*Anastomus oscitans*	Asian Openbill	R	Frogs,crabs, large insects and other small living things.
38	*Ciconia episcopus*	Woolly necked Stork	R	Fish. Frogs. Reptiles, crabs, molluscs, large insects, etc.
39	*Mycteria leucocephala*	Painted Stork	R	Fish, frogs and snakes.
	Family: Jacanidae			
40	*Metopidius indicus*	Bronze Winged Jacana	R	Seeds, roots, etc., aquatic plants, insects and molluscs
41	*Hydrophasianus chirurgus*	Pheasant-tailed Jacana	R	Seeds, roots, etc., aquatic plants, insects and molluscs
	Family: Laridae			
42	*Chlidonias hybridus*	Whiskered Tern	RM	Tiny fishes, tadpoles, crabs, grasshoppers and insects.
43	*Sterna albifrons*	Little Tern	R	Small fish, crustaceans, insects.
44	*Sterna aurantia*	River Tern	R	Fish, crustaceans, tadpoles and water insects.
	Family: Pelecanidae			
45	*Pelecanus philippensis*	Spotbilled Pelican	RM	Fish
46	*Pelecanus crispus Bruch*	Great White Pelican	M	Fish, crustaceans
	Family: Phalacrocoracidae			
47	*Phalacrocorax carbo*	Great Cormorant	RM	Almost exclusively fish
48	*Phalacrocorax fuscicollis*	Indian Cormorant	RM	Almost exclusively fish
49	*Phalacrocorax niger*	Little Cormorant	RM	Exclusively fish
	Family: Phoenicopteridae			
50	*Phoenicopterus ruber*	Greater Flamingo	RM	Crustaceans, worms, insect larvae, seeds of marsh plants.
51	*Phoenicopterus minor*	Lesser Flamingo	RM	Phytoplankton (algae, diatoms, etc.)
	Family: Podicipedidae			
52	*Tachybaptus ruficollis*	Little Grebe	RM	Aquatic insects and larvae, tadpoles, etc.
	Family: Scolopacidae			
53	*Limosa limosa*	Black tailed Godwit	M	Worms, molluscs,crabs, insects.
54	*Limosa lapponica*	Bar-tailed Godwit	M	Marine invertebrates, insects.

Sr. No.	Scientific name	Common name	Migratory Status	Food Habits
55	*Tringa glareola*	Wood Sandpiper	M	Insects, larvae, worms and molluscs.
56	*Tringa hypoleucos*	Common Sandpiper	RM	Insects, worms , molluscs.
57	*Tringa nebularia*	Common Greenshank	M	Insects and other invertebratres, tadpoles, frogs.
58	*Tringa ochropus*	Green Sandpiper	M	
59	*Tringa stagnatilis*	Marsh Sandpiper	M	Insects, invertebrates and small frogs.
	Family: Threskiornithidae			
60	*Platalea leucorodia*	Eurasian Spoonbill	RM	Tadpoles, frogs, molluscs, insects and vegetable matter
61	*Plegadis falcinellus*	Glossy Ibis	RM	Molluscs, crustaceans, insects, etc.
62	*Pseudibis papillosa*	Red-naped/Black Ibis	R	Insects, grain and small reptiles.
63	*Threskiornis melanocephalus*	Oriental White Ibis	R	Tadpoles, frogs, molluscs, insects and vegetable matter
	Order: Coraciiformes			
	Family: Alcedinidae			
64	*Alcedo atthis*	Small Blue Kingfisher	R	Small fish, tadpoles and aquatic insects.
	Family: Cerylidae			
65	*Ceryle rudis*	Pied Kingfisher	R	Fish, tadpoles, frogs and aquatic insects.
	Family: Dacelonidae			
66	*Halcyon smyrnensis*	White-breasted Kingfisher	R	Fish, tadpoles, lizard, grasshoppers and other insects
	Order: Gruiformes			
	Family: Gruidae			
67	*Grus antigone*	Sarus Crane	R	Grain, shoots and other vegetable matter, insects, reptiles.
68	*Grus grus*	Common Crane	M	Largerly vegetarian, tubers, grain, insects and small reptiles
69	*Grus virgo*	Demoiselle Crane	M	
	Family: Rallidae			
70	*Amaurornis akool*	Brown Crake	R	
71	*Amaurornis phoenicurus*	White-breasted Waterhen	R	Insects, worms, molluscs, grain, etc.
72	*Fulica atra*	Common Coot	RM	Grass and Paddy shoots, aquatic weeds, insects, etc.

Sr. No.	Scientific name	Common name	Migratory Status	Food Habits
73	*Gallicrex cinerea*	Watercock	R	Largely vegetarin - seeds and green shoots of rice etc.
74	*Gallinula chloropus*	Common Moorhen	RM	Insects, worms, molluscs, grain, etc.
75	*Porphyrio porphyrio*	Purple Swamphen/Moorhen	R	Shoots and vegetable matter, insects and molluscs.
	Family: Accipitridae (Or.Ciconiformese)			
76	*Circus aeruginosus*	Western Marsh Harrier	M	Frogs, fish small birds, mammals and carrion.
TERRESTRIAL BIRDS				
	Order: Apodiformes			
	Family: Apodidae			
77	*Apus nipalensis*	House Swift	RM	Chiefly dipterous insects.
78	*Cypsiurus balasiensis*	Asian Palm-Swift	R	Tiny winged insects.
	Order: Ciconiiformes			
	Family: Accipitridae			
79	*Accipiter badius*	Shikra	R	Lizards, mice, squirrels, birds etc.
80	*Accipiter virgatus*	Besra	R	Largely small birds, mice, bats, lizards and insects.
81	*Aquila heliaca*	Imperial Eagle	RM	Rodents, ground dwelling birds, reptiles, etc.
82	*Aquila clanga*	Greater Spotted Eagle	RM	Frogs, waterfowl, small birds, etc.
83	*Elanus caeruleus*	Black-shouldered Kite	R	Locusts, crickets, mice, lizards,etc.
84	*Milvus migrans*	Black Kite	R	Offal and garbage, earthworms, mice, lizards etc.
85	*Neophron percnopterus*	Egyptian Vulture	RM	Animal carcasses and freshwater turtles.
86	*Pandion haliaetus*	Osprey	RM	Fish
87	*Pernis ptilorhyncus*	Oriental Honey-buzzard	RM	Honeybee larvae, small birds, reptiles, frogs etc.
88	*Spilornis cheela*	Crested Serpent-Eagle	R	Frogs, lizards, rats, snakes,etc.
	Order: Columbiformes			
	Family: Columbidae			
89	*Columba livia*	Rock Pigeon	R	Cereals, pulses, groundnuts,etc.
90	*Streptopelia chinensis*	Spotted Dove	R	
91	*Streptopelia decaocto*	Eurasian	R	

Sr. No.	Scientific name	Common name	Migratory Status	Food Habits
		Collared-Dove		
92	*Streptopelia orientalis*	Oriental Turtle-Dove	RM	Paddy, cereals, bamboo and grass seeds.
93	*Streptopelia tranquebarica*	Red Collared-Dove	R	
94	*Treron phoenicoptera*	Yellow-footed Green-Pigeon	R	Fruits and berries.
	Order: Coraciiformes			
	Family: Coraciidae			
95	*Coracias benghalensis*	Indian Roller	R	Insects,
96	Merops orientalis	Little Green Bee-eater	R	Insects, chiefly diptera and hymenoptera
97	*Centropus sinensis*	Greater Coucal	R	caterpillars, large insects, snails, lizards young mice etc.
	Family: Cuculidae			
98	*Cuculus micropterus*	Indian Cuckoo	RM	Mainly caterpillars, insects, etc.
99	*Eudynamys scolopacea*	Asian Koel	R	Largely fruits and berries, caterpillars and insects.
	Family: Phasianidae			
100	*Coturnix coturnix*	Common Quail	RM	Grain and grass seeds, termites, etc.
101	*Francolinus pictus*	Painted Francolin	R	Grain, grass seeds, green shoots, white ants and insects.
102	*Francolinus pondicerianus*	Grey Francolin	R	Grain, seeds, termites , beetle larvae, etc.
103	*Pavo cristatus*	Indian Peafowl	R	Grain, Vegetable shoots, insects, lizards, snakes, etc.
	Order: Passeriformes			
	Family: Aegithalidae			
104	*Aegithalos leucogenys*	White-cheeked Tit	R	
	Family: Alaudidae			
105	*Eremopterix grisea*	Ashy-crowned Sparrow-Lark	R	Seeds and insects.
106	*Eremopterix nigriceps*	Black-crowned Sparrow-Lark	R	
	Family: Cisticolidae			
107	*Prinia inornata*	Plain Prinia	R	Insects, caterpillars, ants, small beetles,etc.
108	*Prinia socialis*	Ashy Prinia	R	Insects.
	Family: Corvidae			
109	*Aegithina tiphia*	Common Iora	R	Insects, their eggs and larvae.

Sr. No.	Scientific name	Common name	Migratory Status	Food Habits
110	*Corvus splendens*	House Crow	R	Offal, dead sewe rat, kitchen scraps and refuse, termitesetc
111	*Dendrocitta vagabunda*	Rufous Treepie	R	Fruits, insects, lizards, frogs, centipedes etc.
112	*Dicrurus leucophaeus*	Ashy Drongo	RM	Mainly insects, occasionally reptiles,and small birds.
113	*Dicrurus macrocercus*	Black Drongo	R	Insects, flower nectar, occasionally small birds.
114	*Garrulus glandarius*	Eurasian Jay	R	
115	*Pericrocotus cinnamomeus*	Small Minivet	R	Insects and their larvae.
116	*Rhipidura albicollis*	White-throated Fantail	R	
117	*Rhipidura aureola*	White-browed Fantail	R	Insects, chiefly diptera and hemiptera.
	Family: Hirundinidae			
118	*Delichon urbica*	Northern House-Martin	RM	Midges and other insects.
119	*Hirundo smithii*	Wire-tailed Swallow	R	Midges
	Family: Laniidae			
120	*Lanius vittatus*	Bay-backed Shrike	R	Locusts,lizards , large insects, etc.
	Family: Muscicapidae			
121	*Copsychus saularis*	Oriental Magpie-Robin	R	Insects, flower nectar of Salmalia and Erythrina.
122	*Saxicoloides fulicata*	Indian Robin	R	Insects and their eggs, spiders, etc.
	Family: Nectariniidae			
123	*Nectarinia asiatica*	Purple Sunbird	R	Insects and spiders, very largely flower nectar.
	Family: Passeridae			
124	*Anthus campestris*	Tawny Pipit	RM	
125	*Anthus rufulus*	Paddyfield Pipit	R	Weev and other small insects
126	*Lonchura striata*	White-rumped Munia	R	Grass seeds, etc.
127	*Motacilla cinerea*	Grey Wagtail	M	Tiny insects.
128	*Motacilla flava*	Yellow Wagtail	RM	Insects, spiders and invertebrates, etc.
129	*Passer domesticus*	House Sparrow	R	Grains, insects, fruit buds, flower nectar, etc.
	Family: Pycnonotidae			
130	*Pycnonotus cafer*	Red-vented	R	Insects, fruits and berries, peas

Sr. No.	Scientific name	Common name	Migratory Status	Food Habits
		Bulbul		and vegetables etc.
131	*Pycnonotus leucotis*	White-eared Bulbul	R	Kitchen scraps, berries of peelu and wild caper.
	Family: Sturnidae			
132	*Acridotheres ginginianus*	Bank Myna	R	Grasshoppers and other insects.
133	*Acridotheres tristis*	Common Myna	R	Fruits, insects, kitchen scraps, etc.
134	*Sturnus pagodarum*	Brahminy Starling	R	Chiefly berries, wild figs and insects.
135	*Sturnus roseus*	Rosy Starling	M	Locusts, berries, nectar of Salmalia, etc.
	Family: Sylviidae			
136	*Acrocephalus arundinaceus*	Great Reed Warbler		
137	*Orthotomus sutorius*	Common Tailorbird	R	Tiny insects, their eggs and grubs, flower nectar.
138	*Turdoides caudatus*	Common Babbler	R	Insects, berries, grain and flower nectar.
139	*Turdoides earlei*	Striated Babbler	R	Insects, snails and some vegetable matter.
140	*Turdoides malcolmi*	Large Grey Babbler	R	Insects, berries, grain and flower nectar.
141	*Turdoides striatus*	Jungle Babbler	R	Spiders, cockaroaches, insects and their larvae grain, etc.
	Order: Psittaciformes			
	Family: Psittacidae			
142	*Psittacula krameri*	Rose ringed parakeet	R	Ripening fruits, standing crops of maize and jowar.
	Order: Strigiformes			
	Family: Strigidae			
143	*Athene brama*	Spotted Owlet	R	Chiefly beetle and other insects, mice, lizards, etc.
	Order: Upupiformes			
	Family: Upupidae			
144	*Upupa epops*	Eurasian Hoopoe	RM	Insects, grubs and pupae.

Annexure 2. List of Aquatic & Terrestrial Birds in Thol Bird Sanctuary

Sr. No.	Scientific name	Common name	Migratory Status	Open Water - Deep	Open Water - Shallow	Emergent Aquatic Vegetation	Muddy	Shore Land	Surrounding Environment - Agri	Surrounding Environment - Fallow	Wooded Areas
	Order: Anseriformes										
	Family: Anatidae										
1.	*Anas acuta*	Northern Pintail	M	L	H	L	–	–	–	–	–
2.	*Anas clypeata*	Northern Shoveler	M	–	H	M	M	–	–	–	–
3.	*Anas crecca*	Common Teal	M	–	H	M	M	–	–	–	–
4.	*Anas penelope*	Eurasian Wigeon	M	L	H	L	–	–	–	–	–
5.	*Anas platyrhynchos*	Mallard	RM	L	H	–	L	–	–	–	–
6.	*Anas poecilorhyncha*	Spot billed Duck	RM	L	H	L	L	–	–	–	–
7.	*Anas querquedula*	Garganey	M	L	H	L	–	–	–	–	–
8.	*Anas strepera*	Gadwall	M	L	H	–	L	–	–	–	–
9.	*Anser anser*	Greylag Goose	M	–	M	–	L	M	H	H	–
10.	*Anser indicus*	Bar-headed Goose	RM	–	M	–	L	M	H	H	–
11.	*Aythya ferina*	Common Pochard	M	H	M	L	–	–	–	–	–
12.	*Aythya nyroca*	White-eyed Pochard	RM	H	M	L	–	–	–	–	–
13	*Sarkidiornis melanotos*	Comb Duck	R	L	H	M	–	–	–	–	–
14	*Tadorna ferruginea*	Ruddy Shelduck	RM	L	M	–	L	–	–	–	–
15	*Tadorna tadorna*	Common Shelduck	M	L	L	L	L	L	–	–	–
	Family: Dendrocygnidae										
16	*Dendrocygna javanica*	Lesser Whistling Duck/Teal	R	L	H	M	L	L	L	L	L
	Family: Anhingidae										
17	*Anhinga melanogaster*	Oriental Darter	RM	H	M	–	–	–	–	–	–
	Family: Ardeidae										
18.	*Ardea cinerea*	Grey Heron	RM			L	H	M	–	–	–
19.	*Ardea purpurea*	Purple Heron	RM	–	–	H	M	L	–	–	–
20.	*Ardeola grayii*	Indian Pond Heron	R	–	–	L	H	M	–	–	–
21.	*Bubulcus ibis*	Cattle Egret	R	–	–	L	M	H	H	H	H
22.	*Casmerodius albus*	Great Egret	RM	–	H	M	M	L	L	L	M
23.	*Egretta garzetta*	Little Egret	R	–	M	M	H	M	M	L	M
24.	*Egretta gularis*	Western Reef Egret	RM		H	M	H	M	M	L	M
25.	*Ixobrychus minutus*	Little Bittern	RM	–	–	H	L	–	–	–	–
26.	*Ixobrychus sinensis*	Yellow Bittern	RM	–	–	H	H	L	L	–	–
27.	*Mesophoyx intermedia*	Intermediate Egret	RM	–	–	–	H	–	–	–	–
28.	*Nycticorax nycticorax*	Black crowned Night Hero	R	–	–	M	M	L	L	L	–
	Family: Charadriidae										
29	*Charadrius alexandrinus*	Kentish Plover	RM			L	H	–	–	–	–
30.	*Charadrius dubius*	Little Ringed Plover	RM	–	–	–	H	M	–	–	–
31.	*Vanellus indicus*	Red wattled Lapwing	R	–	–	–	L	H	M	M	–
32	*Vanellus leucurus*	White tailed Lapwing	M	–	–	–	L	H	M	M	–
33	*Vanellus malabaricus*	Yellow wattled Lapwing	R	–	–	–	L	H	M	M	–
34	*Calidris minuta*	Little Stint	M	–	–	L	H	L	–	–	–
	Family: Recurvirostridae										
35	*Himantopus himantopus*	Black winged Stilt	RM	–	M	M	H	M	–	–	–
36	*Recurvirostra avosetta*	Pied Avocet	RM	–	M	L	H	–	–	–	–
	Family: Ciconiidae										
37	*Anastomus oscitans*	Asian Openbill	R	–	H	L	L	–	M	M	M
38	*Ciconia episcopus*	Woolly/White necked Stork	R	–	H	L	M	–	L	L	L
39	*Mycteria leucocephala*	Painted Stork	R	–	H	H	M	–	–	–	–
	Family: Jacanidae										
40	*Metopidius indicus*	Bronze winged Jacana	R	–	M	H	L	L	–	–	–
41	*Hydrophasianus chirurgus*	Pheasant-tailed Jacana	R	–	M	H	L	L	–	–	–
	Family: Laridae										
42	*Chlidonias hybridus*	Whiskered Tern	RM	H	H	–	L	M	–	–	–
43	*Sterna albifrons*	Little Tern	R	H	H	–	L	M	–	–	–

1

44	*Sterna aurantia*	River Tern	R	H	H	–	L	M	–	–	–
	Family: Pelecanidae										
45	*Pelecanus crispus Bruch*	Great White Pelican	RM	H	–	–	–	–	–	–	–
46	*Pelecanus philippensis*	Spot-billed Pelican	M	H	–	–	–	–	–	–	–
	Family: Phalacrocoracidae										
47	*Phalacrocorax carbo*	Great Cormorant	RM	H	–	–	–	–	–	–	–
48	*Phalacrocorax fuscicollis*	Indian Cormorant	RM	H	M	L	–	–	–	–	–
49	*Phalacrocorax niger*	Little Cormorant	RM	H	M	–	–	–	–	–	–
	Family: Phoenicopteridae										
50	*Phoenicopterus ruber*	Greater Flamingo	RM	–	H	–	–	–	–	–	–
51	*Phoenicopterus minor*	Lesser Flamingo	RM	–	H	–	–	–	–	–	–
	Family: Podicipedidae										
52	*Tachybaptus ruficollis*	Little Grebe	RM	H	M	L	–	–	–	–	–
	Family: Scolopacidae										
53	*Limosa limosa*	Black tailed Godwit	M	–	M	L	H	L	–	–	–
54	*Limosa lapponica*	Bar-tailed Godwit	M	–	M	L	H	L	–	–	–
55	*Tringa glareola*	Wood Sandpiper	M	–	–	L	H	L	–	–	–
56	*Tringa hypoleucos*	Common Sandpiper	RM	–	–	L	H	L	–	–	–
57	*Tringa nebularia*	Common Greenshank	M	–	L	L	H	L	–	–	–
58	*Tringa ochropus*	Green Sandpiper	M	–	–	L	H	L	–	–	–
59	*Tringa stagnatilis*	Marsh Sandpiper	M	–	–	–	H	–	–	–	–
	Family: Threskiornithidae										
60	*Platalea leucorodia*	Eurasian Spoonbill	RM	–	H	L	M	L	–	–	M
61	*Plegadis falcinellus*	Glossy Ibis	RM	–	–	M	H	M	L	L	H
62	*Pseudibis papillosa*	Red-naped/Black Ibis	R	–	–	–	M	H	M	M	H
63	*Threskiornis melanocephalus*	Black-headed/White Ibis	R	–	L	L	H	L	L	L	H
	Order: Coraciiformes										
	Family: Alcedinidae										
64	*Alcedo atthis*	Common Kingfisher	R	L	M	H	L	M	–	L	–
	Family: Cerylidae										
65	*Ceryle rudis*	Pied Kingfisher	R	M	H	L	–	L	–	L	–
	Family: Dacelonidae										
66	*Halcyon smyrnensis*	White-throated Kingfisher	R	M	H	L	L	L	L	L	M
	Order: Gruiformes										
	Family: Gruidae										
67	*Grus antigone*	Sarus Crane	R	–	–	–	M	L	H	H	–
68	*Grus grus*	Common Crane	M	–	–	–	L	M	H	H	–
69	*Grus virgo*	Demoiselle Crane	M	–	–	–	L	M	H	H	–
	Family: Rallidae										
70	*Amaurornis akool*	Brown Crake	R	–	–	H	L	L	–	–	–
71	*Amaurornis phoenicurus*	White-breasted Waterhen	R	–	–	H	M	L	L	–	–
72	*Fulica atra*	Common Coot	RM	H	M	L	–	–	–	–	–
73	*Gallicrex cinerea*	Watercock	R	–	–	H	M	–	M	L	–
74	*Gallinula chloropus*	Common Moorhen	RM	–	–	H	M	L	–	–	–
75	*Porphyrio porphyrio*	Purple Swamphen/Moorhen	R	–	–	H	M	L	–	–	–

R: Resident species, M: Migratory species, RM: Resident Migratory species
Habitat: L: Less used habitat, M: Moderately used habitat, H: Highly used habitat

Annexure 3. Habitat Requirement of Waterfowls in Thol Bird Sanctuary

8. References

Ali, Salim (2002). The book of Indian Birds. Oxford University Press, New Delhi.

Ambasht, R. S. (1990). A Text Book of Plant Ecology. Students' Friends & Co. Varanasi.

Anno. (1975) Gujarat State Gazetteers, Mehsana District. Government of Gujarat, Ahmedabad.

Anno. (2001) Management Plan of Thol Bird Sanctuary, Government of Gujarat, Gandhinagar.

Chase M. K., Kristan W. B. III, Lynam A. J., Price M. V. and Rotenberry J. T. (2000) Single species as indicator of species richness and composition in California coastal sage scrub birds and small mammals. Conservation Biology, 14(2) 474-487.

Croll, D. A. et al. (2005) Introduced predators transform subarctic islands from grassland to tundra. Science 307, 888-899

Deepa, R. S, and Ramachandra T. V., (1999) Impact of urbanization in the interconnectivity of wetlands, in proceedings of the National Symposium on Remote Sensing Applications for Natural Resources: Retrospectives and perspective, (Jan 19-21, 1999) organized by Indian Society of Remote Sensing, Bangalore, pp. 343-351.

FSI, (2011). State of Forest Report, Forest Survey of India, Dehradun.

Gaur, R. D. (1982) . Dynamics of vegetation of Garhwal Himalayas pp. 12-15. Vegetational Wealth of the Himalayas (ed. G. S. Paliwal) Delhi.

GEC, (2009). Preparation of Comprehensive Study Report Based on the Analysis of Various Ecological and Environmental Parameters to Implement the Action Plan for the Management and Conservation of Thol Wetland, Gujarat Ecology Commission, Gandhinagar.

GEER, (2002). Ecological study of Thol Lake Wildlife (Bird) Sanctuary, GEER Foundation, Indroda Park, Gandhinagar.

Grimmett, R., Inskipp, C. and Inskipp, T. (1998). Birds of the Indian Subcontinent. Oxford University Press, New Delhi.

Mc Farland, D. (1981). (Ed.) The Oxford Companion to Animal Behaviour, Oxford University Press, New York.

Mitsch, W. J. and Gosselink, J. G. (1986). Wetlands. New York: Van Nostrand Reinhold Comp. Inc., pp. 1-539.

Odum, E. P. (1983). Basic Ecology. Saunders College Publishing, Philadelphia and London.

Patel, Sejal & Dharaiya, Nishith (2008) Marsh Bird Community Index of Biotic Intigrity: A key to study an ecological condition of Wetlands.Proceedings of Tall 2007.

Patel, Sejal, Shingala P., Dave, S.M. and Dharaiya, N. (2006). A comparative study of avifaunal composition in two bird sanctuary of Gujarat. Proceeding of XXI Gujarat Science Congress, 2006.

Post, D.M. et al. (1998) The role of migratory waterfowl as nutrient vectors in a managed wetland. Conserve. Biol. 12, 910-920.

RAMSAR, (1971). Convention on Wetlands of International Importance, Ramsar, Iran.

Ravindranath, S. & Premnath, S. (1997) Biomass studies Field Methods for Monitoring Biomass. Oxford & IBH Publishing Co. Pvt. Ltd. New Delhi. pp 81

Singh, H. S. & Tatu, K. (2000) A study on Indian Sarus Crane (Grus antigon antigone) in Gujarat State (with emphasis on its status in Kheda and Ahmedabad Districts). GEER Foundation, Gandhinagar.

Singh, H. S. (1998). Wildlife of Gujarat (Wildlife and Protected Habitats of Gujarat State).GEER Foundation, Gandhinagar.

Singh, H. S., Patel, B. H., Parvez, R., Soni, V.C., Shah, N., Tatu, K. and Patel, D. (1999). Ecological study of Wild Ass Sanctuary, Little Rann of Kachchh (A Comprehensive study on Biodiversity and Management Issues). GEER Foundation, Gandhinagar.

Singh, R and Joshi, M. C. (1979). Floristic composition and life forms of sand dunes herbaceous vegetation near Pilani, Rajasthan. Indian J. Ecol. 6: 9-19

Singh, R. (1976). Structure and net community production of the herbaceous vegetation in the sand dune regions around Pilani, Rajasthan. Ph. D. Thesis BITS Pilani. 460p.

Vaghela, D. K. (1993) Fishes of Pond and Annual Survey. M. Phill. Dissertation, Department of Zoology, Gujarat University, Ahmedabad.

Vijayan, V. S. (1991). Keoladeo National Park Ecology Study. Final Report (1980-1990). BNHS.

Chapter 2

Species-Diversity Utilization of Salt Lick Sites at Borgu Sector of Kainji Lake National Park, Nigeria

A. G. Lameed and Jenyo-Oni Adetola

Additional information is available at the end of the chapter

http://dx.doi.org/10.5772/51089

1. Introduction

Mineral elements occur in the living tissues or soil in either large or small quantities. Those that occur in large quantities are called macro/major elements while those that occur in small quantities are called micro/minor/trace elements. These macro elements are required in large amount and the micro are required in small amount (Underwood, 1977, Alloway, 1990,) They occur in the tissues of plants and animals in varied concentrations. The magnitude of this concentration varies greatly among different living organisms and part of the organisms (W.B.E, 1995).

Although most of the naturally occurring mineral elements are found in the animal tissues, many are present merely because they are constituents of the animal's food and may not have essential function in the metabolism of the animals. Hence essential mineral elements refer to as mineral elements are those which had been proven to have a metabolic role in the body (McDonald, 1987). The essential minerals elements are necessary to life for work such as enzyme and hormone metabolisms (W.B.E, 1995). Enzymes are activated by trace elements known as metallo enzymes (Mertz, 1996). Ingestion or uptake of minerals that are deficient, inbalanced, or excessively high in a particular mineral element induces changes in the functions, activities, or concentration of that element in the body tissue or fluids. Biochemical defects develop, physiological functions are affected and structural disorder may arise and death may occur under the circumstances (Pethes, 1980).

The influence of wildlife management with the goal of maintaining wildlife population and the entire biodiversity at maximum level and maximum ecosystem utilization depend heavily on the knowledge of mineral elements in the nutrients requirement of animals (Mertz, 1976).

2. Salt lick/mineral lick

Salt licks are deposit of mineral salts used by animals to supplement their nutrition, ensuring enough minerals in their diets. A wide assortment of animals, primarily herbivores use salt licks to get essential nutrients like calcium magnesium, sodium and zinc. Salt licks are natural mineral which are mineral outcrops in the soil which are visited by herbivores for soil eating (biting and chewing) or licking (with tongue). They also supplement mineral that are deficient in animal vegetable diets (Ayeni, 1972).

Animals regularly visit licks in the ecosystem which are composed of primarily common salt (sodium chloride). It provides sodium, calcium, iron, phosphorus etc (Ayeni, 1972). Salt licks occur naturally in certain locations in the forest where mineral salt are found on the ground surface.

Shortage of sodium in the plants which are eaten by wildlife could motivate the game to eat a lot of soil at the lick (Ayeni, 1972). The shortage is as a result of water soluble sodium salts being leached out during heavy rain following long period of desiccation. Some plants even substitute potassium ions for sodium ions uptake from soil without showing mineral deficiency symptom (Buckman and Brandy 1960). Many plants are also rich in sodium and potassium, example are; *Acacia elatoir* (Brenan), *Achyranthes aspera (L), Calyptotheca premna resinosa (Hochst) and Salvadora persica (L) and Echolium amplexicaule* (Moore).

Functions of licks change during rainy seasons to those of natural water holes from which wildlife drank and wallowed. Some animals prefer using the drinking holes example are Buffalo and water-buck, but some still eat the soil lick mineral as well as drinking the lick example, warthog and hartebeest.

When salt lick appears, animals may travel to reach it, the lick become a sort of rally points where lots of wildlife can be observed. The concentration of animals in the area becomes visible, game viewing and photographing for tourist are improved in the area, animal studies and census conduction are also improved.

In an ecosystem, salt lick provides sodium calcium, iron, phosphorus and zinc required in the spring time for bones, muscles and growth for the wildlife. All trace elements like copper, magnesium and cobalt are retain in the salt lick for the metabolism of most mammals. Salt licks are used for hunting purposes by the animals.

Salt lick with a wide range of animals illustrate the ways in which wildlife naturally seek out nutrition which is essential to their survival. It provides nutrition for predicators in the form of conveniently located prey who may be distracted by the salt lick long enough to become a snack.

3. Species utilization of salt lick

The following species of wildlife usually utilized salt licks in Nigeria; Elephant (*Loxodonta africana*), Buffalo (*Syncerus caffer*), Western hartebeest (*Alcelaphus buselaphus*), Roan antelope

(*Hippotragus equimus*), water buck (*kobus defessa),* Kob *(kobus kob),* Bushbuck *(Traqelaphus scriptus),* Oribi *(Ourebi ourebi),* Red flanked duiker *(Cephalophus rufilatus*), Grimm's duikers *(Sylvicapra grimmia),* Warthog *(Phacochoerus aethiopicus) and* Baboon *(Papio Anubis*). The following game species were also observed in east Africa to use salt licks: rhino (*Diceros bicornis*) Zebra (*Equus burchelli*) Impala *(Aepycenus melampus),* Water buck (*Kobns ellipsiprymus*), Eland (*Taurotragus oryx*).

4. Activities of animals using salt lick

Salt licks are used predominantly in the day. Different species used salt licks at different period of the day. Buffalo and Warthog used lick mainly in the morning and afternoon. The shapes of the animal mouths determine the method by which salt licks material is obtained and ingested. Elephant and Warthog dig up the licks content using their tusks and lift up large licks materials with the trunk, whereas baboon used their hands to pick up and throw small pieces into the mouth. Hartbeest cut fresh lick materials by biting deeper into the craters with the incisors while water-buck, buffalo, roan, duiker, oribi and bush buck lick up the powdered lick material with the tongue.

Salt lick is found in abundance in the game reserves that are situated in the drier habitats in Northern Nigeria. Henshaw and Ayeni (1971) postulated that abundance and use of salt licks by wildlife indicates nutritional deficiencies caused by degrading environment or over-population or both. Examples are in the case of Yankari where ten licks are located within 11km along the Gaji River and the frequency of occurrence and intensity of use is very high.

5. Problems of salt licks in game reserved

Three problems have been noticed by the presence of salt lick in the game reserves; soil removal, vegetation destruction and spread of disease (Woodley, personal communication). Over 5000 tons of soils are removed annually. The soil is lost through wallowing, eating or licking and trampling. The areas around the salt lick are always trampled by hoof action, overgrazing and devoid of vegetation cover. Since many or so much drinking, urination, defecation, wallowing and feeding occur at the licks, diseases spread rapidly (Ayeni, 1979, Ogunsanmi, 1997). The use of artificial licks may prevent this.

6. Types of salt lick

Salt licks can be natural or artificial. Artificial salt licks are used by farmers for their cattle, horses and other herbivores to encourage health growth and development. Typically a salt lick in form of block is used in these circumstances. The block may be mounted on a platform so that domesticated animals do not consume dirt from the ground with necessary salt. There is need to medicate shy animals or a large group of animals.

Some people used artificial salt licks to attract wildlife such as deer and moose along with smaller creatures like squirrels. Animals may be attracted purely for pressure of humans who install the salt lick with the goal of watching or photography. Also salt licks are used by hunters to encourage potential prey to frequent an area. Wildlife biologists also used salt lick to assist them in tracking populations and can be a serve as medication in that it is used as birth control to keep animals from proliferating in the area where they are few natural predators, example is the deer.

Artificial salt lick comes in two forms; blocked and bagged. Bagged salt licks are designed to be buried in pits to create a more realistic form of salt lick with the salt and mineral leaching out in wet weather to form a salt deposit which will attract animals. While blocked licks are installed directly or mounted on platforms depending on personal taste. It can also be hanged on a tree in the middle of the farm or ranch house.

The universal popularity of salt licks with wide range of animals illustrates the way in which wildlife naturally seek out nutrition which is essential to their survival and provide nutrition for predators. Salt lick is a natural gathering place for grazing animals, which also attracts the carnivores. Animals ingest it as part of food they eat or they eat it directly

7. Significance of natural salt lick

Natural salt licks are utilizes by wildlife to supplement their mineral requirements (Ayeni, 1979). Wild game especially the herbivores can always identify the spots in their habitat where the essential minerals could be found. The African elephant have been noted to travel great distances to visit areas of saline earth which they swallow in considerable quantities as a purgative (Wari, 1993). Lick utilization is related to the spatial distribution and abundance of environmental minerals. A positive correlation between the spatial distribution of elephants in Wakie National Park, central Africa and the abundance of environmental sodium has been reported (Weir, 1972).

Water holes which usually serve as salt licks for animals during dry season may become heavily contaminated with infectious pathogens which can survive to the dry season (Woodford, 1979, Ogunsanmi, 1997). Droplet of respiratory disease is also made possible when animals crowd together in salt lick spots. Salt lick can be infected with anthrax spores and can act as focal point for the spread of the disease (Woodford, 1979). There is also a high increase in predation illustrated by the frequency of carcasses near licks which often lead to high mortalities (Lasan, 1999).

8. Resources availability and utilization by wildlife

The pattern of utilisation is determined by the growth of different plants and the physiological need of the animals, Benjamin, (2007). The abundant of mineral ions Na^+, K^+, Ca^+ and Mg^+ in salt lick, (Ayeni,1971), cause the concentration of big games along the river during the dry season due to the availability of salt lick, cover, water, and cover and fresh flush of the grasses and browse able materials.

Afolayan (1977), Milligant and Ajayi, (1978), observed that the utilization of salt lick by large mammals in Kainji Lake National park reduced with increased distance from water point particularly streams and rivers. Vertebrates have complex nutritional requirement in the form of chemical elements. Just as water and food supply, salt lick constitute one of the requirements expected in an ecological unit. Ayeni, (1971) observed that not all animals come regularly to salt lick but big games pay much visit e.g. elephants, antelopes, baboons etc. Phosphorus and sodium are believed to be principal trace elements causing animals to use salt lick (Cowan, 1949). Habitat degradation through over grazing and over browsing and soil compaction result from heavy lick use (Ayeni, 1972).

9. Factors influencing wildlife population

The increase in animal population is a function of animal chance to survive and multiply, and that chance to multiply and survive is a function of the environment. Environment is the sum total of all factors that influences the animals' speed of development and expectation of life and fecundity. The component which are claimed to be homogeneous with respect to the way it influences the animals chance of survival, Andrewartha and Birch, (1954) are;

- The resources e.g food, water and cover
- Mate
- Predators, pathogens and aggressors
- Weather condition
- Pollution.

10. Salt lick and trace minerals

Mineral elements play an important role in the nutrition of wild games. Hence a brief discussion on wildlife nutrition is important. Salt is unique in that animals have a much greater appetite for the sodium and chloride in the salt than for other minerals. Because most plants provide insufficient sodium for animals feeding and may lack adequate chloride content, salt supplementation is a critical part of a nutritionally balanced diet for animals. In addition, because animals have definite appetite for salt, it can be used as a delivery mechanism to ensure adequate intake of less palatable nutrient and as a feed limiter.

Sodium plays major roles in nerves and impulse transmission and the rhythmic of heart action. Efficient absorption of amino acids and monosaccharide from the small intestine requires adequate sodium. The other nutrient in salt, chloride, are essential for life. Chloride is a primary anion in blood and the movement of chloride in and out of the red blood cells, is essential in maintaining the acidic-base balance of the blood. Chloride is also a necessary part of the hydrochloric acid produced in the stomach which is required to digest most food. Animals have more defined appetite for sodium chloride than any other compound in nature except water. Ruminants have such a strong appetite for sodium that the exact location of salt source is permanently imprinted in their memory which they can then return to when they become deficient. Horses have been shown to have specific appetite for salt if the diet is deficient in sodium.

Trace elements

There are seven trace minerals that have been shown to be needed in supplementing animal diets. They are iron, copper, zinc, manganese, cobalt, iodine and selenium. They are needed in small amounts, or traces, in the diets and hence their names traces minerals".

Currently trace minerals deficiency is a bigger problem than the acute mineral deficiencies. They are several examples where an area was not recognise to have trace mineral deficient in the past but now has been shown to require supplementation. For example selenium deficiency was not considered to be a problem in the United States until recently.

Salt as a carrier of trace minerals

Salt is known to be a carrier of trace minerals, since all herbivores has natural appetite for salt this could serve as a source of trace minerals for them. Moreover when horses, cattle, sheep and other animals are on pasture with little, or no varying amounts of concentrate feeding, farmer can supply trace mineral salt in the form of a mineral block or loose trace mineral salt in a box.

11. Wildlife nutrition

Nutrition is the study of process by which organic and inorganic substances ingested by living organisms are converted to various needs for life processes such as promoting growth, replacing worn-out and injured tissues and perpetuate life. Wildlife nutrition is concerned with the supply of quality food in an animal environment. The basic requirements of all wildlife are food, water and cover. In general animal with adequate food supply grows larger, produced more young, are more vigorous and healthy, and are more resistant to many form of diseases than those affected by malnutrition (Nancy and Martha, 1995). When wild animals are shipped, moved, migrated, translocated or restocked to destinations where plant food that are typically consumed by freely grazing animals e.g reindeer are not found in sufficient quantities to provide adequate nourishment. Nutritional and digestive disorder set in which could be fatal (Luick, 1968). According to Luick, 1979, more than 60% of reindeer may die in two weeks of departure from their native tundra ranges.

Digestive and nutritional disorders are factors contributing to high mortality rate others include; the stresses of capture and handling, hyperthermia, regurgitation and inhalation of rumen contents, injury and diseases and overdoses of immobilizing agents and tranquilizers, (Luick, 1968, 1976). Nutritional disorders are primarily a result of failure in adjustment of the balance between nutrient input and requirements. These disorders are distinct and specific to a particular nutrient, (Sauvant, 1991)

An increasing amount of information is accumulating to show that many nutrients are needed at higher levels to improve the ability of the animal immune system to cope with infection. Sodium chloride, copper, zinc, Selenium, phosphorus and magnesium already have been shown to be helpful in this regard. Nutrients requirement for growth, gestation and lactation do not necessary mean that the level will be adequate for normal immunity and high resistance to diseases. But nutrient levels higher than those recommended may be needed for maximum productivity and health of the animals.

12. Essential mineral elements in wildlife nutrition

Mineral elements are restricted to mineral elements which have metabolic role in the animal body, (McDonald, 1979). Bowen, (1999), an essential mineral element is necessary to proof that diet lacking the element can cause deficiency symptom in animals. These depend on three basic factors. These are;

1. The organism can neither grow nor complete its life cycle without an adequate supply of the element.
2. The element cannot be replaced by any elements and
3. The element has direct influence on the organism and is involved in its metabolism.

13. Major mineral elements at salt-lick site

a. **CALCIUM (Ca):** Calcium is the most abundant element in the animal body. It is an important constituent of the bone and feet where about 99% of the body total calcium is found. Calcium is also an essential constituent of living cells and tissue fluids. It is essential in the activities of a number of enzymes including those necessary for the transmission of nerve impulses and for contractile properties of muscles. It is also necessary in the coagulation of blood and the normal function of cell membrane, (Clegg and Clegg, 1978; McDonald, 1998).
b. **PHOSPHORUS (P):** It occurs in the bone and is vital for bone formation. It also occurs in protein called phosphor-proteins, nucleic acids and phospholipids. The element plays a vital role in energy metabolism in the formation of sugar phosphate and adenosine di and tri-phosphates. It forms an essential constituent of milk and is necessary for the function of the neuromuscular system (McDonald, 1987).
c. **POTASSIUM (K):** Potassium plays an important role in osmoregulation of the body fluids and in acid-base balance in the animals. It functions principally as the cation of the cell. It plays an important part in muscle and nerve excitability, carbohydrate metabolism and is important blood and interstitial fluid, (McDonald et. al., 1981).
d. **CHLORINE (Cl):** Chlorine is associated with sodium and potassium in acid base balance and osmoregulation. It plays an important role in the gastric secretion where it occurs as hydrochloric acid as well as chloride salts. Chlorine deficiency can lead to an abnormal increase of the alkali reserve of the blood.
e. **SULPHUR (S):** It is an important constituent of protein in animals such as protein containing amino acids cystine, methionine, biotine and thiamine, the hormone, insuline and important metabolite, coenzyme A also contain sulphur.
f. **MAGNESIUM (Mg):** Magnesium is closely associated with calcium and phosphorus in the formation of bones and about 70% is found in the skeleton, the remains distributed

in the soft tissue and fluids. It is the commonest enzyme activator, particularly in the activation of phosphate transferases, decarboxylases and acyl transferases (McDonald *et. al.*, 1987).

14. Trace elements

a. **IRON (Fe):** Iron plays an important role in the synthesis of haemoglobin and enzymes in the foetus. It also occurs in the blood serum in a protein called transfrin which is concerned with the transport of iron from one part of the body to another. Iron is also a component of many enzymes including the cytochrome C reductase, succinic dehydrogenase and furmaric dehydrogenase. Hence, it is vitally important in the oxidative mechanism of all living cells. (Clegg and Clegg, 1978; Pethes, 1980; McDonald, 1987). Symptoms of deficient of iron include paleness of the skin, hypochromic microcyclic anaemia and conjunctiva.

b. **COPPER (Cu):** Copper is found in blood plasma and occurs in various complex forms loosely bound to albumin and amino acids. It is necessary for heamoglobin synthesis and has been found in over 35 enzymes and proteins. These include; red blood cell, cerebrocuprein, and mitochondrocuprein, (Underwood, 1977; Pethes, 1980).

c. **IODINE (I):** Iodine is required by animals for the synthesis of the thyroid hormones, thyroxin and triiodothyroxin produced in the thyroid gland (Pethes, 1980). Deficiency results in endemic goitre.

d. **ZINC (Z):** Zinc is found in every tissue in the animal's body. It plays an important role in enzyme activation and in wound healing. It is important in the fundamental process of RNA, protein synthesis and metabolism. Deficiency results in growth retardation, loss of apetite, alopecia, bone deformation, impaired fertility in both male and female animals and increase occurrence of tetratogenicity and behavioural anomalies.

e. **FLOURINE (F):** It is one of the constituents of bone, teeth, soft tissues and body fluids. It activate adrenal cylase enzyme which is the primary mediator of hormone action (Pethes, 1980). It's function in the prevention of dental canies. Deficiency symptoms include; pain on movement, lameness, arthritis, of the hip, erosion of the tooth enamel and decrease in milk yield, (Pethes, 1980).

15. Study area

The study was conducted in Kainji Lake National Park (KLNP). In Nigeria the role of game reserve in conserving wildlife for various purposes is widely recognised. The flora (plants) constitute only one element of the complex ecosystem which they belong and are not in stable state. The effective management of land as game reserve is the general principle. There is a management plans to be drown. Such plan has been prepared for Kainji Lake National Park (KLNP) (Ayayi and Hall, 1975). The principle purpose of the plan is to provide all available information relevant to the management of the reserve. It also makes provisions for it regular revision and updating and incorporate timetable for these purposes.

16. Location

Kainji Lake National Park extends 80km in an east-west direction and about 60km north-south. It consists of two sectors, the Borgu sector and Zuguruma sector. It lies between 90^0 4^1 to 10^0 30^1 N and 3^0 30^1 to 5^0 50^1 E, covering a total area of 5,340.82km (Tuna, 1983). The Borgu sectors cover an area of 3,970.02km S.E in Borgu Local Government Area of Niger State. Ero, (1985), put the location of Borgu between 10^0 50^1 N latitude and 4^0 19^1 E longitude. The Borgu has the Kainji Lake on the east while the west is by the republic of Benin (Figure 1).

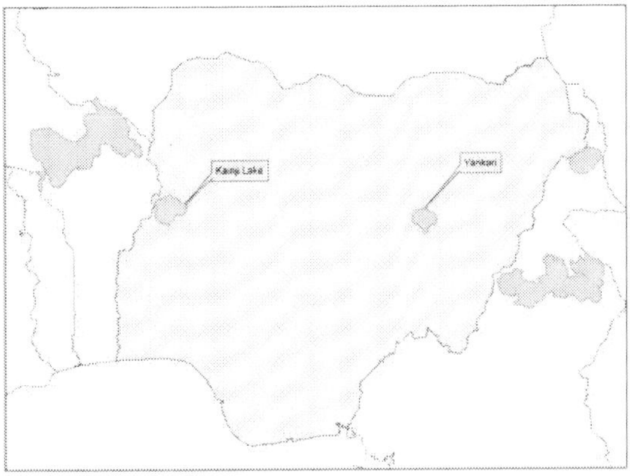

Figure 1. Map of Nigeria showing the location of Kainji Lake National park

Figure 2. Showing the Geo-reference Map of Kainji Lake National Park in respect to the Lake and Salt lick sites.

17. Vegetation of the study area

The vegetation of Nigeria vegetation consists of forest, savannah and Montane. The forest zone comprises mangrove forest, rainforest and dry forest southern and northern guinea savannah zone, Sudan zone and Sahel constitute the savannah vegetation. Montane vegetation is formed by the Montane forest and grassland. The derived savannah forms the boundary between the forest and the true savannah vegetation types.

The Borgu sector of Kainji Lake National park has transitional vegetation which is between the Sudan and the Northern Guinea savannah type (Onyeanusi, 1996). FAO, (1974) also included the vegetation of the study area in the guinea zone in the map of vegetation of Nigeria. FAO, (1974) divides the vegetation into six main vegetations having significance as wildlife habitat types. These are;

a. The Burkea/Detarium microcarpum (Wooded Savannah)
b. The Isoberlinia tomentosa (Woodland)
c. Diospyros mespilliformis (Dry Forest)
d. Terminalia macroptera (Tree Savannah)
e. **The Riparian forest and woodland**
f. **The Oli complex**

Burkea africana/Deterium microcarpum (Wooded savannah)

This consists of the majority of the vegetation of study area. The density and height of woody cover varies with soil and influence of fire. *Afzelia africana* is an associated species which occasionally form patches of woodland. Trees and shrubs common in this area are *Butyrospermum paradoxum, Terminalia avicenniodes, Parinari polyandra, piliostgma thonnigii, Maytenus senegalensis, Gardenia ternifolia, strychnos inocua* (FAO, 1974).

Some of the grasses associated with this community include *Hyperrhenia involucrata, Andropogon gayanus, A. pseudapricus, Genium newtonii, Chlospermum tinctorum, Indigofera bractolats, Hyperthelia dissolute, Brachaiaria jubata, B. brachylopha* and *Aristida kerstingii.* (FAO1974).

Isoberlinia tomentosa (Woodland)

This associates with quartzite ridges and are extensive on higher ground in the study area. The *I. tomentosa* occur in almost pure stands though occasionally *I. doka* is found where stone intrusions occur. *Afzelia africana* and *Ostryoderris stuhlmanni* is found adjacent to *I. tometosa* woodland. Grasses such as *Monocynibium cresiforme, Schizachhyrium sanguineum, Beckeropsis uniseta, andropogon tectorum* and *Andropogon gayanus* occurs extensively on poorer sites.

Diospyros mespiliformis (Dry forest)

This distinctive type occurs as units of a few acres extent at scattered localities in the central part ot the study area. *Diospyros mespiliformis* foms the bulk of the tree layer while

Polysphaeria arbuscula usually comprises of a dense under storey. The grasses of this type are the broadleafed *Opilsmenus hirtellus (*FAO, 1974*).*

Terminalia macroptera (Tree savannah)

This occurs in the low- lying seasonally inundated areas, characterised by dark grey hot-wallowed clays. Typically there are few woody plants in this type. They include *Pseudocedrela kotschyi, Mitrangyna intermis* and *Daniella oliveri* (FAO, 1974). The grasses are those associated with swampy ground. They include *Panicum paucinode, Brachiara jubata, Hyparrhenia glabriuscula, H. rufa, Andropogon perligulatus, A.pseudapricus, Schizachrium schweinfurhii,Hyparrhenia cyanescens and Echinocloa obtussiflora (*FAO,1974*).*

Riparian forest and woodland

This includes the vegetation of all water courses of the reserve with the exception of the Oli and the lower reaches of its two largest tributaries- the Uffa and the Nanu. The riparian/ forest woodland which develops in response to the prevailing high atmospheric water content, resembles the moist savannah bordering the forest zone. The common tree species to this vegetation type include: *Cola laurifolia, Irvingia smithii, Antidesma venosum, Pterocarpus santalinoides, Diospyros mespiliformis, Daniella Oliveni, Gardenia Species, Strychnos spinosa, Terminalia aricennoides* and *Maytenus senegalensis*. The grasses remain green into greater part of the dry season. The grasses include *Acroceras zizanoides.*

The Oli complex

The Oli complex is distinctive in its heterogeneity. It is heterogeneous vegetation which characterises the courses of the Oli River as well as some parts of its main tributaries (e.g. River Nanu, Suna and Suma). Its dominant tree species are assemblages of characteristics rainforest species e.g. *Annogeissus leioxarpus, Vitex doniana, Khaya senegalensis, Mitragyna inernus, Chlorophaora* excels. The oli complex and the riparian forest are closely associated.

The valleys have a scattering of *Terminalia macroptera* in grassland of lots of *Hyperrhenia smithiana*. The upland areas are very distinctive but small unit of *Diospyros ruspiliformis* forest. The evergreen nature of the *Diospyros ruspiliformis* and the shrub makes it difficult for fire to penetrate it. *Oplismenus hitellus* grass also occurs in this unit. In the upland, three woodland units are found they are, *Isoberlinia doka, Isoberlinia tomentosa and monotes kerstingii* woodlands. The largest vegetation is covered by *Burkea Africana-Detarium macrocarpum* wooded savannah.

18. Climate of Kainji Lake National Park

Rainfall: The two major features of the climate of the park are the wet and dry season and they are variable from year to year. The wet season extends from May to November while the dry season extends from December to April. The highest amount of rainfall is always in

	2007	2006	2005	2004	2003	2002	2001	2000
JAN	-	-	-	0.2	-	-	-	-
FEB	-	-	-	-	2.6	-	-	-
MAR	-	-	28	2.8	18.7	-	-	-
APRIL	65.51	30	53.3	67.4	45	84.9	48.5	8.6
MAY	100.7	168.4	58.9	167.6	82.2	34.68	96	70.43
JUNE	119	225.7	179.04	176.5	203.27	103.62	229.9	227.4
JULY	175.8	226.33	184.6	231.3	111.3	154.15	165.23	115.1
AUG	306	166.8	106	158.3	126.56	159.08	111.21	253.8
SEPT	127.2	255.06	329.39	265.76	147.17	122.9	211.65	245.6
OCT	45.8	68.3	110.4	112	110.3	74	119.98	50.12
NOV	-	-	-	12	-	3	-	-
DEC	-	-	-	-	-	-	-	-

Table 1. Kainji Rainfall Data from 2000-2007, Kainji monthly Rainfall (mm)

August (11.89mm) while the lowest rainfall (2.090mm) is in October (Table 1). These values varied yearly. Milligan (1979) indicates that there is a decreased in rainfall from the south to the north and increased rainfall towards the west and east and low condition in central northern regions.

Temperature: The highest is in the dry season just before the rain and lower during the wet season it picks up again towards the end of the wet season and later drops to the lowest value in December and January during harmattan. The minimum temperature during this period ranges from 17.87°C to 19.90°C (Onyeanusi 1998).

Relative humidity: The relative humidity increases gradually from low values at the beginning of the dry season to a peak during the wet season. Generally, the relative humidity follows opposite pattern to that of temperature.

Wind: This influences both incidence and duration of the wet season considering the whole year, southern winds predominate over northern winds in the Borgu sector of the park. There is also a distinct seasonal trend, with the dry, dusty, northern winds prevailing during the beginning of the dry season that is November to February, while the moist southern winds prevail throughout the wet season.

Topography: The landscape of the Borgu Sector of the Kainji Lake National Park is gently undulating. Its features may be related to the lithology and erosion history of the area. The relief is broken in places by quartzite ridges (FAO, 1974). Elevation of the central and Western parts of the park lie between 800 and 1,000ft the highest point in the park is 1135ft in the Northwest. Kubil hill, just outside the Northern boundary of the park has an elevation of 1,684ft (FAO, 1974). The land slope down from the East to the Niger valley. The lowest elevations are along the Kainji shore where the normal maximum high water mark is 465ft contour (FAO, 1974; Afolayan, 1977).

Geology: The Borgu sector of the park is underlain by the Basement Complex helica was considered to be Pre-Cambrian in origin (Afolayan, 1977). Most of the area is composed of geissose rock and other units consist of younger metasediments which are mainly schists and phyllites. According to FAO, (1974), the park is underlain by undifferentiated metasediments in the east and west. The Basement Complex was until recently considered as pre-cambrian. However, Truswell and Cope, (1963) presented age determination which will place it in late Pre-cambrian or early Paleozoic.

Soil: Detailed soil survey of the Borgu sector shows that the soils in the area are low of fertility. It is slightly acid to natural and the acidity increases irregularly with dept. Although the soil nutrient is low, well developed and maintained savanna woodland exists on this soil. Meanwhile, the park has over ferruginous tropical soil and crystalline acid rocks (Anon 1964). Also the soils are shallow.

Drainage: The drainage is to the east because of the slope into River Niger through River Oli which is the largest river in the park. The river Oli and its tributaries drain the western part of the Borgu sector while the eastern part are drained by rivers Doro, Timo and Menai into Kainji lake. The river covers an estimate of about 3.305kms from the Nigeria border with Republic of Benin to where it empties into the Niger River. The river has maximum flow of approximately 600-700mm^3/sec of water at the end of wet season breaks into pools in the dry season. The main tributaries of the Oli river are; Uffer, Koa, Nanu, Suma and Suna. Though other unnamed seasonal rivers also feed the Oli river in the wet season. There are six drainage basins in the western part of Borgu sector, viz the lower and upper Oli, Suna, Nanu, Uffa, and an unnamed basin which covers about three quarters of the sector. In the eastern part are seven drainage basins giving a total of thirteen basins. Eroded water from these rivers contributes to the volume and rate of flow of the Oli river in the early part of the wet season. During dry season, surface flow ceases in all the major rivers particularly the Oli River and only pools remain which provides sources of perennial water for wildlife population.

19. Research methodology

Reconnaissance survey: The reconnaissance survey was carried out in the month of June 2008 to get acquainted with the terrain of the park. Information from the park management and the review of previous literature on salt lick mineral in the park served as a bedrock towards successful reconnaissance survey. This aided in choosing the study side and the ecological survey method used. The study covers between June 2008-February 2009 (6month). The following were predetermined to aid sampling design.

Ecological survey: Based on the methodology adopted from Ayodele,(1988), Lameed,(1995) and Akanbi,(1997), which stated that an ecological survey for an area should be conducted on a comparative bases, particularly the heterogeneous to indicate a long term range. The study was carried out in the Borgu sector of Kainji Lake National Park. Out of twenty salt lick areas in the Kainji Lake National Park, four were selected based on concentration of the salt and species to the spot, management and tourists' preference and proximity to the camp. The inventory of the place was taken which are:

a. Inventory of the salt lick in three places, that is, the middle, the upper and the lower parts to look at the composition of the minerals.
b. The fauna and vegetation were assessed by laying four (4) transect 1km each in each of the salt lick. This transect was taken towards the north, south, east and west which form the radii of the utilized areas.

Point Centre Quarter Method (Vegetation sampling): For vegetation, point centred quarter method were taken at every 10m and the fauna were assessed looking into the direct and indirect survey of fauna species to note the absolute and relative density of the species around the area. The relative density is to asses the significant of the spur rate of the species along transect by identifying the faecal samples, foot prints, death/ carcass and calls of the species available within the transect (Hopkin, 1974).

A transect of 1km was cut from each of the sites with the use of a cutlass and a compass to align the transect line. Each transect was marked at 50m intervals. Daily survey of the transect along the salt lick was carried out as early as 6.30am with the assistance of the research officer and a ranger from the park and backed by 2.pm. The early morning study was designed to survey some of the wild animals that are inactive and sluggish in the afternoon whose detection may be difficult and liable to error. At the beginning of each early outing on the salt lick, the following were recorded;

- Location and the name of the tract
- Date
- Time (beginning and stop)
- Number of the salt lick.

Line Transect Survey: A line transect according to Roger, (1975), in a fixed path independent of external features along which survey will take place. At each outing the following equipments were used;

- Binocular for clearer viewing of animals at a long distance
- Camera for taking photographs
- Field notebook, pen and pencil
- A cutlass
- A wristwatch.

The following information was recorded at each salt lick sites

- Name, number and species of the flora and fauna around the salt lick sites
- Activities of the animals
- Droppings of the named and species of the animal found
- Footprints of the named animals
- Food materials. Carcass of the animal
- Calls of the animal

- Habitat type

20. Search method for direct and indirect methods

Both the direct and the indirect methods were used to study the animals that visit the salt lick. This involved walking along the delineated transect line and looking for signs of species presence. The stop and search method was used as an indirect method of estimating wildlife population through the use of faecal droppings, footprints, feeding/remnant and calls of the animals. The following assumptions for animals that visit salt lick was outlined by Burham et al 1980; Seber, (1982). These are

1. The number of footprints of a particular animal was closely examined and traced to the salt lick.
2. The number of faecal samples was observed along the transect
3. The feed/remnants of animal along the transect were also examined
4. The activities of the animals around the area and along the transect was examined
5. Sighting of animals was carried out
6. The faecal droppings containing salt lick was picked and examined
7. Backward movement along transect to confirmed the samples was limited and observation period not more than 10minutes.
8. Samples positioned directly over the transect line was not missed. It was also understood that not all the sample within the survey area was detected. Some was inevitably missed and possibility of detection declined with increase distance from the transect line or survey path.

21. Population density

To determine the population density for each species of the animal encountered during the survey. Time species count was used as outlined by (Ajayi, 2001). The formula is as follows

$$P = \frac{AZ}{2XY}$$

Where:
A = Area occupied by the species
Z = number of animal seen
2 is constant
X = sighting distance
Y = length of transect

The area (A) is determined by multiplying the transect width by the length of the transect.

The number of the animals seen (Z) = species seen/transect

22. Data analysis

All data collected were subjected to appropriate statistical analysis depending on the nature of the study. Correlation, Analysis of variance, descriptive analysis and T-test to differentiate two variables were used to draw conclusion.

Graphs, photographs and diagrams were used at appropriate places for proper result. Chi-square test was also used to analyse the result.

23. Results and discussion

	A	B	C	D
Kobus kobs	65	51	20	20
Hippotamus	39	21	13	33
Roan antelops	1	10	4	-
Red flank dunker	5	13	1	-
Hares	1	2	-	-
Baboons	10	13	9	4
Silvet cat	2	3	5	2
Bush buck	9	2	-	-
Monkey	4	3	2	1
Lion	1	5	-	1
Flankolin	-	-	-	1
Crocodile	-	-	-	1
Total	137	123	54	63
Average	13.70	12.30	7.71	7.86

Table 2. Table 2: Diversity of Wild animals sampled within the various transects studied.

Table 3 shows record of animal sample within various transects at the four sites (A, B, C and D). It was observed that the highest number of animals is recorded in transect A (137) with an average number of 13.7, followed by transect B (123) with an average number of 12.3. the lowest value is recorded in transect C (54) with an average of 7.7.

Parameters	df	ms	f	p-level
Salt lick1	3	81687.34	10.1038	0.0013+
Error	12	8084.833		
Total	15			

Table 3. ANOVAs for Number of trees/hectare

The table 3 above shows that at p-level there was no significant difference (p>0.05) in the number of trees/hectare. Therefore the alternative hypothesis was accepted that said there was no significant difference in the trees/hectare in the salt lick sites.

Parameters	df	ms	f-cal	p-level
Salt lick2	3	305.58	2.0204	0.1648
Error	12	151.25		
Total	15			

Table 4. ANOVAs for number of different trees at the salt lick sites (tree diversity)

Table 4 above shows that there was significant difference (p<0.05) in the tree diversity across transects at the salt lick sites. Therefore the null hypothesis is accepted while the alternative is rejected.

Parameters	df	ms	f-cal	p-level
Salt lick3	3	1757.67	.390	.7623
Error	12	4702.5		
Total	15			

Table 5. ANOVAs for number of animals/hectare at salt lick sites

The table 5 above shows that at ($p<0.05$) there was significant difference in the animals across the four transects. Therefore a null hypothesis was rejected while the alternative hypothesis was accepted.

Parameters	df	ms	f-cal	p-level
Salt lick4	3	39.3333	.39008	.762310
Error	12	100.8333		
Total	15			

Table 6. Anova for different animals/hectare at the salt lick sites (animal diversity)

The results obtained from the direct and indirect survey shows that f-cal was significantly difference at ($p>0.05$). Alternative hypothesis was accepted that said that there is significant difference in animal species across transects as seen in table 6 above.

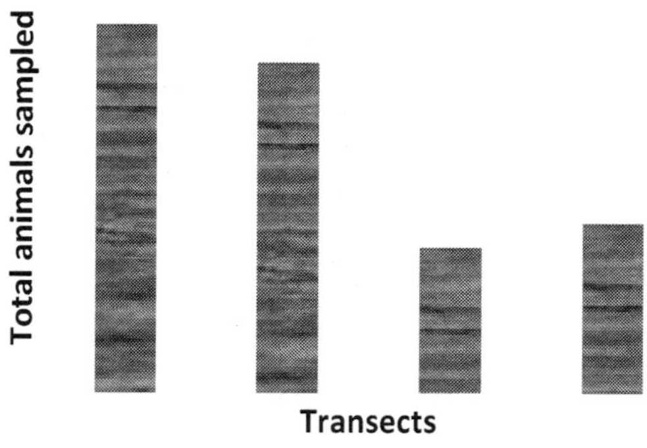

Figure 3. Variation of animal sampling within transects studied.

Figure 3 above and the graph sows that the highest animal sampling was recorded in transect one (13.7), followed by transect two (12.3) and the lowest value was in transect three (7.7).

The table 7 below shows the records of various tree sampled within the transects. Transect three(C) recorded the highest value of trees (213) and the average is 14.20 while the lowest is recorded in transect A (149) and the average of 9.93. Meanwhile *Acasia spp* was found to be dominant species within the four transects.

	Transects			
Trees	A	B	C	D
Acasia spp	42	84	28	81
Annogeissus leiocarpus	30	11	33	6
Daniella Oliverie	6	22	6	14
Terminalia spp	32	8	37	10
Detarium microcarpum	4	4	51	2
Burkea Africana	12	7	12	4
Gardenia sokotolensis	2	7	1	1
G. aquala	8	6	9	6
Strychnos spinosa	1	1	4	14
Bytyrospermum paradoxum	1	3	6	9
Darsperus mispiliformis	1	4	1	1
Tamarindus indica	5	7	3	2
Afzelia Africana	2	3	5	2
Tetrap tetrap	2	4	12	6
Acaria spp	1	4	5	2
Total	149	175	213	160
Average	9.93	11.67	14.20	10.67

Table 7. Diversity of Vegetation sampled within transects studied.

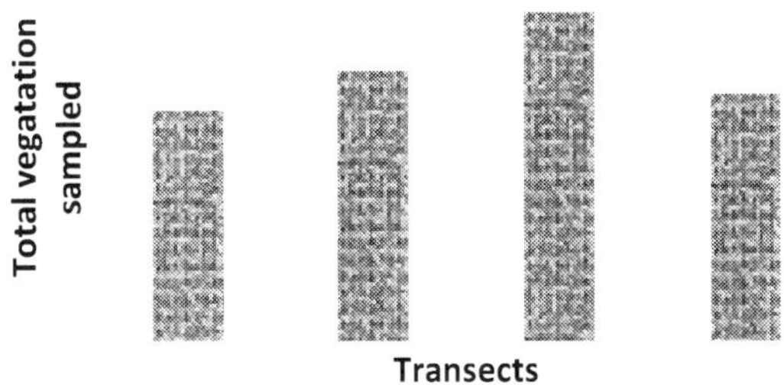

Figure 4. Variation in total vegetation (trees) samples within transects.

Figure 4 above shows that transect three has the highest tree samples (14.2) followed by transect two (11.67) and the lowest was recorded within transect one (9.93).

The results obtained from the direct and indirect survey shows that f-cal from the two survey types was significantly different ($p>0.05$) fro each other. Alternative hypothesis was accepted that said there are significant differences in the trees sampled as seen in table 8 above.

ANOVA: Single Factor

SUMMARY

Groups	*Count*	*Sum*	*Average*	*Variance*
A	15	149	9.933333	179.2095
B	15	175	11.66667	424.9524
C	15	213	14.2	238.3143
D	15	160	10.66667	397.8095

ANOVA

Source of Variation	*SS*	*Df*	*MS*	*F*	*P-value*	*F crit*
Between Groups	156.1833	3	52.06111	0.1679	0.917595	2.769431
Within Groups	17364	56	310.0714			
Total	17520.18	59				

Table 8. ANOVA of trees sampled with the transects groups.

Salt licks	**%Na**	**%Ca**	**%Mg**	**%K**	**Fe (mg/Kg)**	**%PO4**	**%SO4**
1	3.69	1.79	0.88	6.76	1149	0.73	0.35
2	4.18	3.54	1.23	4.46	992.81	1.93	0.55
3	3.79	1.92	0.81	6.81	1168	0.89	0.39
4	4.21	1.85	0.79	6.95	1173	0.87	0.43

Table 9. Chemical analysis of mineral contents in salt licks.

Fig.5 shows that the level of sodium (Na) is highest in salt lick site four (4.21%), followed by salt lick site two (4.18%). There is no much different between the percentage level of sodium salt lick sites one and three (3.695 and 3.765), but the lowest was recorded in salt lick site one.

Salt lick site two has the highest percentage level of calcium (3.54%), while the lowest percentage level of calcium was recorded in salt lick site one (1.79%). Salt lick site three and four has (1.92% and 1.85%) respectively (Fig. 6).

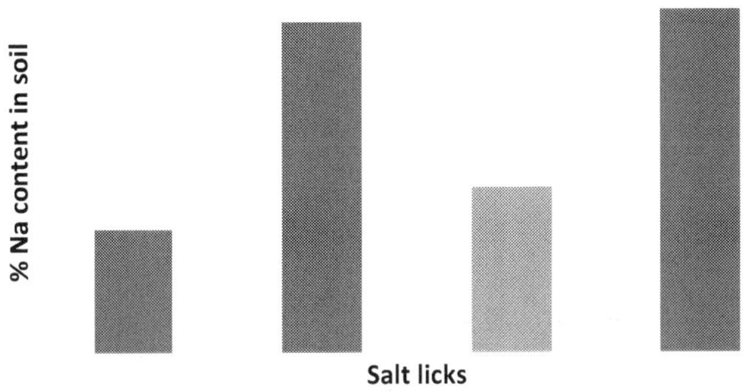

Figure 5. Concentration of Na (%) in salt licks studied.

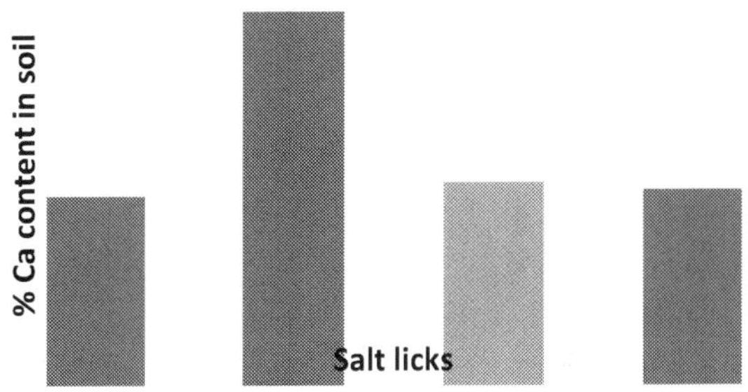

Figure 6. Concentration of Ca (%) in salt licks studied.

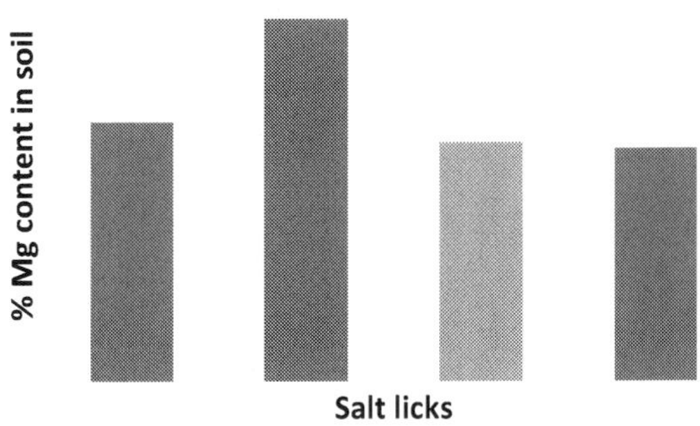

Figure 7. Concentration of Mg (%) in salt licks studied.

Fig.7 above shows that the salt lick site two has the highest level of magnesium (1.23%), followed by salt lick site one (0.88%). The lowest value was recorded in salt lick site four (0.79%).

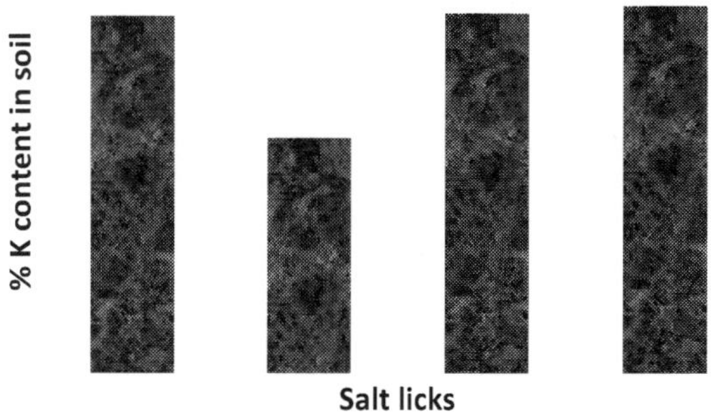

Figure 8. Concentration of K (%) in salt licks studied.

Among all the minerals present in the salt lick sites, potassium was found to be the highest mineral content. Fig. 8 above shows that salt lick site four has the highest level of potassium content (6.95%), followed by salt lick three(6.8%), salt lick one(6.76%) and salt lick two has the lowest (4.46%).

Figure 9. Concentration of PO_4 (%) in salt licks studied.

Figure 9 shows the highest percentage level of phosphate in salt lick two (1.93%). The lowest level is in salt lick one (0.73%). Salt lick three and four have (0.89% and 0.87%) respectively.

Iron is measured in milligramme per kilogramme (mg/kg) and not in percentage. The highest value is in salt lick four (1173mg/kg), while the lowest is in salt lick two (992.81mg/kg). Salt licks one and three have (1149mg/kg and 1168mg/kg) respectively (Fig. 10).

Figure 10. Concentration of Fe (mg/Kg) in salt licks studied.

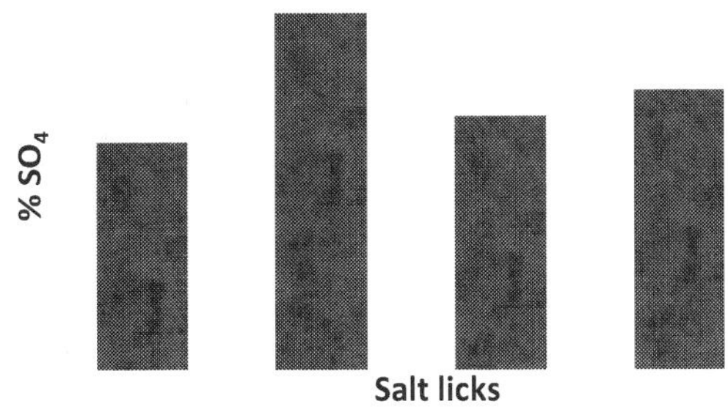

Figure 11. Concentration of SO_4 (%) in salt licks studied.

Sulphate was found to have the highest percentage value in salt lick two (0.55%), and the lowest percentage value in salt lick one (0.35%). Salt lick four has the value of (0.43%) and salt lick three has 0.39% (Fig. 11).

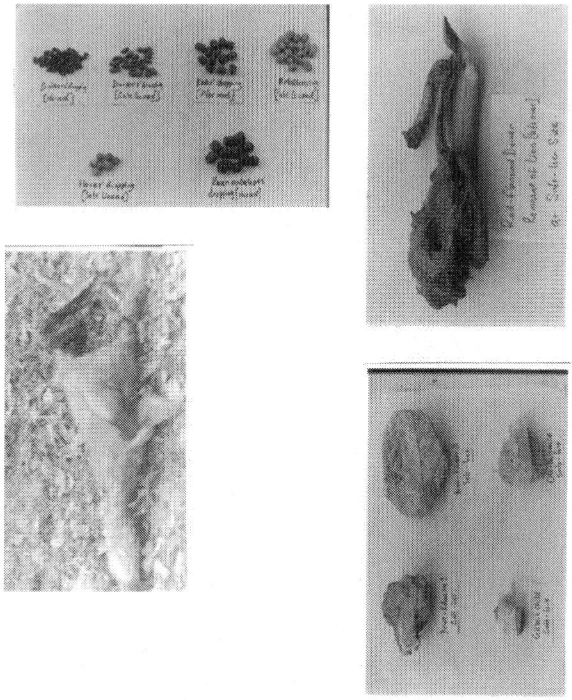

Figure 12. Droppings of wild animals and remains of some species killed by carnivores utilizing the saltlick site within the KLNP.

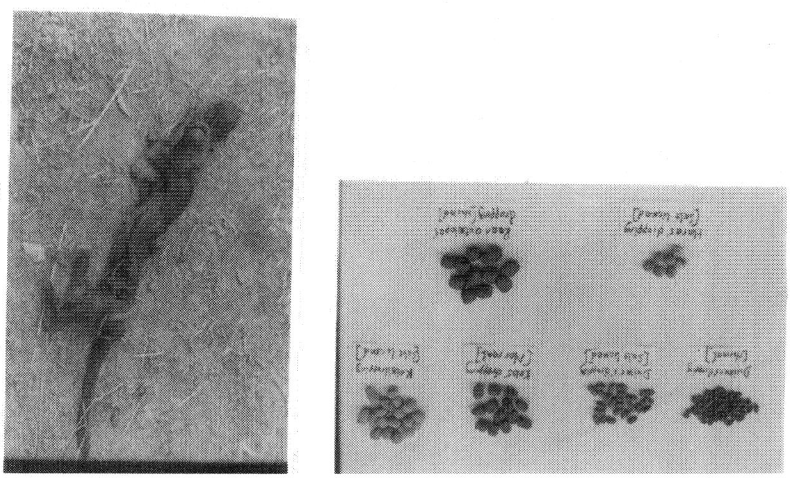

Figure 13. One of the carcass of infant herbivores found at the saltlick site with the collected droppings.

Figure 14. Elephant is one of the common preferred species visiting saltlick site in the Park.

24. Conclusion and recommendations

Kainji Lake National Park is an important reservoir for several wild herbivores. From this study, the results show that the highest population of animal was recorded in salt lick site three (3) and one (1), that is the Oli Entrance and kilometre three Drum. This was significantly different ($p<0.05$) and higher than those in salt licks four (4) and two (2). Therefore it could be deduced that herbivores have more preference to salt licks near river sides than those that are not.

The population density of the animal in the area was established by indirect/relative survey, since it has to do with the observation of some indices like footprints, feed remnants/signs, and faecal droppings. Carnivores were also noticed at the study site like lion by the carcases and call of the animal. It was observed that there was an interrelationship between animals in the lick as some fruits were found at the lick sites. Most of the animals were not seen due to their nocturnal nature and shyness. Therefore indirect/relative method of transects survey was ideal to established their presence, distribution and abundance. The activities of the animals were also assessed.

From the study, the result of the analysis shows that both the macro and trace elements are present in the salt licks of the park. The mineral elements identified in the salt lick are; calcium, sodium, potassium magnesium, phosphate, iron, and sulphate. Potassium is most abundance mineral element in the salt licks while sulphate is the least abundant. The analysis revealed that the salt lick of the Oli complex contain the highest potassium and iron. This is probably due to the accumulation of dead leaves, defecations, urination and soil parent materials. The salt lick one has low calcium and sodium mineral elements due to the fact that there was less abundance of vegetation due to erosion of the upper slope of the reserve. Hence these minerals could have been leached away by erosion.

Salt lick spot are quite abundance in Kainji Lake National Park (KLNP) which is frequently used by different herbivores and carnivores. Hence the lick plays an important role in supplementing minerals lacking in the animal's diets. The licks therefore helps in preventing nutritional diseases and disorders that could result due to lack of these minerals and also in biodiversity conservation of wild fauna species of the park.

25. Recommendations

The results show that different herbivores utilize the salt licks in the park and this cannot be completely exploited on short duration and single handed project of this nature. It required continuous research work to be carried out on many other areas of interrelationship with natural ecosystem where they are found. Effort should be made to establish the proximate population and consistence to monitor the population. This will enable the management of the national park to know if the government and non governmental investment is achieving the desired impact.

Efforts should be made towards the rate of poaching activities in the park by reintroducing the special squad to complement the effort of the forest guards. Communication equipments like walkie-talkie, GSM antenna should be provided in the park, handsets, boots and patrol clots should be provided for the guards to carry out their duty effectively. The numbers of park guards present are grossly not adequate as compared to the increase number of poachers daily. More park stations should be created to boost the morals of the guards and more sophisticated arms given in view of the risky nature of the job. Residential quarters should be built by government for the park guards to live as the situation of rented building is exposing them to high risk and at the same time stand a chance of compromising their work with the poachers. Good incentives like food, medicals allowances should be provided for the park guards. Unannounced visit to the park station by senior officers should be vigorously carried out to ascertain if guards are on patrol regularly.

The Support Zone Development Programme (SZDP) should supervise the disbursement of loans to the community and the use of the loan to the community. The enlightenment on the conservation should be fully understood by the people. Therefore it should go beyond radio, television announcements, pamphlets and seminars, but house -to- house enlightenment should be used for the message to be fully understood. Infrastructures should be provided for the host community to justify the forfeiting of their resources to the government. Indigenes should be given scholarships to enable them go to school to reduce their dependence on hunting and logging which they carried out as a means of livelihood.

Author details

A. G. Lameed

Department of Wildlife and Ecotourism Management, Faculty of Agriculture and Forestry, University of Ibadan, Ibadan, Nigeria

Jenyo-Oni Adetola

Department of Aquaculture and Fisheries Management, Faculty of Agriculture and Forestry, University of Ibadan, Ibadan, Nigeria

26. References

Afolayan, T. A. (1977): Savannah Structure and Productivity in Relation to Burning And Grazing Regime in KLNP, unpublished PhD Thesis. Dept. of Forest Resources Management. University of Ibadan.

Ajayi, S. S. (1979): Utilization of Forest by Wildlife in Africa, F. A. O. Rome 1979

Akanbi, O. A. (1997): An Ecological Basis For Management of Enamania Game Reserve in Benue State, Nigeria. Ph D. Thesis unpublished, Department of Wildlife and Fisheries, University of Ibadan.

Alloway, B. J. (1990): Heavy Metals in Soil. John Wiley and Sons Inc. New York.

Andrewartha, H. G. and Birch, C. C. (1954): The Distribution of Abundance of Animal in Chicago, University of Chicago Press.

Andrewartha, M. C. (1954): Animal Environment and Man in Africa. 3rd Edit. 1974 Pp 120 – 134.

Anon, O. (1991): The Role of Tropical Timber Extraction in Species Extinction. *London Friends of the Earth.*

Ayeni, J. S. O. (1971): Some Aspect of Natural Mineral Licks by Large mammals of Yankari Game Reserve, North Eastern State of Nigeria. B.Sc Desertation in Forest Resources Management. University of Ibadan . Vol. 7 pp 47-53.

Ayeni, S.O. (1972): Chemical Analysis of some soil samples from Natural Licks. Vol, 2 pp 16-23.

Ayeni, S. O. (1975): Utilization if water holes in Natural Parks. *East Africa Wildlife Journals.* Vol. 13 pp 305-323.

Ayeni, S. O. (1979): Utilization of Mineral salt Licks. Wildlife Management in Savana Woodland. Cambridge University Press.

Ayodele, I. A. (1988): An Ecological Basis for Management of Old Oyo National Park, Unpublished Thesis Department of Wildlife and Fisheries Management University of Ibadan.

Benjamin, B. I. (2007): Preservation of Africa's Biodiversity. In Nature Fauna FAO/UNDP publication, Vol. 8, No 1. 1992.

Bowen, H. D. M. (1979): Environmental Chemistry of Mineral Elements. Accademic Press, London.

Buckman, H.O. and Brandy. N. C. (1960): Reaction of Saline and Alkaline Soil. 6th Edition pp 567.

Budd, J. (1973): Infectious Disease of Wild Mammals. (Ed. J. W. Davis, L. H. Karstad and D. O. Trainer). The Iowa State University Press, 36-49pp.

Burnham, A. et al (1980): Estimation of Density from line transect Sampling of Biological population of Wildlife Monogr. 72: pp 202.

Clegg, P. S. and Clegg, A. G. (1978): Biology of Mammals. Williams Heineman Medical Book Limited, London.

Cowam, I. McTaggart and Brink. V.C. (1949): Natural Game Lick in the rocky Mountains National Parks of Canada. Jour of Mammalogy, vol 30 pp 387.

FAO, (1983): Kainji Lake Research Project. An Ecological Survey of Borgu Game Reserve, UNDP/FAO Tech. Report, Rome.

Friend, M. and Trainer, D. O. (1970): Serology of Newcastle Disease in Captured Mallard and Swams. Jour. on Wild Diseases.Vol. 6 pp130-137.

Heady, H.F. and Heady, F.B (1982): Range and Wildlife Management in the Tropics Vol 2. Longman Pub. New York.

Heady, H. F. and Child, R. D. (1994): Rangeland Ecology and Management. Westview Press.

Hedrick, D. W. (1966): What is Range Management? Pp 19-11.

Lasan, A. H. (1979): Habitat Management for Wildlife Conservation with Respect to Fire, Salt Lick and Water Regime. Management in Savannah Woodland (Ed. S. S. Ajayi, and L. B. Halstead). University Press.

Lamed.G. A. (1995): Some Aspect of Ecology of Wildlife in Gashaka Gumti National Park. Unpublished M.Sc. Thesis University of Ibadan.

Luick, J.R. (1979): Nutrition and Metabolism of Reindeer Caribou in Alaska with Special Interest in Nutritional and Environmental Adaptation. Energy Commission 1969, pp 1-32.

Mertz, W. (1976): Trace Elements in Animal Nutrition. Prog. of Symp-Nuclear Techniques in Animal Production and Health, FAO/IAEA, Vienna.

Milligan, K. R. N. (1979): An ecological basis for the management of Kainji Lake National Park, Nigeria. Ph D Thesis. University of Ibadan.

Nancy, A. S. and Haufler, A. A. (1975): Mineral Elements in Animal Nutrition. *Jour. in Animal Nutrition.* 36(5): 169.

NEST, (1991): Nigeria;s Threatened Environment; A National Profile , Nigeria Environmental Action Team, Publication. Ibadan. Pp 1-28.

Okeyoyin, O.A. (1988): Behavioral Ecology of Wildlife in KLNP. A Study of food resources exploitation. Unpublished M. Sc Thesis. Dept of Wildlife and Fisheries Management. University of Ibadan.

Popoola, L. (2000): Practise of Environmental Management. *Forest Production Paper* Presented at International Enabling Conference. Abuja. 9th-10th oct, 2000. Pp 26.

Pethes, G. (1980): The Need for Trace Element Analysis in the Life Sciences. Elemental Analysis of Biological Materials. IAEA, No 197, Vienna.

Roger, O.U. (1975): Ground Census Survey of Wildlife in Woodland Habitat.

Seber, G. A. F. (1982): Estimation of Animal Abundance and Related Parameters 2nd edition, Macmillan. New York.

Society of Range Management (1989): Terms used in Range Management. PP 39.

Stoddart, L.A. (1967): Range Management Techniques. Pp 20-30, 37.

Underwood, E.J. (1977): Trace Elements in Human and Animal Nutrition. Press. New York.

Wari, M. (1993): Know the Natural History of Nigeria Wildlife. Park News, 1(5) 12-15.

World Book Encyclopaedia (1995): Vol. 19. World Book International London.

Chapter 3

Biodiversity and Conservation Status of a Beech (*Fagus sylvatica*) Habitat at the Southern Edge of Species´Distribution

Rosario Tejera, María Victoria Núñez, Ana Hernando,
Javier Velázquez and Ana Pérez-Palomino

Additional information is available at the end of the chapter

http://dx.doi.org/10.5772/51365

1. Introduction

The aim of Habitats Directive (European Council, 1992) is "to contribute towards ensuring biodiversity through the conservation of natural habitats and of wild fauna and flora in the European territory of member states" (Article 2.1). This directive identifies a set of natural habitats and wild species of fauna and flora of Community interest (Annexes I and II of the Directive) and establishes the requirement to maintain a favourable conservation status. Therefore, Member States designate Special Protection Areas (SPAs), which are provisional sites of Community Importance (SCIs).

To ensure its enforcement, Member States should establish the necessary conservation measures involving, if necessary, appropriate management plans (Article 6.1).

According to the Article 1 of the Directive, the state of conservation of natural habitat is considered favourable when:

- its natural range and areas within that range are stable or increasing and
- the specific structure and functions necessary for long-term viability exist and are likely to continue to exist in the foreseeable future and
- the status of its typical species is favourable .

Member states have implemented different strategies for evaluating the conservation status of habitat types and species of Community interest, basing on both the European Commission reports (European Commission, 1995, European Commission, 2006, Shaw and Wind, 1997)) and scientific research (Bock et al., 2005, Dimitriou et al., 2006, Lang and Langanke, 2005, Noss, 1990, Noss, 1999, Roberts-Pichette, 1998, Simboura and Reizopoulou, 2007).

In addition, some Member States have developed their own methodologies for assessing the conservation status, such as Germany, Austria, the Netherlands, Portugal, Spain and the United Kingdom (Velázquez et al., 2010). These previous studies often propose numerical indicators and have been applied at regional or national levels (Cantarello, 2008).

In Spain, in 2009 the Ministry of Rural and Marine Environment issued a set of guidelines at national level to assess the conservation status of habitats and species of Community interest (AUCT. PL. , 2009). The main objectives of these guidelines are to identify and adequately describe the 117 habitat types and typify their conservation status.

2. Objetives

The objective of this study was to determine the conservation status of habitat 9120 - *Atlantic acidophilous beech forests with Ilex and sometimes also Taxus (Quercion robori-petraeae or Ilici-Fagenion)* within the beech forest of "Dehesa del Moncayo" (Spain) by applying the methodology provided by the Spanish Ministry. This implies a revision of the methodology at local level.

3. Methodology

The conservation status of habitats is assessed according to four general factors (European Commission, 2006): range and area occupied by the habitat, typical species, structure and function and future perspectives (Table 1). Each one can take the value of favourable, unfavourable-inadequate, unfavourable-bad or unknown. The overall assessment of the conservation status arises by combining the values obtained in Table 2 with the General Assessment Matrix (European Commission, 2006)

FACTOR	INDICATOR
Range and area occupied	Area (ha) and trend
Typical species	presence and abundance of typical species
Structure and function	Dead wood
	Forest structure
	Fragmentation
	Presence of Picidae
	Degree of defoliation
Future prospects	Current and potential threats

Table 1. Adaptation of the methodology for the habitat 9120.

PARAMETER	CONSERVATION STATUS			
	Favourable (green)	Unfavourable-inadecuate (amber)	Unfavourable-bad (red)	Unknown
Distribution area (range)	The range of habitat is stable (loss and expansion are balanced) or increasing and is not less than the "favourable area of reference"	Any situation other than those described in "green" or "red"	Large decrease in the range (equivalent to a loss of more than 1% per year over a period specified by the EC, other thresholds can be used but should be explained in Annex D Or the range is more than 10% below the "favourable reference range	Not available or insufficient reliable information
Area occupied by the habitat within the range	The area occupied by the habitat is stable (loss and expansion are balanced) or increasing and is not less than the "favourable area of reference " and without major changes in the distribution pattern within the range as a whole (if data are available for evaluation)	Any situation other than those described in "green" or "red"	Large decrease of the surface (equivalent to a loss of more than 1% per year over a period specified by the MS, other thresholds can be used but should be explained in Annex D Or with losses (negative changes) in the pattern of distribution within the range Or the current surface is more than 10% below the "favourable reference range"	Not available or insufficient reliable information
Structure and functions	Structures and functions (including typical species) in good condition and without significant damage/pressure	Any situation other than those described in "green" or "red"	More than 25% of the habitat is unfavourable in terms of its specific structures and functions (including typical species)	Not available or insufficient reliable information

Future prospects (regarding range, area covered and structure and function)	Future prospects are excellent / good, no significant effects of future threats, the long-term viability is guaranteed	Any situation other than those described in "green" or "red"	Future prospects are bad, serious impacts of threats, the long-term viability is not guaranteed	Not available or insufficient reliable information
Overall assessment of conservation status	All "green" or three "green" and one "unknown"	Any situation other than those described in "green" or "red"	Two or more "unknown" combined with "green" or all "unknown"	Not available or insufficient reliable information

Table 2. General Assessment Matrix

3.1. Distribution area and area occupied

The distribution area can be defined as "the current habitat areas " (AUCT. PL. , 2009). It aims to identify changes of distribution patterns of the habitat within the range. This factor makes sense at the biogeographic region scale. However, the range does not apply at the local level.

The area occupied assesses the area covered by the habitat in the study area and its trend:

1. Area occupied by habitat in the study area (in hectares).
2. Date of assessment.
3. Trend of area (stable, increasing, decreasing or unknown).
4. Magnitude of the trend.
5. Period of trend.
6. Reasons for the trend.

The concept of "Favourable Area of Reference" (FAR) shown on the General Assessment Matrix is defined as "the minimum area required within a biogeographic region to ensure long-term viability of a type of habitat" (European Commission, 2006). Neither this concept is of application for the current study, since it is a study at the local scale.

- Measuring procedure

For the present study, the vegetation map of the Moncayo Natural Park has been used. We have distinguished three main types of vegetation: beech (used for extracting charcoal), scot pine reforestated in the 19th century and natural Pyrenean oaks (Gallo Manrique, 2011).

- Assessment of conservation status

The conservation status was assessed based on the trend of the area occupied, giving a value of zero to the status in 2000, as proposed in the methodology.

3.2. Typical species

This factor considers the presence and viability of populations of typical species. That is, those that are indicators of habitat status. They can also be defined as those species relevant to maintain the habitat in a favourable conservation status, either because of their abundance or because of their influence in the ecological functions.

Typical species of the habitat 9120 are:

Flora: Yew (*Taxus baccata L.*), holly *(Ilex aquifolium L.), Lobaria pulmonaria L.*

Amphibians: Salamander (*Salamandra salamandra*)

Mammals: Gray dormouse (*Glis glis*)

Birds: White-backed woodpecker (*Dendrocopus leucotus*), Black Woodpecker (*Dryocopus martius),* Nuthatch (*Sitta Europea*), Treecreeper (*Certhia familiaris*), Pied flycatcher (*Ficedula hypoleuca*), Marsh Tit (*Parus palustris*)

Invertebrates: saproxylic invertebrates: *Elona quimperiana, Rosalia alpina, Osmoderma eremita, Limoniscus violaceus, Cerambyx cerdo, Lucanus cervus, Gnorimus variabilis, Caliprobola speciosa.*

- Measuring procedure

The method used was based on observations of presence/absence of typical species during the field work reinforced with the wildlife catalog of Moncayo Natural Park (Gobierno de Aragón, 2002)

- Assessment of conservation status

It is not imperative that a particular location holds all or most of its typical species for a favourable conservation status (European Commission, 2006). But the set of all the habitats at the national or biogeographic scale must have long-term viable populations of all or many of the typical species of the habitat.

Since this study covers a small area of habitat 9120 in Spain, we assessed the number of typical species present in the forest. The result of this factor must be consistent with the structure and function factors.

3.3. Structure and function

Structure and function define the quality of habitat 9120 through four parameters: dead wood, stand structure, fragmentation, presence of *Picidae* and degree of of defoliation.

To determine the overall status of the structure and function, each indicator takes a value (0: unfavourable-bad, 1: unfavourable-inadequate, 2: favourable). The overall status of the structure and function can be unfavourable-bad — for results below 40% of maximum punctuation —, unfavourable-inadequate — from 40 to 75% —, and favourable — above 75%.

3.3.1. *Dead Wood*

This indicator measures dead wood (m^3/ha), separating it according to: species, standing or fallen, size and level of decomposition.

- Measuring procedure

The inventory of dead wood was done by strip-plots 500 m long and 20 m wide (1 ha), as proposed in the Spanish methodology (Olano and Peralta de Andrés, 2009). In these plots we measured dead wood — both standing and fallen —, diameter, length, species, and degree of decomposition.

The degree of decomposition was assessed according to the following criteria (Table 3).

Degree of decomposition	Description
Level 1	Healthy wood, with bark; wood intact
Level 2	Healthy wood, beginning of the bark loss
Level 3	Wood beginning to rot away. Without bark
Level 4	Very rotten wood, full of holes
Level 5	Completely rotten wood that breaks when touched

Table 3. Degree of d criteria

- Assessment of conservation status

Dead wood in forests ranges from 10 to 150 m^3/ha (Müller and Bütler, 2010). According to these authors, most species linked to dead wood seem to be present in hardwood forests for volumes between 30 and 50 m^3/ha.

- Unfavourable-Bad: less than 10 m^3 of dead wood per hectare.
- Unfavourable-inadequate: 10 to 30 m^3 of dead wood per hectare, with at least 30% of deadwood above 30 cm diameter and 20% of standing dead wood.
- Favourable: more than 30 m^3 of dead wood per hectare, with at least 12 m^3/ha of dead wood above 30 cm diameter and at least 4 m^3/ha of standing dead wood. It is important that dead wood presents all stages of decomposition and it is distributed throughout the habitat.

3.3.2. *Forest structure*

Forest structure is evaluated according to three indicators: abundance of overmature trees (trees with dbh above 45 cm), structural diversity and species diversity. It is necessary to assess the number of stems/ha per diameter class and indicate the proportion of species.

To determine the overall status of forest structure, each indicator has a value (0: unfavourable-bad, 1: unfavourable-inadequate, 2: favourable). The overall status of the structure and function is unfavourable-bad — for results below 40% of the maximum punctuation –, unfavourable-inadequate – from 40 to 75% –, and favourable – above 75%.

- Measuring procedure

We inventoried diameter and species in circular plots 10 m radius, located at the points 100, 300 and 500 m of the strip-plots used for the inventory of dead wood.

- Assessment of conservation status

Overmature tree (dbh> 45 cm):

- Unfavourable-Bad: less than 5 trees/ha
- Unfavourable-inadequate: 6 to 10 trees/ha.
- Favourable: above 10 trees/ha

Species diversity:

- Unfavourable-Bad: less than 5 trees (dbh> 15 cm) /ha of other native tree species.
- Unfavourable-inadequate: 5 to 10 trees (dbh> 15 cm) /ha of other native tree species
- Favourable: above 10 trees (dbh> 15 cm) /ha of other native tree species

Structural diversity:

- Unfavourable-Bad: 90% trees in the same diameter class (classes of 20 cm).
- Unfavourable-inadequate: from 80% to 90% trees in the same diameter class (classes of 20 cm).
- Favourable: less than 80% trees in the same diameter class (classes of 20 cm).

3.3.3. *Fragmentation*

This indicator evaluates whether the habitat is a continuous patch of sufficient extent to ensure species survival or, conversely, is composed of individual patches.

Fragmentation is a very important element for forest communities that affects the quality of habitat and causes loss of species (Telleria and Santos, 2001). In beech forests, typical flora and fauna species are strongly affected by the edge effect due to their dependence on low light and high relative humidity.

- Measuring procedure

Fragmentation is quantified by comparing the total habitat area with the surface free of edge effect (effective area). We considered an edge effect of 30 m from the margins of the patches.

- Assessment of conservation status
    - Unfavourable-Bad: ratio between surface without edge effect and total area less than 80%.
    - Unfavourable-inadequate: ratio between surface without edge effect and total area from 80 to 90%
    - Favourable: ratio between surface without edge effect and total area above 90%.

3.3.4. *Presence of Picidae*

Picidae are known for tapping on tree trunks in order to find insects living in crevices in the bark and to excavate nest cavities. Some of these species require old forests with abundant dead wood, both standing and fallen. The presence of *Picidae* is a good indicator of habitat quality and conservation status.

- Measuring procedure

We performed a visual observation of cavities in the circular plots of the inventory. Additionally we used bibliographic survey (Gobierno de Aragón, 2002).

- Assessment of conservation status
  - Unfavourable-Bad: no *Picidae* nesting.
  - Unfavourable-Inadequate: Only Great Spotted Woodpecker *(Dendrocopos major)* nesting.
  - Pro: woodpecker White-backed Woodpecker (*Dendrocopos leucotos)* or Black Woodpecker (*Dryocopus martius*) nesting

3.3.5. *Degree of defoliation*

This indicator belongs to the group of indicators for the maintenance of health and vitality of forest ecosystems and is considered to be the main indicator of health status (MCPFE, 2002).

In Spain, there is a network of Forest Damage Assessment following the European methodology (International Cooperative Programme on Forests). In Moncayo Natural Park there are 5 plots for that network, but none of them within the beech forest "Dehesa del Moncayo".

- Measuring procedure

We visually assessed the percentage of defoliation in the circular inventory plots

- Assessment of conservation status

We used the thresholds of European Forest Damage Assessment Network (Table 4)

Defoliation class	% defoliation	Description
0	0-10%	No defoliation
1	>10-25%	Minimum
2	>25-60%	Moderate
3	>60-<100%	High
4	100%	Dead tree

Table 4. Criteria for assessing defoliation levels

3.4. Future perspectives

This factor refers to the long-term viability of a habitat considering possible threats, typical species and structure and function factors.

- Measuring procedure

We evaluated the main past and present impacts and the possible future threats that may affect the long term-viability of the habitat.

- Assessment of conservation status
  - Unfavourable: The future scenario does not ensure the long-term viability of habitat 9120
  - Favourable: The future scenario ensures the long-term viability of habitat 9120

4. Case study

4.1. Study area

The study area was the 1494 ha forest " Dehesa del Moncayo" within Moncayo Natural Park (Aragón, Spain) (Fig. 1). It is also included in the Natura 2000 Network as part of SCI ES2430028 "Moncayo" and SPA ES0000297 "Sierra del Moncayo-the-Fayos Sierra Arms" due to six habitats of interest. One of them is habitat 9120 *Atlantic acidophilous beech forests with Ilex and sometimes also Taxus in the shrublayer (Quercion robori-petraeae or Ilici-Fagenion)*.

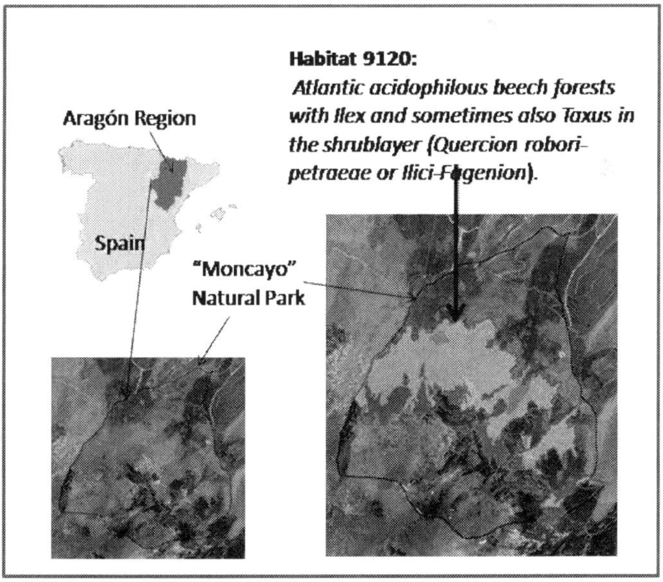

Figure 1. Location scheme

The forest is at the southern edge of the Mediterranean region. However, it is considered an "Atlantic island" due to the altitude and the NW-SE aspect. The Moncayo beech forest is between 1100 and 1900 m a.s.l. facing north or northeast with slopes between 20 and 50%. It has been mainly used for charcoal until 1940, since then it has evolved into a high polewood. It used to be also used for timber but, due to the poor quality of timber, cuttings have been infrequent. In 1978 it was declared Natural Park.

We distinguished four different forest types:

Beech forest (*Fagus sylvatica*)

This is a typical high density beech forest with holly (*Ilex aquifolium*) and blueberry(*Vaccinium myrtillus)* in less dense areas. It covers 360 ha (Table 5).

Fagus sylvatica on screes

Small and branched isolated beech trees on rocky abrupt areas.

Rangeland of *Fagus sylvatica*

It is an area of small size (9.92 ha) used for grazing until 1920. As a result, big trees are accompanied by smaller trees.

Fagus sylvatica with heather (*Erica sp.*)

Beech forest with dense heather and other tree species such as Rebollo oak (*Quercus* pyrenaica) and Scots pine (*Pinus sylvestris*)

Forest types	Area	
	(ha)	(%)
Beech forest (*Fagus sylvatica*)	360,61	76,82
Fagus sylvatica on scree	94,76	20,19
Rangeland of *Fagus sylvatica*	9,92	2,11
Fagus sylvatica with heather (*Erica sp.*)	4,13	0,88
TOTAL	469, 42	-

Table 5. Forests types area (ha and %)

"*Fagus sylvatica* on scree" is assigned to habitat 8130 *Mediterranean and thermophilous screes*. Therefore it was excluded of the conservation status assessment.

4.2. Field survey

We performed a simple random pilot sampling inventory leaning on the network of paths. The pilot sampling was conducted over three consecutive days in July 2010.

The main objective of the pilot sampling was to calculate the variance and to determine whether the error was admissible or the inventory had to be strengthened with new sampling plots.

"Fagus sylvatica with heather" was excluded because it is a small area where the abundance of heather and the low density of trees do not justify the inventory.

4.3. Measuring procedure

4.3.1. Sampling units

We measured dead wood in four strip plots of 500 x 20 m and forest structure in 12 circular plots of 10 m radius (3 in each strip plot), distributed by forest type(Fig. 1).

It was decided to place them on the network of paths since this does not influence significantly the volume estimation of dead wood, due to narrow lanes (less than 1 m wide) and high bandwidth (10 m) both sides the path.

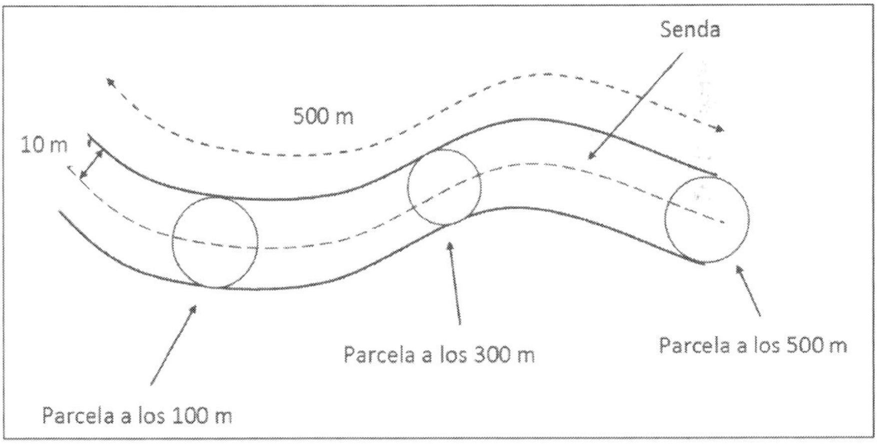

Figure 2. Sampling plots

We measured all dead wood on the ground from a minimum diameter of 10 cm (criterion given by the technical director of the study). Given the abundance of fine twigs on the ground and the large size of the sample plots (1 ha), measurement from 0 cm would have been impossible.

4.3.2. Measured variables

Variables from strip plots:

- Dead wood: Dead wood is classified into several groups: dead wood on the ground, standing dead trees, stumps and dead branches on living trees (Kirby et al., 1998). Diameter of the middle section (diameter at half the length of the fragment), length and level of decomposition was measured for dead wood on the floor.
  Diameter at 1.30m height, total height, and level of decomposition was assessed for standing dead wood.

Variables from circular plots:

- Forest structure: The reference methodology does not establish a minimum diameter for measuring forest structure. Following the technical director criterion, trees below 2.5cm diameter were excluded. Therefore, for the rest of the trees we measured all diameters at 1.30m height and recorded the species.
- Level of defoliation (by visual observation)
- number of cavities (natural or *Picidae*)
- number of trees below 2.5 cm diameter
- Mean height of the stand
- Description of the stand, indicating silvicultural characteristics, non target species, and a sketch/diagram/outline/schema of the vertical forest structure.

We performed a sheet for each plot.

4.4. Field work results

Analysis of variance (ANOVA) was performed for both variables to check significant differences between the two types of beech forest inventoried ("beech *Fagus sylvatica*" and "Rangeland of *Fagus sylvatica*"). The analysis showed no significant differences. So we adopted a single maximum admissible error for these variables.

When sampling dead wood, errors are generally quite high (Kirby et al., 1998, Van Wagner, 1982, Woodall et al., 2006, Woodall and Williams, 2005). Following Van Wagner (1982), in this study we assumed a 20% maximum admissible error.

Error for standing dead trees is higher than admissible (Table 8). However, lack of standing dead trees (Table 7) and heterogeneous distribution are typical of young beech forests.

Furthermore, the error for the variable basal area slightly exceeds the maximum so it was not considered necessary to reinforce the sampling.

strip plot	CIRCULAR PLOT	G (m²)	G (m²/ha)
1	1	0,46	14,60
	2	0,60	18,98
	3	0,75	23,88
2	1	0,79	24,99
	2	1,79	56,90
	3	0,91	29,06
3	1	1,11	35,24
	2	1,15	36,58
	3	0,96	30,48
4	1	0,60	18,96
	2	0,90	28,58
	3	0,61	19,53

Table 6. Basal area (G) by circular plot

	Dead wood volume (m³/ha)		
strip plot	total	Standing dead wood	dead wood on the floor
1	3,71	0,83	2,88
2	5,68	1,70	3,97
3	4,88	0,91	3,97
4	4,23	0,56	3,67

Table 7. Dead wood results by plot

VARIABLE	Mean	Variance	Error (%)
Basal area (m²/ha)	27,79	119,09	21,78
Dead wood on the floor (m³/ha)	3,62	0,27	14,26
Standing dead wood (m³/ha)	1,00	0,24	48,96
Total dead wood (m³/ha)	4,63	0,72	18,36

Table 8. Mean values, variance and error

5. Results for the conservation status assessment

5.1. Range

The results for the area occupied factor according to the methodology are:

Area covered by habitat 9120 within "Dehesa del Moncayo": 374.65 has
Date: 2011.
Trend: stable/increasing.
Trend-period: 1975-2011.
Reasons for the trend: the absence of human influence and good regeneration capacity.

Range	FAVUORABLE

5.2. Typical species

Typical 9120 habitat species present in "Dehesa del Moncayo" are the following:
Flora: Yew (*Taxus baccata*), holly (*Ilex aquifolium*) and *Lobaria pulmonaria*.
Holly is scarce except for some areas of low beech density. Yew and *Lobaria pulmonaria* are scarce or rare.
Birds: Nuthatch (*Sitta European*) and Pied flycatchers (*Ficedula hypoleuca*).
Invertebrates: *Rosalia alpina* and *Cerambyx cerdo*.
During the field work no typical species of fauna were inventoried.
Therefore, the conservation status for this factor is unfavourable-bad.

Typical species	UNFAVOURABLE-BAD

5.3. Structure and function

Snags are scarce and most of them are not large (Fig. 2)

	Volume of deadwood (m^3m^3/ha)		
Strip-plot	Total	Standing dead wood	Dead wood on the floor
1	3.71	0.83	2.88
2	5.68	1.70	3.97
3	4.88	0.91	3.97
4	4.23	0.56	3.67
Mean	**4.63**	**1.00**	**3.62**

Table 9. Volume of total deadwood — standing and on the floor — in each strip-plot

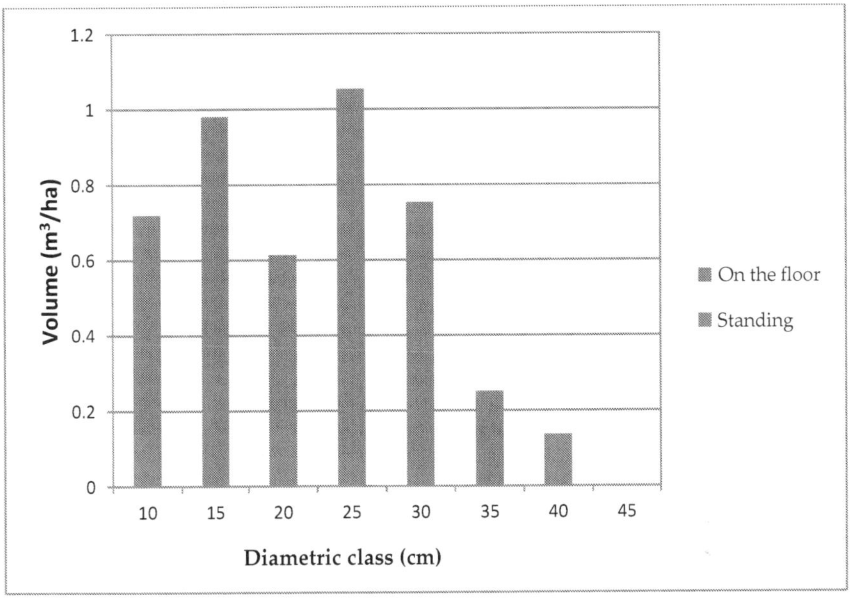

Figure 3. Volume (m^3/ha) by diametric class and type of dead wood

The mean total volume of deadwood is below 10 m^3/ha (Table 9). Therefore, the conservation status is unfavourable-bad.

Deadwood	UNFAVOURABLE-BAD

5.3.1. Forests structure

- Oversized trees: Only 4 oversized trees were sampled (all of them in the "Rangeland of *Fagus sylvatica*" forest type) involving a total of 9.8 tree/ha. An unfavourable-inadequate conservation status was assessed for this component.
- Species diversity: Density of non target species with dbh above 15 cm was 14.7 tree/ha. Based on thresholds proposed in the methodology, a favourable conservation status for this component was assessed.
- Structural diversity:
- Total density reaches 1320 trees/ha. Above 50% of them are trees below 12.5 cm dbh. Regenerated beech (dbh< 2.5 cm)) is the most abundant diameter class, with 32.84% of total trees. Large trees are scarce.
- Less than 80% of trees group into the same diameter class (Fig. 3), so that the status regarding the structural diversity is favourable.

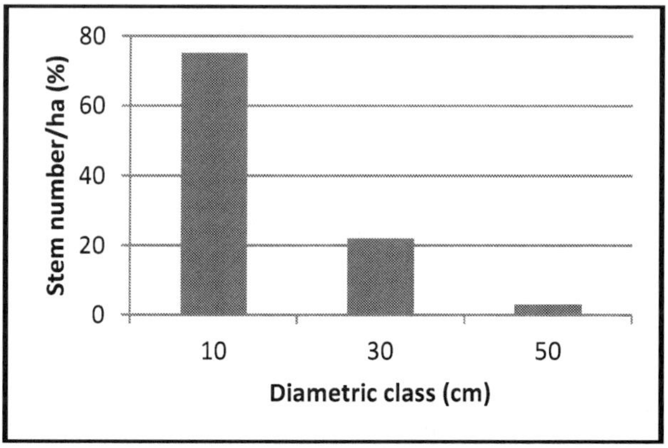

Figure 4. Tree distribution by diameter classes

Individual results (0: unfavourable-bad, 1: unfavourable-inadequate, 2: favourable) reach 5 of the 6 possible points. So the conservation status of forest structure indicator is favourable.

Forest structure	FAVOURABLE

5.3.2. *Habitat fragmentation*

Total area is 374.65 and the effective area, 278.36 ha. This yields a ratio of 74.30% free surface of the edge effect. According to the thresholds in the methodology, this is an unfavourable-bad status.

Fragmentation	UNFAVOURABLE-BAD

5.3.3. *Presence of Picidae*

The only species inventoried are Spotted woodpecker (*Dendrocopos major*), green woodpecker (*Picus viridis*) and Wryneck (*Jynx torquilla*) (experts information).

According to the thresholds established in the methodology, the conservation status is unfavourable-inadequate.

Picidae	UNFAVOURABLE - INADECUATE

5.3.4. Degree of defoliation

Defoliation reaches 12.5%. According to the pan-European forest monitoring criteria, defoliation level is low. Therefore, the conservation status is favourable.

DEFOLIATION	FAVOURABLE

5.3.5. Overall Conservation status of structure and function indicator

The overall conservation status scores 5 points (50% of maximum points), that means unfavourable-inadequate.

INDICATOR	Conservation status	Points
Dead wood	Unfavourable-bad	0
Forest structure	Favourable	2
Habitat fragmentation	Unfavourable-bad	0
Picidae	Unfavourable-inadecuate	1
Degree of defoliation	Favourable	2

Structure and function	UNFAVOURABLE -INADECUATE

5.4. Future Prospects

Given the low grazing pressure of herbivores, the control of public use and the low risk of fire, future prospects are favourable.

Future Prospects	FAVOURABLE

5.5. Global conservation status

Table 10 shows the results of the four general factors used to evaluate the conservation status of the habitat.

FACTOR	Conservation status	Global conservation status
Range	Favourable	Unfavourable-bad
Typical species	Unfavourable-bad	
Structure and function	Unfavourable-inadecuate	
Future Prospects	Favourable	

Table 10. Global Conservation Status for the 9120 habitat within "Dehesa del Moncayo" and final diagnosis

By applying the General Assessment Matrix (Table 2) criteria, we conclude that the conservation status of 9120 habitat within "Dehesa del Moncayo" is unfavourable-inadequate.

6. Discussion

The methodology of this study is an important step for assessing the conservation status of habitats of Community interest. The reference values are based on scientific research which should be adjusted periodically.

When applying this methodology to the 9120 habitat we found some difficulties:

The area of distribution and area occupied had to be adapted locally since the General Assessment Matrix proposes the biogeographic region. This led to only consider the area occupied by the habitat.

Measurement procedure for Typical species has not been standarised yet. We used bibliographic survey that may not accurately represent population and species of our habitat. The conservation status of typical species was unfavourable-bad due to the scarcity of species of fauna. However there may be several abundant and viable populations and more research would be necessary.

Structure and function is the core of the conservation status assessment. This Indicator was unfavourable-inadequate. Only forest structure and level of defoliation parameters had a favourable outcome. Although forest structure result was favourable, tree distribution is far from an uneven-aged forest which is the most suitable structure for biodiversity (Camprodon and Plana, 2007). Therefore, more research studies on thresholds of structure and function should be developed.

Procedures to measure dead wood have not been standarised yet either. Taking into account several studies (Kirby *et al.*, 1998, Woodall and Williams, 2005) we considered that "line transect" method could be more efficient than "strip-plot" method and could allow dead wood on the floor to be measured from 0 cm instead of 2.5 cm.

Finally, the overall conservation status unfavourable-inadequate shows the habitat is far from the favourable status. The lack of typical species of fauna is linked to the scarce dead wood and old trees with cavities. However, this is a young beech forest without productive exploitation since 1975 so the future prospects are favourable.

7. Conclusions

The Preliminary Ecological bases for the conservation of habitat types of Community interest in Spain (AUCT. PL., 2009) assesses the conservation status according to four general factors: range and area occupied, typical species, structure and function, and future prospects.

Although results showed an unfavourable conservation status, the current situation of the beech forest is the best one considering that it was highly harvested in the past. The future prospects are favourable and ensure the capacity of the forest to naturally achieve all the quality thresholds required, with no forest management actions.

Our results indicate that special attention must be paid to thresholds and that more accurate measurement procedures and assessment methods must be developed.

This methodology is an important and comprehensive starting point, however, it requires further applications to identify weaknesses and optimal measurement procedures.

Author details

Rosario Tejera
Research Group for Sustainable Management, Department of Economy and Forest Management - E.T.S.I de Ingenieros de Montes, Technical University of Madrid (U.P.M), Ciudad Universitaria, Madrid, Spain

María Victoria Núñez
Research Group for Sustainable Management, Department of Projects and Rural Planning - E.T.S.I de Ingenieros de Montes, Technical University of Madrid (U.P.M), Ciudad Universitaria, Madrid, Spain

Ana Hernando
Research Group for Sustainable Management, Department of Economy and Forest Management - E.T.S.I de Ingenieros de Montes, Technical University of Madrid (U.P.M), Ciudad Universitaria, Madrid, Spain

Javier Velázquez
Catholic University of Avila. C/ Los Canteros, Ávila, Spain

Ana Pérez-Palomino
E.T.S.I de Ingenieros de Montes, Technical University of Madrid (U.P.M), Ciudad Universitaria, Madrid, Spain

8. Acknowledgement

The research leading to these results has received funding from the Servicio Provincial de Medio Ambiente del Gobierno de Aragón and FEADER. We wish to thank all of the members of the research group "Silvanet" for their support and comments. We extend special thanks to Enrique Arrechea Veramendi for his cooperation and assistance and Miguel Valentín Gamazo for the English review.

9. References

AUCT. PL. (2009). *Preliminary ecological basis for conservation of habitat types of Community interest in Spain* Ministry of Rural and Marine Environment, Madrid.

Bock, M., Rossner, G., Wissen, M., Remm, K., Langanke, T., Lang, S., Klug, H., Blaschke, T. & Vrscaj, B. (2005). Spatial indicators for nature conservation from European to local scale. *Ecological indicators,* 5, 322-338.

Camprodon, J. & Plana, E. (2007). *Conservación de la biodiversidad, fauna vertebrada y gestión forestal,* Publicacions i Edicions Universitat de Barcelona.

Cantarello, E., Newton A.C. (2008). Identifying cost-effective indicators to assess the conservation status of forested habitats in Natura 2000 sites. *Forest Ecology and Management,* 256, 815-826.

Dimitriou, E., Karaouzas, I., Skoulikidis, N. & Zacharias, I. (2006). Assessing the environmental status of Mediterranean temporary ponds in Greece. *Annales De Limnologie-International Journal of Limnology,* 42, 33-41.

European Commission (1995). Standard Data Form for Special Protection Areas (SPA) for sites eligible for identification as Sites of Community Importance (SCI) and for Special Areas of Conservation (SCA).

European Commission (2006). Assessment, monitoring and reporting under Article 17 of the Habitats Directive: Explanatory Notes & Guidelines.

European Council (1992). Directiva 92/43/CEE del Consejo de 21 de mayo de 1992 relativa a la conservación de los hábitats naturales y de la fauna y flora silvestres. DO L 206 de 22.7.1992, p.7, Bruselas.

Gallo Manrique, P. (2011). *4ª Revisión del Proyecto de Ordenación del M.U.P. 251 "Dehesa del Moncayo", Tarazona (Zaragoza).*

Gobierno de Aragón, D.d.M.A. (2002). Plan Rector de Uso y Gestión del Parque Natural del Moncayo. Zaragoza.

Kirby, K.J., Reid, C.M., Thomas, R.C. & Goldsmith, F.B. (1998). Preliminary estimates of fallen dead wood and standing dead trees in managed and unmanaged forests in Britain. *Journal of Applied Ecology*, 35, 148-155.

Lang, S. & Langanke, T. (2005). Multiscale GIS tools for site management. *Journal for Nature Conservation*, 13, 185-196.

MCPFE (2002). Improved Pan-European Indicators for Sustainable Forest Management. Viena.

Müller, J. & Bütler, R. (2010). A review of habitat thresholds for dead wood: a baseline for management recommendations in European forests. *European Journal Forest Resource*, 129, 981-992.

Noss, R.F. (1990). Indicators for monitoring biodiversity - A hierarchichal approach. *Conservation Biology*, 4, 355-364.

Noss, R.F. (1999). Assessing and monitoring forest biodiversity: a suggested framework and indicators. *Forest Ecology and Management*, 115, 135-146.

Olano, J.M. & Peralta de Andrés, J. (2009). 9120 Hayedos acidófilos atlánticos con sotobosque de *Ilex* y a veces de *Taxus* (*Quercinion robori-petraeae* o *Ilici-Fagenion*). En: *Bases ecológicas preliminares para la conservación de los tipos de hábitat de interés comunitario en España*. (Ministerio de Medio Ambiente, y Medio Rural y Marino, Madrid.

Roberts-Pichette, P. (1998). Canada's Ecological Monitoring and Assessment Network with special reference to long-term biodiversity monitoring. *Forest Biodiversity in North, Central and South America, and the Caribbean*, 21, 47-56.

Shaw, P. & Wind, P. (1997). Monitoring the condition and biodiversity status of European Conservation Sites. Conservation, R.t.t.E.E.A.o.b.o.t.E.T.C.o.N.).

Simboura, N. & Reizopoulou, S. (2007). A comparative approach of assessing ecological status in two coastal areas of Eastern Mediterranean. *Ecological indicators*, 7, 455-468.

Telleria, J.L. & Santos, T. (2001). Fragmentación de hábitats forestales y sus consecuencias. En: *Ecosistemas Mediterráneos. Análisis funcional*. (Consejo Superior de Investigaciones Científicas y Asociación Española de Ecología Terrestre, Madrid. pp. 293-317.

Van Wagner, C.E. (1982). Practical aspects of the line intersect method. Canadian Forestry Service, Chalk River, Ontario, Canada.

Velázquez, J., Tejera, R., Hernando, A. & Nuñez, M.V. (2010). Environmental diagnosis: Integrating biodiversity conservation in management of Natura 2000 forest spaces. *Journal for Nature Conservation*, 18, 309-317.

Woodall, C.W., Rondeux, J., Verkerk, P.J. & Stahl, G. (2006). Estimating Dead Wood During National Forest Inventories: a Review of Inventory Methodologies and Suggestions for Harmonization

En: *Proceedings of the Eighth Annual Forest Inventory and Analysis Symposium*.

Woodall, C.W. & Williams, M.S. (2005). Sampling Protocol, Estimation and Analysis Procedures for the Down Woody Materials Indicator of the FIA Program. United States Department of Agriculture (USDA), Forest Service.

Chapter 4

Top-Predators as Biodiversity Regulators: Contemporary Issues Affecting Knowledge and Management of Dingoes in Australia

Benjamin L. Allen, Peter J.S. Fleming, Matt Hayward, Lee R. Allen, Richard M. Engeman, Guy Ballard and Luke K-P. Leung

Additional information is available at the end of the chapter

http://dx.doi.org/10.5772/50246

1. Introduction

Large predators have an indispensable role in structuring food webs and maintaining ecological processes for the benefit of biodiversity at lower trophic levels. Such roles are widely evident in marine and terrestrial systems [1, 2]. Large predators can indirectly alleviate predation on smaller (and often threatened) fauna and promote vegetation growth by interacting strongly with sympatric carnivore and herbivore species (e.g. [3-5]). The local extinction of large predators can therefore have detrimental effects on biodiversity [6], and their subsequent restoration has been observed to produce positive biodiversity outcomes in many cases [7]. Perhaps the most well-known example of this is the restoration of gray wolves *Canis lupus* to the Greater Yellowstone Ecosystem of North America. Since the reintroduction of 66 wolves in 1995 [8], wolf numbers in the area have climbed to ~2000, some large herbivores and mesopredators have substantially declined, and some fauna and flora at lower trophic levels have increased (see [4], and references therein). Similar experiences with some other large predators mean that they are now considered to be of high conservation value in many parts of the world [1, 2, 7], and exploring their roles and functions has arguably been one of the most prominent fields of biodiversity conservation research in the last 10–15 years.

Large terrestrial predators are often top-predators (or apex predators), but not all top-predators are large or associated with biodiversity benefits [5, 9]. For example, feral cats *Felis catus* or black rats *Rattus rattus* may be the largest predators on some islands, but their effects on endemic fauna are seldom positive [10-13]. In geographically larger systems, coyotes (*Canis latrans*) [14] or dingoes (*Canis lupus dingo* and other free-roaming *Canis*) [15], for example, can exacerbate wildlife management problems in highly perturbed ecosystems, where they have the capacity to devastate populations of smaller prey [5, 16-18]. Hence, it is not the trophic position of a predator that determines their ecological effects, but rather their behaviour, impact and function [9]. This is most important for small- and medium-sized predators which can have positive, negative or neutral effects depending on a range of context-specific factors.

Excluding humans, dingoes are the largest terrestrial predator on mainland Australia but, at an average adult body weight of only 15–20 kg [19], are atypical top-predators [20-22]. No other continent has such a small top-predator, and canids have rarely (if ever) been a continent's largest predator, a role typically filled by ursids or felids. Australia's former terrestrial top-predator, a similar-sized marsupial known as the thylacine or Tasmanian Tiger *Thylacinus cynocephalus*, was quickly replaced by dingoes as the largest predator as thylacines became extinct coincident with the introduction of dingoes to Australia about 4000–5000 years ago [23-25]. Like all dogs, dingoes are derived from wolves by human selection [26-29], yet it is a mistake to equate dingoes with wolves (*sensu* [30, 31]) simply because they share a common origin [9, 22, 32] and display some wolf-like behaviours [19]. Hence, the net effects of dingoes on biodiversity might not be readily deduced from studies of other top-predators. Regardless of their derivation and exotic origin, dingoes are common across most of Australia's mainland biomes [33, 34], although their densities have been reduced to very low levels in some regions (<25% of Australia) where sheep *Ovis aries* and goats *Capra hircus* are farmed [15, 34].

Dingoes can have neutral, positive or negative effects (which can be either direct or indirect) on economic, environmental and social values [22, 35]. For example, dingoes can adversely affect livestock production by preying on livestock [36, 37], yet have beneficial effects to livestock producers by preying on livestock competitors [38, 39]. Alternatively, dingoes might help to reduce the impacts of smaller predators (such as introduced red foxes *Vulpes vulpes* or feral cats) on threatened fauna through intraguild predation or exploitative competition [40, 41], yet have detrimental effects on the same fauna through predation [15, 16] and/or disease transmission [42, 43]. Human attitudes towards dingoes are also variable [22, 44-46]. Hence, it should not be surprising to discover evidence for diverse and contrasting functions and values of dingoes in different places and at different times, which adds complexity to their best-practice management [35].

Knowledge of the roles of top-predators on other continents (e.g. [1, 2]) and recent research focus on the positive environmental effects of dingoes (e.g. [41, 47, 48]) has led to calls to cease lethal dingo control (e.g. [31, 49]) and even restore them to sheep and goat production regions (e.g. [23, 50]), actions collectively referred to hereafter as 'positive dingo management'. Serious concerns about the validity and rigour of the science supporting positive dingo management have been raised (e.g. [15, 51, 52], but see also [33, 53, 54]). The issue is further complicated by the changing genetic identity of dingoes [55-58] and the associated ambiguity and misuse of taxonomic terminology ([33]; e.g. compare taxonomic nomenclature between [56], [59], [60], and [55]). The capacity for dingoes to exploit seemingly unsusceptible fauna [61] and the widespread and direct negative effects of dingoes on biodiversity are also overlooked in many cases [15, 16]. There remains, however, a general view that dingoes provide net benefits to biodiversity at continental scales through suppression of foxes (Plate 1), feral cats and herbivores such as kangaroos (*Macropus* spp.) and rabbits (*Oryctolagus cuniculus*) [9, 47], and policy and practice recommendations towards positive dingo management are already occurring (e.g. [49, 62, 63]) despite concerns over the state of the literature and the conflicting roles of the dingo. In most places dingoes are presently managed on the basis of where they occur and what they are (or are perceived to be) doing, not on their genetics or appearance [33, 64].

Out of the confusion arise several knowledge gaps and issues which hamper the informed management of dingoes for biodiversity conservation. In this chapter we discuss critical knowledge gaps about dingo ecology, and highlight the influence of methodological application and design flaws on the reliability of published literature underpinning current knowledge of the ecological roles of dingoes. We offer alternative explanations for the mostly correlative data often mooted as 'clear and consistent evidence' (e.g. [54, 65]) for the fox-suppressive effects of dingoes, and discuss practical obstacles to the accrual of biodiversity benefits expected from positive dingo management. We also discuss the potential consequences of such a management approach for biodiversity and livestock industries, and the management of dingoes at scales which can address their context-specific impacts. Finally, we summarise some surmountable issues presently faced by researchers, land managers and policy makers, and provide recommendations for future research that, when completed, will assist in filling the knowledge gaps required to progress the best-practice management of dingoes for biodiversity conservation in Australia.

2. Knowledge gaps in the literature

Dingoes are one of the most studied animals in Australia, but there is still much to learn about them. Management of dingoes can be advanced by directing researchers towards critical knowledge gaps which require exploration. Unsurprisingly, some gaps need more urgent attention than others. Here, we focus on four key knowledge gaps that we consider to be fundamental to achieving best-practice management of dingoes as biodiversity conservation tools. These are:

1. The relationships between dingoes and biodiversity in relatively intact ecosystems
2. The relationships between dingoes and biodiversity in relatively altered ecosystems characterised by grossly disturbed vegetation structure and composition
3. The effects of current dingo control practices on mesopredators and biodiversity
4. The public's view of what we're trying to conserve (i.e. their pelage, their genetic identity and/or their ecological function)

Dingoes have been studied in many parts of Australia [19], but mostly in relatively intact (i.e. parks, reserves or extensive cattle production regions) and/or arid (Table 1) areas. This is mirrored by international research [2] that primarily comes from a limited number of classic studies conducted in relatively intact ecosystems that do not represent the majority of the earth's surface [66]. Although the relationships between dingoes and biodiversity in these intact areas might be considered well studied, they are not well understood, because the majority of the literature addressing the ecological roles of dingoes in these areas is compromised by a variety of methodological flaws [52]. Even ignoring these flaws, the majority of the relevant literature is only observational and correlative [41], and is therefore subject to plausible alternative explanations [67, 68]. Key among these is the cumulative effects of pastoralism (e.g. [15, 53]), which dramatically transformed pre-European landscapes into those characterised by severely altered vegetation communities [69-71] and a high proportion of now rare and locally extinct native fauna [72-75]. Understanding the roles of dingoes in highly altered ecosystems (i.e. sheep grazing lands and urban ecosystems) may actually be most important, because such systems are those expected to benefit most from positive dingo management [23, 50].

Since the 1960s, when the modern era of dingo research began, most studies have focussed on basic biology, including dingo diet, pack structure, physiology and reproductive biology [19, 76]. The motivation for much of this work has been directed at the negative effects of dingoes on livestock production [19, 64], and dingoes are presently subject to lethal control in many places in attempts to alleviate livestock predation [32, 64, 77]. However, due to the recently reported positive roles of dingoes and other top-predators on biodiversity conservation [1, 2, 7], lethal dingo control has come under increased scrutiny over its perceived indirect effects on biodiversity (e.g. [49]); the idea being that dingo control leads to negative outcomes for faunal biodiversity through trophic effects [23, 78]. Noteworthy however, is that the predicted negative effects of dingo *control* on faunal biodiversity are largely only presumed, and have rarely been demonstrated [79]. Regardless, the conservation and encouragement of dingoes is still being advocated on biodiversity conservation grounds (e.g. [23, 76]). However, *what* exactly requires conservation has not yet been determined for dingoes, which are listed as threatened species [56, 63] not because they are rare (in contrast, there are probably more dingoes now than at any other time in Australia's ecological history [33]), but because their genetic identity is again being altered through hybridisation [55, 57]. Unfortunately, phenotype or pelage is an unreliable indicator of genetic purity [58, 80], though most lay people equate purity with pelage (where only a sandy-coloured dingo is assumed to be pure). Alternatively, it may not be their colour or genetic identity that requires conservation, but their ecological roles [76]. Identifying what is to be conserved is important because most dingoes in Australia are not pure and are expected to become less so with time [55-57].

Understanding the trophic relationships between dingo management practices (i.e. poison baiting, trapping, shooting or no human intervention at all) and the conservation of threatened prey species (R1–R6 in Fig. 1) is the most critical management challenge [22, 41]. A wide variety of taxa may be involved (Plate 1). Ecological relationships between organisms are rarely as simple as those described in Fig. 1, yet they are often assumed to be so in studies of dingoes [32]. The (mostly negative) relationships between exotic mesopredators and threatened prey species (R3) are relatively well understood from other studies [81, 82], as is the relationship between lethal dingo control and dingoes (R1) [64, 83]. The other two relationships (R4 and R6) have received less attention (Table 1), although these are arguably the two relationships most able to address questions relating to the trophic consequences of dingo control. The direct risks dingoes pose to threatened fauna (R5) should also be well established before positive dingo management can be implemented with confidence [22]. Dingoes are highly adaptable and generalist predators capable of threatening many of the species they have also been predicted to protect [16, 17]. Studies that focus on R2 (and report that dingoes are negatively associated with foxes and cats) typically presume that lethal control of dingoes must therefore benefit foxes and cats (R4), though such an assumption is unfounded [22, 32]. Of ultimate importance however, and irrespective of any of the other relationships, understanding the effect of dingo control on threatened prey species (R6) can facilitate the most rapid management progress. The short-term and direct effects of dingo control on threatened fauna were reviewed in [79], which concluded that no studies to date have shown negative effects of dingo control on non-target fauna, a view subsequently ratified in [84]. There remains, however, limited reliable data on the longer term and indirect effects of dingo control faunal biodiversity [41, 85].

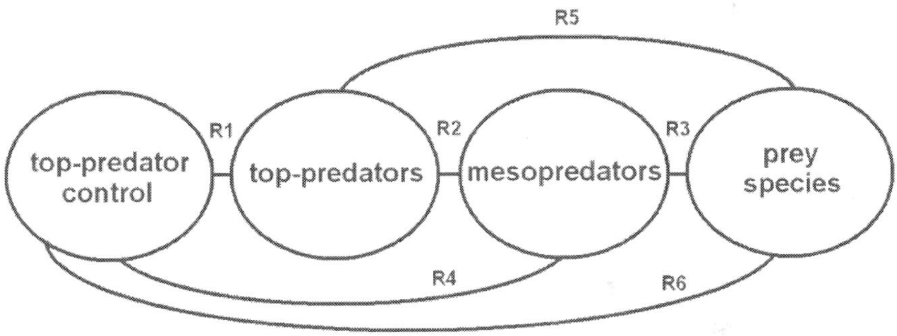

Figure 1. Schematic representation of six relationships (R1–R6) between top-predator control and prey species at lower trophic levels.

Investigating R6 is a 'black box' approach to applied research [86], meaning the observed outcomes of control interventions can enable management progress in the absence of a complete understanding of the mechanisms responsible for the outcomes. For example, [86] summarised the results of 25 years of experimental research on the conservation of threatened black-footed rock-wallabies *Petrogale lateralis*, stating that researchers had found time and again that fox control resulted in more rock-wallabies, but they did not have a good grasp on the mechanisms responsible for it. Thus, if investigations of R6 show that threatened prey populations fluctuate independently of dingo control, lethal control of dingoes might continue to occur without concern from conservationists that such practices inhibit the recovery of threatened fauna through trophic effects. Lethal dingo control may not be incompatible with biodiversity conservation or restoration [32], nor is cattle production always incompatible with dingoes in the absence of dingo control [38, 87, 88]. In a world where resources to manage threatened species are limited, focussing on such applied studies should be of utmost value to land managers and policy makers.

Plate 1. Rufous hare-wallabies *Lagorchestes hirsutus* (bottom right; photo from www.arkive.org), dusky hopping-mice *Notomys fuscus* (bottom left; photo by Reece Pedler) and red foxes *Vulpes vulpes* (top right; photo by Ben Allen) are some of the fauna that are affected both positively and negatively by dingoes (top left; photo by Ben Allen).

3. The state of current evidence for dingoes' ecological roles

Classical manipulative experiments are the best way to advance scientific knowledge [89, 90]. However, performing robust experiments on dingoes at large-enough scales is costly and logistically very difficult or even impossible [41]. Almost all field studies typically sample dingo populations using passive tracking indices (or sand plots) placed along dirt roads and trails. The use of other monitoring techniques, such as camera trapping, are increasingly being used [91, 92]. Although many studies investigating R2 and R5 using passive tracking indices have claimed to provide evidence that dingoes stabilise ecological processes through their top-down effects on sympatric predators and prey, three unresolved issues continue to compromise the reliability of these conclusions for most studies (Table 1):

1. Much of the literature is weakened by methodological flaws (such as seasonal or habitat confounding, or invalid and violated assumptions) which render the reliability of the body of data collected uncertain [52]. In many cases, it is not the technique that is weak, but it is the poor application of otherwise robust techniques that compromise the data collected [51]. This is not to say that the conclusions of such studies are incorrect, but that the reader cannot tell whether they are or not because of the flaws.

2. Regardless of their methodological flaws, most studies are also conducted over small spatial and/or temporal scales. Because of spatiotemporal variation in animal densities [67, 93, 94], behavioural avoidance of top-predators by mesopredators [3, 95, 96], and because most studies sample dingoes along roads (which are favoured by dingoes; [95]), the results of many recent studies may simply be artefacts of sampling biases towards apparent inverse relationships between dingoes and mesopredators.
3. Regardless of methodological flaws or sampling bias, the experimental designs of many studies are still only observational or correlative ([41]), rendering their conclusions subject to a wide variety of plausible alternative explanations [53, 68]. Such studies can only support statements such as 'dingoes *might* perform this role' instead of statements such as 'dingoes *do* perform this role', which can only be made reliably from studies with greater inferential capacity [89].

3.1. Methodological flaws

Critical review has shown that the data in 75% (15 of 20) of recent studies that sampled dingoes using sand plots on roads are potentially confounded by a variety of factors, including (but not limited to) invalid seasonal and habitat comparisons [52]. Dingo activity on roads varies between seasons independent of their actual abundance [52, 97], which can lead to confounding and weakened inferences if not accounted for by the study design. For example, valid comparisons cannot be made between one site sampled in winter and another site sampled in summer, because observed activity differences are likely to be attributable to behavioural changes and not abundance changes. This issue may most easily be understood for reptiles, which usually reduce their activity in winter [98]. For dingoes and foxes, food availability and breeding may drive this variability [19, 99].

Comparisons between different habitats may also be confounded due to varying detection probabilities associated with different habitat types [68, 93]. For example, even if abundance is equal across habitats, animals occupying landscapes with more difficult terrain may utilise roads (i.e. where sampling occurs) more frequently than animals occupying areas which allow more ubiquitous movements (e.g. [100]), with observed activity differences again potentially attributable to behavioural changes and not abundance changes. Moreover, different habitats often have different faunal assemblages, geological and ecological processes (e.g. [101]), which may influence the way some species interact with sand plots placed on roads. Pooling across seasons or habitats may mask differences that could be more easily viewed if separated (e.g. [32]). A variety of assumptions (such as 'footprints of the same species <500m apart and heading in the same direction belong to the same individual' or 'old-looking footprints are *x* days old') are also commonly made (Table 1) and undoubtedly violated ([52]; but see [88, 102-104] for examples). Violation of such assumptions may underestimate dingo distribution or abundance.

Although a wide variety of methodological flaws are evident (Table 1), violation of assumptions and seasonal or habitat confounding may be more important than other flaws, in that they could have greater ecological significance than other methodological errors [52,

93]. Of the 34 studies considered in Table 1, 14 (41%) and 15 (44%) and are potentially weakened by habitat and seasonal confounding, while 12 (35%) made unnecessary assumptions, indicating that multiple studies contain multiple methodological weaknesses. Fundamentally, indices are only useful when they are correlative of abundance [67, 105], and such flaws typically mean that the relationship between observed indices and actual abundances is unknowable. We note however, that accurate knowledge of absolute abundance is near impossible to acquire in the field [67, 105, 106], and we are not aware of any studies of dingoes that have calibrated sand plot activity data with absolute abundance values (because absolute abundance values have not been attainable). However, where the principles outlined in [93, 106] are strictly applied, researchers can acquire reliable estimates of relative abundance, the metric that underpins the vast majority of available field data on dingoes (Table 1).

The use of inappropriate techniques or poor application of otherwise robust techniques reduces the extent to which such data can be used to make reliable statements about ecological processes, and because many studies have made such flaws (Table 1; [52]), much of the available sand plot data on dingoes might be considered unreliable. Overturning this conclusion for any given study requires demonstration that either (1) the methodological flaws described were not made and/or (2) that if made, they did not constitute unreliability [53]. Once collected, it is also rarely possible to un-confound the data using statistical procedures (such as generalised linear modelling) without making the most tenuous of assumptions [52, 105]. The design flaws outlined here are discussed in more detail in [33, 52]. Others [53, 54] have questioned the importance of these flaws, but such methodological flaws are not the only issue undermining evidence for dingoes' ecological roles.

3.2. Sampling bias

An index is a measurement related to the actual variable in question [67, 105, 107] and specific to the circumstances under which the data were collected [93]. Importantly, animal populations are not usually distributed uniformly across the landscape but are instead clumped, producing areas of higher and lower abundance (e.g. [108]). Thus, studies conducted over small spatial scales may acquire severely biased results. For example, the areas sampled in [109] or [110] were very small (<10km2), which likely represented only a fraction of a dingo's home range in such systems [111, 112]. The observed relationships between species within such small areas may have limited applicability outside the areas sampled, where animal abundances may be markedly different (e.g. [108]). Animal activity is also rarely distributed uniformly over temporal scales. Within a 24 hour period, animals may exhibit diurnal, nocturnal or crepuscular behavioural cycles which prevent reliable comparisons of index values from one time period to another. This may be most easily understood for birds, where, for example, observations collected from one area in the early morning should not be compared to observations collected from another area at noon [113, 114]. Many of these considerations essentially amount to issues of detection probability, and have been discussed in greater detail elsewhere [68, 93, 114, 115]. The same principles apply to indexing and population estimation using almost any technique [93, 116].

The highest activity periods for top-predators are also usually optimal, mesopredators usually avoid top-predators during these times, and prey activity usually fluctuates independently of predator activity (e.g. [117-119]). Because mesopredators typically seek to avoid encountering top-predators, mesopredator activity is likely to be lower at times and in places with higher top-predator activity. This has important implications for studies conducted over restricted temporal scales, such as snap-shot or single sample studies (Table 1; e.g. [120-122]). If dingo activity is high on those days, mesopredator activity would be expectedly lower (and vice versa), which means that such temporally limited data is silent on the ability of dingoes to suppress or exclude mesopredator abundances over time, because mesopredators may simply have been avoiding the sampling area on those days. Repeating this snap-shot approach to sampling at any number of multiple sites cannot overcome this issue of bias. Conducting successive surveys over slightly longer timeframes (e.g. three or four surveys over one year) may also be affected by this bias because periods of high or low top-predator activity may endure for several months [52, 97, 111, 123]. Some such studies (e.g. [110, 124]) might been viewed as positive population responses of mesopredators to single dingo control events. Again, however, such observations would be expected given that mesopredator behaviour may change, increasing their use of tracks once the landscape of fear has been altered [96, 125, 126] without necessarily altering their actual abundance (e.g. [110, 124, 127]). Temporally restricted data cannot be reliably used as evidence that dingo control increases the abundance of mesopredators unless the results can be adjusted for seasonal effects by incorporating data from a comparable nil-treatment area. Even over several years, a sampling strategy which focuses on landscape features where dingoes are expected to be more active (such as dirt roads and trails) are also likely to be biased towards dingoes and less sensitive (but not insensitive; e.g. [87]) at detecting foxes or cats [95].

Such issues of bias on sand plots are typically overcome by sampling populations over larger spatial and/or temporal timeframes [93] and means that interspecific comparisons of index values are inappropriate [93, 94]. Other population sampling and analytical techniques might be used (such as estimates derived using photo-mark-recapture [128-131], camera trap rates [132], aerial surveys [133, 134], distance sampling of actual observations or signs [113], occupancy modelling [68] or track transects [135]), but these are all likewise subject to similar issues [114, 116]. Even though magnitudes of index values are meaningless for comparison between species, the population trends defined by the index values over time can be valid given appropriate study design and data analyses [93]. All studies identified in Table 1 have sampled predators for only a few days at a time during each survey, meaning that the results from each individual survey, in isolation, might be artefacts of such bias. This is an important weakness of short-term studies, but when surveys are repeated over several seasons or years, resulting trends may be reliably used to identify relationships between predators. For example, fox activity on sand plots may be much lower than those of dingoes for any (or every) given survey (possibly as a result of sampling bias), but when surveyed repeatedly over longer timeframes, correlations between dingo and fox population trends can be confidently compared. When dingo abundance is further manipulated in an experimental framework, a divergence of activity (or relative abundance) trends between dingoes and foxes would be particularly strong evidence for mesopredator suppression or release. The corollary of this is that non-divergence of dingo and fox population trends over time would be particularly strong evidence that mesopredator suppression by dingoes is not occurring.

Additional to the methodological flaws described earlier, many studies are also conducted over small spatial or temporal scales (Table 1). Thus, their results are likely to be affected by the sampling biases described, giving the potentially mistaken impression of inverse relationships between dingoes and mesopredators. The common presence of this issue throughout the literature further weakens the reliability of data on dingoes' ecological roles. Such biased data might only be suggestive of spatial avoidance between predators, but it cannot demonstrate avoidance. Provided the proper indexing principles are strictly applied and the data analysed appropriately, studies assessing predator population trends over longer timeframes will have a much better ability to identify correlative relationships. However, to identify causal process for observed correlations still requires experimental designs with even greater inferential ability [89, 90].

3.3. Experimental design

Poor application of methods and sampling bias are but two forms of experimental design flaws weakening the reliability of many studies. But even if such issues are overcome through appropriate sampling strategies, different types of experimental designs have inherent limitations to their inferential ability [89]. The implications of these limitations have not been adequately dealt with in most appraisals of the literature on dingoes' ecological roles. In 2007, [41] concluded that the available data on dingoes' ecological roles was mostly observational and correlative, and many studies published since then (e.g. [31, 78, 122, 136-138]) have not improved this situation. It should be understood that 'studies of a more observational nature can make only weak inferences about cause and effect and studies that involve classical experiments can make stronger inferences. Where studies use more observational methods the results should be interpreted and valued as such, and not as equivalent to the results of classical experiments' ([89]; but see also [90]). The replication and randomisation of treatments, along with the use of nil-treatments (or experimental controls) are particularly important design features that can provide a greater ability to demonstrate causal processes – provided methodological flaws and sampling bias are also avoided.

The inferential capabilities of different designs used in 34 studies of dingoes are here ranked between 1 and 16 (1 = highest level of inference, 16 = lowest; from [89]) in Table 1. Without a nil-treatment, the highest rank a study can achieve is a pseudo-experiment type I (Rank 9). Without randomisation, the highest rank possible is a quasi-experiment type I (Rank 5). For studies comparing the effect of contemporary or historical dingo control practices on predators or prey, many researchers cannot randomise their treatments and are constrained to use areas where dingo control is (or is not) already being undertaken (e.g. [83, 139]). In the case of cross-fence comparisons (e.g. [78, 122, 140]), the results of such non-randomised studies may be subject to plausible alternative explanations that cannot be controlled for [15,

101, 121]. Where possible, treatment randomisation offers one way of addressing these constraints, but has only been undertaken by three studies (Table 1). Only one study [32] has involved a classical experiment on dingoes, where treatments and nil-treatments were also replicated (two of each at one site). Thus, almost all of the available literature reports results from experimental designs which cannot reliably demonstrate cause and effect. Each of these three issues (methodological flaws, sampling bias and experimental design limitations) mean that the evidence for dingoes' ecological roles is not as strong as might be supposed, and each of these issues must be overcome in order to change this view.

As an example of how these issues combine to effect the reliability of data, [121] used footprint counts on dirt roads to derive activity indices for dingoes, foxes and cats at three sites on either side of the dingo barrier fence, which was erected in the early 20th century to exclude dingoes from sheep production lands in south-eastern Australia [141-143]). At two sites, fox activity was reportedly ~2–3 times higher in places where dingoes were rare. At a third site, foxes were only detected where dingoes were rare, and cats were reportedly present in equally low abundance on both sides of the fence [121, 138]. The methodological flaws described earlier (and in [52]) mean that the results of [121] could only be considered 'coarse measures'. Although, [53] argued that coarse measures are sufficient in places where the effect sizes are too large to be explained by the methodological shortcomings (such as seasonal confounding), meaning that the quantitative data may be unreliable but the qualitative patterns may still be recognisable. Importantly however, predator activity can naturally vary in excess of 400% in a matter of weeks or months (e.g. [32, 83, 144]), which means that the effect sizes must be enormous for comparisons made between different seasons to not be affected by season. Regardless, sampling occurred only once over a few days at each of the three sites described in [121]. Because, in such habitats, mesopredators typically avoid roads and dingoes do not [95], the low incidence of fox tracks in the presence of greater numbers of dingo tracks could simply be an artefact of spatial avoidance of roads by foxes on the days that footprint counts were collected. This result may not necessarily reflect the relative abundance of foxes at all, because foxes may have been more active in other parts of the landscape on those days – the infrequent detection of mesopredator tracks would be expected at a time of high top-predator activity (or vice versa). Whether the methodological flaws or the potential for sampling bias are considered important or not, [121] was still only a non-randomised correlative quasi-experiment type I [89], with an inferential rank of 5 out of 16 (Table 1). Hence, the observations may equally be explained by alternative factors, such as the cumulative impacts of livestock grazing [15, 121], thus offering only 'inconclusive' support [53] for the functional relationships between the species studied.

We are not trying to argue here that foxes are actually abundant on the same side of the fence as high-density populations of dingoes, or that dingoes are actually abundant on the same side of the fence as high-density populations of foxes. Rather, we seek only to illustrate that the sampling biases inherent to short-term studies prohibit the demonstration of causal relationships. In no way is the preceding discussion on the state of the literature intended to be personally critical of researchers and authors, because achieving robust experiments is logistically very difficult [41] and randomisation of treatments is often impossible. Rather, we simply aim to show that whether it is methodological flaws or sampling bias or experimental design limitations, most studies cannot provide strong evidence for causal factors associated with dingoes' ecological roles. It is also important to remember that because perfect experimental designs can be executed imperfectly and imperfect designs may be executed perfectly, neither may enable reliable inference. In other words, correlative or mensurative studies that avoid the flaws and biases described may be just as inconclusive as experimental studies that contain them. As [145] cautioned, 'don't even start the project if you cant do it right', because if the basics are not right, such projects may 'only represent wasted resources' [115].

Reference	Study topic (climate)	Methodological strengths	Methodological weaknesses	Spatial scale per site & sampling effort	Relation-ships investigat ed^	Experimental design (highest rank of inference)*
Allen B.L. [32]	The effect of dingo control on dingoes (arid)	Manipulative experiment BACI design Random allocation of treatments Treatment replication at some sites Time-series data	Baiting intensity varied within treatments between replicates	50 plots over 50km (x2) 6–10 counts at 4 sites over 2–4yrs	R1	Classical experiment (1) & Unreplicated experiment (3)
Allen L.R. [87]	The effect of dingo control on beef cattle (monsoonal tropics and semi-arid)	Manipulative experiment BACI design Random allocation of treatments Time-series data	No replication at individual sites	50 plots over 50km (x2) 7–19 counts at 3 sites over 3–4yrs	R1, R4, R5, R6	Unreplicated experiment (3)
Allen L.R. [83]	The effectiveness of dingo control campaigns (semi-arid)	Replication of treatments Multiple properties surveyed Temporally intensive sampling Time-series data	Non-random allocation of treatments Non-independence between treatments Baiting intensity varied between properties within-treatments	92–133 plots over 92–133km 16–23 counts at 3 sites over 2–3yrs	R1	Quasi-experiment type I (5)
Augusteyn et al. [146]	The effect of dingo control on dingoes and bridled nailtail wallabies	BACI design Manipulative experiment Time-series data Measured demographic responses of prey	One study site only No nil-treatment	53 plots over 53km 20 counts at 1 site over 5yrs	R1, R2, R5, R6	Pseudo-experiment type VII (15)
Brawata & Neeman [140]	Predator distribution around waterpoints in the arid zone (arid)	Spatial replication of treatments Two indices of predators used	Data confounded by habitat and seasonal effects Used binary observations over potentially continuous measures Two experiments in one, but analysed together Sand plot index data untransformed	15 plots over 20km (x2) and 20 scent stations over 20km (x2) 2 counts at 5 sites over 3yrs	R1, R2, R4	Quasi-experiment type I (5)

Reference	Study topic (climate)	Methodological strengths	Methodological weaknesses	Spatial scale per site & sampling effort	Relation-ships investigat ed^	Experimental design (highest rank of inference)*
Burrows et al. [147]	The effects of dingo control on dingoes, foxes and cats (arid)	BACI design Three indices of predators attempted Time-series data	Non-random allocation of treatments Invalid assumptions when calculating the activity of predators Data confounded by seasonal differences in predator activity Invalid comparisons between species One index technique (cyanide bait uptake) removed individuals from the population	30–60km tracking transects 25 counts at 1 site over 10yrs	R1, R4	Quasi-experiment type III (7)
Catling & Burt [148]	The influence of habitat on small mammals (temperate)	Mensurative study Standardised design	Data confounded by seasonal differences in predator activity Invalid comparisons between habitats Sand plot index data untransformed	20–35 plots over 4–7km 2 counts at 13 sites over 7yrs	R3, R5	Pseudo-experiment type V (13)
Catling et al. [149]	The effects of cane toads on native fauna (monsoonal tropics)	BACI design Three treatments Different indices for some species	Used binary observations over potentially continuous measures Sand plot index data untransformed	25 plots over 5km 4 counts at 1 site over 2yrs	R5	Quasi-experiment type I (5)
Christensen & Burrows [150] (see also [147])	Reintroduction success of native mammals following predator control (arid)	Two measures of predators used	Invalid assumptions when calculating the activity of predators Predators in 'nil-treatment' areas sampled using an index technique (lethal cyanide bait uptake) that removed individuals from the population 'Nil-treatment' area relocated during the course of the study Cyanide sampling technique biased towards dingoes and foxes Only 1 (of 2) treatment was sampled on 7 of the 8 surveys Not all survey results are reported No analyses undertaken	60km tracking transect 8 surveys at 1 site over 4yrs	R1, R2, R3, R4, R5, R6	Quasi-experiment type IV (8)
Claridge et al. [151]	The effect of predator control on activity trends of forest	Mensurative study Spatial replication of treatments and transects Time-series data	Used binary observations over potentially continuous measures Assumed independence between sand plots	75-125 plots over 19-31km 19 counts at 1 site over 9yrs	R1, R4, R6	Quasi-experiment type I (5)

Reference	Study topic (climate)	Methodological strengths	Methodological weaknesses	Spatial scale per site & sampling effort	Relation-ships investigat ed^	Experimental design (highest rank of inference)*
	vertebrates (temperate)					
Corbett [152]	Relationships between dingoes, water buffalo and feral pigs (monsoonal tropics)	BACI design Independent indices of some species Calibrated pig and dingo indices with mark-recapture estimates and total counts Time-series data	Used binary observations over potentially continuous measures	55 plots over 400km 27 counts at 1 site over 7 yrs	R5	Quasi-experiment type I (5)
Edwards et al. [102]	Habitat selection by dingoes and cats (arid)	Mensurative study Standardised design	Invalid assumptions when calculating the activity of predators Data confounded by seasonal and habitat differences in predator activity	25km tracking transects (x4) 9 counts at 1 site over 3yrs	R2	Psuedo-experiment type V (13)
Edwards et al. [153]	The effect of rabbit warren ripping on wildlife (arid)	Spatial replication of treatments	Invalid assumptions when calculating the activity of predators Data confounded by seasonal and habitat differences in predator activity Baiting intensity varied between sites	10km tracking rectangle (x2) 8 counts at 4 sites over 2yrs	R1, R2, R5	Quasi-experiment type I (5)
Edwards et al. [154]	The effect of Rabbit Haemorrhagic Disease on wildlife (arid)	Mensurative study Standardised design	Invalid assumptions when calculating the activity of predators Data confounded by seasonal and habitat differences in predator activity Data influenced by rabbit warren ripping at some sites	10km tracking rectangle (x2 at four sites) 8 counts at 6 sites over 2 yrs	R2, R3, R5, R6	Pseudo-experiment type V (13)
Eldridge et al. [88]	The effect of dingo control on dingoes and wildlife (arid)	Manipulative experiment Two measures of predators used	Invalid assumptions when calculating the activity of predators	10km tracking transects (x6) 7 counts at 3 sites over 3yrs	R1, R4, R6	Unreplicated experiment (3)
Fillios et al. [155]	Relationships between dingoes and kangaroos (arid)	Spatial replication of treatments Independent measures of kangaroos and dingoes	Replication devalued by seasonally staggered indexing Data confounded by seasonal and habitat differences in predator activity	25 plots over 25km (x2) 1 count at 6 sites over 1yr	R5	Quasi-experiment type I (5)

Reference	Study topic (climate)	Methodological strengths	Methodological weaknesses	Spatial scale per site & sampling effort	Relation-ships investigat ed^	Experimental design (highest rank of inference)*
			Sand plot index data untransformed			
Fleming et al [139] (see also [156])	The effects of dingo control on dingoes (temperate)	BACI design Index data transformed Data corrected for detection probability	Non-random allocation of treatments Abundance and activity potentially confounded	120–270 plots over 12–27km (x2) 12 counts at 1 site over 3yrs	R1	Quasi-experiment type 1 (5)
Johnson & VanDerWal [136] (using data from [157, 158])	Dingoes ability to limit fox abundance (temperate)	Source data from mensurative studies Large data set over wide spatial distribution	Source data confounded by seasonal and habitat differences in predator activity Source data used binary observations over potentially continuous measures Invalid comparisons between species Sand plot index data untransformed	From [158]: 45 plots over 18km, 65 plots over 26km and 105 plots over 84km Repeated counts at 3 sites for up to 9yrs From [157]: 20–35 plots over 4–7km 1 or 2 counts at 15 sites over 7yrs	R2	Pseudo-experiment type V (13)
Kennedy et al. [159]	Relationships between dingo control, dingoes and cats (monsoonal tropics)	Mensurative studies and manipulative experiments Spatial replication of treatments Mensurative study temporally replicated Data transformed Time-series data	Site differences not explicitly identified Temporal trends in predator activity not reported	30–50 plots over 30–50km (x10) 3 counts at 2 sites over 3 years, 2 counts at 2 sites over 2–4 weeks	R1, R2, R4	Pseudo-experiment type I (9) & Quasi-experiment type 1 (5)
Koertner & Watson [160]	The impact of dingo control on quolls (temperate)	Uses two measures of efficacy Replication of treatment (individuals exposed)	Used binary observations over potentially continuous measures Index data untransformed	36 plots over 36km 2 counts at 1 site once	R1, R4	Quasi-experiment type I (5) & Pseudo-experiment type V (13)
Letnic et al. [121] (a subset of [122])	Dingoes' role in protecting dusky hopping-mice from predation by foxes and cats (arid)	Spatial replication of treatments Different measures for hopping-mice and dingoes	Replication devalued through seasonally staggered indexing Insensitive measures of grazing pressure used Data influenced by seasonal and habitat differences in predator activity	25–30 plots over 25–30km (x2) 1 count at 3 sites over 1yr	R3, R5	Quasi-experiment type I (5)

Reference	Study topic (climate)	Methodological strengths	Methodological weaknesses	Spatial scale per site & sampling effort	Relation-ships investigated^	Experimental design (highest rank of inference)*
Letnic et al. [122]	Relationships between dingoes and wildlife (arid)	Spatial replication of treatments Different measures for wildlife and dingoes Effect size measured	Replication devalued through seasonally staggered indexing Data influenced by seasonal and habitat differences in predator activity Used binary observations over potentially continuous measures Insensitive measures of grazing pressure used	25–30 plots over 25–30km (x2) 1 count at 8 sites over 2yrs	R3, R5	Quasi-experiment type I (5)
Lundie-Jenkins et al. [110]	Relationships between hare-wallabies and introduced mammals (arid)	Mensurative study Comprehensive dataset collected	Used binary observations over potentially continuous measures Non-independence between plots No details of dingo control program given Very small spatial scale	Intensive plot coverage within a ~10km² area 4 counts at 1 site over 1yr	R1, R2, R3, R4, R5, R6	Simple observations (16)
Moseby et al. [109]	Population dynamics of hopping-mice (arid)	Mensurative study Time-series data	Used binary observations over potentially continuous measures Very small spatial scale	4km transect inside an 8ha grid (x2) 15 counts at 2 sites over 8yrs	R3, R5	Quasi-experiment type II (6) or Pseudo-experiment type VI (14)
Newsome et al. [101]	Fence effect on dingoes and wildlife (arid)	Different measures for wildlife and dingoes	Invalid comparisons between species	Ringed plots around 10 waterpoints (x2) 4 counts at 1 site over 1yr	R3, R5	Quasi-experiment type 1 (5)
Pascoe [161]	Predator ecology and interactions (temperate)	Mensurative study Two measures of dingoes used Spatial replication	Used binary observations over potentially continuous measures for some analyses Sand plot index data untransformed	31 plots over 15km 8 counts at 3 sites over 2yrs	R2, R3, R5	Pseudo-experiment type V (13)
Pavey et al. [162]	Population dynamics of rodents and predators (arid)	Mensurative study Different measures for wildlife and dingoes Two measures of dingo abundance collected	Invalid assumptions when calculating the activity of predators Invalid comparisons between species Merged sandplot and spotlighting data	10km tracking transects (x3) 6 counts at 1 site over 2yrs	R3, R5	Pseudo-experiment type V (13)
Pettigrew [124]	The effect of dingo control on cats (arid)	Demographic data on cats collected Two measures of predators used	Ambiguous description of site and methodology Data from both sampling measures apparently combined Data from some treatments not reported	Spatial scale unknown, but ~100km of transect 12 counts at 1 site over 3yrs	R3, R4, R5	Quasi-experiment type IV (8)

Reference	Study topic (climate)	Methodological strengths	Methodological weaknesses	Spatial scale per site & sampling effort	Relation-ships investigated^	Experimental design (highest rank of inference)*
Purcell [123]	Dingo purity, diet, activity and behaviour (temperate)	Mensurative study Temporally intensive sampling	Used binary observations over potentially continuous measures for some analyses Sand plot index data untransformed	25 plots over 25km (x2) 26 counts at 1 site over 2yrs	R2, R3, R5	Pseudo-experiment type V (13)
Southgate et al. [103, 104]	Bilby and predator distribution and fire (arid)	Three different sampling strategies used Different measures of bilbies and predators	Data influenced by seasonal and habitat differences in predator activity Used binary observations over potentially continuous measures Invalid assumptions when calculating the activity of predators Footprints assumed 'old' were excluded from occupancy analysis	10km rectangle tracking transects (x2) 6–8 counts at 8 sites over 4yrs	R3, R5	Quasi-experiment type I (5)
Wallach & O'Neill [120] (a subset of [31, 78])	Relationship between dingoes and kowaris (arid)	Two measures of dingo abundance collected	Data influenced by seasonal and habitat differences in predator activity Invalid assumptions when calculating the relative abundance, "Index of abundance", and territorial activity of predators Data influenced by the presence of pet dogs and people Multiplication of binary and continuous abundance measures Sand plot index data untransformed Small spatial scale	10–12 strip plots (500m long), and 20 area plots (2ha) 1 count at 2 sites once	R2, R5	Quasi-experiment type IV (8)
Wallach et al. [163] (a subset of [31, 78])	Dingoes' role in protecting yellow-footed rock wallabies and malleefowl from predation by foxes and cats (arid, semi-arid)	Two measures of dingo abundance collected Large data set over wide spatial distribution	Data influenced by seasonal and habitat differences in predator activity Invalid assumptions when calculating the relative abundance, "Index of abundance", and territorial activity of predators Data influenced by the presence of pet dogs and people Multiplication of binary and continuous abundance measures Sand plot index data untransformed	9–25 strip plots (500m long), and 21–39 area plots (2ha) 1–2 counts at 7 sites over 1yr	R2, R5	Quasi-experiment type III (7)

Reference	Study topic (climate)	Methodological strengths	Methodological weaknesses	Spatial scale per site & sampling effort	Relation-ships investigat ed^	Experimental design (highest rank of inference)*
			Small spatial scale			
Wallach et al. [31]	The effect of dingo control on pack structure and social stability (arid)	Two measures of dingo abundance Large data set over wide spatial distribution	Data influenced by seasonal and habitat differences in predator activity Invalid assumptions when calculating the relative abundance, "Index of abundance", and territorial activity of predators Data influenced by the presence of pet dogs and people Multiplication of binary and continuous abundance measures Sand plot index data untransformed Small spatial scale	9–25 strip plots (500m long), and 21–39 area plots (2ha) 1–3 counts at 7 sites over 3yrs	R1	Quasi-experiment type III (7)
Wallach et al. [78]	The effect of dingo control on invasive species (arid)	Two measures of dingo abundance Large data set over wide spatial distribution	Data influenced by seasonal and habitat differences in predator activity Invalid assumptions when calculating the relative abundance, "Index of abundance", and territorial activity of predators Data influenced by the presence of pet dogs and people Multiplication of binary and continuous abundance measures Sand plot index data untransformed Small spatial scale	10–12 strip plots (500m long), and 20–40 area plots (2ha) 1–3 counts at 7 sites over 3yrs	R1, R4	Quasi-experiment type III (7)

Table 1. Methodological details of sand plot studies investigating the relationships between dingoes and faunal biodiversity. ^See Figure 1 for explanation of primary relationships. *See Table 1.2 in [89] for descriptions of experimental designs and rank of inference (rank 1 = highest possible, 16 = lowest possible). Note: different types of experimental design may be possible for some studies depending on the nature of the question/s being investigated, and the designs/rank identified here represent the highest level of design possible from the data collected.

4. The dingo-suppressive effects of foxes

The inability of correlations to describe causation was discussed by [68], and is illustrated here by examining published data on relationships between dingoes and foxes. Intraguild killing and interference competition are the two primary mechanisms given to facilitate the dominance of one predator over another ([1, 2], and references of studies therein). With some noteworthy exceptions (e.g. [144]), observations of intraguild killing are rare, and its occurrence is most often inferred from the remains of one predator in the diet of another (e.g. [164, 165]). Interference competition is typically inferred from studies of dietary overlap between sympatric predators (e.g. [118, 162, 166]), with high levels of dietary overlap used to infer a high level of potential competition. A variety of such studies have been conducted in Australia, which provide compelling correlative evidence that foxes may suppress dingoes through both mechanisms.

Dingo remains have been found in fox scats (e.g. [123, 164, 167, 168]), and even in cat scats (e.g. [169]), suggesting that these mesopredators kill (or at least consume) dingoes on some occasions. Being 2–3 times larger than foxes, dingoes will likely be victors in aggressive encounters between adults of the two species. However, foxes may be a threat to dingo pups, and dingoes may exhibit heightened activity levels during times when their pups are vulnerable [144]. By limiting recruitment of juveniles, foxes have been observed to suppress populations of one of Australia's largest native herbivores, eastern grey kangaroos M. giganteus [170]. Thus, differences in adult body sizes should not automatically discount the potential for foxes to suppress dingoes also. That mesopredators can slow down recruitment of top-predators was precisely the reason why smaller spotted hyaenas Crocuta crocuta were reintroduced with lions Panthera leo in southern Africa [171]. Multiple studies (e.g. [122, 164, 172, 173]) have also shown foxes to have a high level of dietary overlap with dingoes (Fig. 2), or in other words, dingoes and foxes eat the same things. This suggests that interference competition from high-density populations of foxes (which can reportedly be 7–20 times higher than dingoes [101]) reduces the availability of prey that otherwise might be consumed by dingoes; top-predators being primarily limited by bottom-up factors related to their preferred prey [174-176].

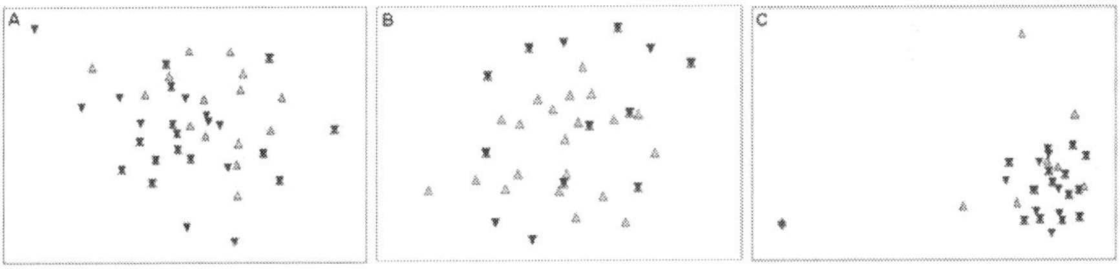

Figure 2. Ordination plot of nonmetric multidimensional scaling analyses showing a high level of dietary overlap between foxes (▼) and dingoes (▲) in the (A) Simpson Desert, (B) Strzelecki Desert and (C) Nullarbor region of arid Australia (from [164]).

Using data from [177], [178] report that dingoes were infrequently detected in places with high fox numbers (Fig. 3). This is further supported by the analyses of [136], which also report that dingo abundance is lower when fox abundance is high (Fig. 4). In contrast, scat indices (or scat collection rates) between dingoes and foxes appeared positively correlated in [123] and foxes (and especially goannas *Varanus varius*) were thought to derive some benefit from dingoes through kleptoparasitism in [173]. Although there are important limitations associated with the use of scats for making inferences about predation and abundance [16, 17, 61, 179], it appears clear from the data published in the aforementioned studies that a substantial and compelling amount of correlative evidence exists to support the hypothesis that foxes suppress dingoes through direct killing and interference competition. In all cases however, alternative hypotheses have been raised. These include the suppression of foxes by dingoes (e.g. [136, 164]) or the cumulative effect of livestock grazing (e.g. [15, 121]). That multiple plausible and competing alternative explanations can be generated is precisely the reason why correlative evidence cannot be trusted to describe causal processes [68] and most of the presently available literature on dingoes' ecological roles is at best inconclusive [52, 53].

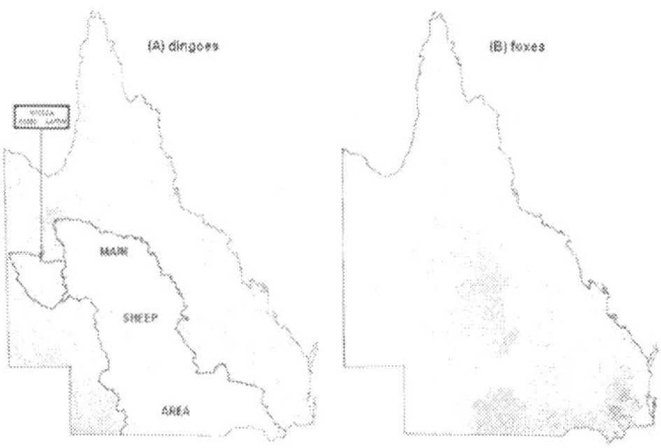

Figure 3. Bounty returns for (A) dingoes and (B) foxes in Queensland for the 1951–52 financial year (from [177], but see also [178]) showing that dingoes were rarely found in the presence of foxes.

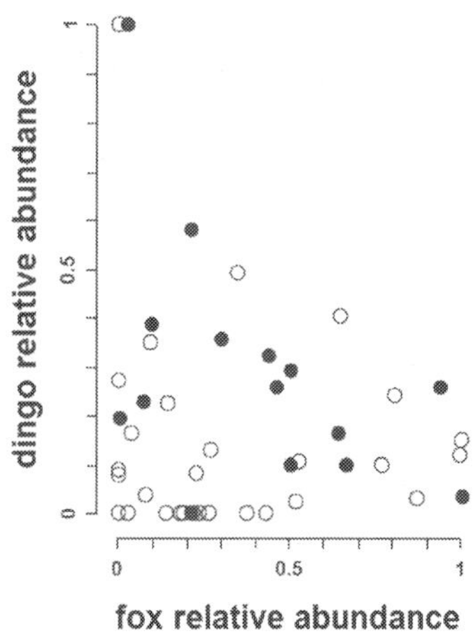

Figure 4. The relationship between dingo and fox abundance in eastern Australian forests (adapted from [136]) showing that the variability in dingo abundance is lower in areas with higher fox abundance (filled circles source data from [101], open circles source data from [157]).

5. What direct risk do dingoes pose to faunal biodiversity?

That dingoes provide net benefits to biodiversity has been almost universally accepted (e.g. [9, 30, 47, 49, 62]) despite the unreliable and inconclusive state of the literature described earlier. Additionally, and disregarded by most, is that dingoes have been implicated in the extinctions of native vertebrates prior to European settlement [23, 180, 181] and the loss of other native vertebrates in the recent past (e.g. [15, 19, 182-185]). Predation by dingoes and other wild-living dogs is therefore identified as a known or potential threat in no less than 14 national threatened species recovery plans listed by the Australian government [17] for species weighing as little as 70 g (i.e. marsupial moles, *Notorycetes* spp. [186]). 'Predation and hybridisation by feral dogs (*Canis lupus familiaris*)' is also a listed Key Threatening Process for 'threatened species, populations, and communities' in New South Wales (see [187] for the listing, see [188] and [57] for the distribution of *Canis* sub-species in Australia, and see [33, 189], [19], [56], [190], [22] for discussion of taxonomy and functional similarities between wild-living sub-species of *Canis*). Dingoes also threaten northern hairy-nosed wombats (*Lasiorhinus krefftii* [184, 191]), bridled nailtail wallabies (*Onychogalea fraenata* [146, 192]) and a range of other species [16, 112, 193, 194] in other areas, where it is predicted that some populations (such as those of koalas *Phascolarctos cinereus* [195, 196], for example) will only persist through the control or absence of canid predators, including dingoes. Not only are many mammals susceptible to exploitation by dingoes, but some bird (e.g. [19, 59, 197]) and reptile (e.g. [112, 198-200]) populations may also be substantially impacted by them. Predation on these less-preferred taxa may increase if mammals become increasingly unavailable [16]. Urgent research focussing on R5 is therefore paramount before positive dingo management is widely adopted in the hope that it will solve our biodiversity conservation problems [16, 17].

Although dingoes and threatened native fauna coexisted sympatrically prior to European settlement, they did not do so in the presence of rabbits, livestock or other landscape-changing effects of pastoralism [23, 70, 201]. Unequivocal data on dingo densities may not have been collected at the time, but post-European provision of virtually unlimited prey and water resources across much of Australia has undoubtedly increased the range and population densities of dingoes in areas outside the dingo barrier fence [19, 112, 202]. Thus, populations of many native fauna have not been exposed to such high and ubiquitous densities of dingoes until modern times. Put simply, the circumstances have changed significantly since dingoes and now-threatened native fauna coexisted sustainably [15, 22], where habitat alteration now enables dingoes (and other predators) to exploit populations that otherwise might have sustained dingo predation. Thus, dingoes clearly present direct risks to threatened fauna that must not be casually overlooked or assumed to be of lesser importance than their indirect benefits [16, 17, 22]. For example, by applying established predation risk assessment methods [50] developed for foxes and cats, [16] showed that up to 94% of extant threatened mammals, birds and reptiles in western New South Wales would be at risk of dingo predation (71% at high risk) should dingoes re-establish there (Table 2). By comparison, only 66% and 81% were predicted to be at risk of cat and fox predation [50].

	Low dingo density			High dingo density		
	No risk	Low risk	High risk	No risk	Low risk	High risk
			EXTANT MAMMALS (n = 16)			
Vulnerable	4	2	2	0	2	6
Endangered	1	5	2	0	0	8
TOTAL	**5**	**7**	**4**	**0**	**2**	**14**
			EXTANT BIRDS (n = 41)			
Vulnerable	16	13	2	4	12	15
Endangered	1	5	4	0	1	9
TOTAL	**17**	**18**	**6**	**4**	**13**	**24**
			REPTILES (n = 23)			
Vulnerable	3	5	4	1	1	10
Endangered	2	5	4	0	2	9
TOTAL	**5**	**10**	**8**	**1**	**3**	**19**
			LOCALLY EXTINCT MAMMALS (n = 17)			
TOTAL	**2**	**6**	**9**	**2**	**0**	**15**
			LOCALLY EXTINCT BIRDS (n = 4)			
TOTAL	**2**	**1**	**1**	**0**	**2**	**2**

Table 2. Summary of overall dingo predation risks to 80 threatened extant and 21 locally extinct mammals, reptiles and birds in western New South Wales (from [16]).

Information on prey important to dingoes seems particularly useful for gauging the potential risks dingoes pose to threatened fauna [16]. While the mere presence of threatened species in dingo diets might be dismissed as uncommon events [169, 203, 204], 71% (33 of 47) of dingo diet studies assess <500 scat or stomach samples [17]. Greater sampling effort and a consideration of additional information has highlighted substantial risks to threatened fauna from dingoes in some cases (e.g. [17, 61, 112]). For example, threatened mammals under 35 g body weight are typically considered to fall outside the primary weight range [75, 205] of preferred prey for dingoes [19], but ([112]; N = 1907 scats) showed that anthropogenic provision of virtually unlimited food and water resources can exacerbate the risk of decline for some such species by facilitating elevated levels of dingo predation (i.e. hyperpredation [10, 206]). In another example, ([17]; N = 4087 scats) reported that although small rodents featured relatively infrequently in dingo scats while rabbits or kangaroos were available, consideration of dingo predation rates on rodents (made possible by knowledge of predator and prey densities) supported earlier assertions by [207] that dingoes alone have the capacity to exterminate rodent (e.g. dusky hopping-mice *Notomys fuscus*, Plate 1) populations within a few months under certain conditions, regardless of any indirect benefit rodents may derive through dingoes' effects on foxes and cats [17]. Even seemingly unsusceptible arboreal and fossorial species (such as sugar gliders *Petaurus breviceps* and beach crabs *Ocypode* spp.) can become important prey for dingoes following the decline of their preferred prey ([61]; $N = 1460$ scats). Using the simple formula:

$$\text{Number of months until population extinction} = \frac{a}{\left(\frac{\left(b \times \left(365 \div 100\right)\right) \times c}{d}\right) \div 12}$$

where a = mean prey density, b = % occurrence of prey in scats, c = mean dingo pack size, and d = mean home range size of a dingo pack, the consideration of predator and prey densities can illuminate the significance of infrequent records of threatened species in dingo diets (Table 3).

Example	Dusky hopping-mice (from [17])	Rufous hare-wallabies (from [110])	Bridled nailtail wallabies (from [146])	Black-footed rock-wallabies (from [182])
Frequency of occurrence in dingo scats (%)	8*	12*	8*	46*
Mean dingo pack size (N=)	10*	10#	8^	5#
Mean dingo home range size (km^2)	25*	50#	40^	50#
Prey density (individuals/km^2)	60*	5#	5*	<1*
Predicted number of months until population extinction by dingoes	**3.08**	**6.85**	**10.27**	**0.71**

Table 3. The hypothetical impact of dingo predation on four threatened species based on the frequency of occurrence in dingo scats and predator and prey densities. (See [17] for rationale and assumptions; *Empirical data reported in original studies; ^L. Allen, unpublished data; #estimated values based on comparable studies).

As an example, [110] report the swift extinction of the small and last remaining mainland population (outside of fenced reserves) of rufous hare-wallabies *Lagorchestes hirsutus* (Plate 1) in 1987 when one or two foxes were detected first on only one occasion in an area that had just been exposed to a dingo control program. A cursory view of this outcome might suggest that dingo control facilitated the mesopredator release of foxes and led to the local extinction of a critically endangered species [41], but this does not explain the driver/s of hare-wallaby decline in the first place. Lethal dingo control had not previously occurred in the area until <100 poisoned baits were distributed along 20–30 km of vehicle tracks within the 10 km^2 area surrounding the hare-wallaby population (G. Lundie-Jenkins, unpublished data), so it could not have been lethal dingo control that caused the decline of the hare-wallabies. Foxes were reportedly absent (or at least uncommon [208]) until the dingo control program occurred [110], so it could not have been foxes which caused the decline either, and cats (which were also in very low abundance [110]) had probably been there for several decades [208, 209]. Notably, artificial water resources had not been established in the area until the 1950s and 1960s when outback mining and pastoralism became established [15, 112]. This undoubtedly increased the density and distribution of dingoes [112, 202] (the primary terrestrial predator of hare-wallabies since the extinction of thylacines [23]), suppressed any extant fox or cat populations, and caused or contributed to the decline of hare-wallabies and other marsupials [15, 19]. Furthermore, hare-wallabies were present in 12% of dingo scats collected prior to the commencement of the study [110]. Hare-wallaby densities were not reported in [110], but considering that the population became extinct just a few months later, there may have been only 50 or so animals (at most) in the population (G. Lundie-Jenkins, pers. comms.). If dingo densities were 0.2/km^2 (or 10 individuals within a home range of 50 km^2) and hare-wallaby densities were 5/km^2 (or 50 individuals within the 10 km^2 study site), and assuming that one scat represents the prey eaten by a dingo in the previous 24 hours, then 12% occurrence in dingo scats could hypothetically represent as many as 438 hare-wallabies consumed by dingoes within the home range of a dingo pack each year. In other words, dingo predation alone had the capacity to exterminate the population of hare-wallabies in <7 months if they could not sustain the loss of that many individuals annually (Table 3). That dingoes were considered to be a limiting factor for their already endangered populations [110] (which is why lethal dingo control was initiated in the first place) suggests that, in association with other causal factors, increased dingo predation over the preceding 30–40 years (a consequence of adding water and dingo prey resources to the area) drove hare-wallabies down to a point where foxes just happened to be the predator to finish the extinction process.

In a somewhat comparable situation, [185] reported that one individual dingo in a dingo-controlled area (which was not detected on sand plots, but from post-mortem evidence on killed animals) was responsible for the surplus killing of 14 (out of 101) reintroduced (and similar sized) burrowing bettongs *Bettongia lesueur* on the first night after release, the rest succumbing to predation by unknown predators within a few months. It should also be noted that the simple calculations described earlier (in Table 3) falsely assume that predation rates remain constant as the prey population declines [17], which limit firm assertions from these considerations. But if the occurrence of a given species in dingo diets is known and a few key assumptions seem reasonable (discussed in [17]), then undertaking this coarse and hypothetical exercise can indicate whether or not dingoes should be considered a potential risk to the population before positive dingo management is implemented. From the preceding discussion, it should be clear that dingoes are certainly not the type of predator that one would want around a population of threatened fauna and should, as a precaution, be considered a significant threat until robust evidence suggests otherwise.

6. Practical issues hampering the realisation of net dingo benefits

Dingo suppression of mesopredators and herbivores are the two primary mechanisms predicted to generate positive biodiversity outcomes for fauna following positive dingo management (e.g. [23, 78]). Herbivore suppression is expected to increase the food and shelter available to threatened species, mesopredator suppression is expected to decrease predation on the same species, and dingoes are simply the tool expected to generate these outcomes. While the ecological theory supporting these mechanisms might be considered sound (e.g. [4, 6]; but see [210, 211] for an alternative considerations), at least two practical factors may prevent the realisation of these expected benefits in the rangelands of south-eastern Australia (where positive dingo management is considered imperative [50]).

6.1. Livestock enterprise switching

Sheep, goats, kangaroos and rabbits may be considered the most widespread and ecologically important herbivores in this area [34, 101, 212], but in places where two or more of them are extant, using dingoes to disentangle their cumulative impacts may be very difficult to achieve. Assuming that dingoes can suppress agriculturally non-productive herbivores (such as rabbits or kangaroos) without also suppressing the livestock with which they coexist, any reduction in undesirable herbivores may be replaced by increased stocking of agriculturally productive herbivores (such as sheep, goats or cattle), thereby maintaining total grazing pressure. For example, sheep populations have suffered precipitous declines in central and southern Queensland over the last decade [213], with no substantial change in the combined grazing pressure of sheep and cattle because of enterprise switching from sheep to cattle (Fig. 5), which are now in much higher densities in the area. Hence, enhancing the prospects for biodiversity conservation by securing improvements in vegetation communities might only be achievable if livestock stocking rates are not increased following the decline of some herbivores. But such may be a trivial consideration anyway, because dingoes are unlikely to kill only livestock competitors without also killing livestock [37, 189]. Importantly though, the positive management of dingoes may be advantageous to livestock producers where dingoes have greater effects on livestock competitors than they do on livestock ([39]; i.e. in arid cattle production regions), but this may not be economically or socially acceptable in places where the impacts of dingoes on smaller livestock species are prohibitive (i.e. sheep and goat production zones).

It should be understood that dingoes can completely eliminate sheep and goat populations [37, 44, 158, 212], and although their extirpation from rangelands might be considered a biodiversity success to some, the global human population need the food and fibre products these livestock produce [214-217]. As the world's largest wool exporter, the largest goat-meat exporter, and the second largest sheep-meat exporter (www.fao.org; www.mla.com.au), the loss of Australia as a globally important supplier of small ruminant products (which dingoes are quite capable of achieving [15, 61, 142, 218]) would need to be countered by an increase in livestock production in other countries. These countries may not be able to produce them as environmentally or economically sustainably as Australia; they may have extant diseases and other pathogens (such as rabies or screwworm flies *Cochliomyia* spp.) that inhibit broad-scale production or export, be forced to clear new land for increased livestock production, or may also have native predators of their own that need controlling in order to viably scale-up their production of livestock. In short, the primary reason for encouraging dingoes in sheep production areas (i.e. to improve biodiversity outcomes) may simply shift the biodiversity conservation problem to other countries where, unlike Australia, the extant top-predators may not be very common and their management may be more complex. These, and other issues will need serious consideration before dingoes are permitted to increase in sheep and goat production areas [22, 219].

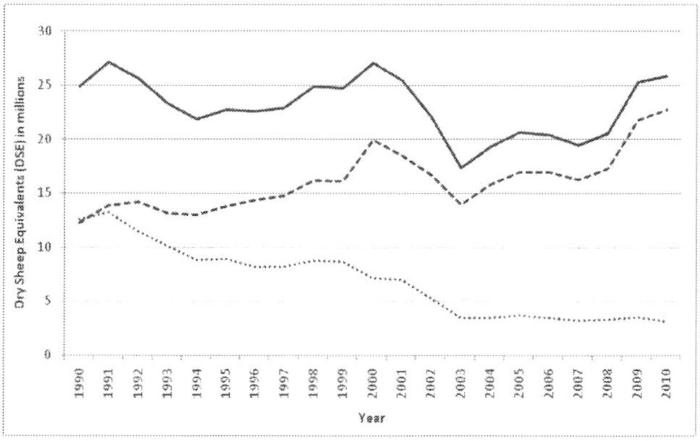

Figure 5. Trends in sheep (dotted line), cattle (dashed line; assuming 8 DSE per cow) and combined (solid line) livestock numbers in southwest and centralwest Queensland 1990–2010 (Australian Bureau of Statistics data, cat. no. 7121.0, Agricultural Commodities Australia, available at www.abs.gov.au).

6.2. Mesopredator release

Although many threatened fauna are indeed at risk of fox and cat predation [50], these fauna may also be equally at risk of dingo predation [16]. Dingoes do not kill only cats, foxes and kangaroos. In fact, these species are relatively uncommon in dingo diets [17, 19, 220], which means that replacing foxes and cats with dingoes (assuming dingoes could achieve this) or simply adding dingoes to an ecosystem might not stem the decline of threatened species [22]. As strongly interactive species, top-predators can have disproportionate effects on mesopredators, where small increases of larger predators dramatically reduce the abundance of smaller ones [1, 2]. Thus it is hypothetically conceivable that small increases in dingo abundances might substantially suppress foxes, leading to a net reduction in predator biomass and predation on threatened species. This does not appear to have been studied in great detail in Australia (Table 1) but may nevertheless prove true in some cases. Even so, the resulting lower levels of predation on threatened species might still be unsustainably high (which is why knowledge of R2 is of lesser value than R5 when considering the positive management of dingoes). In this situation, higher densities of dingoes might simply force threatened species to extinction slower than higher densities of mesopredators – the end result (extinction) being the same no matter which predator is most common (Table 3). Where multiple generalist predators are capable of exploiting the same prey species (as is the case with dingoes, foxes and cats [162, 164, 165, 172]), attempts to identify which predator is worse may be largely unhelpful in securing biodiversity against decline [221,

222]. Rather, identifying the population viability or status of threatened fauna under different management scenarios (R6) may be more useful.

A review of 14 cases of mesopredator release (analysed pairwise [223]) showed positive mesopredator population responses to decreases in higher-order predator abundance, suggesting that increases of dingoes might suppress foxes yet increase populations of cats, which are lower-order predators apparently suppressed by foxes [224]. Some support for this is found in several studies. Cats appeared to be positively associated with dingoes in the Tanami Desert of the Northern Territory [208], which is at the edge of foxes' national distribution [34, 99]. At tropical study sites devoid of foxes, [159] also reported that cats were positively associated with dingoes in the Northern Territory. At similar sites in the Kimberleys, [159] reported that (besides one outlier) cat activity varied little (0.18–0.40 tracks/sand plot/night) despite a nearly four-fold difference in dingo activity (0.80–4.30 tracks/sand plot/night). The cross-fence study of [121] (a subset of the data in [122]) also reported that foxes and cats were negatively and positively correlated with dingo presence, respectively, suggesting that increased dingoes may suppress foxes yet release cats from suppression by foxes. Subsequent analyses of the more comprehensive dataset suggested that cats were in equally low abundance on both sides of the fence [122], suggesting that cat abundance operated independently of the type of top-predator (dingoes or foxes) present.

Although increased populations of dingoes may reduce mesopredator activity they are unlikely to extirpate or exclude them (e.g. [118, 144, 225]). Detailed studies in northern South Australia ([225]; B. Allen, unpublished data from [32]) report the persistence of foxes in the presence of extremely high densities of dingoes, [144] reported that even though dingoes killed foxes they could not exclude them, and [118] showed that dingoes are unable to limit the distribution of foxes at landscape scales. Indeed, the colonisation and subsequent widespread distribution of foxes and cats across Australia [34] would suggest that the presence of dingoes (or the absence of lethal dingo control) neither prevented their establishment or limit their distribution. Rather, dingoes might reduce their densities and alter their behaviour at local scales [118], but whether or not this provides any relief to threatened prey remains unclear.

Given that dingoes are unlikely to extirpate cats, that there is strong overlap in the diets of dingoes, foxes and cats, and that cat predation is listed by the Australian Government as a Key Threatening Process to 18 of the 19 threatened arid-zone mammal species [122], there may be little overall biodiversity conservation benefit to species threatened by both foxes and cats if dingo populations increase [16, 22]. Irrespective of this, the positive management of dingoes would be unnecessary for places with extant (and typically unmanaged [32]) dingo populations, such as areas outside the dingo barrier fence, which are (confusingly) the very areas where some predict their positive management to be of most benefit to threatened fauna [122]. As illustrated earlier for rufous hare-wallabies and in addition to a variety of other important factors (discussed in [71, 72, 74, 226]), at-risk fauna are clearly threatened by predation per se, and not dingo *or* fox *or* cat predation individually (e.g. [221, 222, 227]). The literature is replete with examples of reductions of one pest animal increasing the undesirable impacts of another with no (or worse) overall outcomes for the species of conservation concern (e.g. [12, 110, 228]), and it would be naive to expect the positive management of dingoes across large areas to achieve universally 'good' outcomes for faunal biodiversity at more local scales [16, 22]. Increasing the number of generalist predators may only widen the suite of prey susceptible to predation and subsequent decline [222], and 'one may ask if the faunal biodiversity outcomes are any greater if a species is extinguished by a dingo instead of a fox or feral cat' [22]. Moreover, the biodiversity benefits expected of dingoes are likely to be available only to those prey species which have survived the impacts of cats, foxes and dingoes anyway. Thus, if fox and/or cat impacts are not the limiting factor for threatened species, then encouraging the suppression of foxes and cats by adding dingoes to the ecosystem seems an unlikely prerequisite for their recovery [16].

7. Context-specific management

Dingo impacts, roles and functions are context-specific, and the same is true for other top-predators [5, 229]. For example, the positive effects of wolves on biodiversity in some places may not be as apparent in other places just a few kilometres away, where site-specific factors may affect the strength of influence wolves have in the ecosystem [230, 231]. Such context-specific impacts mean that extreme caution should be exercised when considering using top-predators as biodiversity conservation tools in some new context, based on information collected from another time and place [22, 229]. Bottom-up factors associated with prey availability (such as habitat productivity, structural complexity etc) will affect the density of predators [174-176], the density of prey species [232-234] and their relative vulnerability to predation [221, 222, 227, 235]. Within this diversity, land use also varies from conservation to agriculture, from extensive to intensive livestock enterprises, and from small livestock to cattle production (e.g. [15, 69]). It is, therefore, unreasonable to expect that the goals and outcomes of dingo management will be uniform across Australia, which is why dingoes are presently managed locally for where they are and what they are (or are expected to be) doing [33, 35, 64].

Should positive dingo management to be adopted across large areas, the negative impacts of dingoes expected in some contexts may not be manageable in others. For example, the presence of dingoes has been predicted to benefit some rodents in arid environments [47], but dingo predation alone has the capacity to exterminate local populations of the same rodents under certain conditions (e.g. during droughts; Table 3; [17]) – conditions that are predicted to become more frequent and intense under future climate-change scenarios [236-238]. The negative impacts of dingoes in livestock production areas may also become increasingly unmanageable as dingoes are encouraged in adjacent conservation reserves where their impacts might be positive. Radio and GPS tracking studies indicate that most dingoes are sedentary (e.g. [108, 111, 239, 240]), and a recent continental-scale gene flow study [57] supports this conclusion. But a substantial proportion of dingoes do travel considerable distances (e.g. >550 km in 30 days [97]) for dispersal and exploration (e.g. [97, 123, 239, 241]). Given the capacity for dingoes to disperse, without containment fencing, dingo populations and their impacts (like reintroduced wolves [8]) are unlikely to remain only in reserves.

These issues are outside the capacity of any one individual or agency to manage, and are best addressed through a strategic adaptive management approach that can accommodate differences in situation and objectives [242-244]. The management of dingoes (either positively or negatively) requires adherence to a number of underlying principles including: defining the biological assets to be protected and the people involved, setting measurable goals and timeframes for action, undertaking management actions at a scale appropriate to the enterprise or ecosystem to be enhanced and the wild dog home range and movements, relying on a suite of actions applied in a coordinated sequence, and continuously monitoring in preparation for new incursions or threats [35, 64]. Issues of scale and management unit are particularly important, and the minimum size of the management unit may be determined using the home range size of the animal in the particular environment as a guide. Recorded home range sizes for dingoes vary from 7–2013 km^2 in semi-arid and arid rangeland rangelands, from 2–262 km^2 in mesic environments, and may be <1 km^2 in urban areas [19, 112, 239, 245]. Such variation in scales important to dingoes is likely to preclude management approaches which seek to apply broad-scale solutions to context-dependant problems, such as the widespread prohibition of dingo control for the recovery of an isolated population of threatened mammals.

Although dingo management policies must be general by nature, the process of defining the issue in strategic management ensures that the appropriate scale for actions is decided before commencement. Therefore, where dingoes are determined by reliable experimentation to be important for biodiversity conservation, strategic management can achieve this objective locally or regionally, depending on the minimum size of the management unit required. In short, top-down management approaches which seek to exclude the land manager in favour of government policy intervention (e.g. [70]) and/or apply broad-scale solutions to context-dependant impacts (either positive or negative) are unlikely to succeed in restoring faunal biodiversity [22, 246].

8. Looking forward: surmountable challenges to overcome

Knowing that the available data is lacking rigour and defensible or definite conclusions may seem depressing after the countless hours of hard work expended by many in obtaining it. But all is not lost, and dismissing it completely may be just as dangerous as embracing it uncritically [53]. From the implications of [52], [95] and the present study it seems clear that a greater understanding of the advantages and limitations of sand plot tracking indices are required by many dingo researchers, and it will be difficult in reaching consensus on the state of the available literature until this is achieved. The advantages and limitations of indices and populations estimation procedures have been widely discussed (in [67, 93, 94, 105, 106, 114-116, 247-249]; to cite just a few) to a point where relative abundance indices can be viewed as an incredibly powerful population censusing technique provided appropriate principles and analyses are applied [93, 114]. Moreover, so long as the results of studies with lower inferential ability are valued above those with designs that permit more definitive statements, end-users of the literature may also continue to be confused about the most appropriate dingo and threatened species management strategies. A return to more objective and applied science and management of dingoes is imperative (also suggested by [189]).

Long-term manipulative experiments are able to advance science much more rapidly than other approaches [68, 89, 90], but they are few (Table 1), and more are sorely needed [41, 250]. When conducting such studies, the relationships (Fig. 1) and knowledge gaps being investigated are of utmost importance. Interest in the positive management of dingoes as biodiversity conservation tools is ultimately driven by the desire to improve the status of threatened fauna through trophic effects (e.g. [23, 50]), so should not the threatened faunal response to dingo management be the variable of interest? Demonstration of sustained non-target population responses to predator control can provide 'conclusive proof' [79] for the effects of lethal dingo control on threatened fauna. Hence, in places where dingoes are actively controlled (for whatever reason), it is not the direct or indirect effects of dingoes on fauna that should be of primarily interest, but rather, the effects of dingo management practices on fauna (R6) – the 'black box' approach [86]. Knowledge of the other relationships (R2, R3, R5) is supplementary and may be more important in places where dingoes are typically unmanaged.

In order to focus our collective attention on the questions that matter most, we issue the following challenge. For any given site and population of threatened species:

1. Do contemporary dingo management practices negatively affect the species either directly or indirectly?
2. Do dingoes themselves pose a current or future threat to the species, regardless of their indirect effects on other threatening processes?
3. Is positive dingo management the only practical option to improve conditions for the species?
4. What factors determine which predator becomes ecologically dominant following dingo control programs?

If contemporary dingo management practices (such as poison baiting, trapping or shooting) do not harm threatened species either directly or indirectly (R6), then arguments to cease controlling dingoes remain unjustified on biodiversity conservation grounds. Multiple studies have failed to demonstrate the 'release' of mesopredators following dingo control (R4) (e.g. [87, 88, 159, 251], and no studies to date have shown short-term negative responses from populations of non-target species to dingo or fox control [79]. Hence, lethal dingo control will still be useful in mitigating livestock losses without fear of releasing mesopredators or harming threatened species. If dingoes threaten a particular species to any degree (R5), then researchers must investigate the relative strengths of dingo-prey (R5), mesopredator-prey (R3), and dingo-mesopredator (R2) interactions in order to gauge the likely outcomes of positive dingo management. Positive dingo management is unlikely to benefit the threatened species where the direct effect of dingoes is greater (or may become greater) than their indirect effect on mesopredators.

If dingo control does appear to hinder the conservation of the species, and dingoes do not pose a current or future threat to them, are there any alternative management actions that could improve biodiversity outcomes without compromising livestock production values? For example, livestock guardian dogs might offer a non-lethal approach to reduce the impacts of dingoes on livestock without excluding dingoes from an area [252, 253].

Alternatively, the selective exclusion of agriculturally non-productive herbivores from watering points [254-256] may elicit a greater bottom-up response from threatened species than the top-down suppression of mesopredators by dingoes without threatening the viability of livestock producers. In fact, doing so would probably enhance their viability.

Lastly, the commonly observed presence of foxes in areas free of dingo control suggests that bottom-up factors may largely determine which predator successfully colonises and dominates an area, though these influences remain largely unknown. Foxes appear to be positively associated with disturbed agricultural habitats in a bottom-up manner [257, 258], which may help explain the pattern of fox densities noted by [178] and others (e.g. [50]). Top-predators can also be associated with higher biodiversity in a bottom-up manner [19, 174, 175, 229], and positive correlations between dingoes and greater biodiversity values cannot be immediately interpreted to be the result of top-down processes [52, 68]. When the factors that determine which predator dominates a given area become well understood, our ability to manage predators will be greatly enhanced.

9. Conclusion

Maintaining top-predator function may be an important component of biodiversity conservation initiatives in many places [1, 2]. Although this might be more easily achieved in relatively intact areas, the functions of top-predators may be most needed in the more degraded ecosystems characterised by depleted faunal and floral communities. Importantly though, such systems are typically those used most heavily by humans for agricultural production, and the age-old battle betweens humans and top-predators seems likely to continue into the foreseeable future [214, 259]. Nevertheless, conservative environmental management is required in our efforts to balance the needs of humans with those of the threatened fauna and flora we seek to protect [260]. Evidence-based biodiversity conservation and carefully considered policy approaches are critical to the informed management of top-predators for this purpose [261, 262].

This chapter has discussed the knowledge and management of dingoes for biodiversity conservation. Our overview of the field data underpinning knowledge of dingoes' ecological roles has identified critical knowledge gaps that we believe require the primary attention of researchers and policy makers operating in this area. We have also shown that although dingoes are well-studied, their functional roles may not be well understood. This is because methodological flaws, sampling bias and experimental design limitations inherent to most studies (Table 1; [52]) cannot provide reliable or conclusive evidence for dingoes ecological roles. We therefore agree with [53] that there is inconclusive evidence for the positive roles of dingoes and that cessation of lethal dingo control is presently unjustified on biodiversity conservation grounds. We are cognizant that questioning the conclusions of studies documenting the benefits of fox control on native fauna [263] probably delayed the necessary implementation of broad-scale fox control for biodiversity conservation in many places. Likewise, we acknowledge that questioning the science underpinning the role of dingoes may delay the adoption of positive dingo management in places that might yet be shown to need it. However, we believe there are sufficient concerns regarding the impacts of dingoes on mesopredators and threatened fauna to stress strong caution when considering the positive management of dingoes for biodiversity conservation purposes under current ecological conditions [22].

We therefore challenge researchers and funding agencies to focus on applied science questions that can address the effects of dingo management practices on prey populations of interest. Doing so within an experimental framework that has the capacity to explore and exclude alternative hypotheses will be most useful, and we encourage those with such data to invest time in its analyses and publication. We encourage the continued interest in dingoes as a biodiversity conservation tool, and look forward to the results of future studies on this charismatic and iconic terrestrial top-predator.

Author details

Benjamin L. Allen* and Luke K-P. Leung
School of Agriculture and Food Sciences, the University of Queensland, Gatton, Queensland, Australia

Peter J.S. Fleming
Vertebrate Pest Research Unit, Department of Primary Industries, Orange, New South Wales , Australia

Matt Hayward
Centre for African Conservation Ecology, Nelson Mandela Metropolitan University, Port Elizabeth, South Africa
School of Biological, Earth and Environmental Science, University of New South Wales, Sydney, Australia

Lee R. Allen
Robert Wicks Pest Animal Research Centre, Biosecurity Queensland, Toowoomba, Queensland, Australia

Richard M. Engeman
National Wildlife Research Centre, US Department of Agriculture, Fort Collins, Colorado, USA

Guy Ballard
Vertebrate Pest Research Unit, Department of Primary Industries, Armidale, New South Wales, Australia

Acknowledgement

We thank the editors for the invitation to prepare this report, which was enhanced by discussions, correspondence and input from a variety of researchers, land managers and policy makers involved in dingo and threatened species research and management.

---

* Corresponding Author

10. References

[1] Ray JC, Redford KH, Steneck RS, Berger J. Large carnivores and the conservation of biodiversity. Washington: Island Press; 2005.

[2] Terborgh J, Estes JA. Trophic cascades: Predator, prey, and the changing dynamics of nature. Washington D.C.: Island Press; 2010.

[3] Miller BJ, Harlow HJ, Harlow TS, Biggins D, Ripple WJ. Trophic cascades linking wolves (*Canis lupus*), coyotes (*Canis latrans*), and small mammals. Canadian Journal of Zoology. 2012;90:70-8.

[4] Ripple WJ, Beschta RL. Trophic cascades in Yellowstone: The first 15 years after wolf reintroduction. Biological Conservation. 2012;145(1):205-13.

[5] Soule ME, Estes JA, Miller B, Honnold DL. Strongly interacting species: conservation policy, management, and ethics. BioScience. 2005 2005/02/01;55(2):168-76.

[6] Estes JA, Terborgh J, Brashares JS, Power ME, Berger J, Bond WJ, et al. Trophic downgrading of planet earth. Science. 2011;333:301-6.

[7] Hayward MW, Somers MJ. Reintroduction of top-order predators. Oxford: Wiley-Blackwell; 2009.

[8] Bangs EE, Smith DW. Re-introduction of the gray wolf into Yellowstone National Park and central Idaho, USA. In: Soorae PS, editor. Global re-introduction perspectives: re-introduction case studies from around the globe. Abu Dhabi, UAE: IUCN/SSC Re-introduction Specialist Group; 2008.

[9] Ritchie EG, Elmhagen B, Glen AS, Letnic M, Ludwig G, McDonald RA. Ecosystem restoration with teeth: what role for predators? Trends in Ecology and Evolution. 2012;27(5):265-71.

[10] Courchamp F, Langlais M, Sugihara G. Rabbits killing birds: Modelling the hyperpredation process. Journal of Animal Ecology. 2000;69(1):154-64.

[11] Nogales M, Martin A, Tershy B, Donlan CJ, Veitch CR, Puerta N, et al. A review of feral cat eradication on islands. Conservation Biology. 2004;18:310-9.

[12] Ruscoe WA, Ramsey DSL, Pech RP, Sweetapple PJ, Yockney I, Barron MC, et al. Unexpected consequences of control: competitive vs. predator release in a four-species assemblage of invasive mammals. Ecology Letters. 2011;14(10):1035-42.

[13] Engeman R, Whisson D, Quinn J, Cano F, Quiñones P, White Jr. TH. Monitoring invasive mammalian predator populations sharing habitat with the critically endangered Puerto Rican parrot *Amazona vittata*. Oryx. 2006;40(1):95-102.

[14] Miller B, Reading R, Forest S. Prairie night: Black-footed ferrets and the recovery of endangered species. Washington DC: Smithsonian Institution Press; 1996.

[15] Allen BL. A comment on the distribution of historical and contemporary livestock grazing across Australia: Implications for using dingoes for biodiversity conservation Ecological Management and Restoration. 2011;12(1):26-30.

[16] Allen BL, Fleming PJS. Reintroducing the dingo: The risk of dingo predation to threatened species in western New South Wales. Wildlife Research. 2012;39(1):35-50.

[17] Allen BL, Leung LK-P. Assessing predation risk to threatened fauna from their prevalence in predator scats: dingoes and rodents in arid Australia. PLoS ONE. 2012;7(5):e36426.

[18] Gompper ME. Top carnivores in the suburbs? Ecological and conservation issues raised by colonization of northeastern North America by coyotes. BioScience. 2002;52(2):185-90.

[19] Corbett LK. The dingo in Australia and Asia. Second ed. Marleston: J.B. Books, South Australia; 2001.

[20] Carbone C, Mace GM, Roberts SC, Macdonald DW. Energetic constraints on the diet of terrestrial carnivores. Nature. [10.1038/46266]. 1999;402(6759):286-8.

[21] Carbone C, Teacher A, Rowcliffe JM. The costs of carnivory. PloS Biology. 2007;5(2: e22):0363-8.

[22] Fleming PJS, Allen BL, Ballard G. Seven considerations about dingoes as biodiversity engineers: the socioecological niches of dogs in Australia. Australian Mammalogy. 2012;34(1):119-31.

[23] Johnson C. Australia's mammal extinctions: A 50 000 year history. Melbourne: Cambridge University press; 2006.

[24] Johnson CN, Wroe S. Causes of extinction of vertebrates during the Holocene of mainland Australia: Arrival of the dingo, or human impact? Holocene. 2003 Nov;13(6):941-8.

[25] Letnic M, Fillios M, Crowther MS. Could direct killing by larger dingoes have caused the extinction of the thylacine from mainland Australia? PLoS ONE. 2012;7(5):e34877.

[26] Oskarsson MCR, Klutsch CFC, Boonyaprakob U, Wilton A, Tanabe Y, Savolainen P. Mitochondrial DNA data indicate an introduction through Mainland Southeast Asia for Australian dingoes and Polynesian domestic dogs. Proceedings of the Royal Society B. In press;xx(xx):xx-xx.

[27] Saetre P, Lindberg J, Leonard JA, Olsson K, Pettersson U, Ellegren H, et al. From wild wolf to domestic dog: Gene expression changes in the brain. Molecular Brain Research. 2004 2004/7/26;126(2):198-206.

[28] Savolainen P, Leitner T, Wilton AN, Matisoo-Smith E, Lundeberg J. A detailed picture of the origin of the Australian dingo, obtained from the study of mitochondrial DNA. Proceedings of the National Academy of Sciences of the United States of America. 2004 August 17, 2004;101(33):12387-90.

[29] vonHoldt BM, Pollinger JP, Lohmueller KE, Han E, Parker HG, Quignon P, et al. Genome-wide SNP and haplotype analyses reveal a rich history underlying dog domestication. Nature. 2010;464:898-903.

[30] Bowman D. Bring elephants to Australia? Nature. 2012;482(7383):30.

[31] Wallach AD, Ritchie EG, Read J, O'Neill AJ. More than mere numbers: The impact of lethal control on the stability of a top-order predator. PloS ONE. 2009;4(9):e6861.

[32] Allen BL. The effect of lethal control on the conservation values of *Canis lupus dingo*. In: Maia AP, Crussi HF, editors. Wolves: Biology, conservation, and management. New York: Nova Publishers; 2012. p. 79-108.

[33] Allen BL, Engeman RM, Allen LR. Wild dogma II: The role and implications of wild dogma for wild dog management in Australia. Current Zoology. 2011;57(6):737-40.

[34] West P. Assessing invasive animals in Australia 2008. Canberra: National Land and Water Resources Audit, The Invasive Animals Cooperative Research Centre2008.

[35] Fleming PJS, Allen BL, Allen LR, Ballard G, Bengsen AJ, Gentle MN, et al. Management of wild canids in Australia: free-ranging dogs and red foxes. In: Glen AS, Dickman CR, editors. Carnivores of Australia: past, present and future. Collingwood: CSIRO Publishing; In press.

[36] Hewitt L. Major economic costs associated with wild dogs in the Queensland grazing industry. Brisbane: Agforce2009.

[37] Thomson PC. Dingoes and sheep in pastoral areas. Journal of Agriculture. 1984;25:27-31.

[38] Allen BL, editor. The effect of regional dingo control on calf production in northern South Australia, 1972-2008. Queensland Pest Animal Symposium; 2010; Gladstone, Queensland.

[39] Wicks S, Allen BL, editors. Returns on investment in wild dog management: beef production in the South Australian arid lands. 56th Australian Agricultural Resource Economics Society conference; 2012; Fremantle, Western Australia: AARES.

[40] Glen AS, Dickman CR. Complex interactions among mammalian carnivores in Australia, and their implications for wildlife management. Biological Reviews. 2005 August;80(3):387-401.

[41] Glen AS, Dickman CR, Soule ME, Mackey BG. Evaluating the role of the dingo as a trophic regulator in Australian ecosystems. Austral Ecology. 2007;32(5):492-501.

[42] Barnes TS, Goldizen AW, Morton JM, Coleman GT. Cystic echinococcosis in a wild population of the brush-tailed rock-wallaby (*Petrogale penicillata*), a threatened macropodid. Parasitology. 2008;135:715-23.

[43] Jenkins DJ, Macpherson CNL. Transmission ecology of *Echinococcus* in wild-life in Australia and Africa. Parasitology. 2003 2003;127(Supplement):S63-S72.

[44] Allen LR, Sparkes EC. The effect of dingo control on sheep and beef cattle in Queensland. Journal of Applied Ecology. 2001;38:76-87.

[45] Atkinson SA, editor. Dingo control or conservation? Attitudes towards urban dingoes (*Canis lupus dingo*) as an aid to dingo management. 23rd Vertebrate Pest Conference; 2008; California: University of California, Davis.

[46] Hytten KF. Dingo dualisms: Exploring the ambiguous identity of Australian dingoes. Australian Zoology. 2009;35(1):18-27.

[47] Letnic M, Ritchie EG, Dickman CR. Top predators as biodiversity regulators: the dingo *Canis lupus dingo* as a case study. Biological Reviews. 2012;87(2):390-419.

[48] Visser RL, Watson JEM, Dickman CR, Southgate R, Jenkins D, Johnson CN. Developing a national framework for dingo trophic regulation research in Australia: Outcomes of a national workshop. Ecological Management and Restoration. 2009;10(2):168-70.

[49] Carwardine J, O'Connor T, Legge S, Mackey B, Possingham HP, Martin TG. Priority threat management to protect Kimberley wildlife. Brisbane: CSIRO Ecosystem Sciences; 2011.

[50] Dickman C, Glen A, Letnic M. Reintroducing the dingo: Can Australia's conservation wastelands be restored? In: Hayward MW, Somers MJ, editors. Reintroduction of top-order predators. Oxford: Wiley-Blackwell; 2009. p. 238-69.

[51] Allen BL. Did dingo control cause the elimination of kowaris through mesopredator release effects? A response to Wallach and O'Neill (2009). Animal Biodiversity and Conservation. 2010;32(2):1-4.

[52] Allen BL, Engeman RM, Allen LR. Wild dogma I: An examination of recent "evidence" for dingo regulation of invasive mesopredator release in Australia. Current Zoology. 2011;57(5):568-83.

[53] Glen AS. Enough dogma: seeking the middle ground on the role of dingoes. Current Zoology. In press;xx(xx):xx-xx.

[54] Letnic M, Crowther MS, Dickman CR, Ritchie E. Demonising the dingo: How much wild dogma is enough? Current Zoology. 2011;57(5):668-70.

[55] Jones E. Hybridisation between the dingo, *Canis lupus dingo*, and the domestic dog, *Canis lupus familiaris*, in Victoria: A critical review. Australian Mammology. 2009;31:1-7.

[56] Corbett LK. *Canis lupus ssp. dingo*. IUCN 2010. IUCN Red List of Threatened Species. Version 2010.4. www.iucnredlist.org. 2008:Downloaded on 20 April 2011.

[57] Stephens D. The molecular ecology of Australian wild dogs: hybridisation, gene flow and genetic structure at multiple geographic scales [PhD]. Perth: The University of Western Australia; 2011.

[58] Elledge AE, Allen LR, Carlsson B-L, Wilton AN, Leung LK-P. An evaluation of genetic analyses, skull morphology and visual appearance for assessing dingo purity: implications for dingo conservation. Wildlife Research. 2008;35:812-20.

[59] Boland CRJ. Breeding biology of Rainbow Bee-Eaters (*Merops ornatus*): A migratory, colonial, cooperative bird. Auk. 2004 Jul;121(3):811-23.

[60] Smith BP, Litchfield CA. A review of the relationship between indigenous Australians, dingoes (*Canis dingo*) and domestic dogs (*Canis familiaris*). Anthrozoös. 2009;22(2):111-28.

[61] Allen LR, Goullet M, Palmer R. The diet of the dingo (*Canis lupus dingo* and hybrids) in north-eastern Australia: a supplement to Brook and Kutt. The Rangeland Journal. In press;xx(xx):xx-xx.

[62] Baker L, Letnic M, Nesbitt B. Ecological functional landscapes: the role of native top-order predators as trophic regulators. New South Wales: Public submission to the Northern Rivers Catchment Management Authority; 2011.

[63] Clarke M. Final recommendation on a nomination for listing: *Canis lupus* subsp. *dingo*. Victoria: Scientific Advisory Committee, Department of Sustainability and Environment; 2007.

[64] Fleming P, Corbett L, Harden R, Thomson P. Managing the impacts of dingoes and other wild dogs. Bomford M, editor. Canberra: Bureau of Rural Sciences; 2001.

[65] Ritchie E, editor. Dingoes: managing for ecosystem resiliance and pastoral productivity. Queensland Pest Animal Symposium; 2010; Gladstone.

[66] Linnell JDC. The relative importance of predators and people in structuring and conserving ecosystems. Conservation Biology. 2011;25(3):646-7.

[67] Caughley G. Analysis of vertebrate populations. London: John Wiley and Sons; 1977.

[68] MacKenzie DI, Nichols JD, Royle JA, Pollock KH, Bailey LL, Hines JE. Occupancy estimation and modelling: Inferring patterns and dynamics of species occurrence. London: Academic Press (Elsevier); 2006.

[69] Hamblin A. Land, Australia State of the environment report (Theme report). Canberra: CSIRO Publishing, on behalf of the Department of Environment and Heritage; 2001.
[70] Letnic M. Dispossession, degradation, and extinction: environmental history in arid Australia. Biodiversity and Conservation. 2000;9:295-308.
[71] McKenzie NL, Burbidge AA, Baynes A, Brereton RN, Dickman CR, Gordon G, et al. Analysis of factors implicated in the recent decline of Australia's mammal fauna. Journal of Biogeography. 2007;34(4):597-611.
[72] Dickman CR, Pressey RL, Lim L, Parnaby HE. Mammals of particular conservation concern in the Western Division of New South Wales. Biological Conservation. 1993 1993;65:219-48.
[73] Lunney D. Causes of the extinction of native mammals of the western division of New South Wales: An ecological interpretation of the nineteenth century historical record. Rangeland Journal. 2001;23(1):44-70.
[74] Smith PJ, Pressey RL, Smith JE. Birds of particular conservation concern in the western division of New South Wales. Biological Conservation. 1994;69:315-38.
[75] Burbidge AA, McKenzie NL. Patterns in the modern decline of Western Australia's vertebrate fauna: Causes and conservation implications. Biological Conservation. 1989;50:143-98.
[76] Purcell BV. Dingo. *Australian Natural History Series*. Collingwood: CSIRO Publishing; 2010.
[77] Fleming PJS, Allen LR, Lapidge SJ, Robley A, Saunders GR, Thomson PC. Strategic approach to mitigating the impacts of wild canids: Proposed activities of the Invasive Animals Cooperative Research Centre. Australian Journal of Experimental Agriculture. 2006;46(6-7):753-62.
[78] Wallach AD, Johnson CN, Ritchie EG, O'Neill AJ. Predator control promotes invasive dominated ecological states. Ecology Letters. 2010;13:1008-18.
[79] Glen AS, Gentle MN, Dickman CR. Non-target impacts of poison baiting for predator control in Australia. Mammal Review. 2007;37(3):191-205.
[80] Elledge AE, Leung LK, Allen LR, Firestone K, Wilton AN. Assessing the taxonomic status of dingoes *Canis familiaris dingo* for conservation. Mammal Review. 2006;36(2):142-56.
[81] Denny EA, Dickman CR. Review of cat ecology and management strategies in Australia. Canberra: Invasive Animals Cooperative Research Centre; 2010.
[82] Saunders GR, Gentle MN, Dickman CR. The impacts and management of foxes *Vulpes vulpes* in Australia. Mammal Review. 2010;40(3):181-211.
[83] Allen LR. Best-practice baiting: Evaluation of large-scale, community-based 1080 baiting campaigns. Toowoomba: Robert Wicks Pest Animal Research Centre, Department of Primary Industries (Biosecurity Queensland); 2006.
[84] APVMA. Review findings for sodium monofluoroacetate: The reconsideration of registrations of products containing sodium monofluoroacetate and approvals of their associated labels, Environmental Assessment. Canberra: Australian Pesticides and Veterinary Medicines Authority; 2008.

[85] Visser RL, Watson JEM, Dickman CR, Southgate R, Jenkins D, Johnson CN. A national framework for research on trophic regulation by the Dingo in Australia. Pacific Conservation Biology. 2009;15:209-16.

[86] Kinnear JE, Krebs CJ, Pentland C, Orell P, Holme C, Karvinen R. Predator-baiting experiments for the conservation of rock-wallabies in Western Australia: A 25-year review with recent advances. Wildlife Research. 2010;37:57-67.

[87] Allen LR. The impact of wild dog predation and wild dog control on beef cattle production, PhD Thesis. Brisbane: Department of Zoology, The University of Queensland; 2005.

[88] Eldridge SR, Shakeshaft BJ, Nano TJ. The impact of wild dog control on cattle, native and introduced herbivores and introduced predators in central Australia, Final report to the Bureau of Rural Sciences. Alice Springs: Parks and Wildlife Commission of the Northern Territory2002.

[89] Hone J. Wildlife damage control. Collingwood, Victoria: CSIRO Publishing; 2007.

[90] Platt JR. Strong inference: Certain systematic methods of scientific thinking may produce much more rapid progress than others. Science. 1964;146(3642):347-53.

[91] Meek PD. Refining and improving the use of camera trap technology for wildlife management and research in Australia and New Zealand Final report to the Winston Churchill Memorial Trust of Australia. 2011.

[92] Robley A, Woodford L, Lindeman M, Ivone G, Beach M, Campbell I, et al. Assessing the effectiveness of buried baiting for the control of wild dogs in Victoria. Heidelberg, Victoria: Department of Sustainability and Environment; 2011.

[93] Engeman R. Indexing principles and a widely applicable paradigm for indexing animal populations. Wildlife Research. 2005;32(3):202-10.

[94] Wilson GJ, Delahay RJ. A review of methods to estimate the abundance of terrestrial carnivores using field signs and observation. Wildlife Research. 2001;28:151-64.

[95] Mahon PS, Banks PB, Dickman CR. Population indices for wild carnivores: A critical study in sand-dune habitat, south-western Queensland. Wildlife Research. 1998;25:11-22.

[96] Creel S, Christianson D. Relationships between direct predation and risk effects. Trends in Ecology and Evolution. 2008;23(4):194–201.

[97] Allen LR. Best practice baiting: dispersal and seasonal movement of wild dogs (*Canis lupus familiaris*). Technical highlights: Invasive plant and animal research 2008–09. Brisbane: QLD Department of Employment, Economic Development and Innovation; 2009. p. 61-2.

[98] Cogger H. Reptiles and amphibians of Australia (Sixth edition). Florida: Ralph Curtis Publishing; 2000.

[99] Saunders G, McLeod L. Improving fox management strategies in Australia. Canberra: Bureau of Rural Sciences; 2007.

[100] Evangelista P, Engeman R, Tallents L. Testing a passive tracking index for monitoring the endangered Ethiopian wolf. Integrative Zoology. 2009;4:172-8.

[101] Newsome AE, Catling PC, Cooke BD, Smyth R. Two ecological universes separated by the dingo barrier fence in semi-arid Australia: Interactions between landscapes,

herbivory and carnivory, with and without dingoes. Rangeland Journal. 2001;23(1):71-98.

[102] Edwards GP, de Preu N, Crealy IV, Shakeshaft BJ. Habitat selection by feral cats and dingoes in a semi-arid woodland environment in central Australia. Austral Ecology. 2002;27:26-31.

[103] Southgate R, Paltridge R, Masters P, Carthew S. Bilby distribution and fire: a test of alternative models of habitat suitability in the Tanami Desert, Australia. Ecography. 2007 December;30(6):759-76.

[104] Southgate R, Paltridge R, Masters P, Ostendorf B. Modelling introduced predator and herbivore distribution in the Tanami Desert, Australia. Journal of Arid Environments. 2007 Feb;68(3):438-64.

[105] Sutherland WJ. Ecological census techniques. Cambridge, UK: Cambridge University Press; 1996.

[106] Pollock KH. The challenge of measuring change in wildlife populations: A biometrician's perspective. In: Grigg GC, Hale PT, Lunney D, editors. Conservation through the sustainable use of wildlife. The University of Queensland: Centre for Conservation Biology; 1995. p. 117-21.

[107] Krebs CJ. Ecology: The experimental analysis of distribution and abundance. 6 ed. San Francisco: Benjamin-Cummings Publishing; 2008.

[108] Thomson PC. The behavioural ecology of dingoes in north-western Australia: IV. Social and spatial organisation, and movements. Wildlife Research. 1992;19(5):543-63.

[109] Moseby KE, Owens H, Brandle R, Bice JK, Gates J. Variation in population dynamics and movement patterns between two geographically isolated populations of the dusky hopping mouse (*Notomys fuscus*). Wildlife Research. 2006;33:223-32.

[110] Lundie-Jenkins G, Corbett LK, Phillips CM. Ecology of the rufous hare-wallaby, *Lagorchestes hirsutus* Gould (Marsupialia: Macropodidae), in the Tanami Desert, Northern Territory. III. Interactions with introduced mammal species. Wildlife Research. 1993;20:495-511.

[111] Allen BL. Do desert dingoes drink daily? Visitation rates at remote waterpoints in the Strzelecki Desert. Australian Mammalogy. In press;xx(xx):xx-xx.

[112] Newsome TM. Ecology of the dingo (*Canis lupus dingo*) in the Tanami Desert in relation to human-resource subsidies [PhD]. Sydney: The University of Sydney; 2011.

[113] Buckland ST, Anderson DR, Burnham KP, Laake JL, Borchers DL, Thomas LN. Introduction to distance sampling: estimating abundance of biological populations. Oxford: Oxford University Press; 2001.

[114] Johnson DH. In defense of indices: the case of bird surveys. Journal of Wildlife Management. 2008;72(4):857-68.

[115] Anderson DR. The need to get the basics right in wildlife field studies. Wildlife Society Bulletin. 2001;29(4):1294-7.

[116] Engeman RM. More on the need to get the basics right: population indices. Wildlife Society Bulletin. 2003;31:286-7.

[117] Atwood TC, Fry TL, Leland BR. Partitioning of anthropogenic watering sites by desert carnivores. The Journal of Wildlife Management. 2011;75(7):1609-15.

[118] Mitchell BD, Banks PB. Do wild dogs exclude foxes? Evidence for competition from dietary and spatial overlaps. Austral Ecology. 2005;30(5):581-91.

[119] Hayward MW, Slotow R. Temporal partitioning of activity in large African carnivores: tests of multiple hypotheses. South African Journal of Wildlife Research. 2009;39:109-25.

[120] Wallach AD, O'Neill AJ. Threatened species indicate hot-spots of top-down regulation. Animal Biodiversity and Conservation. 2009;32(2):127-33.

[121] Letnic M, Crowther M, Koch F. Does a top-predator provide an endangered rodent with refuge from a mesopredator? Animal Conservation. 2009;12(4):302-12.

[122] Letnic M, Koch F, Gordon C, Crowther M, Dickman C. Keystone effects of an alien top-predator stem extinctions of native mammals. Proceedings of the Royal Society of London B. 2009;276:3249-56.

[123] Purcell BV. Order in the pack: Ecology of *Canis lupus dingo* in the southern Greater Blue Mountains World Heritage Area [PhD]. Sydney: University of Western Sydney, School of Natural Sciences; 2009.

[124] Pettigrew JD. A burst of feral cats in the Diamantina: a lesson for the management of pest species? In: Siepen G, Owens C, editors. Cat Management Workshop Proceedings. Brisbane: Queensland Department of Environment and Heritage; 1993. p. 25-32.

[125] Brown JS, Laundre JW, Gurung M. The ecology of fear: optimal foraging, game theory, and trophic interactions. Journal of Mammalogy. 1999; 80:385-99.

[126] Ripple WJ, Beschta RL. Wolves and the ecology of fear: Can predation risk structure ecosystems? Bioscience. 2004;54(8):755-66.

[127] Engeman RM, Pipas MJ, Gruver KS, Allen LR. Monitoring coyote population changes with a passive activity index. Wildlife Research. 2000;27:553-7.

[128] Karanth KU, Chundawat RS, Nichol JD, Kumar NS. Estimation of tiger densities in the tropical dry forests of Panna, Central India, using photographic capture-recapture sampling. Animal Conservation. 2004 Aug;7:285-90.

[129] Karanth KU, Nichols JD. Estimation of tiger densities in India using photographic captures and recaptures. Ecology Letters. 1998;79:2852-62.

[130] Karanth KU. Estimating tiger *Panthera tigris* populations from camera-trap data using capture-recapture models. Biological Conservation. 1995;71(3):333-8.

[131] Karanth KU, Kumar NS, Nichols JD. Field surveys: Estimating absolute densities of tigers using capture-recapture sampling. Monitoring tigers and their prey: A manual for researchers, managers and conservationists in tropical Asia. Karnataka: Center for Wildlife Studies; 2002.

[132] Rowcliffe JM, Field J, Turvey ST, Carbone C. Estimating animal density using camera traps without the need for individual recognition. Journal of Applied Ecology. 2008;45:1228-36.

[133] Dawson MJ, Miller C. Aerial mark–recapture estimates of wild horses using natural markings. Wildlife Research. 2008;35(4):365-70.

[134] Grigg GC, Beard LA, Alexander P, Pople AR, Cairns SC. Aerial survey of kangaroos in South Australia 1978-1998: A brief report focusing on methodology. Australian Zoologist. 1999 June;31(1):292-300.

[135] Stephens PA, Zaumyslova OJ, Miquelle D, Myslenkov AE, Hayward GD. Estimating population density from indirect sign: track counts and the Formozov-Malyshev-Pereleshin formula. Animal Conservation. 2006;9:339-48.

[136] Johnson C, VanDerWal J. Evidence that dingoes limit the abundance of a mesopredator in eastern Australian forests. Journal of Applied Ecology. 2009;46:641-6.

[137] Johnson CN, Isaac JL, Fisher DO. Rarity of a top predator triggers continent-wide collapse of mammal prey: Dingoes and marsupials in Australia. Proceedings of the Royal Society, Biological Sciences Series B. 2007 February 7;274(1608):341-6.

[138] Letnic M, Koch F. Are dingoes a trophic regulator in arid Australia? A comparison of mammal communities on either side of the dingo fence. Austral Ecology. 2010;35(2):267–175.

[139] Fleming PJS, Thompson JA, Nicol HI. Indices for measuring the efficacy of aerial baiting for wild dog control in north-eastern New South Wales. Wildlife Research. 1996;23(6):665-74.

[140] Brawata RL, Neeman T. Is water the key? Dingo management, intraguild interactions and predator distribution around water points in arid Australia. Wildlife Research. 2011;38(5):426-36.

[141] Chippendale JF. The Queensland barrier fence - A preliminary economic analysis. Proceedings of the Vertebrate Pest Control Conference. Adelaide1991. p. 143-7.

[142] Payne WLR, Fletcher JW, Tomkins B. Report on the Royal Commission on rabbit, dingo, and stock route administration. Queensland: Queensland Government; 1930.

[143] Yelland L. Holding the line: A history of the South Australian Dog Fence Board, 1947 to 2000. Adelaide: Primary Industries and Resources South Australia; 2001.

[144] Moseby KE, Neilly H, Read JL, Crisp HA. Interactions between a top order predator and exotic mesopredators in the Australian rangelands. International Journal of Ecology. 2012;Article ID 250352:15 pages.

[145] White GC. Why take calculus? Rigour in wildlife management. Wildlife Society Bulletin. 2001;29:380-6.

[146] Augusteyn J, editor. Determining the effectiveness of canine control at Taunton National park (Scientific) and its impact on the population of bridled nailtail wallabies. Queensland Pest Animal Symposium; 2010; Gladstone, Queensland.

[147] Burrows ND, Algar D, Robinson AD, Sinagra J, Ward B, Liddelow G. Controlling introduced predators in the Gibson Desert of Western Australia. Journal of Arid Environments. 2003;55:691-713.

[148] Catling PC, Burt RJ. Studies of the ground-dwelling mammals of eucalypt forests in south-eastern New South Wales: The effect of habitat variables on distribution and abundance. Wildlife Research. 1995;22:271-88.

[149] Catling PC, Hertog A, Burt RJ, Wombey JC, Forrester RI. The short-term effect of cane toads (Bufo marinus) on native fauna in the Gulf Country of the Northern Territory. Wildlife Research. 1999;26(2):161-85.

[150] Christensen P, Burrows N. Project desert dreaming: experimental reintroduction of mammals to the Gibson Desert, Western Australia. In: Serena M, editor. Reintroduction biology of Australian and New Zealand fauna. Sydney: Surrey Beatty & Sons; 1995. p. 199-207.

[151] Claridge AW, Cunningham RB, Catling PC, Reid AM. Trends in the activity levels of forest-dwelling vertebrate fauna against a background of intensive baiting for foxes. Forest Ecology and Management. 2010;260(5):822-32.
[152] Corbett L. Does dingo predation or buffalo competition regulate feral pig populations in the Australian wet-dry tropics? An experimental study. Wildlife Research. 1995;22(1):65-74.
[153] Edwards GP, Dobbie W, Berman DM. Warren ripping: Its impacts on European rabbits and other wildlife of central Australia amid the establishment of rabbit haemorrhagic disease. Wildlife Research. 2002;29(6):567-75.
[154] Edwards GP, Dobbie W, Berman DM. Population trends of European rabbits and other wildlife of central Australia in the wake of rabbit haemorrhagic disease. Wildlife Research. 2002;29:557-65.
[155] Fillios M, Gordon C, Koch F, Letnic M. The effect of a top predator on kangaroo abundance in arid Australia and its implications for archaeological faunal assemblages. Journal of Archaeological Science. 2010;37(5):986-93.
[156] Fleming PJS. Aspects of the management of wild dogs (*Canis familiaris*) in north-eastern New South Wales. Armidale: The University of New England; 1996.
[157] Catling PC, Burt RJ. Why are red foxes absent from some eucalypt forests in eastern New South Wales? Wildlife Research. 1995;22:535-46.
[158] Newsome A. The biology and ecology of the dingo. In: Dickman C, Lunney D, editors. A symposium on the dingo: Royal Zoological Society of New South Wales; 2001.
[159] Kennedy M, Phillips B, Legge S, Murphy S, Faulkner R. Do dingoes suppress the activity of feral cats in northern Australia? Austral Ecology. 2011;37(1):134-9.
[160] Koertner G, Watson P. The immediate impact of 1080 aerial baiting to control wild dogs on a spotted-tailed quoll population. Wildlife Research. 2005;32(8):673-80.
[161] Pascoe JH. Apex predators in the Greater Blue Mountains World Heritage Area [PhD]. Sydney: The University of Western Sydney; 2011.
[162] Pavey CR, Eldridge SR, Heywood M. Population dynamics and prey selection of native and introduced predators during a rodent outbreak in arid Australia. Journal of Mammalogy. 2008 June;89(3):674-83.
[163] Wallach AD, Murray BR, O'Neill AJ. Can threatened species survive where the top predator is absent? Biological Conservation. 2009;142:43-52.
[164] Cupples JB, Crowther MS, Story G, Letnic M. Dietary overlap and prey selectivity among sympatric carnivores: could dingoes suppress foxes through competition for prey? Journal of Mammalogy. 2011;92(3):590-600.
[165] Glen AS, Pennay M, Dickman CR, Wintle BA, Firestone KB. Diets of sympatric native and introduced carnivores in the Barrington Tops, eastern Australia. Austral Ecology. 2011;36(3):290-6.
[166] Letnic M, Dworjanyn SA. Does a top predator reduce the predatory impact of an invasive mesopredator on an endangered rodent? Ecography. 2011;34(5):827-35.
[167] Triggs B, Brunner H, Cullen JM. The food of fox, dog and cat in Croajingalong National Park, south-eastern Victoria. Australian Wildlife Research. 1984;11:491-9.

[168] Lunney D, Law B, Rummery C. Contrast between visible abundance of the brush-tailed rock wallaby, *Petrogale penicillata*, and its rarity in fox and dog scats in the gorges east of Armidale, New South Wales. Wildlife Research. 1996;23(3):373-80.

[169] Paltridge R. The diets of cats, foxes and dingoes in relation to prey availability in the Tanami Desert, Northern Territory. Wildlife Research. 2002 2002;29:389-403.

[170] Banks PB, Newsome AE, Dickman CR. Predation by red foxes limits recruitment in populations of eastern grey kangaroos. Austral Ecology. 2000;25:283-91.

[171] Hayward MW, Adendorff J, O'Brien J, Sholto-Douglas A, Bissett C, Moolman LC, et al. The reintroduction of large carnivores to the Eastern Cape Province, South Africa: an assessment. Oryx. 2007;41:205-14.

[172] Glen AS, Dickman CR. Niche overlap between marsupial and eutherian carnivores: Does competition threaten the endangered spotted-tailed quoll? Journal of Applied Ecology. 2008 April;45(2):700-7.

[173] Pascoe JH, Mulley RC, Spencer R, Chapple R. Diet analysis of mammals, raptors and reptiles in a complex predator assemblage in the Blue Mountains, eastern Australia. Australian Journal of Zoology. 2012;59(5):295-301.

[174] Carbone C, Gittleman JL. A common rule for the scaling of carnivore density. Science. 2002;295:2273-6.

[175] Hayward MW, O'Brien J, Kerley GIH. Carrying capacity of large African predators: predictions and tests. Biological Conservation. 2007;139:219-29.

[176] Jedrzejewska B, Jedrzejewski W. Predation in vertebrate communities: the Bialowieza Primeval Forest as a case study. Berlin, Germany: Springer; 1998.

[177] Anon. Annual report on the operations of 'The Stock Routes and Rural Lands Protection Acts, 1944-1951' for the year 1951-52. Queensland Parliamentary Papers. 1952;1951-52.

[178] Letnic M, Greenville A, Denny E, Dickman CR, Tischler M, Gordon C, et al. Does a top predator suppress the abundance of an invasive mesopredator at a continental scale? Global Ecology and Biogeography. 2011;20(2):343-53.

[179] Mitchell B, Balogh S. Monitoring techniques for vertebrate pests: Wild dogs. Orange: NSW Department of Primary Industries, Bureau of Rural Sciences; 2007.

[180] Archer M. New information about the Quaternary distribution of the thylacine (Marsupialia, Thylacinidae) in Australia. Journal of the Proceedings of the Royal Society of Western Australia. 1974; 57:43-50.

[181] Baird RF. The dingo as a possible factor in the disappearance of the *Gallinula mortierii* from the Australian mainland. Emu. 1991;91:121-2.

[182] Moseby K, Read J, Gee P, Gee I. A study of the Davenport Range black-footed rock wallaby colony and possible threatening processes. Final report to Wildlife Conservation Fund, Department for Environment and Heritage. 1998:Adelaide.

[183] Kerle JA, Foulkes JN, Kimber RG, Papenfus D. The decline of the brushtail possum, *Trichosurus vulpecula* (Kerr 1798), in arid Australia. The Rangeland Journal. 1992;14(2):107-27.

[184] Horsup A. Recovery plan for the northern hairy-nosed wombat *Lasiorhinus krefftii* 2004-2008 Canberra: Report produced by the Environmental Protection

Agency/Queensland Parks and Wildlife Service for the Department of Environment and Heritage; 2004.

[185] Moseby KE, Read JL, Paton DC, Copley P, Hill BM, Crisp HA. Predation determines the outcome of 10 reintroduction attempts in arid South Australia. Biological Conservation. 2011;144(12):2863-72.

[186] Benshemesh J. Recovery plan for marsupial moles *Notoryctes typhlops* and *N. caurinus*, 2005-2010. Alice Springs: Northern Territory Department of Infrastructure, Planning and Environment; 2004.

[187] Major R. Predation and hybridisation by feral dogs (*Canis lupus familiaris*) - Key threatening process listing: New South Wales Department of Environment, Climate Change, and Water; 2009.

[188] Corbett LK. The conservation status of the dingo *Canis lupus dingo* in Australia, with particular reference to New South Wales: threats to pure dingoes and potential solutions. In: Dickman CR, Lunney D, editors. Symposium on the dingo. Mossman: Royal Zoological Society of New South Wales, Australian Museum; 2001. p. 10-9.

[189] Coman B, Jones E. The loaded dog: On objectivity in the biological sciences and the curious case of the dingo. Quadrant. 2007;1 Nov 2007.

[190] Claridge A, Hunt R. Evaluating the role of the Dingo as a trophic regulator: Additional practical suggestions. Ecological Management and Restoration. 2008;9(2):116-9.

[191] Banks SC, Horsup A, Wilton AN, Taylor AC. Genetic marker investigation of the source and impact of predation on a highly endangered species. Molecular Ecology. 2003 June 2003;12(6):1663-7.

[192] Lundie-Jenkins G, Lowry J. Recovery plan for the bridled nailtail wallaby (*Onychogalea fraenata*) 2005-2009: Report to the Department of Environment and Heritage (DEH), Canberra. Brisbane: Environmental Protection Agency/Queensland Parks and Wildlife Service2005.

[193] Coutts-Smith AJ, Mahon PS, Letnic M, Downey PO. The threat posed by pest animals to biodiversity in New South Wales. Canberra: Invasive Animals Cooperative Research Centre2007.

[194] Newsome A, Pech R, Smyth R, Banks P, Dickman C. Potential impacts on Australian native fauna of rabbit calicivirus disease. Canberra: Biodiversity Group, Environment Australia1997.

[195] Lunney D, Gresser S, O'Neill LE, Matthews A, Rhodes J. The impact of fire and dogs on koalas at Port Stephens, New South Wales, using popualtion viability analysis. Pacific Conservation Biology. 2007;13(3):189-201.

[196] Mifsud G. Hansard excerpt of the transcript from the Commonwealth senate enquiry into the status, health and sustainability of Australia'a koala population, 19 May 2011. Canberra: Official Committee Hansard, the Commonwealth of Australia; 2011.

[197] Benshemesh J. National recovery plan for malleefowl. South Australia: Department for Environment and Heritage; 2007.

[198] Heard GW, Robertson P, Black D, Barrow G, Johnson P, Hurley V, et al. Canid predation: a potentially significant threat to relic populations of the Inland Carpet Python *Morelia spilota metcalfei* (Pythonidae) in Victoria. The Victorian Naturalist. 2006;123(2):68-74.

[199] Somaweera R, Webb JK, Shine R. It's a dog-eat-croc world: dingo predation on the nests of freshwater crocodiles in tropical Australia. Ecological Research. 2011;26(5):957-67.

[200] Whiting SD, Long JL, Hadden KM, Lauder ADK, Koch AU. Insights into size, seasonality and biology of a nesting population of the Olive Ridley turtle in northern Australia. Wildlife Research. 2007;34(3):200-10.

[201] James CD, Landsberg J, Morton SR. Provision of watering points in the Australian arid zone: A review of effects on biota. Journal or Arid Environments. 1999;41:87-121.

[202] Davies KF, Melbourne BA, James CD, Cuningham RB. Using traits of species to understand responses to land use change: Birds and livestock grazing in the Australian arid zone. Biological Conservation. 2010;143:78-85.

[203] Brook LA, Kutt AS. The diet of the dingo (*Canis lupus dingo*) in north-eastern Australia with comments on its conservation implications. The Rangeland Journal. 2011;33:79–85.

[204] Claridge AW, Mills DJ, Barry SC. Prevalence of threatened native species in canid scats from coastal and near-coastal landscapes in south-eastern Australia. Australian Mammalogy. 2010;32(2):117-26.

[205] Johnson CN, Isaac JL. Body mass and extinction risk in Australian marsupials: The 'Critical Weight Range' revisited. Austral Ecology. 2009;34(1):35-40.

[206] Smith AP, Quin DG. Patterns and causes of extinction and decline in Australian conilurine rodents. Biological Conservation. 1996;77:243-67.

[207] Newsome AE, Corbett LK. Outbreaks of rodents in semi-arid and arid Australia: Causes, preventions, and evolutionary considerations. In: Prakash I, Gosh PK, editors. Rodents in desert environments. The Hague, The Netherlands: Dr. W. Junk; 1975.

[208] Paltridge RM. Predator-prey interactions in the spinifex grasslands of central Australia [PhD]. Wollongong: School of Biological Sciences, University of Wollongong; 2005.

[209] Abbott I. Origin and spread of the cat, *Felis catus*, on mainland Australia, with a discussion of the magnitude of its early impact on native fauna. Wildlife Research. 2002;29:51-74.

[210] Wilson DS. The adequacy of body size as a niche difference. American Naturalist. 1975;109:769-84.

[211] Hutchinson GE. Homage to Santa Rosalina, or why are there so many kinds of animals? American Naturalist. 1959;93:145-59.

[212] Pople A, Froese J. Distribution, abundance and harvesting of feral goats in the Australian rangelands 1984-2011. Final report to ACRIS Management Committee. Brisbane: Department of Employment, Economic Development and Innovation; 2012.

[213] East IJ, Foreman I. The structure, dynamics and movement patterns of the Australian sheep industry. Australian Veterinary Journal. 2011;89(12):477-89.

[214] Gordon IJ, Acevedo-Whitehouse K, Altwegg R, Garner TWJ, Gompper ME, Katzner TE, et al. What the 'food security' agenda means for animal conservation in terrestrial ecosystems. Animal Conservation. 2012;15(2):115-6.

[215] Kearney J. Food consumption trends and drivers. Philosophical Transactions of The Royal Society B. 2010;365:2793-807.

[216] Thornton PK. Livestock production: recent trends, future prospects. Philosophical Transactions of The Royal Society B. 2010;365:2853-67.

[217] Wright HL, Lake IR, Dolman PM. Agriculture—a key element for conservation in the developing world. Conservation Letters. 2012;5(1):11-9.

[218] Allen LR. Control of dingoes inside the Barrier Fence. Rural Lands Protection Board meeting minutes. 1987;May:40-2.

[219] Rural Management Partners. Economic assessment of the impact of dingoes/wild dogs in Queensland. Queensland: Commissioned by the Department of Natural Resources2004. Report No.: Project LP02/03NRM.

[220] Corbett L. Does dingo predation or buffalo competition regulate feral pig populations in the Australian wet-dry tropics? An experimental study. Wildlife Research. 1995 1995;22:65-74.

[221] Holmes JC. Population regulation: A dynamic complex of interactions. Wildlife Research. 1995;22:11-9.

[222] Holt RD, Lawton JH. The ecological consequences of shared natural enemies. Annual Review of Ecology and Systematics. 1994;25: 495-520.

[223] Ritchie EG, Johnson CN. Predator interactions, mesopredator release and biodiversity conservation. Ecology Letters. 2009;12(9):982-98.

[224] Risbey DA, Calver MC, Short J, Bradley JS, Wright IW. The impact of cats and foxes on the small vertebrate fauna of Heirisson Prong, Western Australia. II. A field experiment. Wildlife Research. 2000;27(3):223-35.

[225] Bird P. Improved electric fences and baiting techniques: A behavioural approach to integrated dingo control. Adelaide: Animal and Plant Control Commission, Department of Primary Industries South Australia; 1994.

[226] Woinarski JCZ, Legge S, Fitzsimons JA, Traill BJ, Burbidge AA, Fisher A, et al. The disappearing mammal fauna of northern Australia: context, cause, and response. Conservation Letters. 2011;4(3):192-201.

[227] Sinclair ARE, Pech RP, Dickman CR, Hik D, Mahon P, Newsome AE. Predicting the effects of predation on conservation of endangered prey. Conservation Biology. 1998;12(3):564-75.

[228] de Tores PJ, Marlow NJ. A review of the relative merits of predator exclusion fencing and repeated 1080 fox baiting for protection of native fauna: five case studies from Western Australia. In: Somers MJ, Hayward MW, editors. Fencing for conservation. New York: Springer; 2011. p. 21-42.

[229] Sergio F, Caro T, Brown D, Clucas B, Hunter J, Ketchum J, et al. Top predators as conservation tools: Ecological rationale, assumptions, and efficacy. Annual Review of Ecology, Evolution and Systematics. 2008;39:1-19.

[230] Garrott RA, Gude JA, Bergman EJ, Gower C, White PJ, Hamlin KL. Generalizing wolf effects across the Greater Yellowstone Area: a cautionary note. Wildlife Society Bulletin. 2005 2005/12/01;33(4):1245-55.

[231] Vucetich JA, Hebblewhite M, Smith DW, Peterson RO. Predicting prey population dynamics from kill rate, predation rate and predator–prey ratios in three wolf-ungulate systems. Journal of Animal Ecology. 2011;80(6):1236-45.

[232] Dickman CR, Letnic M, Mahon PS. Population dynamics of two species of dragon lizards in arid Australia: the effects of rainfall. Oecologia. 1999;119:357-66.

[233] Dickman CR, Mahon PS, Masters P, Gibson DF. Long-term dynamics of rodent populations in arid Australia: The influence of rainfall. Wildlife Research. 1999;26:389-403.

[234] Pople AR, Grigg GC, Phinn SR, Menke N, McAlpine C, Possingham HP. Reassessing the spatial and temporal dynamics of kangaroo populations. In: Coulson G, Eldridge M, editors. Macropods: The biology of kangaroos, wallabies and rat-kangaroos. Melbourne: CSIRO Publishing; 2010. p. 197-210.

[235] Gese EM, Knowlton FF. The role of predation in wildlife population dynamics. In: Ginnetr TF, Henke SE, editors. The role of predator control as a tool in game management. San Angelo, Texas: Texas Agricultural Research and Extension Center; 2001. p. 7-25.

[236] Whetton PH, Fowler AM, Haylock MR, Pittock AB. Implications of climate change due to the enhanced greenhouse effect on floods and droughts in Australia. Climatic Change. 1993;25(3):289-317.

[237] Moise AF, Hudson DA. Probabilistic predictions of climate change for Australia and southern Africa using the reliability ensemble average of IPCC CMIP3 model simulations. Journal of Geophysical Research. 2008;113:D15113.

[238] Hughes L. Climate change and Australia: trends, projections and impacts. Austral Ecology. 2003;28:423-43.

[239] Claridge AW, Mills DJ, Hunt R, Jenkins DJ, Bean J. Satellite tracking of wild dogs in south-eastern mainland Australian forests: Implications for management of a problematic top-order carnivore. Forest Ecology and Management. 2009;258:814-22.

[240] Robertshaw JD, Harden RH. The ecology of the dingo in north-eastern New South Wales, I. Movement and home range. Australian Wildlife Research. 1985;12(1):25-37.

[241] Thomson PC, Rose K, Kok NE. The behavioural ecology of dingoes in north-western Australia: VI. Temporary extraterritorial movements and dispersal. Wildlife Research. 1992;19(5):585-95.

[242] Braysher M. Managing vertebrate pests: Principles and strategies. Canberra: Bureau of Rural Sciences, Australian Government Publishing; 1993.

[243] Chapple RS, Ramp D, Bradstock RA, Kingsford RT, Merson JA, Auld TD, et al. Integrating science into management of ecosystems in the Greater Blue Mountains. Environmental Management. 2011;48:659-74.

[244] Holling CS. Adaptive environmental assessment and management. London: John Wiley and Sons; 1978.

[245] Allen BL, editor. The spatial ecology and zoonoses of urban dingoes, and the use of Traversed Area Polygons (TAPs) to calculate home range sizes. workshop on the remote monitoring of wild canids and felids; 2007 21-22 March 2007; Australian National University, Canberra: Invasive Animals Cooperative Research Centre.

[246] Sodhi NS, Butler R, Raven PH. Bottom-up conservation. Biotropica. 2011;43(5):521-3.

[247] Caughley G, Sinclair ARE. Wildlife ecology and management. Cambridge, Massachusetts: Blackwell Sciences; 1994.

[248] Engeman R, Pipas M, Gruver K, Bourassa J, Allen L. Plot placement when using a passive tracking index to simultaneously monitor multiple species of animals. Wildlife Research. 2002;29(1):85-90.

[249] Engeman RM, Allen LR, Zerbe GO. Variance estimate for the activity index of Allen *et al.* Wildlife Research. 1998;25(6):643-8.

[250] Lindenmayer DB, Likens GE, Andersen A, Bowman D, Bull M, Burns E, et al. Value of long-term ecological studies. Austral Ecology. In press;xx(xx):xx-xx.

[251] Fleming PJS. Ground-placed baits for the control of wild dogs: Evaluation of a replacement-baiting strategy in north-eastern New South Wales. Wildlife Research. 1996;23(6):729-40.

[252] van Bommel L, Johnson CN. Good dog! Using livestock guardian dogs to protect livestock from predators in Australia's extensive grazing systems. Wildlife Research. 2012;39(3):220-9.

[253] Allen LR, Byrne D. How do guardian dogs 'work'? 15th Australasian Vertebrate Pest Conference; Sydney, Australia: Invasive Animal Cooperative Research Centre; 2011. p. 158.

[254] Norbury GL. An electrified watering trough that selectively excludes kangaroos. The Rangeland Journal. 1992;14(1):3-8.

[255] Berman D, Brennan M, Elsworth P. How can warren destruction by ripping control European wild rabbits (*Oryctolagus cuniculus*) on large properties in the Australian arid zone? Wildlife Research. 2011;38(1):77-88.

[256] Russell BG, Letnic M, Fleming PJS. Managing feral goat impacts by manipulating their access to water in the rangelands. The Rangeland Journal. 2011;33(2):143-52.

[257] Killengreen ST, Lecomte N, Ehrich D, Schott T, Yoccoz NG, Ims RA. The importance of marine vs. human-induced subsidies in the maintenance of an expanding mesocarnivore in the arctic tundra. Journal of Animal Ecology. 2011;80(5):1049-60.

[258] Elmhagen B, Rushton SP. Trophic control of mesopredators in terrestrial ecosystems: top-down or bottom-up? Ecology Letters. 2007;10:197-206.

[259] Treves A, Karanth KU. Human-carnivore conflict and perspectives on carnivore management worldwide. Conservation Biology. 2003;17(6):1491-9.

[260] Cooney R. The precautionary principle in biodiversity conservation and natural resource management: An issues paper for policy-makers, researchers, and practicioners. Gland, Switzerland and Cambridge: IUCN; 2004.

[261] Trouwborst A. Managing the carnivore comeback: international and EU species protection law and the return of lynx, wolf and bear to western Europe. Journal of Environmental Law. 2010;22(3):347-72.

[262] Pullin AS, Knight TM. Doing more good than harm: building an evidence-base for conservation and environmental management. Biological Conservation. 2009;142:931-4.

[263] Hone J. Fox control and rock-wallaby population dynamics - assumptions and hypotheses. Wildlife Research. 1999;26:671-3.

Chapter 5

Characterization and Biological Activity of *Bacillus thuringiensis* Isolates that Are Potentially Useful in Insect Pest Control

Analía Alvarez and Flavia del Valle Loto

Additional information is available at the end of the chapter

http://dx.doi.org/10.5772/48362

1. Introduction

Bacillus thuringiensis (*Bt*) is a spore-forming bacterium well-known for its insecticidal properties associated with its ability to produce crystal inclusions during sporulation. These inclusions are proteins encoded by *cry* genes and have shown to be toxic to a variety of insects and other organisms like nematodes and protozoa [1]. The primary action of Cry proteins is to lyse midgut epithelial cells through insertion into the target membrane and form pores [2]. Once ingested, crystals are solubilized in the alkaline environment of midgut lumen and activated by host proteases [3]. On the other hand, the involvement of *Bt* proteases in processing inactive protoxins is also reported [3]. These toxins are also highly specific and completely biodegradable, hence no toxic products are accumulated in the environment. In fact, Calderón et al. [4] suggest the potential use of some crystal proteins as adjuvants for the administration of heterologous antigens.The activity spectrum of *Bt* toxins continually increases as result of the ongoing isolation of new strains around the world.

The fall armyworm, *Spodoptera frugiperda* (*S. frugiperda*) (Lepidoptera: Noctuidae), and the variegated cutworm, *Peridroma saucia* (*P. saucia*) (Lepidoptera: Noctuidae), are two lepidopteran pests that cause severe damage to a variety of crops. While the first one mainly attacks corn, rice, peanuts, cotton, soybeans, alfalfa and forage grasses [5], the second one targets peanuts, sunflowers, soybeans and grapevines, among others [6]. Currently, control of this pest relies on chemical insecticides. Nevertheless, the rapid increases in resistance to insecticides together with the potential adverse environmental effects produced by these chemicals have encouraged the development of alternative methods for Lepidoptera control [7,8]. Among these methods the use of *Bt* as a biocontrol agent has shown to be extremely valuable. The diversity of Cry toxins produced by *Bt* allows the formulation of a variety of bioinsecticides by using the bacteria themselves or by expressing their toxin genes in transgenic plants. To date, many plant species have been genetically modified with *cry* genes, resulting in transgenic plants with a high level of resistance to insect [9]. However, it has been reported that several pests have developed resistance against Cry proteins [9, 10]. The current approach used to delay evolution of resistance to transgenic crops uses a "high dose" and "refuge" strategy [9, 11]. In addition, it is important to use a combination of *cry* genes and/or other genes encoding insecticidal proteins within the same transgenic crop [12, 13]. Due to extensive use of transgenic crops in developing countries based on *cry*-type genes, there is a need for alternative *cry* gene sequences to meet the challenge of novel insect resistance [7]. Crucial to this development is the identification of novel and more active strains with respect to insect pests of economically important crops [14].

The *cry* genes of *Bt* strains are known to be related to their toxicity [15, 16] and identification of these genes by means of PCR has been used to characterize and predict insecticidal activity of the strains [17, 18]. Nevertheless, a more complete characterization should include alternative methods. Phenotypical analysis such as protein profile determination provides useful information for typing and comparative studies [19]. The literature data report the possibility of using the whole-cell protein profile as a discriminating method with potency similar to RAPD with combined DNA patterns [1]. However, there is not always a good correlation between these factors and insecticidal activity of *Bt* strains [20, 21]. In addition, there is a need to develop knowledge about the biological properties and diversity of *Bt* isolates since these data allow a better understanding of the biological factors that determine insecticidal properties. Extracellular factors such as phospholipases, proteases and chitinases have shown to contribute to insecticidal activity of *Bt* [22].

During a screening programme of *Bt* isolates native to Argentina and toxic against Lepidoptera, several strains were characterized according to different biological parameters. In addition, promising isolates regarding their useful in biological control programmes -an environmentally safe technology of pest control- were exhaustively studied [14, 19, 23]. The present work showed most relevant results obtained during a course of those investigations. The discovery of highly pathogenic isolates against devastating insect pests reveals the usefulness of screening studies for novel *Bt* strains.

2. Biochemical characterization of *B. thuringiensis* isolates and assessment of toxicity

Crystalliferous spore-forming bacteria were isolated from both *S. frugiperda* larvae showing disease symptoms and soil samples collected in Argentina [19]. These samples came from maize, sorghum, wheat, grape or sugarcane cultivated fields. Briefly, larvae and soil sample suspensions were made in distilled water, heated at 80 °C for 15 min and then plated onto LB-agar. Plates were incubated at either 30 or 55 °C for 24 h. Colonies that did not grow at 55 °C were then analyzed for the presence of parasporal crystals by microscopic examination [24]. From a total of 254 colonies isolated from 490 different environmental samples, 14 were identified as crystal producer strains, giving a mean *Bt* index of 0.05. This result suggested that samples analyzed contained a high background level of other spore-forming bacteria. One crystalliferous strain came from sorghum cultivated field, while the others came from maize cultivated field. Concerning the source of isolation, 50% of crystal producers came from soil samples and the other 50% came from ill larvae. Interestingly, the last source provided the most pathogenic strains (Table 1).

Bacteria -characterized by conventional microbiological methods- possessed typical cellular and colonial morphologies, as well as physiological, biochemical and nutritional features that resembled *Bacillus* spp. They were motile and produced ellipsoidal endospores, located at sub-terminal position in the sporangia, and formed cream-colored colonies with irregular or circular edges on LB agar.

Phenotypical and molecular characterization	Strains														
	TRC11*	TMAN2*	THM8*	NN1**	TRC10*	RT**	TSA2*	TRC12*	N28**	MAN8**	MAN1**	THM30*	*Bt* 4D1***	LSM**	LQ**
Central spore	+	+	+	+	+	+	+	-	+	+	+	-	+	+	+
Sub terminal spore	-	-	-	-	-	-	-	+	-	-	-	+	-	-	-
Growth at pH 9	+	+	-	+	+	+	+	+	+	+	+	-	+	+	+
[b e] Growth in 0.2 % chitin	-	+	+	+	-	+	+	+	-	+	-	-	+	+	+
[b c e] CMC hydrolysis (0.5 %)	+	+	+	+	+	-	-	+	+	+	-	+	+	-	-
[b e] Chitin hydrolysis (0.2 %)	-	-	+	-	-	+	+	-	-	-	-	-	+	+	+
[b] Gelatin hydrolysis (12 %)	+	-	-	-	-	-	+	-	+	-	-	+	-	-	-
[b] Starch hydrolysis (2 %)	-	-	+	+	+	+	+	-	+	-	+	+	+	+	+
Gas production in glucose	-	+	-	+	-	+	-	+	-	-	-	-	+	+	+
[d] clindamycin	+	-	+	+	+	+	+	+	+	+	-	+	+	+	+
[d] gentamicin	+	+	-	-	+	-	-	+	+	-	-	-	+	-	-
[d] rifampicin	+	+	+	+	-	+	-	+	-	+	+	+	+	+	+
cry1	-	+	-	+	-	+	-	-	-	-	+	+	+	+	ND
cry2	-	+	+	-	+	+	-	+	+	-	-	+	+	+	ND
cry1Aa						-							+	+	
cry1Ab						+							+	+	
cry1Ac						+							+	+	
cry2Aa						-							+	+	
cry2Ab						+							+	+	

[a] *Asterisks indicate the source of the isolates: *soil, **ill larvae and ***Bacillus Genetic Stock Center*

[b] Expressed in w/v

[c] CMC: carboxy methyl cellulose

[d] Sensitivity to antibiotics was determined by using the routine diffusion plate technique. (+): sensitive and (-): resistant

[e] Growth on chitin and chitin hydrolysis were determined using colloidal chitin according to Kaur et al. (2005). This protocol was also used to analyze CMC hydrolysis

*Strains isolated from soil

**Strains isolated from ill larvae

***B. *thuringiensis* var. *kurstaki* 4D1 was provided by the Bacillus Genetic Stock Center (BGSC) (Columbus, Ohio) as well as the others *Bt* reference strains (see below).

ND: no determined (without amplification)

Table 1. Biochemical characteristics that presented variable response among the bacterial isolates. Molecular characterization is also showed (see below).

From a biochemical point of view, the 14 strains were catalase-positive, reduced nitrate and produced acetyl methyl carbinol in Voges-Proskauer broth; growth was observed at pH 7 on LB agar supplemented with 2, 3 and 5% NaCl and on LB agar at 30, 37 and 45 ºC. The strains also hydrolyzed casein and were motile on soft LB agar. Negative results for all strains were obtained in several tests: no growth was observed on LB agar at pH 4 or at 50 ºC and none of the strains hydrolyzed carboxymethyl cellulose (CMC) and urea. Antibiotic sensitivity tests revealed a resistance profile to penicillin, oxacillin, trimethoprim and a sensitive profile to erythromycin, vancomycin, levofloxacin, minocycline, chloramphenicol and teicoplanin. Phenotypic features that presented variability among the strains are showed in Table 1. The positive or negative result of each biochemical assay was entered in a 1-0 matrix. These data were subsequently analyzed through correspondence multivariate analysis, using Multivariate Statistical Package (MSVP) software (version 3.13). A cluster diagram based on these variable biochemical properties (that represented 54% of data variability) revealed that the strains formed two main groups (Figure 1). Group A comprised nine crystalliferous isolates which were clustered together with the reference strain *B. thuringiensis kurstaki* 4D1 (*Bt* 4D1). A second group (B) included three *Bt* strains while the remaining two strains presented more divergent features and hence were not included in any group. Isolates from the same sample and/or the same geographic region differed in their phenotypic features and consequently were not grouped together. This indicates that there was no clear association between *Bt* strains biochemical profile and the environments from which they were obtained [23].

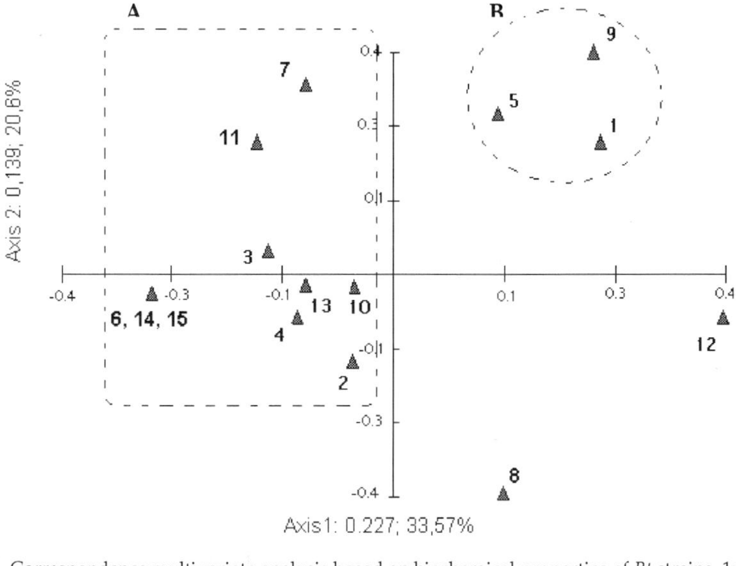

Figure 1. Correspondence multivariate analysis based on biochemical properties of *Bt* strains. 1: TRC11; 2: TMAN2; 3: THM8; 4: NN1; 5: TRC10; 6: *Bt* RT; 7: TSA2; 8: TRC12; 9: N28; 10: MAN8; 11: MAN1; 12: THM30; 13: **Bt* 4D1 (reference strain); 14: LQ and 15: LSM

Although most of the native isolates presented similar biochemical and phenotypical characteristics compared with reference strain *Bt* 4D1 (Group A) (Figure 1), they differed in their toxicity to *S. frugiperda*. Our results showed that the mortality on *S. frugiperda* neonate larvae was variable [19], ranging among values corresponding to Bt strains of bioinsecticides action low to moderate (Figure 2). However, strains named RT, LSM and LQ were found to be highly pathogenic, two of them, even more than reference strain *Bt* 4D1 which was selected for this analysis given it is the most widely used microorganism to control lepidopteran pests [25] (Table 2). This strong biological effect was represented by both a shorter LT_{50} and a higher mortality, which reached 100% in the case of RT strain on *S. frugiperda,* after five days of treatment. This result is extremely relevant considering that *S. frugiperda* is believed a pest with low sensitivity to *Bt* toxins [26]. In addition, when this strain was assayed against first instar larvae of *P. saucia,* reached 93% of mortality (Figure 3) suggesting that RT strain native to Argentina could possibly be employed in biological control of lepidopteran pests [19, 23]. It is important to stress that the high levels of mortality in the present work were obtained with a concentration of a spore-crystal suspension that was lower than some commercial *Bt* formulations; while our crystal spore suspensions presented a dose of 10^7 c.f.u. ml^{-1}, *Bt kurstaki* preparations generally present a dose of 10^9 c.f.u. ml^{-1} [27].

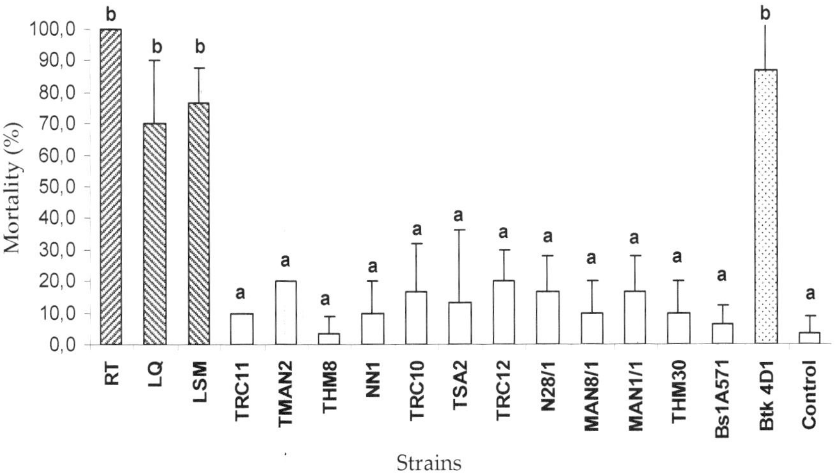

Figure 2. Insecticidal activity of crystalliferous native strains isolated from soil (□) and ill larvae (▧).Bars sharing the same letter were not significantly different ($P > 0.05$, Tukey post-test). Reference strain: *Bt* 4D1 (⬚).Mortality was measured at the 7th day of assay. Ten individuals per treatment were observed and each treatment was repeated 10 times. *B. subtilis* 1A571 *(Bs* 1A571*)* was provided by the Bacillus Genetic Stock Center (BGSC).

Bt strain	*,[a] Mortality (%) ± SD	[b] LT_{50} (h) (95% fiducial limits)	*Specific biomass bound protease activity (± SD) (U g dry wt $^{-1}$)
RT	100 ± 0 a	9.2 (10.4 –16.0)	1.98 ± 98 b
LSM	90.0 ± 7.3 a	37.7 (27.8 – 46.2)	1.80 ± 93 b
LQ	73.0 ± 5.7 c	79.6 (68.2 – 90.7)	1.14 ± 25 a
Bt 4D1	86.0 ± 15.1 b	58.7 (50.4 – 66.0)	946 ± 14 a
control	1.0 ± 3.1 d		

*Values followed by different letters were significantly different ($P<0.05$, Tukey post-test)
[a] Ten individuals per treatment were observed and each treatment was repeated ten times
[b] 50 % lethal time (LT_{50}) was determined by Probit analysis. Mortality was scored every 24 h during seven days

Table 2. Comparison of mortality and 50% lethal time (LT_{50}) of first instar larvae of *S. frugiperda* among native and reference *Bt* strains. Biomass-bound protease activity of pathogenic isolates are also showed.

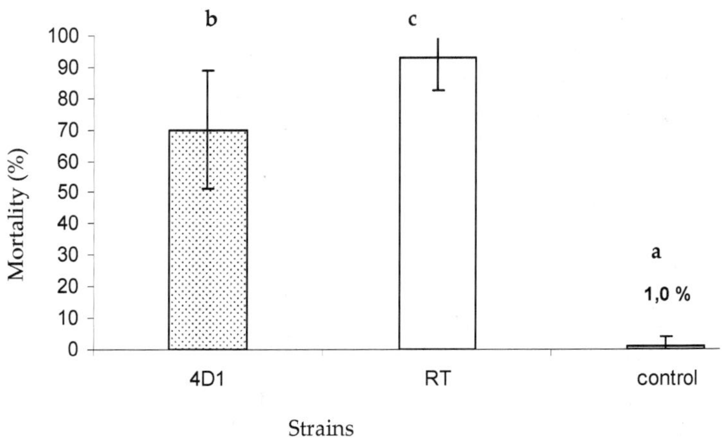

Figure 3. Comparison of insecticidal activity of *Bt* RT and the reference strain *Bt* 4D1 against the first instar larvae of *Peridroma saucia*. Mortality was measured at the 7th day of assay. Ten individuals per treatment were observed and each treatment was repeated 10 times. Mortality differed significantly between all treatments ($P < 0.05$, Tukey post-test).

3. Numerical analysis of insecticidal crystal proteins of *B. thuringiensis*

In order to differentiate native crystalliferous isolates and to evaluate the relationship between the toxicity assays against *S. frugiperda,* isolated bacteria and parasporal crystals were characterized by the whole-cell protein profiles in SDS-PAGE [19]. Protein bands were individually identified by their specific migration rates in the gels. Once bands were properly and distinctively identified, binary (0/1) matrices were constructed to compare the patterns. Electrophoretic analysis revealed the presence of 53 distinct bands with molecular weights ranging from 266 to 20 kDa (Figure 4). Numerical analysis clearly showed two distinct clusters (Figure 5). Cluster A comprised 11 isolations and the reference strain *B. thuringiensis* var. *thuringiensis* 4A4 (*Bt* 4A4) as well as those crystalliferous isolations that had no or very low toxicity against *S. frugiperda* first instar larvae. Interestingly, this group of native microorganisms produced proteins from 28 to 31 kDa but not proteins of ~135 and ~65 kDa. These lower molecular mass could correspond to Cyt toxins, entomocidal crystal proteins highly active against Diptera larvae [28]. On the other hand, all isolations with high toxic activity against *S. frugiperda* (RT, LSM and LQ strains) (Table 2) were located in cluster B, as well as the reference strains *Bt* 4D1 and *B. thuringiensis* var. *kurstaki* 4D3 (*Bt* 4D3). The isolation RT had a protein profile similar to *Bt* 4D1 with proteins of ~140 and ~70 kDa. Strains LSM and LQ showed protein bands of ~100 and ~81 kDa. These results demonstrate that the whole cell protein profiling not only allowed the differentiation of *Bt* at strain level but also revealed a possibility to apply protein profile analysis in classification of toxicity patterns.

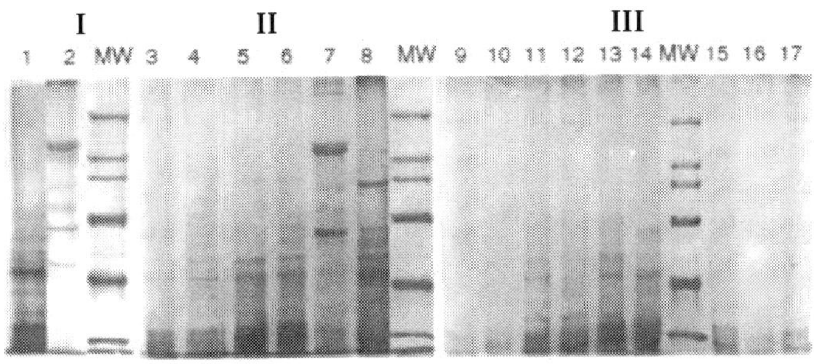

Figure 4. SDS-PAGE of whole-cell protein of crystalliferous strains. Gel I. Lines: 1: *Bt* 4D3, 2: *Bt* 4D1. Gel II. Lines: 3: N28, 4: *Bt* 4A4, 5: LSM, 6: LQ, 7: RT, 8: MAN8. Gel III. Lines: 9: THM8, 10: TMAN2, 11: THM30, 12: NN1, 13: TSA2, 14: MAN1, 15: TRC12, 16: TRC11, 17: TRC10. MW: Molecular weight marker Sigma-Aldrich were rabbit skeletal myosin (200 kDa), *E. coli b*-galactosidase (116.25 kDa), rabbit muscle phosphorylase b (97.4 kDa), Bovine serum albumin (66.2 kDa), hen egg white ovalbumin (45 kDa) and bovine carbonic anhydrase (31 kDa). Gels were stained with silver reagent. *B. thuringiensis* var. *kurstaki* 4D3 and *B. thuringiensis* var. *thuringiensis* 4A4 were also provided by the Bacillus Genetic Stock Center (BGSC).

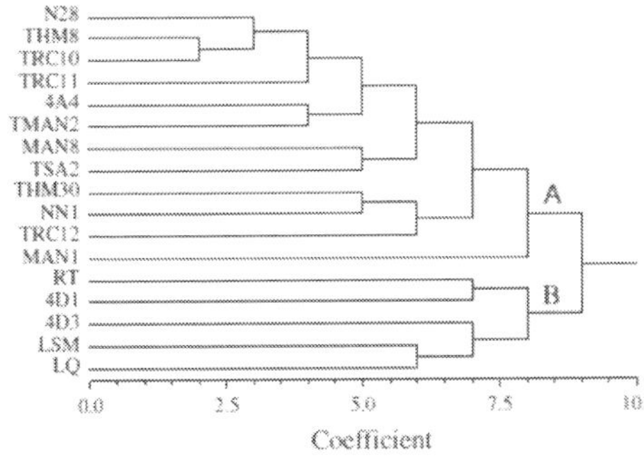

Figure 5. Dendrogram showing the relationship among *Bt* isolates based on electrophoretic whole cell protein patterns. Associations were produced using the simple matching coefficient and the neighbor-joining clustering method.

4. Molecular characterization of *B. thuringiensis* strains and crystal morphology

Although the presence of parasporal crystals is a diagnostic characteristic of *Bt* strains [1], the taxonomic identity of the toxic crystalliferous isolates was confirmed by amplification and partial sequencing of their 16S rDNA genes [19] (Table 3). The partial 16S rDNA sequences were tested by BLAST analysis against the GenBank data base. *Bt* LSM strain showed exact BLAST matches with the sequence from *Bacillus thuringiensis* var *kurstaki* (1 hit, 100% of identity, accession number EF638796). *Bt* LQ strain produced 4 hits (99% of identity, accession number EF638798), all of these corresponding to *Bt* species. Similarly, *Bt* RT (best hits, 13; 99% of identity, accession number EF638795) also shared a close relationship with others *Bt* strains, including *Bt* LSM.

Generally, *B. thuringiensis* insecticidal protein toxin genes (*cry*) reside on large self-transmissible plasmids, and individual *B. thuringiensis* strains can harbor a diverse range of plasmids that can vary in number from 1 to 17 and in size from 2 to 80 MDa [29,30], although it has also been suggested that they are present in the chromosome [31]. In this context, to study the plasmid profiles of *Bt* strains is an important parameter to determine their identity, since the number and size of these is associated with a particular *Bt* strain. Comparison between strains belonging to the same serotype showed a great difference in variability [30]. Some serotypes (e.g., israelensis) showed the same basic pattern among all its strains, while other serotypes (e.g., morrisoni) showed a great diversity of patterns. These results indicate that plasmid patterns are valuable tools to discriminate strains below the serotype level [30]. The profile of extrachromosomal elements in *Bt* is influenced by a number of stressful growth conditions, which determine its stability and heritability (e. i. high temperatures determine the plasmid loss), therefore it is neccesary to take some care . In this study, cultures were routinely grown at 30 °C to avoid this phenomenon. Detection and isolation of plasmid DNA was conducted following the method of Kado and Liu [32]. DNA plasmid samples were electrophoresed on 0.8 % (wt vol^{-1}) agarose gel. Our results showed that selected *Bt* strains present a complex plasmid profile (Figure 6).

In this experiment, the plasmid DNA was not linearized and therefore the same plasmid can produce as many as three different bands in the agarose gel. This made it difficult to determine the precise number of plasmids present in each complex plasmid profile. For this reason, we will refer to the number and size of plasmidic bands rather to plasmids themselves. An intense band above the chromosomal band (C) was observed in *Bt* RT, *Bt* LSM and *Btk* 4D1 suggesting that a large plasmid or plasmids is/are found in this strains which might be responsible for production of parasporal bodies. Compared with the other bacteria, *Bt* LQ presented a very different profile array, suggesting a different *cry*-genotype.

Identification of *cry* genes by means of PCR has been used to predict insecticidal activity of the strains [17,18] and to determine the distribution of *cry* genes within a collection of *B. thuringiensis* strains [20, 33]. In this context, our crystalliferous strains were characterized in terms of presence of *cry1* and *cry2* genes by amplification with general primers. The most toxic *Bt* strains RT, LSM and LQ were characterized through additional PCR with specific

C

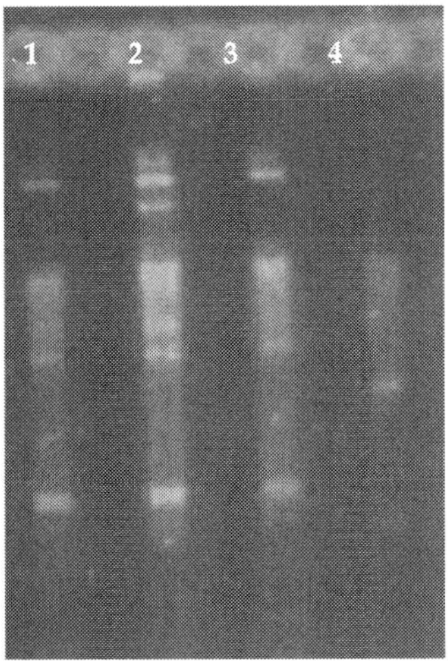

Figure 6. Plasmid profiles of *Bacillus thuringiensis* strains. Lanes: 1: *Bt* RT; 2: *Bt* 4D1; 3: *Bt* LSM; 4: *Bt* LQ. "C" indicate chromosomal DNA.

primers to identify the presence of *cry1Aa, cry1Ab, cry1Ac, cry1Ad, cry2Aa, cry2Ab* and *cry2Ac* genes (Table 3). PCR analysis showed presence of *cry1* and/or *cry2* genes in most of the isolates (Table 1). Specific PCR showed identical *cry* gene profile in both *Bt* LSM and the reference *Bt* 4D1, while *cry* gene content of *Bt* RT was different from them. DNA of *Bt* LQ was not amplified under the current reaction conditions (Table 1).

In addition, amplified fragments corresponding to *cry1* and *cry2* genes from *Bt* RT were sequenced and compared with *cry* genes sequences available from GenBank. This sequences had 99 and 95% identity with *cry1Ab* (EU220269) and *cry2Ab* (EU094885) genes, respectively. As shown in Figure 7, *cry1* and *cry2* partial sequences from *Bt* RT and *Bt* 4D1 were also aligned with five and six GenBank published *cry* sequences, respectively. The phylogenetic analysis revealed that *cry*1 partial sequences from *Bt* RT and *Bt* 4D1 possess almost the same level of evolutionary distance (Figure 7A), while *cry2* partial sequence from *Bt* RT lies on a separate diverse branch not only of *cry2* from *Bt* 4D1 but also of the others analyzed *cry2* sequences (Figure 7B). Considering the phylogenetic analysis, it could be expected toxicity mediated by Cry1 rather than Cry2 crystal protein. In fact, *cry2* partial sequence from *Bt* RT shared a 95% homology with *cry2* sequence from a Colombian native *Bt* strain active against *Tecia solanivora* (Lep:Gelechiidae) (EU094885).

As mentioned before, *cry* genes are a family of genes associated with the toxicity of *Bt* against insects. While *cry1* encodes for proteins forming bipyramidal crystals and are related to toxicity to Lepidoptera [29] *cry2* encodes for cuboidal proteins, toxic to Lepidoptera and Diptera [39]. Our molecular and electron microscopy analyses of *Bt* RT are in agreement with all this evidence, since this highly pathogenic strain has both genes (Table 1) and both kinds of proteins (Figure 8A). In contrast, and although *Bt* LSM showed amplification products with *cry2* general and specific primers (Table 1) no cuboidal proteins were

Primer pairs	Nucleotide sequence	Reference
16S: 27F 1492R	5′-AGAGTTTGATCCTGGCTCAG-3′ 5′-GGTTACCTTGTTACGACTT-3′	[34]
ITS: ISR-1494 ISR-35	5′-GTCGTAACAAGGTAGCCGTA -3′ 5′-CAAGGCATCCACCGT-3′	[35]
Gral-*cry1*	5′-CTGGATTTACAGGTGGGGATAT-3′ 5′-TGAGTCGCTTCGCATATTTGACT-3′	[36]
Gral-*cry2*	5′-GAGTTTAATCGACAAGTAGATAATTT-3′ 5′-GGAAAAGAGAATATAAAAATGGCCAG-3′	[37]
Spe-*cry1*Aa	5′-TTATACTTGGTTCAGGCCC-3′ 5′-TTGGAGCTCTCAAGGTGTAAA-3′	[38]
Spe-*cry1*Ab	5′-AACAACTATCTGTTCTTGAC-3′ 5′-CTCTTATTATACTTACACTAC -3′	
Spe-*cry1*Ac	5′- GTTAGATTAAATAGTAGTGG-3′ 5′- TGTAGCTGGTACTGTATTG-3′	
Spe-*cry1*Ad	5′-GTTGATACCCGAGGCACA-3′ 5′-CCGCTTCCAATAACATCTTTT-3′	
Spe-*cry2*Aa1	5′-GTTATTCTTAATGCAGATGAATGGG-3′ 5′-GAGATTAGTCGCCCCTATGAG-3′	[17]
Spe-*cry2*Ab2	5′-GTTATTCTTAATGCAGATGAATGGG-3′ 5′-TGGCGTTAACAATGGGGGAGAAAT-3′	
Spe-*cry2*Ac	5′-GTTATTCTTAATGCAGATGAATGGG-3′ 5′-GCGTTGCTAATAGTCCCAACAACA-3′	

Table 3. Primer sequences used in this study.

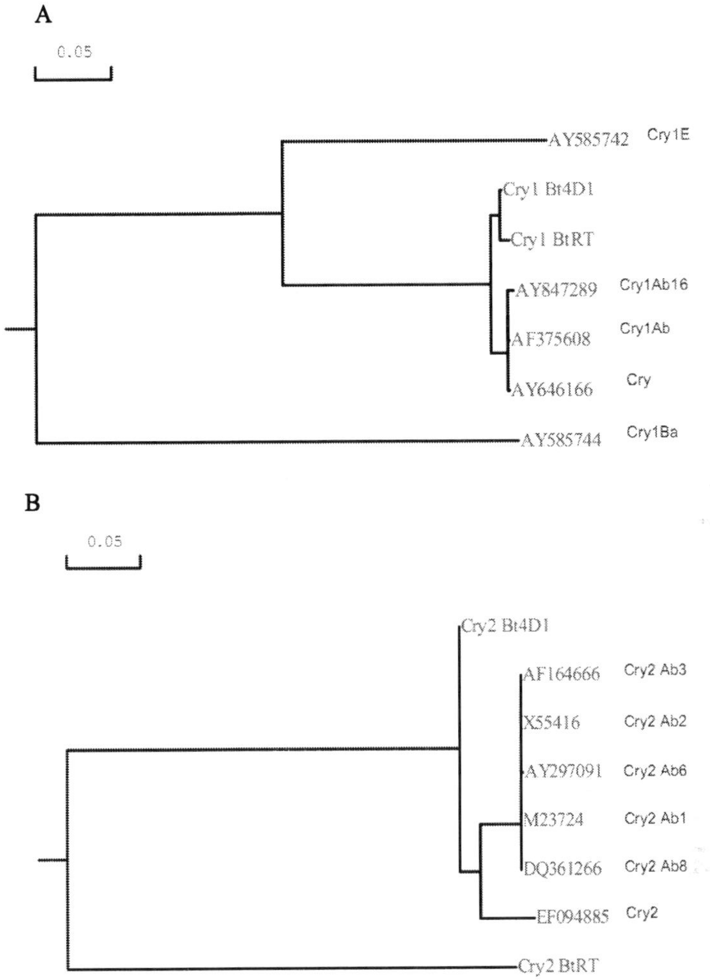

Figure 7. Phylogenetic rooted tree of *cry1* (A) and *cry2* (B) partial sequences from *B. thuringiensis* strains.

identified (Figure 8B). This suggests that a modification in the regulation of the gene would be responsible for the lack of protein product of this gene. Although the experimental growth conditions employed could also explain the lack of cuboidal proteins, the production of these proteins by *Bt* RT under identical experimental conditions argue against this possibility [14]. Cloning and sequencing the putative toxins with surrogate production made help clarify this issue as well as to confirm toxicity. In addition, *Bt* LQ showed no amplification products of *cry1* and *cry2* gene in several attempts (Table 1), despite the presence of bipyramidal crystals (Figure 8C). Noguera e Ibarra, [40] found that *cry* genes of a *Bt* strain isolated in Argentina that showed elongated bipyramidal crystals [41] presented 98% identity with *cry5Ba* genes. Therefore, *Bt* LQ may have Cry proteins other than Cry1 that form bipyramidal crystals.

From a methodological point of view, washing of crystal suspensions with absolute ethanol/distilled water (Figure 9B) was more appropriate for microscopic observation than washing with distilled water (Figure 9A).

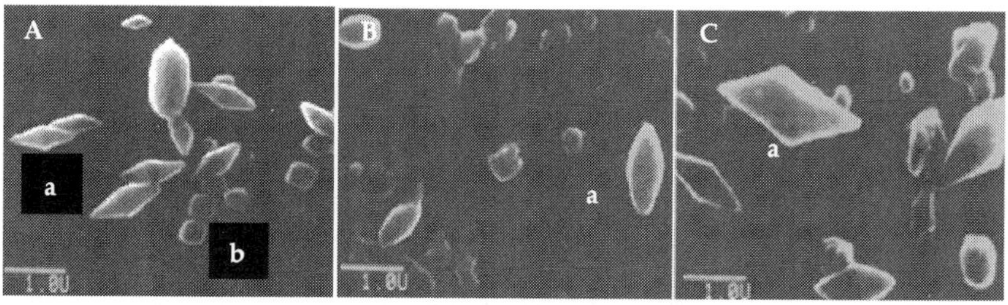

Figure 8. Scanning electron microscopy (SEM) of spore-crystal proteins from *Bt* strains. Concentrated spore-crystal suspensions were placed on a microscope lid and air-dried overnight. Samples were then coated with gold and examined using a scanning electron microscope. A) *Bt* RT; B) *Bt* LSM; C) *Bt* LQ. Both bipyramidal (a) and cuboidal (b) pesticidal crystal proteins are observed. Scale bar: 1 μm.

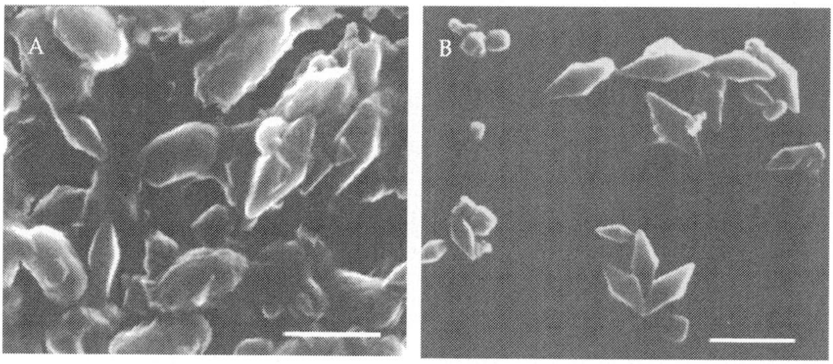

Figure 9. Scanning electron microscopy of spore-crystal proteins from *Bt* RT washed twice either with ethanol/water (1:1, v/v) (A) or with water (B). Scale bar: 1 μm.

As mentioned above, identification of *cry* genes by means of PCR has been used to predict insecticidal activity of the strains [17,18]. However, but since the primers are designed against known genes, the technique presents limitations in the search of novel *cry* genes. Moreover, the reliability of the prediction of insecticidal activity based on PCR results is dependent on the expression of the genes. In this context, a more complete characterization of *Bt* strains should include alternative PCR fingerprinting methods. Among them, assessment of length polymorphism of intergenic transcribed spacers (ITS) between the 16S and 23S rDNA genes has been shown to be an important tool for differentiating bacterial species and even prokaryotic strains [42]. In this context, ITS-PCR was performed as previously described by Daffonchio et al.[31]. Used primer pairs are showed in Table 3. Evaluation of ITS length polymorphism revealed an identical pattern among *Bt* RT, LSM and LQ strains and also with *Bt* 4D1 suggesting that ITS exhibited no polymorphism among the strains (Figure 10). In connection with this, Reyes-Ramirez and Ibarra [43] studied ITS profiles of 31 *Bt* strains and found them to be insufficient to discriminate between isolates.

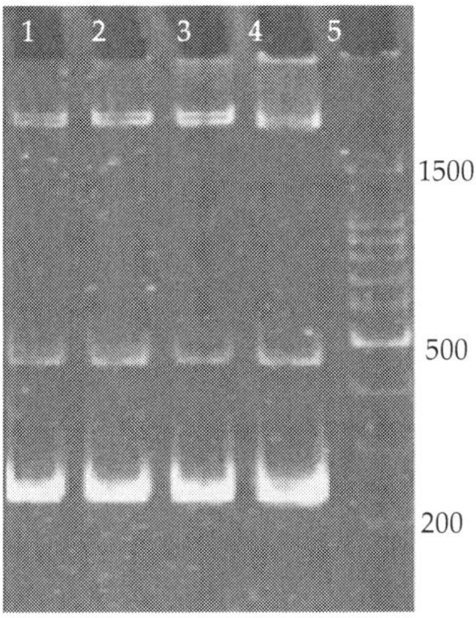

Figure 10. ITS-PCR of *B. thuringiensis* strains. Lanes 1: *Bt* LSM; 2: *Bt* LQ; 3: *Btk* 4D1; 4: *Bt* RT; 5: 100 bp DNA Ladder.

5. Assessment of enzyme activities in *B. thuringiensis*

Phenotypic characterization of selected strains allows identification of properties that are relevant at the moment of selecting bacteria for their use in environmental and agricultural microbiology. Synthesis of lytic enzymes by *Bacillus* species during the early sporulation phase is one of these properties. Secreted microbial enzymes may function as virulence factor that are essential for survival and spread in the host [44,45]. Our results indicate that the native *Bt* strains responded diversely regarding proteolytic, cellulolytic and/or chitinolytic activity (Table 1). Chitinolytic activity is a contributing factor in *Bt* pathogenicity [21] which our most pathogenic strains possessed. The enzymes involved would act on the peritrophic membrane of the host, which facilitates the entry of pathogens into the haemocoel of susceptible insects [21]. In addition, these strains showed no cellulolytic activity in medium supplemented with carboxymethyl cellulose (CMC), one of the products used as a matrix to protect *Bt* spores against high temperatures and UV exposure prevailing in natural environments [46]. This lack of cellulolytic activity is a desirable property given that gelled CMC will not be degraded at the time of *Bt* formulation, and therefore it can be employed for this purpose [13].

In *Bt*, high levels of protease activity are associated with both crystal and spore formation [47] and this activity may contribute in processing inactive Cry protoxin to active toxin [3].

While there is a reasonable understanding of soluble midgut proteases in toxin activation, little is known about the role of *Bt* protease in entomotoxicity. In this connection, during our investigations, biomass-bound protease and extracellular protease activities were determined in the toxic strains, which were process according to [48]. Proteolytic activity was assayed by using azocasein as substrate [49]. Table 2 shows that *Bt* RT, *Bt* LSM and *Bt* LQ displayed high biomass-bound protease activity. To our knowledge, the presence of this naturally immobilized enzyme activity has not been reported in *Bt* [18]. On the other hand, extracellular protease activity was observed when crude extracts of *Bt* strains were electrophoresed on SDS-PAGE containing gelatin powder [50]. The gels were then processed according to [51] for proteolysis to occur. Gel was stained with 0.1% (wt vol^{-1}) Coomasie Blue R-250. Proteinase K (10 mg ml^{-1}) was used for comparative analyze. All strains presented a clear zone of proteolytic activity which were larger in both *Bt* RT and *Bt* LSM (Figure 11). Although no correlation between protease activities and mortality values was initially detected, this result could be complementary information to consider in commercial *Bt* formulations, since the cell structure may act as a natural matrix able to protect the biomass-bound enzymes from the possible negative action of external agents; and therefore it could be that an increased percentage of *Bt* protease may actually reach the larvae midgut. Finally, it would be useful to explore the role of the extracellular and biomass-bound protease activities in crystal protein modification during *Bt* fermentation, the synergy of this protease source with insect entomotoxicity and the possible addition of vegetative cells in the final *Bt* formulation [18].

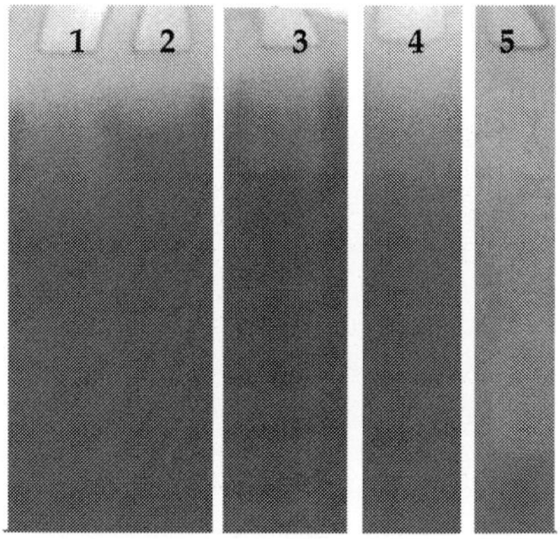

Figure 11. Identification of extracellular protease activity on 10% (wt vol^{-1}) SDS-PAGE containing 0.1% (wt vol^{-1}) gelatin powder. Lanes: 1: *Bt* RT; 2: *Bt* 4D1; 3: *Bt* LQ; 4: *Bt* LSM 5: Proteinase K.

Protein profiles are a useful tool to discriminate among strains, as they provide information about the proximity between species, subspecies and biovars [18, 52]. Considering this, characterization of microorganisms by means of their extracellular isoenzymes showing high polymorphism, as is the case of esterase, is particularly appealing. To determine extracellular esterase profiles, *Bt* strains were processed according to [44]. Briefly, strains were cultured on LB plates during 48 h at 30 ºC and crude extracts were recovered from solid media. Then, extracts were separated by native-PAGE. Esterase activity was assayed using 1.3 mM of α-naphthyl acetate (C2) derivative as substrate. Known electrophoretic esterase profiles of *Bacillus pumilus* A55 (EF638794.1) (*Bp* A55) were used for comparative analysis. The electrophoretic profiles of esterase activity showed differences among strains (Figure 12). *Bt* LQ showed a unique band/enzyme of 40 kDa as well as *Bt* LSM, but of about 60 kDa while *Bt* RT presented two bands of 95 and 60 kDa. Our results are in accordance with those by Norris [53], since it was possible to differentiate *Bt* strains by comparing the electrophoretic migration profiles of esterase produced during the vegetative growth phase (Figure 12) [13].

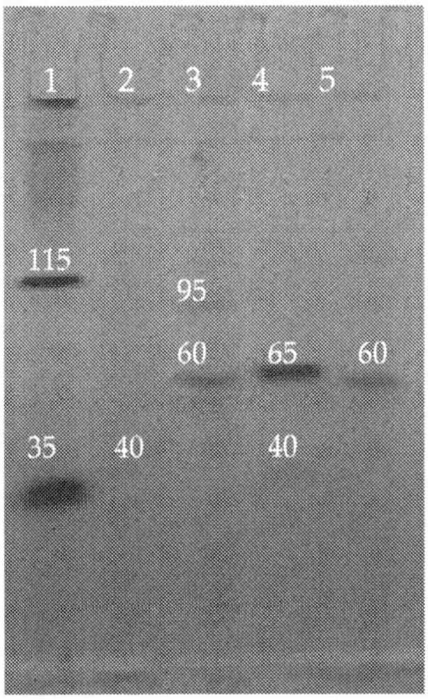

Figure 12. Enzymatic profile of esterase activity in native-PAGE 10%(wt vol^{-1}). Lanes: 1: *Bp* A55; 2: *Bt* LQ; 3: *Bt* RT; 4: *Bt* 4D1; 5: *Bt* LSM. Molecular weight of each band/enzyme is showed in kDa.

6. Conclusion

Lepidoptera causes some of the most devastating insect pests in important crops in America. Since economy of these regions depends largely of agriculture, their control is a priority as well as a necessity. In this context, use of environmentally safe technology to reduce crop damage like *B. thuringiensis* would be extremely valuable. Consequently, we set out to establish and characterize a collection of *Bt* isolates from samples collected in different Argentinean localities in order to find novel strains toxic against insect pests of economically important crops (like soybean and maize).

Fourteen *Bt* strains were isolated and phenotypically, genetically and biologically characterized. Analysis of larvicidal activity indicated that three strains exhibited high toxicity against lepidopteran larvae; this toxicity was in most, higher than that of the reference strain *Bt* 4D1.

The discovery of a highly toxic isolates reveals the usefulness of screening studies for novel *Bt* strains. The future application of these strains in biological control programmes requires optimization of the production conditions of the microorganisms using low-cost substrates. In this context, characterization of phenotypic and biochemical properties as evaluated in this study is highly relevant.

Author details

Analía Alvarez and Flavia del Valle Loto
Pilot Plant of Industrial and Microbiological Processes (PROIMI), CONICET, Tucumán, Argentina

Analía Alvarez
Natural Sciences College and Miguel Lillo Institute, National University of Tucumán, Tucumán, Argentina

7. References

[1] Konecka E, Kaznowski A, Ziemnicka J, Ziemnicki K (2007) Molecular and phenotypic characterization of *Bacillus thuringiensis* isolated during epizootics in *Cydia pomonella* L. J. Invertebr. Pathol. 94: 56-63.

[2] Bravo A, Gill S, Soberón M (2007) Mode of action of *Bacillus thuringiensis* Cry and Cyt toxins and their potential for insect control. Toxicon 49: 423-435.

[3] Brar S, Verma M, Tyagi R, Surampalli, R Bernabé S, Valéro J (2007). *Bacillus thuringiensis* proteases: Production and role in growth, sporulation and synergism. Process Biochem. 42: 773-790.

[4] Calderón, M, Alcocer González J, Molina M, Tames Guerra R, Rodríguez Padilla (2007). Adjuvant effects of crystal proteins from a Mexican strain of *Bacillus thuringiensis* on the mouse humoral response. Biologicals *35*, 271-276.

[5] Virla E, Alvarez A, Loto F, Pera L, Baigorí M (2008) Fall Armyworm strains (Lepidoptera: Noctuidae) in Argentina, their associate host plants and response to different mortality factors in laboratory. Fla. Entomol. 91: 63-69.

[6] Moreno Fajardo O, Serna Cardona F (2006). Biología de *Peridroma saucia* (Lepidoptera: Noctuidae: Noctuinae) en flores cultivadas del híbrido comercial de *Alstroemeria* spp. Rev. Fac. Nal. Agr. Medellín. 2: 3435-3448.

[7] Berón C, Salerno G (2006) Characterization of *Bacillus thuringiensis* isolates from Argentina that are potentially useful in insect pest control. BioControl 51: 779-794.

[8] Gomes Monnerat R, Cardoso Batista A, Telles De Medeiros P, Soares Martins E, Melatti VM, Praca IB, Dumas VF, Morinaga C, Demo C, Menezes Gomes AC, Falcao R, Siqueira CB, Silva-Werneck JO, Berry C (2007) Screening of Brazilian *Bacillus thuringiensis* isolates active against *Spodoptera frugiperda, Plutella xylostella* and *Anticarsia gemmatalis*. Biol. Control 41: 291-295.

[9] Gassmann A, Carrière Y, Tabashnik E (2009) Fitness Costs of Insect Resistance to *Bacillus thuringiensis*. Annu. Rev. Entomol. 54: 147-63.

[10] Tabashnik B, Gassmann A, Crowder D, Carriére Y (2008) Insect resistance to *Bt* crops: evidence versus theory. Nat. Biotechnol. 26: 199-202.

[11] Tabashnik B, Carrière Y (2009) Environmental Impact of Genetically Modified Crops. In: Ferry N, Gatehouse A, editors. Insect Resistance to Genetically Modified Crops. Cambridge: CAB International.74-101pp.

[12] Christou P, Capell T, Kohli A, Gatehouse J, Gatehouse A (2006) Recent developments and future prospects in insect pest control in transgenic crops. Trends Plant. Sci. 11: 302-308.

[13] Sauka D, Benintende G (2008) *Bacillus thuringiensis*: generalidades. Un acercamiento a su empleo en el biocontrol de insectos lepidópteros que son plagas agrícolas. Rev. Argent. Microbiol. 40: 124-140.

[14] Alvarez A, Virla E, Pera L, Baigorí M (2011) Biological characterization of two *Bacillus thuringiensis* strains toxic against *Spodoptera frugiperda*. World J. Microbiol. Biotechnol. 27: 2343-2349.

[15] Carozzi N, Kramer V, Warren G, Evola S, Koziel MG (1991) Prediction of insecticidal activity of *Bacillus thuringiensis* strains by polymerase chain reaction product profiles. Appl. Environ. Microbiol. 57: 3057-3061.

[16] Padidam M (1992) The insecticidal crystal protein Cry1A (c) from *Bacillus thuringiensis* is highly toxic for *Heliothis Armigera*. J. Invertebr. Pathol. 59: 109-111.

[17] Ben-Dov E, Zaritski A, Dahan E, Barak Z, Sinai R, Manasherob R, Khamraeb A, Troitskaya E, Dubitsky A, Berezina N, Margalith Y (1997) Extended screening by PCR for seven *cry*-group genes from field-collected strains of *Bacillus thuringiensis*. Appl. Environ. Microbiol. 63: 4883-4890.

[18] Hansen B, Damgaard P, Eilenberg J, Pedersen JC (1998) Molecular and phenotypic characterization of *Bacillus thuringiensis* isolated from leaves and insects. J. Invertebr. Pathol. 71: 106-114.

[19] Alvarez A, Virla E, Pera L, Baigorí M (2009a). Characterization of native *Bacillus thuringiensis* strains and selection of an isolate active against *Spodoptera frugiperda* and *Peridroma saucia*. Biotechnol. Lett. 31: 1899-1903.

[20] Porcar M, Juarez-Perez V (2003) PCR-based identification of *Bacillus thuringiensis* pesticidal crystal genes. FEMS Microbiol. Rev. 26: 419-432.

[21] Martínez C, Ibarra J, Caballero P (2005) Association analysis between serotype, *cry* gene content, and toxicity to *Helicoverpa armigera* larvae among *Bacillus thuringiensis* isolates native to Spain. J. Invertebr. Pathol. 90: 91-97.

[22] Soberón M, Bravo A (2001) *Bacillus thuringiensis* y sus toxinas insecticidas. In: Microbios en línea. UNAM México. Available vía DIALOG http://www.biblioweb.dgsca.unam.mx/libros/microbios /Cap12 Accessed 1 March 2011.

[23] Alvarez A, Pera L, Loto F, Virla E, Baigori M (2009b) Insecticidal crystal proteins from native *Bacillus thuringiensis*: Numerical analysis and biological activity against *Spodoptera frugiperda*. Biotechnol. Lett. 31: 77-82.

[24] Sharif F, Alaeddinoğlu N (1988) A rapid and simple method for staining of the crystal protein of *Bacillus thuringiensis*. J. Ind. Microbiol. 3: 227-229.

[25] Arango J, Romero M, Orduz S (2002) Diversity of *Bacillus thuringiensis* strains from Colombia with insecticidal activity against *Spodoptera frugiperda* (Lepidoptera: Noctuidae). J. Appl. Microbiol. 92: 466-474.

[26] Del Rincón-Castro M, Méndez-Lozano J, Ibarra J (2006) Caracterización de cepas nativas de *Bacillus thuringiensis* con actividad insecticida hacia el gusano cogollero del maíz *Spodoptera frugiperda* (Lepidoptera: Noctuidae). Folia Entomol. Mex. 45: 157-164.

[27] Seligy V, Rancourt J (1999) Antibiotic MIC/MBC analysis of *Bacillus*-based commercial insecticides: use of bioreduction and DNA-based assays. J. Ind. Microbiol. Biotechnol. 22: 565-574.

[28] Gough J, Kemp D, Akhurst R, Pearson R, Kongsuwan K (2005) Identification and characterization of proteins from *Bacillus thuringiensis* with high toxic activity against the sheep blowfly, *Lucilia cuprina*. J. Invertebr. Pathol. 90: 39-46.

[29] González J, Carlton B. (1980) Patterns of plasmid DNA in crystalliferous and acrystalliferous strains of *Bacillus thuringiensis*. Plasmid 3:92–98.

[30] Reyes-Ramírez IJ (2008) Plasmid patterns of *Bacillus thuringiensis* type strains. Appl. Environ. Microbiol. 74: 125-129.

[31] Kronstad J, Schnepf H, Whiteley H (1983) Diversity of locations for *Bacillus thuringiensis* crystal protein genes. J. Bacteriol. 154:419–428.

[32] Kado C, Liu S (1981) Rapid procedure for detection and isolation of large and small plasmids. J. Bacteriol. *145:*1365-73.

[33] Chak K, Chao D, Tseng M, Kao S, Tuan, S Feng, T (1994) Determination and distribution of *cry*-type genes of *Bacillus thuringiensis* isolated from Taiwan. Appl. Environ. Microbiol. 60:2415-2420.

[34] Weisburg W, Barns S, Pelletier D, Lane D (1991) 16S Ribosomal DNA amplification for phylogenetic study. J. Bacteriol. 173, 697-703.

[35] Daffonchio D, Borin S, Frova G et al (1998) PCR fingerprinting of whole genomes: the spacers between the 16S and 23S rRNA genes and of intergenic tRNA gene regions reveal a different intraspecific genomic variability of *Bacillus cereus* and *Bacillus licheniformis*. Int. J Syst Bacteriol 48:107–116.

[36] Bravo A, Sarabia S, Lopez L, Ontiveros H, Abarca C, Ortiz A, Ortiz M, Lina L, Villalobos FJ, Peña G, Nuñez-Valdez ME, Soberón M, Quintero R (1998) Characterization of *cry* genes in Mexican *B. thuringiensis* strain collection. Appl. Environ. Microbiol. 64: 4965-4972.

[37] Ibarra J, del Rincón M, Ordúz S, Noriega D, Benintende G, Monnerat R, Regis L, Oliveira C, Lanz H, Rodriguez M, Sánchez J, Peña G, Bravo A (2003) Diversity of

Bacillus thuringiensis strains from Latin America with insecticidal activity against different mosquito species Appl. Environ. Microbiol. 69:5269–5274.

[38] Cerón J, Covarrubias L, Quintero R, Ortiz M, Aranda E, Lina L, Bravo A (1994) PCR Analysis of the *cryI* Insecticidal Crystal Family Genes from *Bacillus thuringiensis*. Appl. Environm. Microbiol. 60:353-356.

[39] Al-Momani F, Saadoun I, Obeidat M (2002) Molecular characterization of local *Bacillus thuringiensis* strains recovered from Northern Jordan. J. Basic Microbiol. 2: 156-161.

[40] Noguera P, Ibarra J (2010) Detection of New *cry* Genes of *Bacillus thuringiensis* by Use of a Novel PCR Primer System. Appl. Environ. Microbiol. 76: 6150-6155.

[41] Benintende G, López-Meza J, Cozzi J, Ibarra JE (1999) Novel non-toxic isolates of *Bacillus thuringiensis*. Lett .Appl. Microbiol. 29: 151-155.

[42] Daffonchio D, Raddadi N, Merabishvili M (2006) Strategy for identification of *Bacillus* cereus and *Bacillus thuringiensis* strains closely related to *Bacillus anthracis*. Appl. Environ. Microbiol. 72:1295–1301.

[43] Ramírez R, Ibarra J (2005) Fingerprinting of *Bacillus thuringiensis* type strains and isolates by using *Bacillus cereus* group-specific repetitive extragenic palindromic sequence-based PCR analysis. Appl. Environ. Microbiol. 71: 1346-1355.

[44] Stefanova M, Leiva M, Larrinaga L, Coronado M (1999) Metabolic activity of *Trichoderma* spp. isolates for a control of soilborne phytopatogenic fungi. Rev. Fac. Agron. (LUZ) 16: 509-516.

[45] Lengyel MJ, Pekár S, Felföldi G, Patthy A, Gráf L, Fodor A, Venekei I (2004) Comparison of Proteolytic Activities Produced by Entomopathogenic *Photorhabdus* Bacteria: Strain- and Phase-Dependent Heterogeneity in Composition and Activity of Four Enzymes. Appl. Environ. Microbiol. 70: 7311-7320.

[46] Cokmus C, Elcin M (1995) Stability and controlled release properties of carboxymethylcellulose-encapsulated *Bacillus thuringiensis* var. israelensis. Pest. Sci. 45: 351-355.

[47] Ibarra JE, Del Rincón Castro MC, Galindo E, Patiño M, Serrano L, García R, Carrillo JA, Pereyra-Alférez B, Alcázar-Pizaña A, Luna-Olvera H, Galán-Wong L, Pardo L, Muñoz-Garay C, Gómez I, Soberón M, Bravo A (2006) Los microorganismos en el control biológico de insectos y fitopatógenos. Rev. Latinoam. de Microbiol. 48: 113-120.

[48] González C, Martínez A, Vázques F, Baigorí M, Figueroa L (1996) New method of screening and differentiation of exozymes from industrial strains. Biotechnol. Tech. 10: 519-522.

[49] Secades P, Guijarro A (1999) Purification and Characterization of an Extracellular Protease from the Fish Pathogen *Yersinia ruckeri* and Effect of Culture Conditions on Production. Appl. Environ. Microbiol. 65: 3969-3975.

[50] Heussen C, Dowdell E (1980) Electrophoretic analysis of plasminogen activators in polyacrilamide gels containing sodium dodecyl sulfate and copolymerized substrates. Anal. Biochem. 102: 196-202.

[51] Ferrero M, Castro G, Abate M, Baigorí M, Siñeriz F (1996) Thermostable alkaline proteases of *Bacillus licheniformis* MIR 29: Isolation, Production and Characterization. Appl. Microbiol. Biotechnol. 45: 327-332.

[52] Berber I (2004) Characterization of *Bacillus* species by numerical analysis of their SDS-PAGE protein profiles. J. Cell Mol. Biol. 3: 33-37.

[53] Norris J (1964) The classification of *Bacillus thuringiensis*. J. Appl. Bacteriol. 27:39- 447.

Chapter 6

Decreased Epiphytic Bryophyte Diversity on Mt. Odaigahara, Japan: Causes and Implications

Yoshitaka Oishi

Additional information is available at the end of the chapter

http://dx.doi.org/10.5772/48357

1. Introduction

When wildlife populations grow excessively, they affect other flora and fauna within their ecosystems (Fuller & Gill, 2001; Pellerin et al., 2006; Rooney, 2001; Schütz et al., 2003; Stewart & Burrows, 1989; Stockton et al., 2005; Takatsuki, 2009; Webster et al., 2005). For example, the recent increase in the sika deer population in Japan has led to the degradation of ecosystems in many areas. From 1979 to 2002, the range of this species expanded by as much as 70% (Nakajima, 2007). Although stripping of bark, grazing on grass, and browsing on tree understories are normal foraging behaviors in deer, these activities in excess can cause severe damage. Excessive bark stripping causes wood decay, leading to a decline in the forest cover (Akashi & Nakashizuka, 1999; Miquelle & van Ballenberghe, 1989; Yokoyama et al., 2001), and excessive browsing and/or grazing may alter the structure and composition of vegetation on the forest floor (Kumar et al., 2006; Rooney & Waller, 2003; Schütz et al., 2003; Stockton et al., 2005; Webster et al., 2005). These environmental changes indirectly affect other organisms in the forest ecosystem (Allombert et al., 2005; Feber et al., 2001; Flowerdew & Ellwood, 2001; Rooney, 2001).

To protect forest vegetation from further damage by the increased sika deer population, protective management, for example, wrapping tree trunks in wire mesh, have been implemented in addition to deer population control via culling and the erection of deer-proof fences (Ministry of the Environment-Kinki Regional Environment Office, 2009; Takatsuki, 2009). However, although these measures are effective for the protection of vegetation, they sometimes negatively affect other organisms.

This chapter describes the effects of protective management activities on epiphytic diversity at Mt. Odaigahara in central Japan, which is a hotspot for bryophyte diversity. It also discusses the best practices for biodiversity conservation in this scenario on the basis of a previously published article (Oishi, 2011).

2. Study site

2.1. Location and characteristics of the study site

Mt. Odaigahara (34°N, 136°E; altitude, ca. 1,500 m) is located in Yoshino Kumano National Park, which is in the southeastern part of the Nara Prefecture in Japan (Fig. 1). The climate in this region is relatively mild (annual mean temperature, 5.7 °C), with high levels of precipitation (annual mean precipitation, 4,500 mm; Nara Local Meteorological Observatory 1997). The vegetation on Mt. Odaigahara is classified into 2 main types: (1) the dominant tree species on the eastern part of the mountain is *Picea jezoensis* (Sieb. et Zucc.) Carriere var. *hondoensis* (Mayr) Rehder, and (2) those on the western part are *Fagus crenata* Blume and *Abies homolepis* Sieb. et Zucc (Ide & Kameyama, 1972).

2.2. Deer population

The population density of sika deer in Mt. Odaigahara has rapidly increased from the 1960s to the 1990s. To be specific, it has increased from approximately 12.0–22.2 individuals per square kilometer in the 1980s to 17.5–39.5 individuals per square kilometer in the 1990s (Ando & Goda, 2009). This increase led to serious damage to the forest vegetation in this

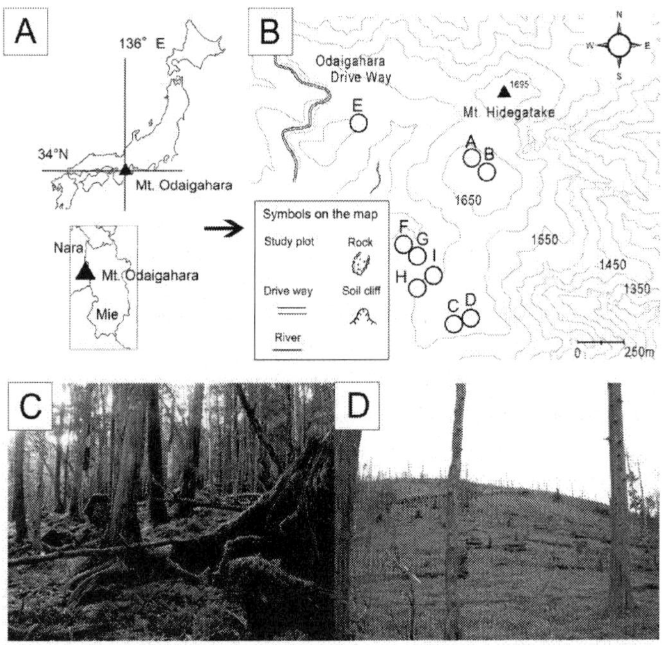

A; Location of Mt. Odaigahara in Japan. B: Location of study plots. C and D: Views of forests in Mt. Odaigahara. Photo C shows a forest of *P. jezoensis* var. *hondoensis* trees, whereas D shows a heavily declined forest.

Figure 1. Mt. Odaigahara and study plots

mountain region; for example, extensive bark stripping by the deer resulted in the dieback of damaged trees, and excessive browsing/grazing led to the loss of vegetation on the forest floor (Fig. 2). The Ministry of the Environment initiated a forest protection program in 1986 to conserve the forest ecosystem in this region. This program was executed in a part (ca. 703 hectare) of Mt. Odaigahara (Ministry of the Environment, Kinki Regional Environment Office, 2009). To prevent bark stripping by the deer, the trunks of around 32,500 trees were wrapped with wire mesh composed of zinc-coated galvanized iron (Ministry of the Environment, Kinki Regional Environment Office, 2009) (Fig. 3).

The photo on the left shows *Sasa nipponica* Makino et Shibata. browsed by sika deer. In this photo, the upper parts of the plants were browsed by sika deer. The right photo shows a deer fence and the effects of protection from browsing by sika deer: the height of the plants within the deer fence (back) is greater than that of the plants outside the fence.

Figure 2. Deer fence and the influence of browsing by deer on vegetation

The middle part of the tree trunk that does not have wire mesh protection has been debarked by deer.

Figure 3. Examples of tree trunks without (left) and with (right) wire mesh protection

2.3. Bryophyte diversity

In addition to the rapidly increasing population of sika deer on Mt. Odaigahara, this region is recognized for its bryophyte diversity. In fact, Mt. Odaigahara is home to approximately 30% (>620 species) of the bryophytes in Japan (Doei, 1988), including several nationally endangered species that are listed in the Red Data Book of Japan (e.g., *Iwatsukia jishibae* (Steph.) N. Kitag). The rich diversity of epiphytic bryophytes in this region is attributed to the high humidity of this region (Doei, 1988) (Fig. 4). The major species on Mt. Odaigahara include *Pogonatum japonicum* Sull. & Lesq., *Dicranum japonicum* Mitt., *Hylocomium splendens* (Hedw.) Schimp., and *Pleurozium schreberi* (Brid.) Mitt., which grow on the forest floor; *Heterophyllium affine* (Hook.) M. Fleisch, *Bazzania yoshinagana* (Steph.) S. Hatt., *Mylia verrucosa* Lindb., and *Scapania ampliata* Steph., which grow at the base of trees; and *Pterobryon arbuscula* Mitt., *Hypnum tristo-viride* (Broth.) Paris, and *Bazzania denudata* (Torr. ex Lindenb.) Trevis., which grow on tree trunks (Fig. 5). Bryophytes contribute to species diversity as well as the ecological integrity of Mt. Odaigahara because they function as microhabitats for the seedbeds of vascular plants and are involved in rainfall interception and nutrient cycling (Coxson, 1991; Nadkarni, 1984; Nakamura, 1997; Pypker et al., 2006) (Fig. 6).

2.4. Changes in bryophyte diversity and the deer population since the 1960s

In the 1980s, epiphytic bryophyte flora in *P. jezoensis* var. *hondoensis* were surveyed in 2 parts of Mt. Odaigahara—Masakitoge and Masakigahara. These areas were dense forests in the 1960s, but by 2008, they had become open forests because the changes in the environmental conditions after a severe typhoon in 1959 gradually resulted in dieback. Because many trees were blown down by the typhoon, the light conditions in the forests improved, and *S. nipponica* grew vigorously on the forest floor. This expansion resulted in an increase in the population of sika deer, which in turn resulted in dieback due to debarking.

The lower parts of many tree trunks (left) and fallen logs (right) are extensively covered with epiphytic bryophytes because of the high humidity.

Figure 4. Epiphytic bryophytes in Mt. Odaigahara

These species frequently occurred in the study plots: A, *D. japonicum*; B, *P. arbuscula*; C, *H. affine*; D, *H. tristo-viride*; E, *H. splendens*; F, *M. verrucosa*

Figure 5. Major species in the study plots

In 2008, we surveyed the epiphytic bryophyte flora in almost the same places as those examined in a previous study by using 20 × 20 m quadrants (plots A, B, C, and D in Fig. 1), and we examined the changes in these areas that had occurred over in the last 30 years. Table 1 summarizes the environmental conditions in these plots.

The species richness of individual *P. jezoensis* var. *hondoensis* trees decreased over 30 years from 18.0 ± 3.5 to 5.7 ± 3.4 in Masakitoge and from 18.0 to 7.5 ± 5.3 in Masakigahara (mean or mean ± SD) (Fig. 7). Thus, in direct contrast to the increasing deer population, epiphytic bryophyte diversity significantly declined over time.

Possible reasons for the decrease in bryophyte diversity are that (1) the decline in the forest cover indirectly affected bryophyte diversity because of the changes in the environmental conditions (e.g., air humidity), and (2) the protection of trees using wire mesh directly affected bryophyte diversity because of metal pollution.

To determine the reasons for the decline in bryophyte diversity, we examined the correlation between the diversity of epiphytic bryophytes and environmental variables, including wire mesh protection.

Figure 6. Bryophyte function in the ecosystem

Bryophytes provide safe microhabitats for the seedbeds of vascular plants (left). B: Bryophytes absorb water from rain drops and mist and therefore function in water storage in forests (right).

Plot	Year	No. of *Picea jezoensis* var. *hondoensis* trees surveyed		DBH (mean ± S.D.)
		Total	With wire mesh	
Masakitoge	1980s	2	0	23.7
	2008	10	8	23.8 ± 4.2
Masakigahara	1980s	13	0	34.4 ± 7.6
	2008	10	9	25.0 ± 11.2

Table 1. Summary of the characteristics of the study plots sampled for comparing the changes in the species richness from the 1980s to 2008

The bars represent the mean value of species richness and epiphyte cover on a single tree, and the error bars represent the corresponding standard deviations.

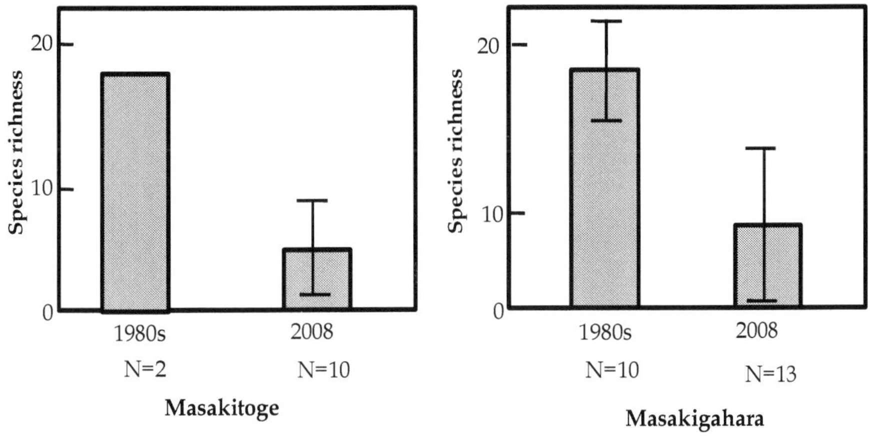

Figure 7. Species richness of epiphytic bryophytes on a single *P. jezoensis* var. *hondoensis* tree in the 1980s and 2008

3. Bryophyte diversity and environmental variables

3.1. Major environmental factors influencing bryophyte diversity

3.1.1. Site selection

A preliminary survey was conducted to identify plots of forest that were dominated by *P. jezoensis* var. *hondoensis* trees, including those with and without a protective wire mesh. In total, 9 plots (each 20 × 20 m in size) were selected (Fig. 1 A–I; Table 2) to examine the influence of wire mesh protection on epiphytic bryophyte diversity. Plots A–D were identical to those mentioned in section 2.4. We observed that the tree trunks were completely wrapped with wire mesh, from the ground up to a height of 150–180 cm. The mesh was composed of zinc-coated galvanized iron, a commonly used material for wire meshes (Japan Society of Corrosion Engineering, 2000). In each plot, the tree density (m^2/plot) was measured on the basis of the total basal area of the trunks. Further, the fraction of the trunk area of the *P. jezoensis* var. *hondoensis* trees that had been debarked by sika deer was also recorded in each plot.

3.1.2. Bryophyte sampling

The epiphytic bryophyte flora on the trunks of *P. jezoensis* var. *hondoensis* trees in the study plots were surveyed from October to November 2008. The bryophyte species covering the tree trunks from ground level to a height of 1.5 m were examined.

Bryophyte nomenclature followed that reported by Iwatsuki (2001). The proportion of bryophyte cover, as a percentage of the total available bark area being investigated, was divided into 6 levels: 1 (<1%), 2 (≥1% to <10%), 3 (≥10% to <25%), 4 (≥25% to <50%), 5 (≥50% to <75%), and 6 (≥75%).

3.1.3. Analysis of the correlation between bryophyte diversity and environmental variables

A simple generalized linear model (GLM) using R software for Windows 2.11.0 (R Development Core Team, 2010) was used to identify the correlations between species richness and the bryophyte cover with respect to environmental variables. To identify the most parsimonious model, we performed automated stepwise model selection using the Akaike information criterion (AIC) using the minimum AIC as the best-fit estimator. Bryophytes that had been identified only up to the genus level were not included in the calculation of species richness if any species of that genus was sampled. The environmental variables used in the GLMs were tree density, host tree diameter at breast height (DBH), percentage of debarked area, and percentage of tree trunks with wire mesh protection.

3.1.4. Results & discussion

Associated with the 110 tree trunks in the sampling plots, 68 species were identified in the bryophyte flora survey: 29 mosses and 39 liverworts (Appendix). Fig. 8 shows the species richness and cover in the study plots. The species richness on a single tree ranged from no species to 34 species (mean = 9.1, SD ± 9.0), while the bryophyte cover ranged from 0 to level 5 (mean = 2.0, SD ± 1.8).

The GLMs constructed using the environmental variables are presented in Table 3. These models showed that the species richness and bryophyte cover were significantly correlated with DBH (height, 1.5 m) and tree density ($p < 0.01$) but negatively correlated with the presence of wire mesh protection ($p < 0.01$). The GLMs for species richness and bryophyte cover explained 70.1% and 80.4% of the variance, respectively ($p < 0.01$ for both models).

High tree density and host tree DBH have been suggested to be beneficial for bryophyte diversity, as they provide better microclimates, e.g., humid conditions (Hazell et al., 1998; Ojala et al., 2000; Thomas et al., 2001). Further, the species richness and bryophyte cover may be positively correlated with high host tree DBH because DBH is correlated with bark features (e.g., bark thickness and bark roughness) (Boudreault et al., 2008; Ojala et al. ,2000).

The results of this study raise the question of how wire mesh protection negatively affects bryophyte diversity. Considering that the galvanized iron, which is the primary component of the wire mesh, is coated with zinc, it is likely that the zinc affects the bryophytes. In the next section, we compare the zinc concentrations in bryophytes on trees with and without wire mesh protection.

3.2. Effect of wire mesh protection on bryophytes)

3.2.1. Bryophyte samples

To examine the influence of wire mesh protection on the bryophytes, inductively coupled plasma-mass spectrometry (ICP-MS) was used to compare the concentration of zinc in bryophyte samples, since zinc coats the wire mesh surface. For this evaluation, 2 species of bryophyte that are commonly found on the trunks of *P. jezoensis* var. *hondoensis* trees of this region, both with and without wire mesh, were sampled: *H.tristo-viride* and *S. ampliata*. For each species, 3 sets of samples each were collected from trees with and without wire mesh.

3.2.2. *Analysis of zinc concentration*

Dry samples (0.05–0.10 g) were placed in polytetrafluoroethylene vessels and weighed. Subsequently, 5 mL of nitric acid was added to the samples, and they were digested using a microwave system (MLS-1200 MEGA; Milestone General, Tokyo, Japan) before ICP-MS analysis. The samples were then analyzed using a 7500CX ICP-MS system (Agilent Technologies, Wilmington, DE, USA). Spectral interference was minimized or eliminated using the octopole reaction system, with helium as the reaction gas at a flow rate of 2.5 mL/min. The ICP-MS analysis was repeated twice for each sample, and the mean values were used in one-sided Student's t-test comparisons of the zinc concentration from bryophyte samples on tree trunks with and without wire mesh.

Plot	Altitude	Tree density (m^2/plot)	No. of *Picea jezoensis* var. *hondoensis* trees Total	With wire mesh
Plot A	1676	3.1	3	2
Plot B	1672	4.0	7	6
Plot C	1621	5.5	2	2
Plot D	1619	4.3	8	7
Plot E	1597	13.9	12	0
Plot F	1572	11.9	15	0
Plot G	1576	19.9	11	11
Plot H	1590	9.9	30	30
Plot I	1597	12.7	22	0

Table 2. Summary of the characteristics of the study plots sampled, including altitude, tree density, and number of trees surveyed

	Cover			Species richness		
Variables	coefficients	*t*-value	*p*	coefficients	*t*-value	*p*
Intercept	7.30	3.81	< 0.01	2.26	7.30	< 0.01
Tree density	3.51×10^{-4}	3.10	< 0.01	6.30×10^{-5}	3.43	< 0.01
Host tree DBH	2.01×10^{-1}	3.12	< 0.01	2.29×10^{-2}	2.18	< 0.05
Wire mesh	-1.43×10^{-1}	–14.2	< 0.01	-3.19×10^{-2}	–19.5	< 0.01
Adjusted R squared	0.701			0.804		

Table 3. Generalized liner models showing the association of species richness and bryophyte cover with environmental variables

The significance level of the coefficients and adjusted R^2 values are shown.

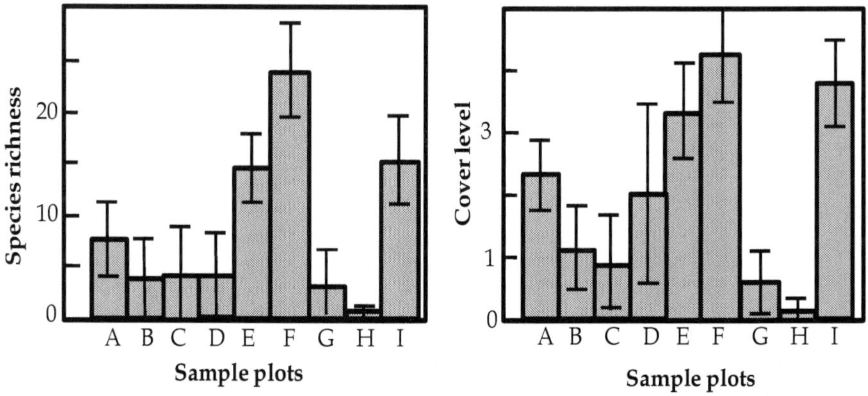

The bars represent the mean value of species richness and epiphyte cover on a single tree, and the error bars represent the corresponding standard deviations. Most tree trunks in plots A, B, C, D, G, and H had wire mesh protection.

Figure 8. Comparison of bryophyte species richness between trees with and without wire mesh protection

3.2.3. Results & discussion

ICP-MS analysis showed a significant 3- to 6-fold higher concentration of zinc in bryophytes inhabiting the bark of trees with wire mesh protection than in those without wire mesh protection (Fig. 9). Previous studies have shown that a considerable amount of zinc is leached from the zinc coating of galvanized iron by rain and dew (Harris, 1946; Seaward, 1974). Research has also shown that zinc is highly toxic to bryophytes (Tyler, 1990). Consequently, from the decreased diversity and increased zinc concentration of bryophytes on trees with wire mesh protection, it is reasonable to conclude that the loss of bryophyte cover and species richness has primarily occurred because of the toxicity of the zinc in the wire mesh. Additionally, other heavy metals in the wire mesh (e.g., iron) may affect bryophytes, with different heavy metals exerting varying levels of toxicity for bryophytes (Tyler, 1990).

4. Implications for biodiversity conservation

The results show that epiphytic bryophyte diversity is positively influenced by tree density and host tree DBH but negatively influenced by wire mesh protection, because of zinc toxicity (Fig. 10). The decline in bryophyte abundance and diversity on the lower parts of the tree trunks may be a cause for concern for biodiversity conservation on Mt. Odaigahara. This is because bryophytes contribute significantly to the species richness and biomass of tree trunks (Fritz, 2009; Lyons et al., 2000), as well as for ecosystem functions.

Furthermore, in addition to bryophytes, tree bark also provides important habitats for lichens and vascular epiphytes (Williams & Sillett, 2007). However, as heavy metals are toxic to these plants (Tyler et al., 1989), wire mesh protection may also contribute towards decreasing their levels of diversity and ecosystem functions. Unfortunately, considering that wire mesh protection is generally used against mammalian pests due to its direct effectiveness (Salmon et al. 2006; Vercauteren et al. 2006), this negative impact on bryophyte diversity may be widespread.

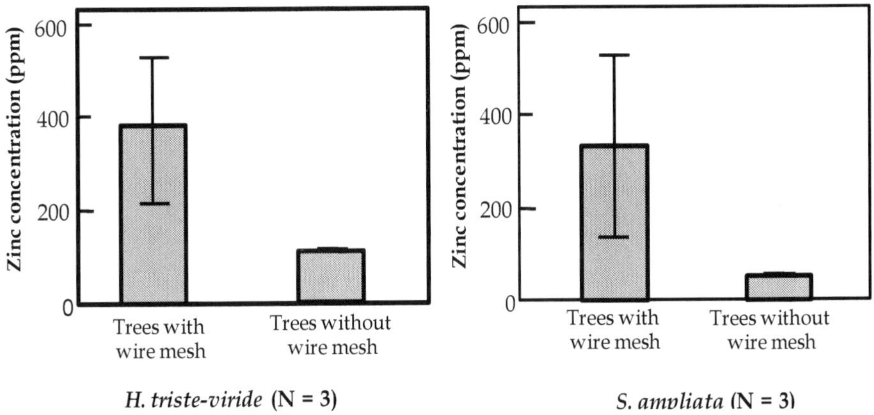

Figure 9. Zinc concentration

The zinc concentration was significantly higher in bryophytes on trees with wire mesh than in those without wire mesh ($p < 0.05$; t-test).

Therefore, to establish best practices for biodiversity conservation that includes bryophytes, we should not only protect trees against bark stripping by deer but also focus on the materials used for protection. Alternative techniques for plant protection include the use of tree shelters in which trees are enclosed in plastic tubes (Ward et al., 2000) and forest enclosures using plastic mesh fencing (Vercauteren et al., 2006). However, these alternatives may also affect biodiversity conservation. For example, tree shelters decrease light transmission (Ward et al., 2000), which might alter the composition of bryophyte species on tree trunks. Further, Shibata et al. (2008) reported that forest enclosures sometimes hamper tree regeneration within the fenced areas because of serious seed predation by increased mouse populations.

5. Conclusion

The difficulties faced in minimizing the effect of plant protection methods on ecosystems that have complex community interactions are shown in this chapter. To establish best practices for biodiversity conservation, adaptive management should be adopted. Within such frameworks, we should examine and revise protective management practices on the basis of scientific data assimilated from regular monitoring of such ecosystems, while also preferentially using metal-free plant protection materials.

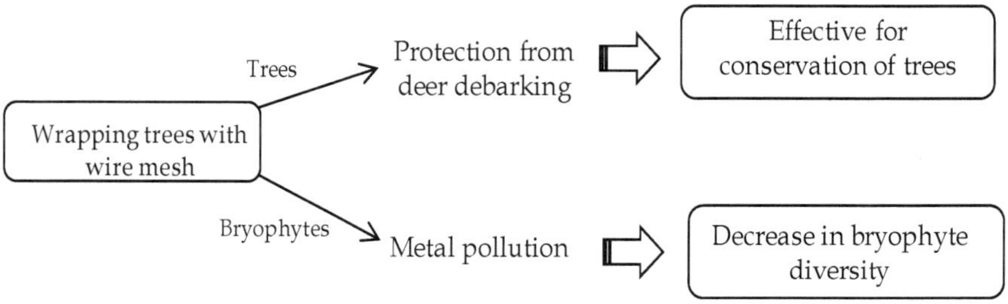

Figure 10. Positive and negative effects of wire mesh protection on the biodiversity

Author details

Yoshitaka Oishi
Department of Forest Science,
Faculty of Agriculture,
Shinshu University,
Japan

Acknowledgement

The author thanks Professor Yukihiro Morimoto and Associate Professor Hiroyuki Akiyama for providing critical comments and suggestions for improvements to this chapter, in addition to Dr. Kosaku Yamada for assistance with bryophyte identification and Kaori Kuriyama for the bryophyte survey and photos. This research was supported by a Grant-in-Aid for Scientific Research (A) (No. 18201008) from the Japan Society for the Promotion of Science and the Global COE Program "Global Center for Education and Research on Human Security Engineering for Asian Megacities," MEXT, Japan.

Appendix

This list shows the average cover of each species on a single *P. jezoensis* var. *hondoensis* tree. The cover levels are indicated for reference to the text. The bryophytes nomenclature follows that reported by Iwatsuki (2001).

Plots	A	B	C	D	E	F	G	H	I
Mosses									
Pogonatum alpinum (Hedw.) Röhl.	0	0	1.0	0	0.1	0.5	0	0	0.4
Pogonatum contortum (Brid.) Lesq.	0	0	0	0	0.1	0.1	0	0	0
Pogonatum japonicum Sull. & Lesq.	0	0	0	0	0.2	0	0	0	0
Dicranum japonicum Mitt.	0	0	0	0	0	0.3	0	0	0
Dicranum nipponense Besch.	0	0	0	0	0	0.2	0	0	0.2
Dicranum hamulosum Mitt.	0.3	0.1	0.5	0	1.2	0.7	0.1	0	0.6
Dicranum leiodontum Card.	1.0	0.6	1.0	0.3	0.8	0.6	0	0	0.6
Dicranum mayrii Broth.	0	0	0	0	0	0.1	0	0	0
Dicranum scoparium Hedw.	0	0.1	0	0	0.8	0.5	0	0	0.2
Dicranum viride (Sull. & Lesq.) Lindb. var. *hakkodense* (Card.) Takaki	1.0	0.4	1	0.5	0.9	0.9	0.4	0	0.4
Dicranodontium denudatum (Brid.) E. G. Britt. ex Williams	0	0.1	0.5	0	0.5	0.4	0	0	0.3
Dicranoloma cylindrothecium (Mitt.) Sakurai	0	0	0	0	0.3	1.2	0	0	0.8
Leucobryum bowringii Mitt.	0	0	0	0	0	0	0	0	0
Leucobryum juniperoideum (Brid.) Müll.Hal.	0	0	0	0	0.1	0.1	0	0	0
Trachycystis flagellaris (Sull. & Lesq.) Lindb.	0	0	0	0	0	0.3	0	0	0
Pterobryon arbuscula Mitt.	0	0	0	0	0	0.1	0	0	0
Neckera konoi Broth.	0	0	0	0	0	0	0	0	0
Fauriella tenuis (Mitt.) Card.	0	0	0	0	0.1	0.1	0	0	0
Thuidium tamariscinum (Hedw.) Shimp.	0	0	0	0	0	0	0	0	0
Plagiothecium euryphyllum (Card. & Thér.) Z.Iwats.	0	0	0.5	0	0	0.2	0	0	0
Heterophyllium affine (Hook.) M.Fleisch.	1.3	0.1	0.5	0	3.3	3.4	0.3	0	1.5
Brotherella fauriei (Card.) Broth.	0	0	0	0	0.1	0.3	0	0	0.2
Brotherella henonii (Duby) M.Fleisch.	0	0	0	0	0.1	0.2	0	0	0
Herzogiella turfacea (Lindb.) Z.Iwats.	0	0	0	0	0	0.1	0	0	0
Hypnum tristo-viride (Broth.) Paris	1.3	0.4	1.0	0.4	3.8	3.5	0.5	0	2.7
Hypnum fujiyamae (Broth.) Paris	0.3	0.1	0	0	0.5	0.5	0	0	0.5
Pseudotaxiphyllum pohliaecarpum (Sull. & Lesq.) Z.Iwats.	0	0	0	0	0.1	0.1	0	0	0
Hylocomium splendens (Hedw.) Schimp.	0	0	0	0	0	0.7	0	0	0.1
Pleurozium schreberi (Brid.) Mitt.	0	0	0	0	0	0.4	0	0	0.1

Liverworts									
Herbertus aduncus (Dicks.) Gray	0.7	0.1	0.5	0	0.4	0.4	0	0	0.4
Trichocolea tomentella (Ehrh.) Dumort.	0	0	0	0	0	0	0	0	0
Blepharostoma trichophyllum (L.) Dumort.	0	0	0	0	0.7	0.9	0	0	0.4
Lepidozia reptans (L.) Dumort.	0	0	0	0	0.1	0.5	0	0	0.4
Lepidozia subtransversa Steph.	0	0	0	0	0	0.1	0	0	0
Lepidozia vitrea Steph.	0.3	0	0	0	0.8	1.0	0.1	0	0.7
Bazzania bidentula (Steph.) Steph.	0	0	0	0	0	0.1	0	0	0.1
Bazzania denudata (Torr. ex Lindenb.) Trevis.	0.3	0.1	0.5	0.1	1.1	1.0	0.1	0	0.9
Bazzania yoshinagana (Steph.) S.Hatt.	0	0	0	0	0.3	1.2	0.1	0	0.8
Cephalozia leucantha Spruce	0	0	0	0	0.1	0.1	0	0	0
Cephalozia lunulifolia (Dumort.) Dumort.	0	0	0	0	0.1	0	0	0	0
Chephaloziella sp.	0	0	0	0.1	0	0.2	0	0	0
Nowellia curvifolia (Dicks.) Mitt.	0	0	0	0	0.1	0.2	0	0	0
Odontoschisma denudatum (Mart.) Dumort.	0	0.1	0.5	0.1	0.8	0.6	0.1	0	0.2
Odontoschisma grosseverrucosum Steph.	0	0	0	0	0	0.1	0	0	0
Jamesoniella autumnalis (DC.) Steph.	0	0	0	0	0.3	0	0	0	0
Jungermannia subulata A.Evans	0	0	0	0	0	0.1	0	0	0
Anastrophyllum michauxii (F.Weber) H. Buch	0	0	1.0	0	0.6	0.4	0.1	0	0.3
Lophozia longiflora (Nees) Schiffn.	0	0	0	0	0.3	0.6	0.1	0	0.2
Lophozia incisa (Schrad.) Dumort..	0	0	0	0	0	0.5	0.1	0	0.2
Mylia verrucosa Lindb.	0	0	0	0	0	0.3	0	0	0.4
Scapania ampliata Steph.	0.3	0.3	0	0	0.4	1.9	0.4	0	1.2
Scapania bolanderi Austin	0	0	0	0	0	0.1	0	0	0.1
Scapania ciliata Sande Lac.	0	0	0	0	0.1	0	0	0	0
Scapania hirosakiensis Steph. ex Müll. Frib.	0	0	0	0.1	0.3	0.3	0	0	0
Heteroscyphus planus (Mitt.) Schiffn.	0	0	0	0	0.1	0	0	0	0
Plagiochila gracilis Lindenb. & Gottsche	0	0.1	0.5	0.3	0.2	0.5	0.2	0	0.1
Plagiochila ovalifolia Mitt.	0	0	0	0	0	0.1	0	0	0
Plagiochila semidecurrens (Lehm. & Lindenb.) Lindenb.	0	0	0	0.1	0.4	0.9	0.2	0	0.1
Radula cavifolia Hampe ex Gottsche, Lindenb. & Nees	0	0	0	0	0	0.1	0	0	0
Radula brunnea Steph.	0	0	0	0	0	0.4	0	0	0
Ptilidium pulcherrimum (Weber) Vain.	0	0.1	0	0	0.1	0	0	0	0
Frullania tamarisci (L.) Dumort. subsp. *obscura* (Verd.) S.Hatt.	0	0	0	0	0.6	1.6	0.2	0	1
Cololejeunea macounii (Spruce ex Underw.) A.Evans	0	0	0	0	0	0.1	0	0	0
Drepanolejeunea angustifolia (Mitt.) Grolle	0	0	1	0.1	0	0.4	0	0	0.1
Drepanolejeunea ternatensis (Gottsche) Steph.	0	0	0	0	0.1	0	0	0	0
Drepanolejeunea teysmannii Steph.	0	0	0	0	0	0.2	0.1	0	0
Nipponolejeunea pilifera (Steph.) S.Hatt.	1.0	0.7	0	0.4	1.0	0.9	0.2	0	0.9
Nipponolejeunea subalpina (Horik.) S.Hatt.	0.3	0	0	0.1	0.1	0.1	0	0	0.1
Lejeunea ulicina (Taylor) Gottsche, Lindenb. & Nees	0.3	0	0.5	0.1	0	0.5	0.1	0	0

6. References

Akashi. N. & Nakashizuka, T. (1999). Effects of bark-stripping by Sika deer (*Cervus nippon*) on population dynamics of a mixed forest in Japan. *Forest Ecolgy and Management*, Vol.113, No.1, (January 1999), pp. 75–82, ISSN 0378-1127

Allombert, S.; Gaston, A. J. & Martin, J.-L. (2005). A natural experiment on the impact of overabundant deer on songbird populations. *Biological Conservation*, Vol.126, No.1, (November 2005), pp. 1–13, ISSN 0006-3207

Ando, M. & Goda, R. (2009). Sika deer (*Cervus nippon* Temminck) of Mount Ohdaigahara, In: *Ecology of sika deer and forest ecosystem of Mt. Ohdaigahara,* Shibata, E. & Hino, T. (Eds.), 46-52 ,Tokai University Press, ISBN 978-4-486-01830-8, Kanagawa, Japan

Boudreault, C.; Coxson, D. S.; Vincent, E.; Bergeron, Y. & Marsh, J. (2008). Variation in epiphytic lichen and bryophyte composition and diversity along a gradient of productivity in *Populus tremuloides* stands of northeastern British Columbia, Canada. *Ecoscience*, Vol.15, No.1, (March 2008), pp. 101–112, ISSN 1195-6860

Coxson, D. S. (1991). Nutrient release from epiphytic bryophytes in tropical montane rain forest (Guadeloupe). *Canadian Journal of Botany*, Vol.69, No.10, (October 1991), pp. 2122–2129, ISSN 1480-3305

Doei, H. (1988). Bryophytes of Mt. Ohdaigahara in the Kii Peninsula, Central Honshu, Japan, I. *Nanki Biology*, Vol.30, pp. 14–23

Feber, R. E.; Brereton, T. M.; Warren, M. S. & Oates, M. (2001). The impacts of deer on woodland butterflies: the good, the bad and the complex. *Forestry*, Vol.74, No.3, pp. 271–276, ISSN 1464-3626

Flowerdew, J. R. & Ellwood, S. A. (2001). Impacts of woodland deer on small mammal ecology. *Forestry*, Vol.74, No.3, pp. 277–287, ISSN 1464-3626

Fritz, Ö. (2009). Vertical distribution of epiphytic bryophytes and lichens emphasizes the importance of old beeches in conservation. *Biodiversity and Conservation*, Vol.18, No.2, (October 2008), pp. 289–304, ISSN 1572-9710

Fuller, R. J. & Gill, R. M. A. (2001). Ecological impacts of increasing numbers of deer in British woodland. *Forestry*, Vol.74, No.3, pp. 193–199, ISSN 1464-3626

Harris, T. M. (1946). Zinc poisoning of wild plants from wire netting. *New Phytologist*, Vol.45, No.1, (June 1946), pp. 50–55, ISSN 1469-8137

Hazell, P.; Kellner, O.; Rydin, H. & Gustafsson, L. (1998). Presence and abundance of four epiphytic bryophytes in relation to density of aspen (*Populus tremula*) and other stand characteristics. *Forest Ecology and Management*, Vol.107, No.1–3, (August 1998), pp. 147–158, ISSN 0378-1127

Ide, H. & Kameyama, A. (1972). Vegetation of Ohdaigahara. *Applied phytosociology*, Vol.1, pp. 1–48.

Iwatsuki, Z. (Ed). (2001). *Mosses and liverworts of Japan*, Heibonsha, ISBN 978-4582535075, Tokyo, Japan

Japan Society of Corrosion Engineering. (Ed.) (2000). *Corrosion handbook*, Maruzen, ISBN 978-4621046487, Tokyo, Japan

Kumar, S.; Takeda, A. & Shibata, E. (2006). Effects of 13-year fencing on browsing by sika deer on seedlings on Mt. Ohdaigahara, central Japan. *Journal of Forest Research,* Vol. 11, No.5, (April 2006), pp. 337–342, ISSN 1610-7403

Lyons, B.; Nadkarni, N. M. & North, M. P. (2000). Spatial distribution and succession of epiphytes on *Tsuga heterophylla* (western hemlock) in an old-growth Douglas-fir forest. *Canadian Journal of Botany* , Vol.78, No.7, (June 2000), pp. 957–968, ISSN 1480-3305

Ministry of the Environment-Kinki Regional Environment Office. (2009). *Nature restoration plan for Mt. Ohdaigahara: second period.* Ministry of the Environment- Kinki Regional Environment Office, Osaka, Japan

Miquelle, D. G. & Van Ballenberghe, V. (1989). Impact of bark stripping by moose on aspen spruce communities. *The Journal of Wildlife Management,* Vol.53, No.3, (July 1989), pp. 577–586, ISSN 1937-2817

Nadkarni, N. M. (1984). Epiphyte biomass and nutrient capital of a neotropical elfin forest. *Biotropica,* Vol.16, No.4, (December 1984), pp. 249–256, ISSN 1744-7429

Nakajima, N. (2007). Changes in distributions of wildlife in Japan, In: *Rebellion of wildlife and collapse of forest,* Forest and Environment Research Association, Japan (Ed.), 57-68, Shinrinbunka Association, ISBN 978-4021001284, Tokyo, Japan

Nakamura, T. (1997). Effect of bryophytes on survival of conifer seedlings in subalpine, pp.forests of central Japan. *Ecological Research* Vol.7, No.2, pp. 155–162, ISSN 1440-1703

Nara local meteorological observatory. (1997). *Climate of Nara Prefecture in the last 100 years.* National printing bureau, Tokyo, Japan

Oishi, Y. (2011). Protective management of trees against debarking by deer negatively impacts bryophyte diversity. *Biodiversity and Conservation,* Vol.20, No.11, (June 2011), pp. 2527-2536, ISSN 1572-9710

Ojala, E.; Mönkkönen, M. & Inkeröinen, J. (2000). Epiphytic bryophytes on European aspen Populus tremula in old-growth forests in northeastern Finland and in adjacent sites in Russia. *Canadian Journal of Botany,* Vol.78, No.4, (April 2000), pp. 529–536, ISSN 1480-3305

Pellerin, S.; Huot, J. & Côté, S. D. (2006). Long term effects of deer browsing and trampling on the vegetation of peatlands. *Biological Conservation,* Vol.128, No.3, (March 2006), pp. 316–326, ISSN 0006-3207

Pypker, T. G.; Unsworth, M. H. & Bond, B. J. (2006). The role of epiphytes in rainfall interception by forests in the Pacific Northwest II. Field measurements at the branch and canopy scale. *Canadian Journal of Forest Reserch,* Vol.36, No.4, (April 2006), pp. 819–832, ISSN 1208-6037

R Development Core Team. (2010). *R 2.11.0, A language and environment for statistical computing computer program,* R Development Core Team, ISBN 3-900051-07-0, Vienna, Austria

Rooney, T. P. (2001). Deer impacts on forest ecosystems: a North American perspective. *Forestry,* Vol.74, No.3, pp. 201–208, ISSN 1464-3626

Rooney, T. P. & Waller, D. M. (2003). Direct and indirect effects of white-tailed deer in forest ecosystems. *Forest Ecology and Management,* Vol.181, No.1-2, (August 2003), pp. 165–176, ISSN 0378-1127

Salmon, T. P.; Whisson, D. A. & Marsh, R. E. (2006). *Wildlife Pest Control around Gardens and Homes* (2nd edition), University of California, Agricultural and Natural Resources Publication, ISBN 978-0931876660, California

Schütz, M.; Risch, A. C.; Leuzinger, E.; Krüsi, B. O. & Achermann, G. (2003). Impact of herbivory by red deer (*Cervus elaphus* L.) on patterns and processes in subalpine grasslands in the Swiss National Park, *Forest Ecolgy and Management* Vol.181, No.1-2, (August 2003), pp. 177–188, ISSN 0378-1127

Seaward, M. R. D. (1974). Some observations on heavy metal toxicity and tolerance in lichens. *The Lichenologist,* Vol.6, No.2, pp. 158–164, ISSN 0024-2829

Shibata, E.; Saito, M. & Tanaka, M. (2008). Deer-proof fence prevents regeneration of Picea jezoensis var.hondoensis through seed predation by increased wood mouse populations. *Journal of Forest Research,* Vol.13, No.2, (April 2008), pp. 89–95, ISSN 1610-7403

Stewart, G. H. & Burrows, L. E. (1989). The impact of white-tailed deer *Odocoileus virginianus* on regeneration in the coastal forests of Stewart Island, New Zealand. *Biological Conservation,* Vol.49, No.4, pp. 275–293, ISSN 0006-3207

Stockton, S. A.; Allombert, S.; Gaston, A. J. & Martin, J.-L. (2005). A natural experiment on the effects of high deer densities on the native flora of coastal temperate rain forests. *Biological Conservatoin,* Vol.126, No.1, (November 2005), pp. 118–128, ISSN 0006-3207

Takatsuki, S. (2009). Effects of sika deer on vegetation in Japan: A review. *Biological Conservation,* Vol.142, No.9, (september 2009), pp. 1922-1929, ISSN 0006-3207

Thomas, S. C.; Liguori, D. A. & Halpern, B. C. (2001). Corticolous bryophytes in managed Douglas-fir forests: habitat differentiation and responses to thinning and fertilization. *Canadian Journal of Botany,* Vol.79, No.8, (August 2001), pp. 886–896, , ISSN 1480-3305

Tyler, G.; Balsberg Påhlsson, A.-M.; Bengtsson, G.; Bååth, E. & Tranvik, L. (1989). Heavy metal ecology of terrestrial plants, microorganisms and invertebrates. *Water, Air, and Soil Pollution,* Vol.47, No.3-4, pp. 189–215, ISSN 1573-2932

Tyler, G. (1990). Bryophytes and heavy metals: a literature review. *Botanical Journal of the Linnean Society,* Vol.104, No.1-3, (September 1990), pp. 231–253, ISSN 1095-8339

Vercauteren, K. C.; Michael, J. L. & Hygnstrom, S. (2006). Fences and deer-damage management: A review of designs and efficacy. *The Wildlife Society Bulletin,* Vol.34, (August 2006), pp. 191–200, ISSN 0091-7648

Ward, J. S.; Gent, M. P. N. & Stephens, G. R. (2000). Effects of planting stock quality and browse protection-type on height growth of northern red oak and eastern white pine. *Forest Ecology and Management,* Vol.127, No.1-3, (March 2000), pp. 205–216, ISSN 0378-1127

Webster, C. R.; Jenkins, M. A. & Rock, J. H. (2005). Long-term response of spring flora to chronic herbivory and deer exclusion in Great Smoky Mountains National Park, USA. *Biological Conservavion,* Vol.125, No.3, (October 2005), pp. 297–307, ISSN 0006-3207

Williams, C. B. & Sillett, S. C. (2007). Epiphyte communities on redwood (Sequoia sempervirens) in northwestern California. *The Bryologist*, Vol.110, No.3, pp. 420– 452, ISSN 0007-2745

Yokoyama, S.; Maeji, I.; Ueda, T.; Ando, M. & Shibata, E. (2001). Impact of bark stripping by sika deer, Cervus nippon, on subalpine coniferous forests in central Japan. *Forest Ecolgy and Management*, Vol.140, No.2-3, (January 2001), pp. 93–99, ISSN 0378-1127

Chapter 7

Tropical Forest and Carbon Stock's Valuation: A Monitoring Policy

Thiago Metzker, Tereza C. Spósito, Britaldo S. Filho,
Jorge A. Ahumada and Queila S. Garcia

Additional information is available at the end of the chapter

http://dx.doi.org/10.5772/48355

1. Introduction

Carbon is the fourth most abundant element on Earth. It is estimated that the world's forests store 283 gigatonnes (1Gt = 1 billion tons) of carbon in their biomass alone and 638 Gt of carbon in the ecosystem as a whole (to a soil depth of 30 cm). Thus, forests contain more carbon than the entire atmosphere. Carbon is found in forest biomass and dead wood, as well as in soil and litterfall [1]. Consequently, changes in forest carbon storage, resulting from a shift in land use, have a significant impact on global climate change [2].

Changes in climate occur naturally, through processes operating on a geologic time scale. For example, the main species presently inhabiting the planet have survived climate changes during the Pleistocene, adjusting their geographical distribution to weather conditions. However, the speed and magnitude of changes that have been occurring in the Earth's climate system since the Industrial Revolution are currently of great concern. In 1991, the Intergovernmental Panel on Climate Change (IPCC) published a first report about global temperature increases caused by the intensification of the greenhouse effect. After this official announcement, the IPCC has established different working groups with scientists from various parts of the world in order for them to meet and compile as much information as possible and to update scientific predictions about the climatic future of the planet. The reports that have been produced by the international scientific community are considered as the main reference for global climate change.

Currently, scientific societies question the capacity of the present biota to tolerate such changes, in an environment that has been highly fragmented by human intervention and where what is still left intact is confined within protected areas. Changes within biota can result in changes in the ecosystem services they provide. Human well-being depends directly and indirectly on the environmental services provided for free by the natural world, including climate regulation, soil formation, erosion control, carbon storage, nutrient cycling, provision of water (both quality and quantity), maintenance of hydrological cycles, preservation of genetic resources, scenic beauty, among others [3]. Furthermore, tropical forests contain 50% of all world species and are considered mega-diverse environments. Therefore, changes in any of these services can have serious consequences for biodiversity, for the natural carbon cycle and the hydrological cycle, which may in turn alter the world economy and affect the everyday life of humans and other species on the planet.

How can these changes be monitored? One way to monitor biodiversity and carbon stocks over large areas is through the establishment of forest inventories. These are effective tools for estimating the type, amount and condition of forest resources over large areas [4]. The regular collection of measurements within Permanent Monitoring Plots (PMPs), combined with the use of statistical techniques, provide a baseline for assessing changes in the structure and dynamics of a forest and permit the construction of predictive models [5]. In the last decade, there has been a large increase in the installation of PMPs in different tropical forest sites around the world, especially in the Amazon Rainforest, where large monitoring networks (TEAM, PELD, CTFS, RAINFOR, LBA, REDEFLOR, PDBFFE and CIFOR) have been established. These programs increase the level of understanding of ecological systems, transforming the knowledge base [6]. However, there are still serious deficiencies in estimating carbon stocks and other components of other types of tropical forests, types and others components of tropical forests.

Within the current political and environmental international situation it is vital that all countries, whether or not signatories of the Kyoto Protocol, do promote initiatives to monitor their biodiversity and their carbon stocks. These data are strategic for each country because they indicate where and how the management of natural resources can bring benefits to local people (local scale), they support the creation of public policies that can become part of the country's legislation (regional scale) and promote policies for adaptation to an increased vulnerability to climate change (global scale).

This chapter, *"Tropical Forest and Carbon Stock's 1 Valuation: A Monitoring Policy"*, incorporates parts of the TEAM (Tropical Ecology Assessment Monitoring) protocol [7] and the knowledge generated over six years of monitoring permanent plots in an area of the Atlantic Rainforest in Brazil. It aims to discuss the importance of planning and implementation of PMPs, the main techniques used, and the errors associated with them. Biomass, carbon stock calculation techniques and data analysis will also be discussed, among other topics. Data collection and analysis have a greater value when incorporated into natural resource management policies, such as Payment for Environmental Services (PES), which are provided by nature. A comprehensive approach involving stakeholders at all levels, from the local to the global scale, is essential for the success of integrated policies. Each of the topics listed below will be presented with the aid of practical examples, figures and tables, in order to allow readers the opportunity to fully engage with the subject matter and, most importantly, to begin to understand how to apply these practices in their own social and environmental contexts.

2. Inventory Data

2.1. Methods for establishing Permanent Monitoring Plots (PMPs)

The establishment of vegetation monitoring networks is a strategy that aims to develop an integrated database through systematized collections using a single monitoring protocol on various sites. In the vegetation network implementation, it is extremely important that the database management team be clear about the questions to be asked and the objectives for the collection of field data. This systemization has implications directly related to the method of collection and the definition of the protocol for implementation and monitoring. The primary analyses to be conducted also must be predefined as they too have a direct impact on the sample design and the means of data collection.

During the planning of a monitoring network, it is important to keep in mind that the key objective is to conduct large-scale analyses that can speak to physiognomy, biomes and wider generalizations. This scale of work is fundamental in order to accomplish robust analyses and to study broad-scale ecological processes. However, it should be noted that local and regional data and publications are also part of this network as they promote the development of local scientific knowledge, along with the participation of the team responsible for the collection of field data. These initiatives encourage cooperation and sharing of experience, in addition to motivating those who are responsible at the local level to continue the work of monitoring once the objectives and results of the initiative are made clear to all involved.

The means of disseminating results should also be defined in the planning phase. For example, during this phase, contact can be made with the editors of scientific journal where there is an intention to publish, in order to establish a connection with the journal and develop credibility for a strong relationship. The sharing of the monitoring protocol, the initial results and the key conclusions at national and international conferences provides visibility for the project and stimulates ongoing discussions with other researchers in the topic area. This interaction and sharing of experience always benefit the project as they increase quality and strengthen key elements. The network planning team should also identify other forms of communication for scientific dissemination, such as specialized documentaries, news networks, community sites and scientific blogs. These promote dissemination and constructive discussion of the conclusions and methods of the published initiative. Another tactic that can make a significant contribution to successful monitoring over the long term by strengthening relationships with local teams is the development of news releases in the local language where the data was collected.

As with any good plan, the protocol must be rigorous. Several protocols for monitoring tropical forests are available including RAINFOR's [8], TEAM's [7] and the Smithsonian's Center for Tropical Forest Science [9]. However, it must also be flexible enough to be adapted and to evolve naturally according to the knowledge generated during the planning process, as well as to the local reality of each site. Ongoing workshops with the local team guarantee that acquired experience is formally recorded, in addition to facilitating the continuous improvement of the protocol by applying experience acquired through its execution *in situ*.

2.2. Geoprocessing techniques for area selection

Many field procedures involve high costs due to transportation and logistics. Therefore, prior to any field procedure, errors in area selection can be minimized by careful planning using GIS techniques. In addition to playing an important role in the preliminary phase (planning), these tools are also very useful in the data analysis phase. When these instruments are used extensively by a qualified professional, significant economies of time and financial resources can be achieved.

After clearly defining the objectives for the implementation of the monitoring network, the next phase is the selection of potential areas to house the plots. The use of GIS allows for a more confident selection of the target areas since it works with georeferenced bases and shapes which allow for simulation of PMPs implemented in practically any location in the world. These areas can be selected by process of elimination from those that, for example, do not have the required attributes or by selection of multiple criteria that involves interpolation of various bases. Through experience acquired in the implementation and monitoring of PMPs, we understand that the minimal criteria for exclusion of target areas for monitoring include:

- Areas that possess accentuated declivity;
- Areas that are not easily accessible and complicate field logistics;
- Areas with creeks, swamps, lakes and rivers;
- Areas that have significant spatial heterogeneity;
- Areas that have variations in the type of soil.

Assuming that the objective of monitoring is to evaluate the temporal dynamics of primary vegetation areas, the areas that are not located in Conservation Units can be excluded first. It is understood that forested areas protected by law in any part of the world represent the highest percentage of protected primary areas. After this first filter, the layers or shapes that meet the exclusion criteria cited above are applied. This type of cut is made relatively quickly, while still in the office, but can reduce a universe of potential samples by more than 90% in certain regions of the world, thus optimizing the accuracy and use of the project's financial resources.

Following elimination of the areas not selected for the sample, the professional responsible for the GIS technology should create polygons capable of housing the future PMPs so that random samples can be selected from within the universe of possible options, thus establishing statistical confidence for the sample. Another important point is that the PMPs should be replicated in areas where there is similar physiognomy, so that means, errors and reliable statistics can be obtained.

It is of fundamental importance for the field team that thematic maps be developed by the GIS team. These maps should be easy to visualize and understand, with current satellite images and superimposed colored sketches of the PMPs in various layers. Essential factors for successful field work include the standardization of symbols, language and scale of work, as well as pre-definition of a standard datum, and being in a system of unique coordinates compatible with the use of local GPSs. The field maps should also be plasticized to avoid stains and tears which can often occur with the use of these materials in the middle of the forest.

2.3. Choosing target areas

The field team should also be very clear about the objective of monitoring. When the project's primary issue is related to the dynamics of areas in recovery or to the differences between primary and secondary vegetation areas, area selection involves different parameters. When the question is focused on temporal variations in areas of intact vegetation in the climactic stage, area selection will be directed primarily toward areas protected by legal mechanisms in each region, ensuring that there will be no interference in the plot throughout the years of monitoring. Depending on the objective, criteria for inter-site analysis can also be established, such as a latitudinal gradient temperature or rainfall gradient, soil gradient, etc.

Once all of the criteria have been established, the field team should depart in order to locate and validate the target areas *in situ*. In addition to being accompanied by local guides, the team should be supplied with basic field supplies as well as thematic maps developed by the GIS team, a GPS, a compass, and a camera for the validation or invalidation of areas previously defined by the GIS team. Additionally, the field team should have in your GPS all points and layers that were previously prepared by the GIS team. For example, see [10] for a complete data transfer protocol.

It is important that the field team be fully trained on the monitoring protocol and have the ability to independently decide at any given moment if an area truly possesses the defined selection criteria or if it would be better to search for a new area. This decision is a key since all monitoring throughout the years ahead will depend on the correct choice and demarcation of these plots. In order to select the best areas for PMPs to be implemented, various factors should be taken into consideration, including the homogeneity of the forest typology to be sampled, the existence of water courses, logistics, access, type of soil and inclination of the terrain.

Due to difficulties of orientation and localization in interior bush areas, the geographical coordinates should be checked and the location of the field team confirmed upon arrival at the target area. Once the location has been verified, a marker should be placed in the ground (a PVC tube of about 1.3 m can be used) to be the point of coordinates 0,0 (X, Y), which will serve as a reference point for the validation of the area as well as for future plot implementation. This point will be used to evaluate the area to decide whether or not it will be selected for PMP implementation. Thus, using a compass, the direction of the course should be read, so that the angle of the directions has a difference of 90° (straight angle). The course is followed in the first direction (X), remaining aligned with the lead angle on the compass, stopping every 20 meters to check the coordinates and the direction of the course. In the field, detours are very common during a walk/hike due to natural obstacles such as fallen trees and branches, the presence of lianas or holes in the ground, or large trees that have to be circumvented. It is important in this verification phase, as well as in the PMP implementation phase, that knives and scythes are not to be used to open trails or forest passages as they can have a long term impact with significant implications on the dynamics of vegetation. Thus, when faced with a natural obstacle, the ideal would be for the team to circumvent it and return to the defined course in order to continue with area verification.

The team should be aware of sudden changes in the type of soil, the existence of accentuated declivity that was not possible to identify in the satellite images, or any other element that strongly differentiates the landscape and that could negatively impact the monitoring or the homogeneity of the plot. This should be recorded in a designated worksheet in order to justify the decision not to use the area in question. Once line X has been verified, the same procedure is conducted with line Y beginning from ground zero. If an area does not possess significant heterogeneity, the selection of the plot must be validated, assigning a number and a syllable to be used throughout the entire period of monitoring and analysis of that area (e.g. 01-LP).

2.4. Implementation of PMPs in the field

Once the entire validation process is complete, the actual marking of the PMP in the field is undertaken. On the day prior to departure, a checklist should be reviewed of all equipment required for field implementation, such as PVC tubes, rubber hammer, colored tape, polypropylene cord, compass, GPS, binoculars, clipboard, collection worksheets, plastic bags, masking tape, pencils, erasers and pens. In addition to support materials, specialized clothing must also be taken, such as boots, leggings and field jackets (with many pockets). The PMP implementation team should be comprised of at least 4 people, primarily to divide the weight of materials to be taken to the selected PMP area, as the tubes or stakes used to mark the chosen spots are very heavy and bulky.

Upon arrival at the PMP location previously marked as 0,0, a suitable location to leave all of the equipment should be identified, as well as an appropriate place to have snacks or lunch while in the field. This location, named "Support Station - SS" should be located in the outlying area of the PMP so that it does not interfere with the vegetation to be monitored on the plot. The ground should be covered by a light blue tarp (or any color that strongly contrasts the forest floor), upon which all of the equipment should be placed to avoid loss. Again, it is imperative that the team be careful not to allow any type of vegetation (lianas, branches or shrubs) to be cut during plot implementation.

In the following example, we simulate the implementation of a 1 ha PMP (10.000 m^2) according to the TEAM protocol for vegetation monitoring [7]. The size of the PMP will depend on the initial objective outlined by the team responsible for managing the project. The size of 1 ha is widely used in permanent plots whose objectives are related to monitoring the dynamics and carbon stocks for the site in question.

Starting at 0,0, two baselines (X, Y) should be projected, at 90° perpendicular angles, which will serve as reference points throughout the PMP implementation. Each baseline should be spiked every 20 meters, with their distance verified using a measuring tape and direction verified by reading the course angle on the compass. After the 6 spikes for each baseline have been duly marked and inserted into the ground, the entire line should be measured to confirm its length, which should be a total of 100 meters. Each spike placed every 20 meters should be sequentially numbered, as well as having its Cartesian coordinates on the plot recorded (e.g. 20, 0; 40, 0; 60, 0;...). Once line X has been completed, the formation of line Y can be undertaken using the same procedures previously followed.

Once the two baselines have been formed, the internal squares of the PMP can be developed. In order to close a PMP, two basic methods can be used: creating 5 lines parallel to baseline Y (Figure 1-B) or creating small 400 m^2 squares, forming sequential lines until the entire PMP is closed (Figure 1-A).

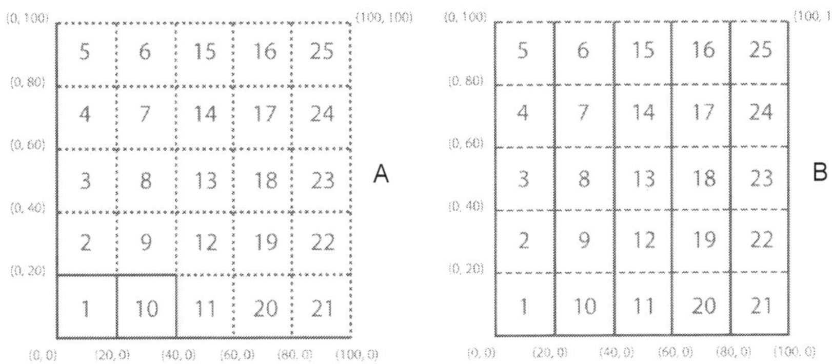

Figure 1. Sample structure of the Permanent Monitoring Plot (PMP), with 25 sub-plots. A – Means of closure using squares; B – Means of closure using lines. Adapted from TEAM (2010).

2.5. Marking trees

After marking the PMP, the individuals to be monitored are marked and data collection is undertaken. For studies related to long-term monitoring of the structure and dynamics of vegetation, it is common for the sample to include all individuals in the forest that have DBH ≥ 10 cm (Diameter at Breast Height). For studies of biomass and carbon stocks, individuals with DBH < 10 cm are not included, due to their low contribution to the total stocks of the PMP. In general, if the objective is to monitor changes in floristic composition and the biodiversity of the PMP, these smaller individuals should be incorporated into the monitoring.

In this case, all of the trees palms and lianas with a DBH greater than or equal to 10 cm should be marked and measured. The POM (Point of Measurement) is the point on the tree or liana where their respective diameters are measured. The POM is marked at 1.30 m with the help of a PVC tube graded at 1.60 m and 1.30 m to avoid error related to the different heights of the field markers. However, for individuals with tabular roots, sapopemas or buttress roots, the POM should be identified at 50 cm above the highest root (Figure 2). This is a valid change since it is common in forest inventories to find all stems with their DBHs measured at 1.30 m. When these data are inserted into allometric equations to calculate biomass, they overestimate biomass, increasing the standard error of these calculations [9-13].

In the case of trees that have many deformities at the POM, a modular ladder up to 12 meters (4 modules of 3 meters each to make it easy to transport in the forest) should be used so that the best location on the tree can be selected for diameter measurement (Figure 2 and Figure 3). Leaning or fallen trees should have their DBH measured following the methodology above; however, the distance from the base should be measured from the underside of the tree (Figure 2) in order to obtain an accurate distance. For trees with multiple trunks, where forking occurs below 1.30 m, each trunk should be considered a separate individual (Figure 2), with the number of measurements matching the number of trunks for the tree.

Once the best area for DBH measurement has been selected, it should be painted with yellow paint. This can also be done with a type of stamp (stencil) that can be made out of a sheet of hard plastic that is cut in the center in the following dimension: 20 x 3 cm. After selecting the location to be painted, the stamp (stencil) is placed on the tree and the POM is painted (Figure 3). In addition to facilitating field work, this stamp also standardizes the width of the paint marking on the trees, thus reducing the possibility of errors in future tree measurements.

This marking should be re-done every two years so that the specific POM is not lost. In order to avoid errors related to POM marking, the height at which the POM is marked should be recorded in a designated field worksheet. This procedure, along with painting the POM, guarantees that the measurement will be done at the exact same point during re-census throughout the monitoring period.

All of the individuals selected should be marked with nails and aluminium tags using increasing numbers according to the layout within the PMP. The nail should always be a distance of 40 cm from the POM so that the nail hole does not damage the trunk and consequently alter the POM. It is very common to see trees in the forest that have significant deformities resulting from a small nail hole. Bacteria and pathogens can enter through this small orifice and cause significant stress to tree trunks. Another important point is that the nail should be pointed downward whereby the tag is touching the head of the nail, since it is common to see trees that envelop around the tags over time when the tags had been touching the trees themselves.

After numeration and marking are complete, each individual should be identified at the highest taxonomic level possible in the field. It is highly recommended that photos be taken of the collected branches and that a collection of each species within the PMP be maintained as a botanical collection specific to each region. The data should be recorded in field worksheets and branch samples that are not identified should be taken for laboratory activities, herbarium consultations and completion of taxonomic identification by specialists. All field collections should be labelled with masking tape, recording their PMP number and reference code.

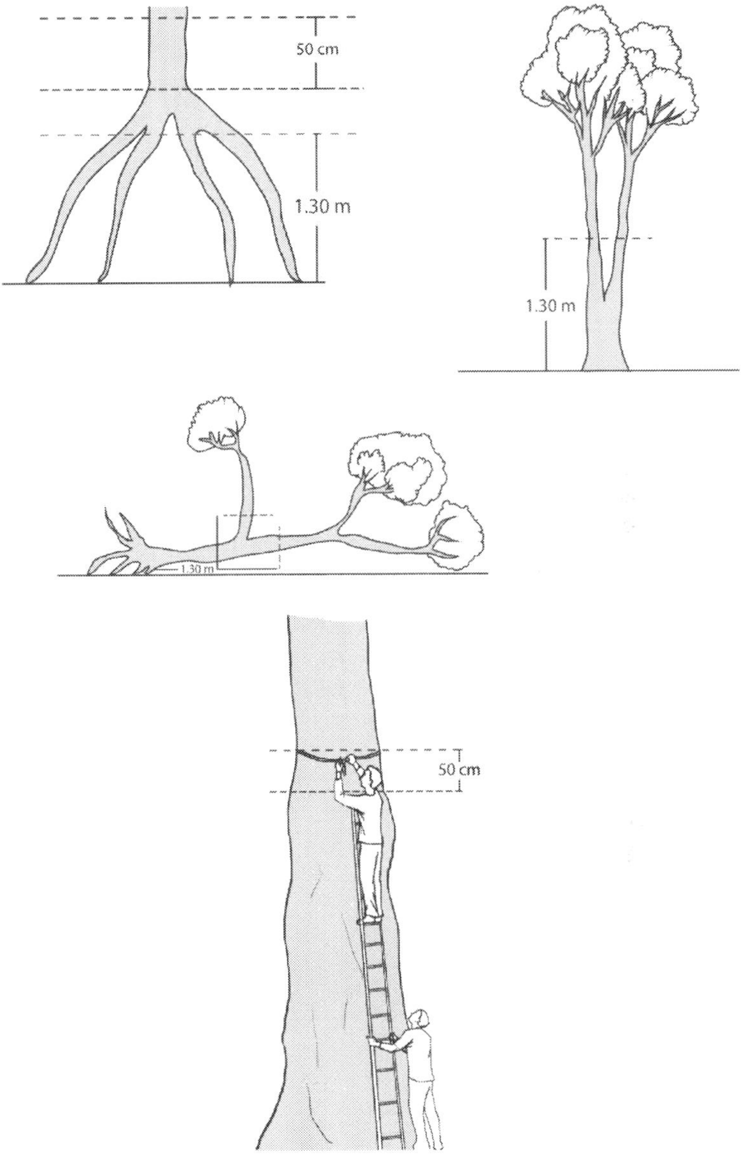

Figure 2. Details for marking trees with deformities in the field. A – For tabular roots, the POM is measured 50 cm above the last root; B – For multiple trunks, each is measured as a separate individual, provided the forking is below 1.30 m; C – For fallen trees, the distance is taken from the underside; D – For tall trees, the measurement should be done with the support of modular ladders.

Figure 3. Details of marking big trees at Rio Doce State Park: Use of a modular ladder up to 12 meters and POM painting process using a stencil. Source: Metzker, T.

With the collection and identification of botanical material, local guidebooks can be developed for the identification of trees registered within the PMPs. The guidebook could include photos of dried plants, taxonomic identification, location of the species, whether or not there are medicinal purposes, and details about flowers or fruits. In collaboration with local experts, the production of this type of material strengthens relationships between project managers and the execution team, in addition to producing registered material that is easily understood by the local population.

2.6. Calibration of diameter tape

As a result of the measurement process, diameter tape can become stretched or it may come from the factory already with small defects. Considering that the annual growth rate of a tree stratum in the forest is ~0.2 cm/year [14] small measurement errors can have a strong impact on the final results. In order to avoid this type of error, the diametric tape should be calibrated using an aluminium ruler prior to each census, thus maximizing the level of precision in the results.

2.7. Measurement calibration

Errors in reading the diametric tape or errors in the position of the tape on the tree can be common during the census, negatively impacting the processing and analysis of data.

Therefore, prior to each census, it is also necessary to calibrate the technician responsible for measuring the trees.

On the first day of the census, all possible measurements should be completed within a given PMP. One or two days later, the same person who measured the trees on day one should return to the same area and re-measure all of the previously measured trees. The results are considered good if the one measuring the trees obtains a minimum of 70% accuracy, or 90% with less than 1 mm of error. If these parameters are not reached, the procedure is repeated, even with others doing the measuring, until the required precision is obtained. The objective in each phase is to minimize potential errors that generally occur in field activities and which substantially impact data analysis.

2.8. Census and re-census

The measurement of the individuals located in a PMP is the heart of the entire initiative. The measurements conducted during the first census should be done with careful attention so that the complete methodology for measurement and marking is constantly being verified and validated. Despite the fact that there are technicians responsible for data collection who are fully trained in the methodology, a copy of the measurement protocol and its specifications should be available for consultation in the field.

It is important to remember that the period for plot measurement (completion of the first census or re-censuses) should be defined by the analysis of a series of rainfall in the region under study so that the measurements can always be done at the same time of year, that is, in the month that has the least amount of precipitation. This strategy seeks to take advantage of the best transportation logistics, generally by ground, and to avoid the influence of rain in the diameter measurements since tree bark can become saturated with water, thus affecting/falsifying growth data.

For the individual measurement of trees, it is recommended that diametric tape (e.g. Diameter Tape – Forest Suppliers) be strictly used. The use of tapes that measure the circumference of individual trees, in order to later convert to diameter, increases estimation errors. The technician responsible for measurement should note, tree by tree, any loose bark, lichens, lianas or other factors that could impact diameter measurement. The technician cleans the measurement area by passing his/her hand along the trunk and then runs the diametric tape around it. Also responsible for worksheet data, the technician should seek to assist the one who is measuring the trees, primarily during the evaluation of large trees, in order to verify the correct position of the tape and to determine if there is anything between the tape and the tree.

During the annual re-censuses, the technician responsible for recording data in the worksheets should pay even greater attention to the data that are found to be divergent from the previous year's records, which could likely be due to an error in reading the diametric tape. If an error is found, the technician should ask for a re-measurement and a re-reading of the diameter for recording in the worksheet.

Another important activity undertaken during the re-censuses is an active search throughout the PMP for new individuals to be included in the sample (recruits) and individuals that no longer exhibit vegetative activity (dead). All of the new trees, palms and lianas that have met the inclusion criteria (DBH ≥ 10 cm) are included in the sample and the same marking methodology is followed. Individuals marked in the first census but which, during the re-census, did not exhibit vegetative activity or were not found after a detailed sweep of the plot, should be considered dead.

It is also possible that trees that had died in the previous year show activity through diametric growth or new growth. In this case, the processing worksheet should be modified, correcting the data recorded the previous year and including this individual once again in the sample since it was not actually dead.

3. Analyzing data

3.1. Tabulation of data

For all field activities related to planning, implementation and monitoring of PMPs, there should be specific worksheets. The standardization of the entry of information that will be generated is of fundamental importance to guarantee the quality of the data. Each worksheet should include the following information at minimum:

- PMP name and abbreviation;
- Complete date when the collection was done;
- Names of each of the team members;
- Number of each individual;
- Registration number of the sub-plot to which each individual belongs;
- Data related to the POM and DBH;
- Pertinent observations.

Upon completion of the field work, all of the worksheets used should be digitized, scanned, and saved in a digital file and then stored in a dry, safe place. These procedures assure that the original worksheets can be consulted in the case of duplicate or conflicting information, when typing errors occur, or when mistakes are made in noting information in the field. After digitizing the worksheet data, the new worksheets should be printed and evaluated by pairs for accuracy, followed by the correction of any confirmed errors.

3.2. Spatial mapping

Spatial mapping of the individuals marked in the PMPs allows for the possibility of analyses of the distribution of species or guilds in the forest. For these analyses, indices of aggregation, such as Morisita [15] or McGuinness [16], can be used, thus defining the spatial distribution of the individuals as aggregate, random or regular. This knowledge is fundamental to ecological analyses as it facilitates an understanding of how a certain species uses available resources in the forest. While the aggregation factor can vary within a species, in different diametric classes, it shows how the life stages of an individual can change the way it uses an available resource.

For mapping, each individual should have its Cartesian coordinates X and Y measured in the PMP. The distances can be measured using a 50 meter measuring tape or a digital measuring stick. It is important that a compass always be used to support the measurements so that the distances are consistently taken in a straight line with respect to the position within each sub-plot. In the example below (Figure 4), the individual marked in the PMP has Cartesian coordinates of X = 56.2 meters and Y = 74.3 meters.

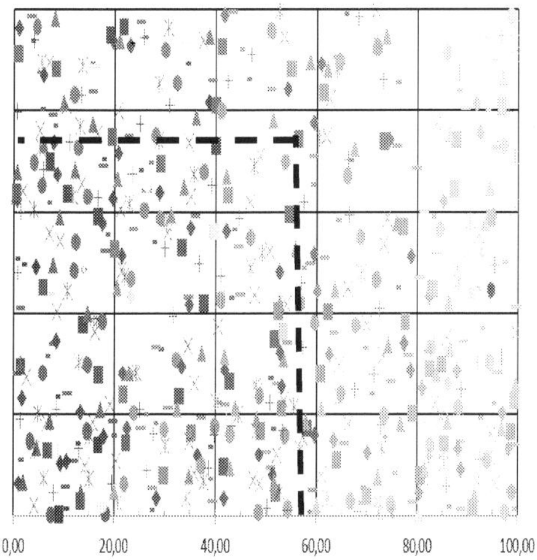

Figure 4. Example of the result of spatial mapping of the field individuals within the PMP at Rio Doce State Park – Minas Gerais, Brazil.

3.3. Estimates of biomass and carbon stocks

The estimates of aboveground live biomass and the resulting carbon stocks can be obtained using two key methods. The first, based on destructive sampling (direct method), involves cutting, drying and weighing separately (roots, trunk and leaves) all of the trees in a specific area. This technique becomes unviable in the case of monitoring since it can damage the sample over the life of the vegetation. The second method (indirect method) consists of estimating biomass and carbon stocks by measuring field variables without having to fall the tree. In this case, DBH data and/or total height of the trees (Ht) and/or specific density of the wood (p) are inserted into previously developed allometric equations in order to estimate the biomass and carbon stocks of the PMP.

Table 1 shows examples of allometric equations already developed and that can be used to calculate biomass. The selection of the best equation should be based on the objective of the project and on the questions to be answered. Allometric models that offer greater precision should be given preference [17].

Types	Allometric Equations	R^2
Wet Forest [18]	$EXP(-2.557 + 0.940 * LN(p * DBH^2 * Ht))$	0.99
Moist Forest [18]	$EXP(-2.977 * LN(p * DBH^2 * Ht))$	0.99
Dry Forest [18]	$EXP(-2.187 + 0.916 * LN(p * DBH^2 * Ht))$	0.99
Palms [19]	$\frac{EXP\left((5.7236 + 0.9285 * LN(DBH^2)) * 1.05001\right)}{10^3}$	0.82
Lianas [20]	$EXP\left(0.07 + 2.17 * (LN(DBH))\right)$	0.95
Amazon [21]	$EXP(-1.754 + 2.665 * LN(DBH))$	0.92
Amazon [21]	$EXP(-0.151 + 2.17 * LN(DBH))$	0.90
Tree ferns [22]	$-4266348/(1 - (2792284\, EXP(0.313677 * Ht)))$	0.88
Wet Forest [23]	$EXP(21.297 - 6.953(DBH) + 0.74(DBH^2))$	0.91

[18] – Chave *et. al.* (2005);
[19] – Nascimento & Laurance (2002);
[20] – Gerwing & Farias (2000);
[21] – Higuchi *et. al.* (1998);
[22] – Tiepolo *et al.* (2002);
[23] – Brown *et al.* 1997.

Table 1. Example of allometric equations used to estimate the aboveground biomass (kg) of trees, palms and lianas in different tropical forest types. DBH – Diameter at breast height; Ht – Total height; and p – Wood mean density g/m^3.

In order to conduct accurate comparisons with other areas or to serve as a potential indicator of carbon stocks for a specific region, simpler allometric equations with only one variable – DBH can be used [17]. In this case, it is not necessary to collect data related to the height or wood density of individuals, resulting in the inventory being completed much faster. An important detail regarding the selection of the equation is that the results for some are fresh biomass data, while for others they are dry biomass data, and still others provide results as carbon quantity.

As previously mentioned, the ideal would be to use an allometric model that provides the highest degree of confidence. The best model has to explain most of the variation in the data or has the lowest AIC (Akaike Information Criterion). In the most cases, equations that use multiple entries with 3 variables per individual (DBH, Ht and *p*) are better. DBH data are easily collected as previously outlined. The data related to tree height are generally complicated to collect due to error associated with height estimations, in addition to the need for greater time in the field, which results in inventories having higher costs. In order to optimize this work, an estimate of tree height can be used by creating an allometric equation adjusted by the diametric and height measurements of a specific number of trees in the plot (Figure 4). This requires the collection of height data for a certain part of the plot. These data should be collected with the greatest precision possible, using cords, ladder or equipment such as a rangefinder. It is recommended that height be measured for a random sample of 20% of the individuals of a PMP in order to later relate them to the diameters, producing an equation for site-specific heights (Figure 4). In order to collect data for specific wood density, there are some protocols for extracting and obtaining these values for each tree in a PMP. With a view to obtaining economies of time and project resources, existing databases can be used, for example, Global Wood Density Database [24, 25], which makes available a series of wood density values for species that exist in almost every part of the world.

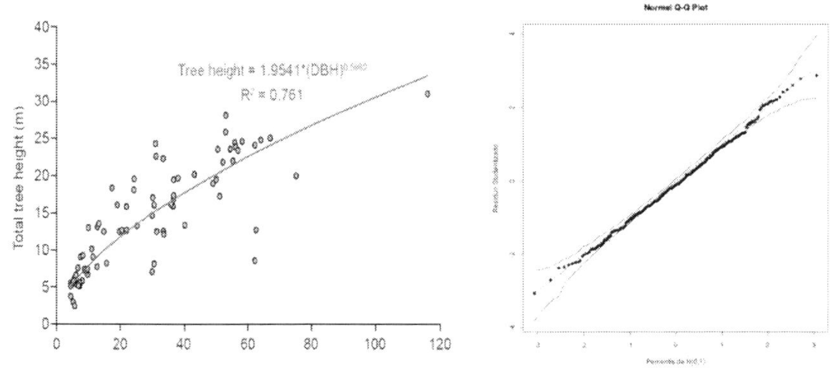

Figure 5. Examples of the development of a site-specific equation for the calculation of height using tree diameters [17] and of equation adjustment using the observance of the normality of residues.

With biomass calculated, many different possibilities for analysis become available. For example, comparisons of biomass can be done between primary and secondary areas, between one year and another, and total biomass can be calculated for the PMP and extrapolated to large forested areas of the same typology. In addition to comparing the relative data to the average annual increment of biomass (or of carbon) of a PMP, analyses of the change in biomass between years can also be conducted. This can be achieved by subtracting the biomass in year one from the biomass in year 0, remembering that this biomass value should include the biomass of recruits in year 1 while the biomass of individuals considered to be dead is subtracted. Another factor that can be considered is the number of days between each census in order to standardize the calculations for a specific period. For example, for 1 year, the following equation would be used:

$$\left(\frac{AGBt2-AGBt1}{DTt2-DTt1}\right) * 365; \quad (1)$$

where AGBt2 refers to biomass in year 2, and AGBt1 to biomass in year 1. DTt2 refers to the date the census was taken in year 2, and DTt1 to the date the census was taken in year 1 (D. Clark personal communication).

Table 2 shows the data for aboveground biomass (AGB) for different neotropical forest sites (adapted by Alves, 2010 [26]). The highest values were primarily found in the Brazilian Amazon (Manaus and Santarém).

Sites	AGB (Mg.ha^{-1})	Reference
Submontane moist semideciduous secondary forest, Marliéria, Brazil	92.0	[27] Metzker et al. (2011)
Submontane moist semideciduous secondary forest, Marliéria, Brazil	107.0	[27] Metzker et al. (2011)
Lowland seasonally dry forest, Mexico	109.0	[28] Vargas et al. (2008)
Lowland wet forest, La Selva, Costa Rica	148.7	[11] Clark and Clark (2000)
Seasonally flooded forest (Restinga), Ubatuba, Brazil	154.0	[26] Alves et al. (2010)
Montane wet forest, Venezuela	157.0	[29] Delaney et al. (1997)
Montane moist forest, Venezuela	173.0	[29] Delaney et al. (1997)
Submontane moist semideciduous primary forest, Marliéria, Brazil	174.0	[27] Metzker et al. (2011)
Lowland moist forest, Venezuela	179.0	[29] Delaney et al. (1997)
Lowland moist forest, BCI, Panama	179.1	[30] DeWalt and Chave (2004)
Submontane moist semideciduous primary forest, Marliéria, Brazil	179.8	[27] Metzker et al. (2011)
Lowland moist forest, Ubatuba, Brazil	198.4	[26] Alves et al. (2010)
Submontane moist semideciduous primary forest, Marliéria, Brazil	201.0	[27] Metzker et al. (2011)
Lowland wet fores, La Selva, Costa Rica	203.2	[30] DeWalt and Chave (2004)
Lowland forests, SW Amazonia (Bolivia, Peru)	206.7	[31] Baker et al. (2004)
Lowland forests, NW Amazonia (Peru, Ecuador)	220.8	[31] Baker et al. (2004)
Submontane semideciduous forest, La Chonta, Bolivia	236.6	[32] Broadbent et al. (2008)
Submontane moist forest, Ubatuba, Brazil	239.3	[26] Alves et al. (2010)
Lowland wet forest, Manaus, Brazill	240.2	[30] DeWalt and Chave (2004)
Lowland moist forest, Rio Branco, Brazil	244.1	[33] Vieira et al. (2004)
Lowland moist forest, BCI, Panama	260.2	[34] Chave et al. (2003)
Montane moist forest, Ubatuba, Brazil	262.7	[26] Alves et al. (2010)

Lowland forests, Central & Eastern Amazonia (Brazil)	277.5	[31] Baker et al. (2004)
Lowland moist forest, Santarem, Brazil	281.2	[33] Vieira et al. (2004)
Lowland semideciduous forest, Roraima, Brazil	292.1	[35] Nascimento et al. (2007)
Lowland moist forest, Santarem, Brazil	294.8	[36] Rice et al. (2004)
Lowland moist forest, Santarem, Brazil	298.0	[37] Pyle et al. (2008)
Lowland moist forest, Rondonia, Brazil	306.8	[38] Cummings et al., 2002
Lowland wet forest, Manaus, Brazil	307.6	[39] Castilho et al. (2006)
Lowland wet forest, Nouragues, French Guiana	317.0	[40] Chave et al. (2001)
Lowland wet forest, Manaus, Brazil	325.5	[19] Nascimento and Laurance (2002)
Lowland moist Cocha Cashu, Peru	332.8	[30] DeWalt and Chave (2004)
Lowland wet forest, Manaus, Brazil	334.0	[37] Pyle et al. (2008)
Lowland semideciduous forest, Linhares, Brazil	334.5	[41] Rolim et al. (2005)
Lowland wet forest, Manaus, Brazil	360.2	[33] Vieira et al. (2004)

Table 2. Estimates of aboveground biomass in different forest typologies on neotropical sites. Adapted by (Alves et al. 2010 [26]). AGB data ($Mg.ha^{-1}$).

3.4. Recruitment and mortality rates

Calculations of the annual rates of Recruitment (Eq. 2) and Mortality (Eq. 3) can be done using the equations by Sheil and Mail [42]. These rates are an excellent indicator of forest dynamics, providing a solid understanding of forest behaviour as it is affected by seasonal events causing variations in water availability, or by extreme climatic events or to conduct multiple comparisons. Since, in reality, everything depends on the proposed objective, forest dynamics can be compared, for example, between those individuals that belong to the higher diametric classes and those who belong to the lower, or the behaviour between different species, among others.

$$R = \left[\left(\left(\frac{No + Nr}{No} \right)^{\left(\frac{1}{t}\right)} \right) - 1 \right] * 100 \qquad (1)$$

$$M = 1 - \left[\left(\left(\frac{No - Nm}{No}\right)^{\left(\frac{1}{t}\right)}\right)\right] * 100; \quad (2)$$

where: *No* equals the number of individuals at time 0; *Nm* is the number of dead individuals between the interval; and *Nr* is the number of individuals recruited in the same time interval *(t)*.

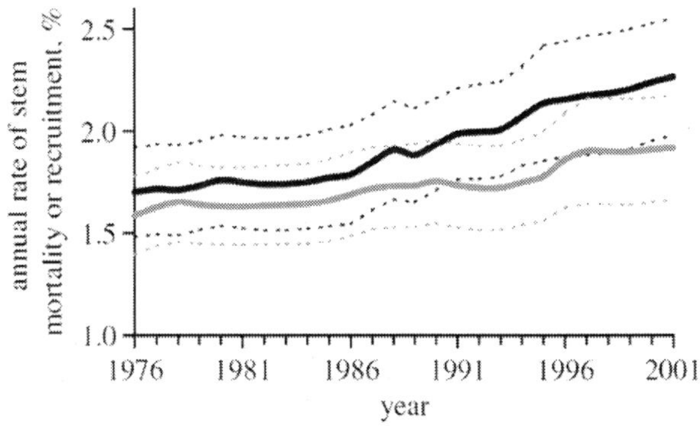

Figure 6. Example taken from Phillips et al., (2008) [43] referring to the analysis of mortality rates (grey lines) and recruitment rates (black lines), using a monitoring time period of 25 years. Solid lines are means and dotted lines are 95% CIs.

4. The valuation of tropical forests

In this section, we will explain how to assign value to carbon stock estimates taken from collected data on PMPs. We will also discuss issues regarding the Payment for Environmental Services (PES), which is provided by tropical forests that are connected with major international protocols and signed agreements.

4.1. Assigning value

Forest conservation strategies to be effective, local communities must first be significantly involved and they must believe in the importance of biodiversity to guarantee quality of life. These communities are the key to a conservationist network. The second step is to invest financially in these initiatives. The project should clearly demonstrate that forest conservation efforts are more economical lucrative when compared with the opportunity costs of using the soil in a given region, for example, for cattle-raising. Therefore, investing in the protection of biodiversity in order to encourage the social and economic development of local communities is one of the best long-term conservation strategies for biodiversity and the ecosystemic services it generates.

One of the difficulties in assigning value to biodiversity and the services it offers is how to specifically quantify this value. First, the value of its natural attributes is immeasurable, such as the services offered by bees when pollinating plantations throughout the world or the atmospheric regulation offered by forests (see, [44]). Thus, the carbon valuation and commercialization market has an advantage, since the prices per ton are already known by the market. Despite being affected by countries' economic changes, a ton of carbon (sequestered or saved) has its own regulations derived from international agreements, such as the Kyoto Protocol or by mechanisms such as the CDM (Clean Development Mechanism) and REDD (Reducing Emissions from Deforestation and Forest Degradation). Therefore, projects that seek to assign economic value to environmental services can include "carbon valuation" as a more precise indicator of the technical reliability of the project.

Forest projects began to participate in the global carbon credit market when companies partnered in order to preserve forests and plant trees with the goal of neutralizing their greenhouse gas emissions [3]. Due to the initial difficulty of negotiating these credits within a regulated market (compliance market), many of these initiatives looked for the voluntary market [3] and other financial transactions that could neutralize their emissions by trees capturing carbon. These new mechanisms opened the door for a wide variety of carbon projects that include voluntary initiatives as payment for the recovery of degraded areas as a means of neutralizing emissions and even responsibility for conserving existing forest areas.

4.2. Development of public policies

These widely diverse ongoing carbon projects have one objective in common: to take advantage of existing market mechanisms in order to assign economic values to rainforests. Today, the REDD+ mechanisms is considered one of the most interesting since it focuses on creating an institutional structure and economic incentives required for developing countries to substantially reduce their CO_2 emissions resulting from deforestation and forest degradation [45].

A practical example of implementing public policies connected to carbon projects is the program called Bolsa Floresta (Forest Fund), created by the state of Amazonas through Law no. 3135 on 05/06/2007. Through this initiative, the Government pays R$50 (~USD $30) per month to registered families who live in State Conservation Units and who have signed a collective agreement to stop deforestation [45]. In the state of Minas Gerais, the Government created an initiative called Bolsa Verde (Green Fund) (Law no. 17.127 in 2008), whose objective is to help conserve native vegetation cover in the State by paying property owners for environmental services if they already preserve or are committed to restoring native vegetation on their properties. In this case, the financial incentive is relative to the size of the protected area, which is a priority for family farms and rural producers. Thus, the REDD+ has a comprehensive rural planning strategy that values rainforests and their recovery, as well as supporting the sustainable development of rural livelihoods [45] and facilitating true socio-environmental gains.

For all of these initiatives works, there must also be reliable data on existing carbon stocks to serve as a baseline for the projects. Permanent Monitoring Plots are technically considered to be the best way to obtain these data. For forest recovery projects, where it is not possible to implement PMPs, they can be implemented using adjacent areas or existing data can even be used to extrapolate biomass values. During the monitoring of carbon projects, the random distribution of PMPs serves as a statistically equivalent sample area for forest recovery monitoring. As an example of other monitoring sites using a standardized methodology we can cite the TEAM network (http://teamnetwork.org/) which has more than 15 monitoring sites in tropical forests. In Brazil we can cite two of these sites: Manaus, Caxiuanã, which have 05 PMPs each. Another success case in the monitoring area is the LBA project (http://lba.inpa.gov.br/lba/), which has a vast network of PMPs in the Amazon forest, that in ten years been able to train more than 500 masters and doctors in Brazil, publishing ~1000 articles in specialized journals.

Regardless of the type of project or the mechanism that is used to implement it, projects that use a ton of carbon (sequestered or saved) as the base, guarantee the long-term presence of these stocks in nature. But most importantly, these projects require the assured quality of the data they propose to collect. These data should have Measurement, Reporting and Verification (MRV) to guarantee the technical quality of the project (e.g. see the Standard CCBA – Climate Community and VCS - Voluntary Carbon Standard). In order to guarantee viability, these projects should also have local community involvement as a goal, whether in the implementation phase or during monitoring, in order to facilitate the improvement of quality of life and the resulting socio-environmental gains. In addition to facilitating the socio-environmental benefits already outlined, the implementation of local PMPs has a powerful differential: calibrating the calculation of international methodologies with highly reliable data, collected locally and using a standardized methodology [27].

Author details

Thiago Metzker
Programa de Pós-Graduação em Ecologia, Conservação e Manejo da Vida Silvestre, Brazil

Tereza C. Spósito
Departamento de Botânica, Universidade Federal de Minas Gerais (UFMG), CP 486, CEP 31270-970, Belo Horizonte, Brazil

Britaldo S. Filho
Departamento de Geociências, UFMG, Belo Horizonte, Brazil

Jorge A. Ahumada
Tropical Ecology Assessment and Monitoring Network, Science and Knowledge Division, Conservation International, Arlington, VA, USA

Queila S. Garcia
Departamento de Botânica, Universidade Federal de Minas Gerais (UFMG), CP 486, CEP 31270-970, Belo Horizonte, Brazil

Acknowledgement

Support for this research was received through FAPEMIG (Fundação de Amparo à Pesquisa no Estado de Minas Gerais, Process APQ-02183-09), PELD (Long Term Ecological Research – CNPq, Process 520031/98-9), the Tropical Ecology Assessment and Monitoring (TEAM) Network, a collaboration between Conservation International, the Missouri Botanical Garden, the Smithsonian Institution, and the Wildlife Conservation Society, and partially funded by these institutions, the Gordon and Betty Moore Foundation, and other donors, and USF&WS. We thank Edgar Paiva for the illustrations in this chapter, Orbifish Global Solutions for linguistic review and the whole community of Rio Doce State Park. We thank UFMG, ECMVS and IEF (State Forestry Institute) for logistical support. T.M. received a Doctor fellowship from CAPES (Brazil) and Q.S.G. received a scholarship from CNPq (Brazil).

5. References

[1] FAO. Global Forest Resoucers Assessment (2005). Progress towards sustainable forest management. FAO Forestry Paper 147: Food and Agriculture Crganization of the United Nations. 350 p.

[2] Keeling, H.C. & Phillips, O.L. (2007) The global relationship between forest productivity and biomass. Global Ecol. Biogeogr. Vol. 16: 618–631.

[3] MMA (2011). Pagamento por Serviços Ambientais na Mata Atlântica: lições apreendida e desafios. Guedes, F.B. & Seehusen, S.E. (Eds.). Série Biodiversidade 42. 276 p.

[4] Corona, P. et al. (2011). Forest Ecology and Management Contribution of large-scale forest inventories to biodiversity assessment and monitoring. Forest Ecology and Management, Vol. 262: n 11, p. 2061-2069.

[5] Lund, H.G. et al. (1998). Plots, pixels, and partnerships: potential for modeling, mapping and monitoring biodiversity. In: Forest Biodiversity Research, Monitoring and Modeling: Conceptual Background and Old World Case Studies, F. Dallmeier, J.A. Comiskey, (Eds.). Man and the Biosphere Series. Vol. 20: UNESCO & The Parthenon Publishing Group. Carnforth, Lancashire, UK.

[6] Andelman, S.J. and Willig, M. R. (2004). Networks by Design: A Revolution in Ecology. Science. Vol: 305. 2004.

[7] TEAM Network (2010). Vegetation Protocol Implementation Manual. Tropical Ecology, Assessment and Monitoring Network, Science and Knowledge Division, Conservation International. Vol. 1.5.1: 75 p.

[8] Malhi, Y. et al. (2002). An international network to understand the biomass and dynamics of Amazonian forests (RAINFOR). Journal of Vegetation Science. Vol 13: 439-450.

[9] Condit, R. (1998). Tropical Forest Census Plots. Springer-Verlag, Berlin, and R. G. Landes Company, Georgetown, Texas.

[10] TEAM Network. (2011). TEAM Network Sampling Design Guidelines. Tropical Ecology, Assessment and Monitoring Network, Science and Knowledge Division, Conservation International, Arlington, VA, USA.

[11] Clark, D.B. & Clark, D.A. (2000). Landscape-scale variation in forest structure and biomass in a tropical rain forest. Forest Ecology and Management, Vol: 137. 185–198.

[12] Clark, D.A. (2002). Are tropical forests an important carbon sink? Reanalysis of the long-term plot data. Ecological Applications. Vol 12: 3–7.

[13] Chave, J. et al. (2004). Error propagation and scaling for tropical forest biomass estimates. Phil. Trans. R. Soc. Lond. B. Vol 359: 409–420.

[14] Schaaf L. B. et al. (2205). Incremento diamétrico e em área basal no período 1979-2000 de espécies arbóreas de uma floresta ombrófila mista localizada no sul do Paraná. Floresta. Vol 35:2.

[15] Morisita, M. (1959). Measuring of the dispersion of individuals and analysis of the distributional patterns. Men. Fac.Sci. Kyushi Univ., Ser. E (Biol.) Vol. 2: n. 4, p. 215 - 235.

[16] McGuinnes, W.G. (1934). The relationship between frequency index and abundance as applied to plant populations in a semi-arid region. Ecology. Washington. Vol 16: p. 263-282.

[17] Vieira, S.A. (2008). Estimation of biomass and carbon stocks: the case of the Atlantic Forest. Biota Neotrop Journal. Vol. 8: n 2. ISSN 1676-0603. Available at: http://www.biotaneotropica.org.br/v8n2/pt/fullpaper?bn00108022008+en

[18] Chave, J. et al. (2005) Tree allometry and improved estimation of carbon stocks and balance in tropical forests. Oecologia. Vol 145: 87–99.

[19] Nascimento, H. E. & Laurance, W.F. (2002). Total aboveground biomass in central Amazonian rainforest: a landscape-scale study. For. Ecol. Manage. Vol 168: 311–321.

[20] Gerwing, J. J. & Farias, D. L. (2000). Integrating liana abundance and forest stature into an estimate of aboveground biomass for an eastern Amazonian forest. J. Trop. Ecol. Vol 16: 327–336.

[21] Higuchi, N. et al. (1998). Biomassa da parte aérea da vegetação da floresta tropical úmida de terra firme da Amazônia Brasileira. Acta Amazônica. Vol 28(2): 153-166.

[22] Tiepolo, G. et al. (2002). Measuring and monitoring carbon stocks at the Guaraquec¸aba Climate Action Projetct, Paraná, Brasil. In: International Symposium on Forest Carbon Sequestration and Monitoring. Extension Series Taiwan Forestry Research Institute. Available at:
http://www.spvs.org.br/download/monitoramento ingles.pdf, pp. 98– 115.

[23] Brown, S. (1997). Estimating Biomass and Biomass Change of Tropical Forests: a Primer. For the food and agriculture organization of the united nations. Rome, FAO Forestry Paper 134.

[24] Chave J. et al. (2009). Towards a worldwide wood economics spectrum. Ecology Letters. Vol 12: 351-366.

[25] Zanne A. E. et al. (2009). Data from: Towards a worldwide wood economics spectrum. Dryad Digital Repository.

[26] Alves, L. F. (2010). Forest Ecology and Management Forest structure and live aboveground biomass variation along an elevational gradient of tropical Atlantic moist forest (Brazil). Forest Ecology and Management. Vol 260(5): 679-691. doi:10.1016/j.foreco.2010.05.023. http://dx.doi.org/10.1016/j.foreco.2010.05.023.

[27] Metzker, T. et al. (2011). Forest dynamics and carbon stocks in Rio Doce State Park – an Atlantic rainforest hotspot. Current Science. Vol 100(12): 2093-2098. Available at: http://cs-test.ias.ac.in/cs/Volumes/100/12/1855.pdf

[28] Vargas, E. et al. (2008). Biomass and carbon accumulation in a fire chronosequence of a seasonally dry tropical forest. Glob. Change Biol. Vol 14, 109–124.

[29] Delaney, M. et al. (1997). The distribution of organic carbon in major components of forests located in five life zones in Venezuela. J. Trop. Ecol. Vol 13, 697–708.

[30] DeWalt, S.J. & Chave, J., 2004. Structure and biomass of four lowland Neotropical forests. Biotropica. Vol 36: 7–19.

[31] Baker, T.R. et al. (2004). Variation in wood density determines spatial patterns in Amazonian forest biomass. Glob. Change Biol. Vol 10: 545–562.

[32] Broadbent, E. et al. (2008). Spatial partitioning of biomass and diversity in a lowland Bolivian forest: linking field and remote sensing measurements. Forest Ecol. Manag. Vol 255: 2602–2616.

[33] Vieira, S.A., et al. (2004). Forest structure and carbon dynamics in Amazonian tropical rain forests. Oecologia. Vol 140: 468–479.

[34] Chave, J. et al. (2003). Spatial and temporal variation of biomass in a tropical forest: results from a large census plot in Panama. J. Ecol. Vol 91: 240–252.

[35] Nascimento, M.T. et al. (2007). Above-ground biomass changes over an 11-year period in an Amazon monodominant forest and two other lowland forests. Plant Ecol. Vol 192: 181–191.

[36] Rice, A.H. et al. (2004). Carbon balance and vegetation dynamics in an old-growth Amazonian forest. Ecol. Appl. Vol 14: 55–S71.

[37] Pyle, E.H. et al. (2008). Dynamics of carbon, biomass, and structure in two Amazonian forests. J. Geophys. Res. Vol 113: G00B08.

[38] Cummings, D.L. et al. (2002). Aboveground biomass and structure of rainforests in the southwestern Brazilian Amazon. Forest Ecol. Manag. Vol 163: 293–307.

[39] Castilho, C.V.et al. (2006). Variation in aboveground tree live biomass in a central Amazonian Forest: effect of soil and topography. Forest Ecol. Manag. Vol 234: 85–96.

[40] Chave, J. et al. (2001). Estimation of biomass in a Neotropical forest of French Guiana: spatial and temporal variability. J. Trop. Ecol. Vol 17: 79–96.

[41] Rolim, S.G. et al. (2005). Biomass change in an Atlantic tropical moist forest: the ENSO effect in permanent sample plots over a 22-year period. Oecologia. Vol 142: 238–246.

[42] Sheil, D. & May, R.M. (1996). Mortality and recruitment rate evaluations in heterogeneous tropical forests. Journal of Ecology. Vol 84: 91–100.

[43] Phillips, O. L. et al. (2008). The changing Amazon forest. Philosophical Transactions of the Royal Society B Biological Sciences. Vol 363: 1819-1827. Available at: http://www.pubmedcentral.nih.gov/articlerender.fcgi?artid=2374914&tool=pmcentrez&rendertype=abstract.

[44] Artaxo, P. et al. (2005). Química atmosférica na Amazônia: a floresta e as emissões de queimadas controlando a composição da atmosfera amazônica. Acta Amazônica. Vol 35:185-196.

[45] Soares Filho, B. et al. (2012). Challenges for Low-Carbon Agriculture and Forest Conservation in Brazil. IDB Publications 64798, Inter-American Development Bank.

Chapter 8

Effect of *Pseudotsuga menziesii* Plantations on Vascular Plants Diversity in Northwest Patagonia, Argentina

I. A. Orellana and E. Raffaele

Additional information is available at the end of the chapter

http://dx.doi.org/10.5772/50340

1. Introduction

Forests biodiversity conservation is a global concern because they are home to 80% of the biodiversity of terrestrial environments [1, 2, 3, 4]. The replacement of all or part of these ecosystems with monocultures creates mosaics of vegetation, contributing to habitat fragmentation [5]. The new landscape includes more homogeneous vegetation units and may differ in patterns and processes from the original landscape formed by primary or secondary forests.

During the second half of the twentieth century there was an increase of areas subjected to forestations with exotic species. It is estimated that about 187 millions ha were planted worldwide, which represents 5% of global forests [2]. The annual growth of forest plantations worldwide is estimated at 2-3 million ha per year [6]. Sixty percent (60%) of forest plantations are located in four countries: China, India, Russian Federation and United States. In the southern hemisphere, emerging forestry countries are: Brazil, New Zealand, Chile, South Africa, Argentina, Uruguay, Venezuela and Australia [7]. In non-tropical areas, a third of the area of native forest destroyed is used for forest plantations [3].

In the West of Chubut, Río Negro and Neuquén provinces in Argentina, the forested area with exotic species reached 70,000 ha in 2007 [8]. The species used are *Pinus ponderosa, Pseudotsuga menziesii, Pinus radiata* and *Pinus contorta*. In the late 1990's it was estimated that forestation had a rates of 10,000 ha per year [9, 10]. Exotic plantations effects on biological diversity in *Austrocedrus chilensis* forests and mixed shrublands in Temperate Forests of South America are still poorly understood.

Several studies have shown that species diversity decreases in areas of forestations, and seems to depend on the proximity of plantations to native environments and treatments prior to afforest. In Congo, diversity of vascular plants was compared in mixed stands of *Eucalyptus* - *Acacia* - *Pinus*, with secondary forests and the African savannah which showed a reduction in the number of species in plantations understory at compared to the understory of secondary forest. At the plantations edges, however, the loss of species richness was lower, so that proximity to the pre-existing forest help to maintain diversity over short distances [11]. In a conservation rainforest study, the richness of species was found to be lower in *Coffea arabica* and *Elettaria cardamomum* plantations than in the native forest [12]. On the other hand, at landscape scale, there has been that *E. cardamomum* plantations where shrubs and herbs strata were retained, connectivity was maintained among fragmented forest patches. Similar results were found when analyzing the feasibility of employing commercial pine plantations as complementary habitat to conserve threatened species in Chile [13, 14].

Besides the proximity to native environments, the diversity in forestations it was found related to age. For example, in 20 years old *P. radiata* plantations in New Zealand, *Rubus fruticosus* and other generalist species were frequently found while in later succession (40 years), *R. fruticosus* is replaced by several species of shade-tolerant native ferns and shrubs [15]. Similar results were found in other older plantations of *P. radiata* in New Zealand, where diversity of vascular plants was similar in plantations and nearby native forests, which also was confirmed that plantations provided habitat for some species of birds such as *Apteryx mantelli* [16, 17]. In some cases, older plantations increase the supply of habitat, increases in spatial and vertical heterogeneity, increases in light levels, development of organic soil layers and the associated fungal flora [4]. However, there are some studies that show that diversity does not increase with age of forest plantations. The diversity of the beetles ensemble was lower in older plantations of *Picea abies* than in younger plantations [18].

In Chilean temperate forests, where the replacement of native forests by exotic forestations was important, there was a decrease in the distribution of endangered and vulnerable vascular plants [19, 20]. There was also a loss of *Nothofagus* native forest structure, with the disappearance of strata, as well as decreases in species richness of vascular plants [21, 22]. Similar patterns with decreases in vascular plants, beetles and birds species were found in mixed plantations of *P. menziesii*, *P. radiata* and *Pinus sylvestris* installed in *Nothofagus* forests and *P. ponderosa* plantations instaled in the steppe in Argentina [23, 24]. In other studies it was found that the richness and composition of birds was more affected by the structure of plantations than by their tree species composition [25, 26]. While a study on ant assemblages showed that in plantations there are decreases in abundance and changes in the composition of species respect to the nearby steppe [27]. All these studies support the hypothesis that high individual density in forest plantations affect biodiversity within them, and propose lower density of trees as an alternative to improve biodiversity.

Diversity loss in South America temperate forests, is a topic of great interest since these ecosystems are characterized by high levels of endemism, a product of a deep biogeographic isolation with common ancestry biota, as well as its extremely heterogeneous floristic composition, derived from various biogeographical sources (e.g. Gondwana, Neotropical, Boreal) [28, 29]. The vascular flora has about 34% of woody genera endemic. Most of the endemisms are monotypic with only one species per genus [28]. *Austrocedrus chilensis* is an endemic monotypic species of South America Temperate Forests, with a smaller distribution area than that occupied in ancient geological times [30, 31]. At the present, this species has serious conservation problems due to multiple anthropogenic disturbances, and is included in the IUCN Red List in the "vulnerable" status [32]. This chapter presents some results related to vascular plants diversity in *A. chilensis* forests, and mixed shrublands when they are replaced by the exotic conifer *P. menziesii* plantations.

1.1. Hypothesis

In *P. menziesii* forestations in Patagonia there is reduced vascular plant diversity compared with the natural communities they replace.

1.2. General aim

Study and compare the vascular plants diversity in *P. menziesii* forestations and *A. chilensis* forests and mixed shrublands adjacent in the Northwest of Chubut Province and Southwest of the Río Negro Province, Argentina.

1.3. Specific aims

1. To estimate the alpha and beta diversity of vascular plants in *P. menziesii* plantations and contiguous *A. chilensis* forests and mixed shrubland.
2. To analyze the similarity in the composition and abundance of herbaceous and woody shrubs growing in *P. menziesii* forestations, *A. chilensis* forests and mixed shrublands.

2. Methods

2.1. Study system

The study area includes the West of Chubut and the Southwest of the Río Negro Provinces, in Argentina, between the localities of Corcovado 43º 32' 36.54" South, 71° 26' 37.5" West and San Carlos de Bariloche 41º 8' 16.83 " South, 71° 17' 12.09" W (Fig. 1). In this area there are about 103 *P. menziesii* plantations. The planted surfaces vary between 0.5 and 12 ha, and initial densities are 1.000 trees per ha [9, 33]. The age of the plantations of *P. menziesii* studied for 2006, ranged between 17 and 35 years old, and all had reached reproductive maturity. *P. menziesii* is native to North America where it is distributed between 55° and 19º N, in temperate climates [34]. In Patagonia, Argentina, *P. menziesii* plantations were installed in a range of precipitations between 1500 mm to 600 mm. In this area, various native plant communities were replaced by afforestations, but the best growths are associated with the natural range of the forests of *A. chilensis* and the mixed shrublands, so that these environments have a higher substitution pressure.

The mixed shrublands are characterized by a shrub stratum of 5 to 7 m high, in which the most abundant species are: *Diostea juncea, Lomatia hirsuta, Embothrium coccineum, Schinus patagonicus, Fabiana imbricata* and some isolated trees of *A. chilensis* and *Maytenus boaria*. It is also distinguished is a shrub sub-stratum with similar species composition, and a herbaceous stratum, dominated by species of the families *Poaceae, Asteraceae* and *Rosaceae*. In *A. chilensis* forests a distinguished tree stratum of 15 m in height is found, in which *A. chilensis* is the dominant species. The shrub stratum was mainly composed by *S. patagonicus, L. hirsuta*, and *E. coccineum*, among others. *Asteraceae* and *Poaceae* dominate the herbaceous stratum.

2.2. Sampling design

Four sites were selected where *P. menziesii* plantations were adjacent to mixed shrublands and eleven sites of *P. menziesii* plantations adjacent to *A. chilensis* forests (Fig. 1).

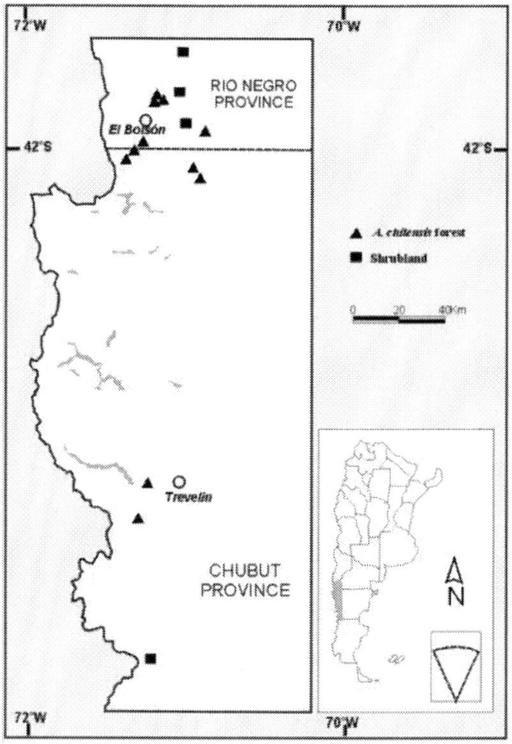

Figure 1. Map of the study area and locations of study sites.

On each site (plantation-native community edge area) a transect perpendicular to the edge line was established (Fig. 2). Each transect was subdivided into 11 plots of 100 m^2, three plots were installed in plantations at -30, -20 and -10 m from the edge line, and eight in the native communities at 10, 20 , 30, 40, 50, 60, 70 and 80 m from the edge line.

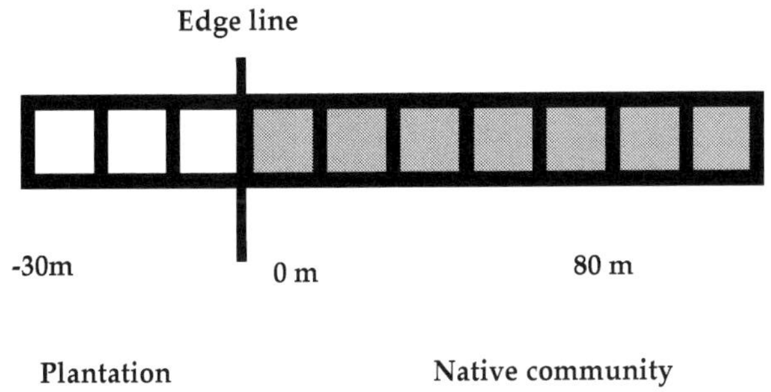

Figure 2. Scheme of a sampling transect established in a plantation-native community edge area, sampling units (plots) and edge line are indicated.

In each 100 m^2 plots, the composition and abundance of vascular plants was registered and mesured, according to the following classification:

1. herbaceous stratum: included herbaceous and woody species seedlings below 10 cm height. The measurements were performed using four 0.5 m^2 circular plots randomly selected within each sampling unit of 100 m^2. The species cover percentage was measured, and individual numbers for woody seedlings counted.
2. shrub stratum: included shrubs, woody vines and woody species saplings above 10 cm in height and below 5 cm in diameter at breast height (DBH). The number of individuals by species was counted on a plot of 25 m^2 randomly selected in each 100 m^2 plot.
3. tree stratum: includes trees and shrubs greater than 5 cm DBH. All individuals were counted in the 100 m^2 plot.

Vascular plant field samples were collected and identified in the laboratory. The reference of support was Patagonian Flora collection volume VIII, Parts I, II, III, IVa and b, V, VI and VII [35].

2.3. Alpha diversity data analysis

The Alpha diversity analysis was made by species accumulation curves, by using EstimateS software [36, 37]. The vascular plants diversity was analyzed in *P. menziesii* plantations and in the adjacent mixed shrublands, as in plantations and *A. chilensis* forests contiguous. The Clench equation was used, which has demonstrated good fits for multiple sampling of taxa of species [$S_n = a*n/ (1+b*n)$], where:

S_n = mean number species for each sampling unit
a = increase rate of new species at the beginning of samplings
b = is a parameter related to the shape of the curve
n = number of sampling units

This function was applied using a nonlinear estimation, through the iterative adjustment of Simplex and Quasi-Newton algorithm with Statistica 7 [38, 39]. The fit of the equation to the accumulation curve, was analyzed by calculating the Determination Ccoefficient: R^2. The slope of the curve (S) when the number of samples is maximum was used to assess sampling quality [$S = a / (1 + bn)^2$] [39]. The flora proportions recorded provided further information on vegetation sampling quality [[$S_{obs.}$/(a/b)] * 100] [39].

2.4. Beta diversity data analysis

The diversity between habitats was analyzed by using the Jaccard similarity index [37]. This index calculated the species replacement degree across environmental gradients. Where environments are very different, and there are no shared species a 0 value occurs. If all species are shared the index value is set to 1.

This study the Jaccard similarity index [$I_J = c / a+b-c$] was obtained for the following pairs of communities: *P. menziesii* plantations (A)-mixed shrublands (B), *P. menziesii* plantations (A) – *A. chilensis* forests (B).

Where:

a = vascular plant species number in A community
b = vascular plants species number in B community
c = vascular plants species number in both A and B communities.

2.5. Analysis of similarity in composition and abundance of herbaceous and shrubs species strata by ANOSIM and MDS

In order to determine if plantations affected the species composition of native communities, the similarity in composition and abundance of herbaceous and shrubs in plantations and native communities were analized by using the multivariate ANOSIM method [40]. This analysis performs permutations on similarity matrices and produces a statistic (R) which is an absolute measure of distance between groups. An R value close to 1 indicates that the assemblages are very different, while a R value close to 0 indicates that the assemblages are similar [40]. To illustrate the assembly of herbaceous and shrub species found in plantations and native communities the Non-Metric Multidimensional Scaling (NMDS) method was applied with the similarity index of Bray-Curtis. These analyzes were performed using the statistical program PRIMER-E Ltd. [41].

3. Results

3.1. Alpha diversity analysis in plantations and native communities

The species accumulation curves obtained from mixed shrublands and *P. menziesii* plantations are presented in Figure 3. Both curves show how species richness increases with increasing number of sampling units. They also show that species richness is higher in mixed shrublands that in contiguous *P. menziesii* plantations.

Model **(1)** describes the species accumulation curve in the mixed shrublands, the model **(2)** describes species accumulation curve in adjacent *P. menziesii* plantations (n = number of sampling units).

$$S_{n\ \text{mixed shrublands}} = 18{,}4 * n / \left(1 + 0{,}2 * n\right) \quad R^2 = 0{,}99 \tag{1}$$

$$S_{n\ P.\ menziesii\ \text{plantations}} = 3{,}3 * n / \left(1 + 0{,}22 * n\right) \quad R^2 = 0{,}99 \tag{2}$$

The Determination Coefficients (R^2) of both models indicate that adjustments to the models accumulation curves are highly representative. Those models were used to calculate the slope of the tangent line when it reaches the maximum number of sampling units. For the species accumulation curve obtained in *P. menziesii* plantations, the slope was **S** = 0.25 (n = 12), so it would have been possible to add new species by increasing the sampling units number. The proportion of vascular plants recorded in *P. menziesii* plantations was 73%. In mixed shrublands, the slope was **S** = 0.33 (n = 32) with a proportion of 86% of vascular plants recorded.

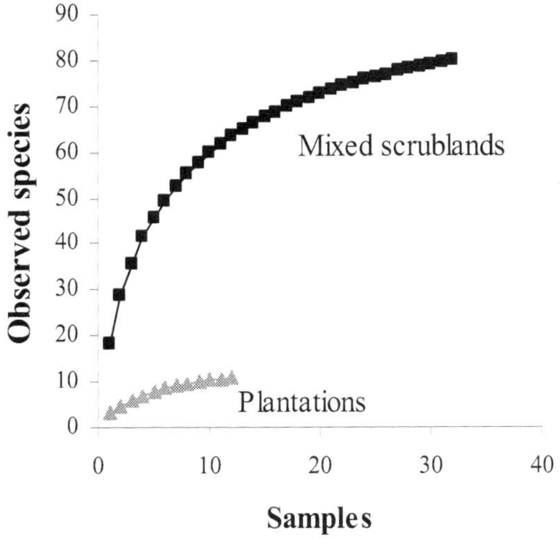

Figure 3. Species accumulation curves obtained from mixed shrublands and the adjacent *P. menziesii* plantations.

The species accumulation curves obtained from *A. chilensis* forests and *P. menziesii* plantations also show a greater species richness in *A. chilensis* forests that in *P. menziesii* plantations (Fig. 4).

Model **(3)** describes the species accumulation curve in *A. chilensis* forests, model **(4)** describes the species accumulation curve in adjacent *P. menziesii* plantations (n = number of sampling units).

$$S_{n\ A.chilensis\ \text{forests}} = 13{,}05{*}n / (1 + 0{,}07{*}n)\ R^2{=}0{,}99 \tag{3}$$

$$S_{n\ P.menziesii\ \text{plantations}} = 5{,}71 * n / \left(1 + 0{,}05 * n\right) R^2 = 0{,}99 \tag{4}$$

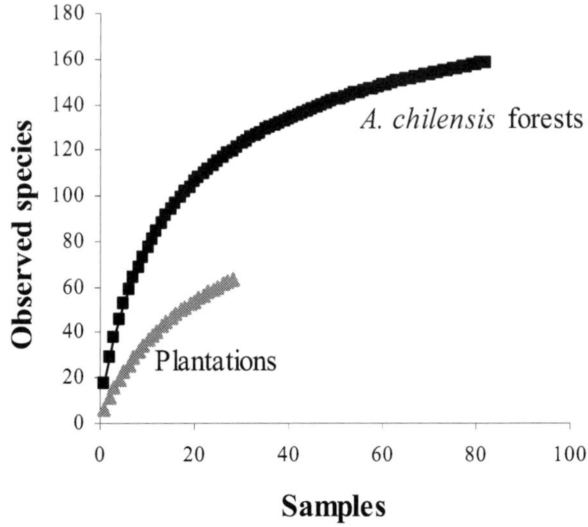

Figure 4. Species accumulation curves obtained from *A. chilensis* forests and the adjacent *P. menziesii* plantations.

The Determination Coefficients (R^2) of both models indicate that adjustments to the models accumulation curves are highly representative. For the species accumulation curve obtained in *P. menziesii* plantations, the slope was **S** = 0.88 (n = 28), so it would have been possible to add new species by increasing the number of sampling units. The sampling registered only 60% of flora of *P. menziesii* plantations (n = 28). In the *A. chilensis* forests contiguous afforestations, for n = 82, **S** = 0.28, so that also have been possible to obtain a greater number of species by increasing the number of sampling units. There, sampling showed 85% of flora of *A. chilensis* forests.

3.2. Beta diversity

When similarity was analyzed between *P. menziesii* plantations and mixed shrublands, the Jaccard index was 0.14 ± 0.02 (± SE), indicating that plantations and mixed shrublands contiguous are dissimilar in species composition. In the same way *P. menziesii* plantations and *A. chilensis* forests showed that both adjoining communities were dissimilar in vascular plants composition, the Jaccard index reached a value of 0.17 ± 0.04 (± SE).

3.3. Floristic similarity between *P. menziesii* plantations, *A. chilensis* forests and mixed shrublands

According to the R value obtained, herbaceous species composition was similar in the three communities studied (R = -0.106, p = 4.3%). However, the confidence level for the analysis was not significant (p <0.1%). The greatest similarities in herbaceous species composition were recorded from *A. chilensis* forests and adjoining plantations, while communities with greater differences in herbaceous species composition are plantations and mixed shrublands (Table 1).

Groups	R	Significance levels (%)
F-S	0,256	95,6
F-P	-0,02	65,4
S-P	-0,366	**99,6**

Table 1. Analysis of similarity (ANOSIM) of herbaceous species found in *P. menziesii* plantations (P), *A. chilensis* forests (F), and mixed shrublands (S). R value is a measure of similarity of species in different communities, R values close to 1 indicate differences, and 0 indicate similarities in species composition. Significant values highlighted in bold ($p < 0.1\%$).

Non-Metric Multidimensional Scaling (NMDS) analysis, showed greater variability in the composition and abundance of herbaceous species in the differente sites corresponding to *P. menziesii* plantations. Most points (sites) were located towards the periphery of the graph and away from each other (Fig. 5). While in the center of the graph the cloud of points corresponding to mixed shrublands and *A. chilensis* forest sites. showed a more similar species composition between both native plant communities. Stress = 0.13, equivalent to a good level of confidence.

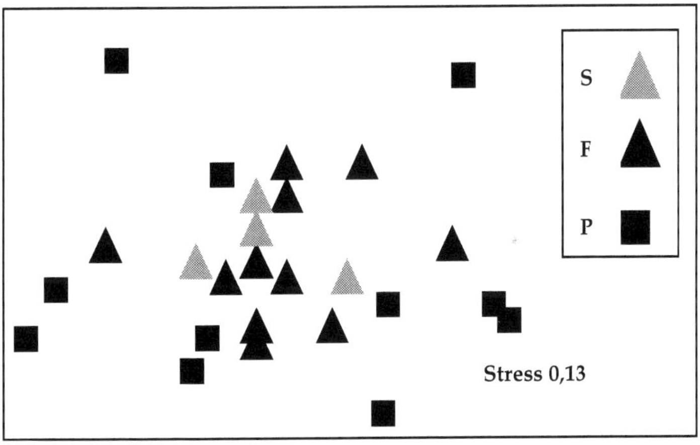

Figure 5. Graphical representation of herbaceous composition and abundance (NMDS) in different communities: F = *A. chilensis* forests, P = *P. menziesii* plantations, and S = mixed shrublands.

Shrub species composition was similar between the communities (R = 0.2, p = 0.4%). The more similar communities to each other on shrub species composition were mixed shrublands and *A. chilensis* forests, but as the above analysis, the confidence level was not significant. The least similar communities were forests and plantations (Table 2).

The NMDS analysis shows two clouds of points, one to the right and the other one to the left of the graph (Fig. 6).The right cloud includes points corresponding to *P. menziesii* plantations. This pattern indicates that composition and abundance of shrub species assemblages is more similar between plantations than between plantations and neighboring native communities. To the left, the points cloud is less scattered, includes sites of *A. chilensis* forests and mixed shrublands. This analysis shows greater similarity in the assembly of vascular plants in *A. chilensis* forests and mixed shrublands.

Groups	R	Significance levels (%)
F-S	-0,13	78,6
F-P	0,34	0,1
S-P	0,20	10,2

Table 2. Analysis of similarity (ANOSIM) of shrubs species found in *P. menziesii* plantations (P), *A. chilensis* forests (F), and mixed shrublands (S). R value indicates the species similarity between the different communities, R values close to 1 indicate differences, and 0 indicate similarities in species composition.

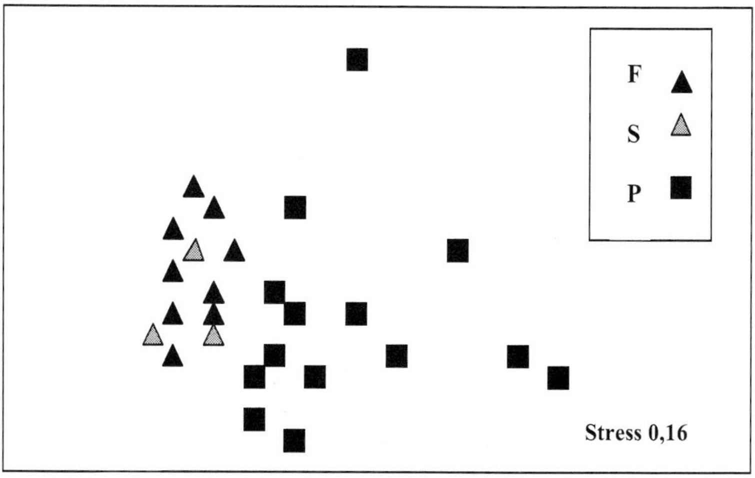

Figure 6. Graphical representation of the ordination (NMDS) of shrub stratum species in different communities: F = *A. chilensis* forests, P = *P. menziesii* plantations, and S = mixed shrublands.

4. Discussion and conclusions

In the plant communities studied, the greatest vascular plant diversity was found in *A. chilensis* forests, where 168 vascular plants species were recorded, while in adjacent *P. menziesii* plantations were 37.5% of this number of species. In mixed shrublands 86 vascular plants species were recorded, and only 13% of this species number in contiguous *P. menziesii* plantations. Species accumulation curves allowed comparison of species richness in adjacent communities: *P. menziesii* plantations- *A. chilensis* forests, and *P. menziesii* plantations- mixed shrublands. Greater species richness was observed in native communities than in plantations, confirming that there is a loss of vascular plant diversity in *P. menziesii* plantations.

While the similarity analysis (ANOSIM) showed a similar species composition between all pairs of communities, these analyses was not significant. The similarity analysis based on the composition and abundance (NMDS) for herbaceous and shrub strata, showed that both mixed shrublands and *A. chilensis* forests are more similar to each other than with the contiguous *P. menziesii* plantations. It is striking that different native communities (forests and shrublands) separated by distances of up to 400 km presented the highest similarity of vascular plants. If there was no effect of plantations on vascular plants composition and abundance, a greater similarity between native community and their corresponding neighboring plantation would be expected. In general terms, this analysis confirms that *P. menziesii* plantations changed the native communities with a noticeable loss of diversity and changes in the abundance and composition of vascular plant.

The vascular plants loss in plantations adjacent to mixed shrublands can be explained by decreases in radiation and environmental heterogeneity in plantations [42]. The environmental heterogeneity is an important factor that promotes biodiversity [23, 43]. Differences in radiation between shrublands and plantations, are stronger than differences between *A. chilensis* forests and *P. menziesii* plantations. So understory species of *A. chilensis* forests, could find a more similar environment in adjacent plantations, than species in mixed shrublands [42]. Furthermore, *A. chilensis* forests are a major source of vascular plants diversity, which may spread to adjoining plantations. While the native communities proximity to plantations promotes species dispersal towards plantations, environmental conditions in plantations prevent the establishment.

The diversity loss found in *P. menziesii* plantations that replaced *A. chilensis* forests and mixed shrublands is similar to the results found in Chile, South Africa, New Zealand and Argentina [11, 19-23, 44, 45]. All these studies support the idea that there is a loss of diversity in forest plantations. This study shows the effects on vascular plants diversity when native ecosystems are replaced by exotic species forestations. In addition to changes in vascular plant diversity in forested areas, there are other edge effects that alter vascular plant structure in the native communities surrounding *P. menziesii* plantations [42]. One of the most important processes recorded in the edge areas is the establishment and dispersion of *P. menziesii* seedlings and saplings from *P. menziesii* plantations [46, 47]. These invasion processes together with diversity loss contribute to native ecosystems degradation.

Author details

I. A. Orellana
Andean Patagonian Forest Research and Extension Center (CIEFAP),
Patagonia San Juan Bosco National University (UNPSJB), Esquel, Chubut, Argentina

E. Raffaele
Ecotone Laboratory, Comahue National University - INIBIOMA/ CONICET,
S. C. de Bariloche, Río Negro, Argentina

Acknowledgement

This manuscript was supported by Universidad Nacional de la Patagonia San Juan Bosco PI N° 560, Centro de Investigación y Extensión Forestal Andino Patagónico (CIEFAP), and PICTO Forestal Nº36879 MinCyT Argentina. We would like to thank J. Monges for his field assistance and Javier Puig, Guillermo Defosse and Mark Austin for their help with the language.

5. References

[1] Naciones Unidas (1992) Convention on Biological Diversity. New York: Environmental Policy and Law.

[2] FAO (2001) State of the World's forests, 2001. Roma, Italia: Food and Agriculture Organization of the United Nations.

[3] Carnus JM, Parrotta J, Brockerhoff EG, Arbez M, Jactel H, Kremer A Lamb D, O'Hara K, Walters B (2003) Planted Forests and Biodiversit. UNFF Intersessional Experts Meeting on the Role of Planted Forests in Sustainable Forest Management, New Zealand.

[4] Brockerhoff EG, Jactel H, Parrotta JA, Quine CP, Sayer J (2008) Plantation forests and biodiversity: oxymoron or opportunity? Biodiversity Conservation, 17, 925–951.

[5] SchelhasJ, Greenberg R (1996) The value of forest patches. In: J. Schelhas y R. Greenberg (Eds.), Forest patches in tropical landscapes (pp. xv – xxxvi). Washington DC: Island Press.

[6] FAO (2006). Global Forest Resources Assessment 2005—Progress towards sustainable forest management. FAO Forestry Paper 147. Roma, Italia: Food and Agriculture Organization of the United Nations.

[7] Sedjo RA, Lyon KS (1990) The long term adequancy of wordl timber supply. Resource for the future.Washington DC

[8] Rusch V, Vila A (2007) La conservación de la biodiversidad en ambientes bajo uso productivo. I reunión sobre forestación en la Patagonia. Esquel, Argentina: CIEFAP.

[9] Davel M (2008) Estimación de la productividad de sitio. In: M. Davel (Ed.), Establecimiento y manejo del pino oregón en Patagonia (pp. 30-42). Esquel, Argentina: CIEFAP.

[10] Schlichter T, Laclau P (1998) Ecotono estepa-bosque y plantaciones forestales en la Patagonia norte. Ecología Austral, 8, 285-296.

[11] Huttel C, Loumeto JJ (2001) Effect of exotic tree plantations and site management on plant diversity. In: F. Bernhard-Reversat (Ed.), Effect of Exotic Tree Plantations on Plant Diversity and Biological Soil Fertility in the Congo Savanna: With Special Reference to Eucalypts, (pp. 9-18). Bogor, Indonesia: Center for International Forestry Research.

[12] Raman TRS (2006) Efects of habitat structure and adjacent habitats on birds in tropical rainforest fragments and shaded plantations in the Western Ghats, India. Biodiversity Conservation, 15, 1577–1607.

[13] Saavedra B, Simonetti JA (2005) Small mammals of Maulino forest remnants, a vanishing ecosystem of south-central Chile. Mammalia, 69 (3-4), 337-348.

[14] Vergara PM, Simonetti JA (2006) Abundance and movement of understory birds in a maulino forest fragmented by pine plantations. Biodiversity and Conservation, 15 (12), 3937-3947.

[15] Brockerhoff EG, Ecroyd CE, Leckie AC, Kimberley MO (2003) Diversity and succession of adventive and indigenous vascular understorey plants in *Pinus radiata* plantation forests in New Zealand. Forest Ecology and Management, 185, 307–326.

[16] Allen .B, Platt KH, Coker REJ (1995) Understorey species composition patterns in a *Pinus radiata* D. Don plantation on the central North Island volcanic plateau, New Zealand. New Zealand Journal of Forestry Science, 25, 301-317.

[17] Kleinpaste R (1990) Kiwis in a pine forest habitat. In: E. Fuller (Ed.), Kiwis, a Monograph of the Family Apterygidae (pp. 97-138). Auckland, New Zealand: Seto Publishing.

[18] Magura T, Tothmeresz B, Elek Z (2006) Changes in carabid beetle assemblages as Norway spruce plantations age. Community Ecology, 7(1), 1-12.

[19] Benoit I (1989) Libro Rojo de la Flora Terrestre de Chile. Corporación Nacional Forestal, Ministerio de Agricultura, Santiago de Chile

[20] Lara A, Donoso PJ, Cortés M (1991) Development of conservation and management alternatives for native forest in south-central Chile. Santiago, Chile: Reporte Final WWF-US-CODEFF.

[21] Von Buch MW, Osorio M (1987) Probleme um die Pinus radiata-Monokulturen in Südchile. Forstarchiv, 58, 249-253.

[22] Frank D, Finckh M (1997) Impactos de las plantaciones de pino oregón sobre la vegetación y el suelo en la zona centro-sur de Chile. Revista Chilena de Historia Natural, 70, 91-211.

[23] Raffaele, E. y Schlichter, T. (2000). Efectos de las plantaciones de pino ponderosa sobre la heterogeneidad de micrositios en estepas del noroeste patagónico. Ecología Austral, 10, 151-158.

[24] Paritsis J, Aizen MA (2008) Effects of exotic conifer plantations on the biodiversity of understory plants, epigeal beetles and birds in Nothofagus dombeyi forests. Forest Ecology and Management, 255, 1575–1583.

[25] Lantschner MV, Rusch V (2007) Impacto de diferentes disturbios antrópicos sobre las comunidades de aves de bosques y matorrales de *Nothofagus antarctica* en el NO Patagónico. Ecología Austral 17, 99-112.

[26] Lantschner MV, Rusch V, Peyrou C (2008) Bird Assemblages in Pine Plantations Replacing Native Ecosystems of N.W. Patagonia, Argentina. Biodiversity and Conservation 17(5), 969-989.

[27] Corley JC, Sackmann P, Rusch V, Bettinelli J, Paritsis J (2006) Effects of pine silviculture on the ant assemblages (Hymenoptera: Formicidae) of the Patagonian steppe. Forest Ecology and Management 222, 162-166.

[28] Arroyo MK, Cavieres L, Peñaloza A, Riveros M, Faggi AM (1996) Relaciones fitogeográficas y patrones regionales de riqueza de especies en el bosque templado lluvioso de Sudamérica. In: J. Armesto, C. Villagrán y M.K. Arroyo (Eds.), Ecología de los bosques nativos de Chile (pp. 71-92). Santiago, Chile: Editorial Universitaria.

[29] Armesto JJ, Lobos PL, Arroyo MK (1996) Los bosques templados del sur de Chile y Argentina: Una Isla Biogeográfica. In: J. Armesto, C. Villagrán y M.K. Arroyo (Eds.), Ecología de los bosques nativos de Chile (pp. 23-27). Santiago, Chile: Editorial Universitaria.

[30] Cabrera AL (1971) Fitogeografía Argentina. Boletín de la Sociedad Argentina de Botánica, 14(1), 36-38.

[31] Donoso C, Escobar B, Pastorino M, Gallo L, Aguayo J (2006) *Austrocedrus chilensis* (D. Don) Pic. Ser. Et Bizzarri Ciprés de la Cordillera, Len. In: C. Donoso (Ed.), Las especies arbóreas de los bosques templados de Chile y Argentina: Autoecología (pp.54-67). Valdivia, Chile: Mariza Cuneo Ediciones.

[32] IUCN Red List of Threatened Species (2012). Available: www.iucnredlist.org

[33] Gonda H, Mohr Bell D, Sbrancia R, Lencinas J, Bava J, Monte C, Mortero A, Sieber A (2009) Inventario del bosque implantado de la Provincia del Neuquén. Argentina. Patagonia Forestal, 14 (4), 13-16.

[34] Hermann RK, Lavender DP (1990) *Pseudotsuga menziesii* (Mirb.) Franco. In: R.M. Burns y B.H. Honkala (Eds.) Silvics of Noth America Volume 1 Conifers. Agriculture Handbook 654 (pp. 527-540). Washington, USA: Forest Service.

[35] Correa MN (1971-1999) Flora Patagónica VIII (I, II, III, IVa, IVb, V, VI, VII). Buenos Aires, Argentina: INTA.

[36] Gotelli NJ, Colwell RK (2001) Quantifying biodiversity: procedures and pitfalls in the measurement and comparison of species richness. Ecology Letters, 4, 379-391.

[37] Moreno CE (2001) Métodos para medir la biodiversidad. (Manuales y Tesis 1). Zaragoza, España: Sociedad Entomológica Aragonesa.

[38] Soberón J, Llorente J (1993) The use of species accumulation functions for the prediction of species richness. Conservation Biology, 7, 480-488.

[39] Jiménez Valverde A, Hortal J (2004) Las curvas de acumulación de especies y la necesidad de evaluar la calidad de los inventarios biológicos. Revista Ibérica de Aracnología, 8(31), 151-161.

[40] Clarke KR, Gorley RN (2001) PRIMER V5: User manual tutorial. Plymouth, United Kingdom: Plymouth Marine Laboratory.

[41] Clarke KR (1993) Non-parametric multivariate analyses of changes in community structure. Australian Journal of Ecology, 18, 117-143.

[42] Orellana I (2011) Efecto de borde de las plantaciones de *Pseudotsuga menziesii* sobre comunidades vegetales naturales en el noroeste patagónico. PhD thesis Universidad Nacional del Comahue, Argentina.

[43] Rosenzweig ML, Abramsky Z (1993) How are diversity and productivity related?. In: R.E. Ricklefs y D. Schuler (Eds.), Species diversity in ecological communities (pp. 52-65). Chicago: University of Chicago Press.

[44] Ogden J, Braggins J, Stretton K, Anderson S (1997) Plant species richness under *Pinus radiata* stands on the central north island volcanic plateau New Zealand. New Zealand Journal of Ecology, 21(1), 17-29.

[45] Barbaro L, Rossi JP, Vetillard F, Nezan J,Jactel H (2007) The spatial distribution of birds and carabid beetles in pine plantation forests: the role of landscape composition and structure. Journal of Biogeography 34:652–664.

[46] Sarasola MM, Rusch VE, Schlichter TM, Ghersa CM (2006) Invasión de coníferas forestales en áreas de estepa y bosques de ciprés de la cordillera en la Región Andino Patagónica. Ecología Austral, 16(2), 143-156.

[47] Orellana IA, Raffaele E (2010). The spread of *Pseudotsuga menziesii* in the *Austrocedrus chilensis* forest and shrubland, in the North Western Patagonia. New Zealand Journal of Forest Science, 40: 199-209.

Section 2

Health Related Matter

Chapter 9

Biodiversity and Mental Health

Hector Duarte Tagles and Alvaro J. Idrovo

Additional information is available at the end of the chapter

http://dx.doi.org/10.5772/48345

1. Introduction

Humans depend on goods and services provided by natural environments for a decent, healthy, and secure life [1]. There is an increasing evidence of the health benefits to the people exposed to natural environments [2]. Physical health improvement by exposure to natural environments has been attributed mainly to the access and motivation of people to engage in physical activities (the so-called "green exercise" [3]), although, some controversies still remain [4]. It is well-known the positive association between physical activity and health by improving the physical fitness of people [5], and some studies have reported the beneficial impact of exercising to mental health as well [6].

Other studies have found the association between improved mental health and natural environment exposure by psychological mechanisms of restoration, rather than through mere physical exertion [7], or by enhancing social cohesion [8]. However, most of the studies investigating the association between natural environment exposure and mental health had focused on urban settings (green areas) of developed countries, where social, demographic, and geographic contexts, may be different from those of less developed economies. There is a lack of studies about the association between exposure to natural environments or green spaces and mental health in medium to low-income countries. A report of a cross-national prevalence of major depressive episodes, showed a significant higher lifetime prevalence in high-income countries than medium to low-income countries [9]. However, no significant difference in 12-month prevalence of major depressive episode was found.

The chapter begins with definitions about biodiversity and provides some arguments of concern for its current status. Then, from a theoretical and empirical perspective, it is explained the general relationship of biodiversity with human health, focusing on the association with mental health. A special part of the chapter will be the explanation of the underlying theories that give support to the plausible association between biodiversity and mental health. The particular mental health problem being analysed and explained is depression. Along the chapter, all relevant information for the association of biodiversity with depression will be referred as to what is found or being done in Mexico. The chapter will end with a conclusion about the need for the conservation of the different forms of biodiversity, not only for aesthetic purposes but for the positive impact on human health, despite the gaps in attributing causal effects.

2. Biodiversity

The association between physical environment and health has been known for a long time. In fact, the health and disease process is the result of a permanent interaction of human beings with the environment where they live [10]. The living and physical components of the environment, and the relationships that take place among them, define a particular ecosystem which, when it is disturbed, may produce direct and indirect alterations to the entire set of integrating elements [11]. An ecosystem then, is a complex dynamic group of various living organisms acting as a whole functioning unity [12]. The diverse group of ecosystems, the species living within those ecosystems and the genetic variations within each population, in addition to the process involving their functioning, constitutes what is called biodiversity [13].

Biological diversity or biodiversity refers to the sum of the total biotic variability present in any ecosystem; therefore, it may be estimated in different ways. Although the most common measure is by counting the number of species identified within a time and space frame (known as species richness), there are also other forms of biodiversity measurements. The multidimensional aspect of the concept allows the quantification of biodiversity using three non-exclusive criteria: a) species richness (numeric values of abundance), b) the evenness of their spatial distribution (using biodiversity index), or c) the phenotypic differentiation and genetic variability of the living organisms (at different taxonomic levels) [14]. Approximately, 1.75 millions of species have been identified in the planet, but it is estimated that the real number could be 10 times higher [15]. Ecosystems provide the supporting vital systems for any form of life on Earth, including humans. Not only provide resources for nourishing and fuel, but also they permit the air and water purification, clear and retain toxic substances, degrade waste and recycle nutrients, allow natural and crop pollination, improve soil fertility, buffer out climate change effects, among many other functions and services [1].

With more than 81,000 identified species, and a vast heterogeneity of terrestrial and aquatic ecosystems, Mexico is placed fourth world-wide in biodiversity records. Closed to 10% of the Planet biodiversity lives in Mexican territory, ranking first in reptile diversity and second in mammals, sharing with Brazil the first place in number of ecosystems [16]. In an attempt to estimate the number of species of different taxa (e.g. plants, angiosperms, amphibians, reptiles, birds, mammals, etc.) R Mittermeier created a list of the 17 countries in the world with the greatest diversity, which represents less than 10% of the Planet's surface but host seven out of ten recognised species (Table 1).

CONTINENT	COUNTRIES
Africa	Congo, Madagascar, South Africa
Asia	China, India, Indonesia, Malaysia, Philippines
Australia	Australia, Papua New Guinea
America	USA, Mexico, Brazil, Colombia, Peru, Venezuela, Ecuador

Table 1. Megadiversity countries and the Continents where they are located[17].

Biodiversity, as an important feature of ecosystems, may threaten the continuity of any form of life within when it is affected or diminished. It is estimated that 27,000 species of living organisms are lost annually (about one specie every 20 minutes), which is high above the expected rate of 3 species per year [18]. The global environmental impact due to biodiversity loss has been extensively addressed, but only recently, the focus has been centred on the health consequences of biodiversity loss.

2.1. Biodiversity and health

Human health relies in many ways on biodiversity conservation [19]. When biodiversity is affected, the entire ecosystem destabilizes reducing its resilience capacity, altering the abundance and distribution of living organisms and modifying the interactive relationships among them and with the physical environment as well. In addition, the productivity of the ecosystem is also affected, reducing the benefits that products and services may provide to humans, such as drug biosynthesis from plants and animals [20]. When natural areas are deforested for agricultural use purpose or for new urban settlements, human population becomes exposed to many vectors and species carrying communicable diseases, while limiting the population of natural predators that could exert control over the dispersion of pathogen populations [21].

The main relationship between biodiversity and human health is food provision. However, biodiversity also has direct influence on human health through other pathways not linked to food production [22]. This type of benefits has been observed in urban green spaces, where people reported more psychological benefits and better recovery capacity of mental fatigue as they were exposed to green areas with greater plant diversity [23]. The study conducted in Sheffield, UK, estimated biodiversity as species richness measured by the Gotelli-Colwell index of species density for plants. Total plant richness was the logarithmic-transformed sum estimates for woody and herbaceous plants. Butterflies and bird species were also monitored within the green space, covering a surface of 13 km^2. Psychological well-being was measured by the administration of a questionnaire to 312 peasants about green space usage for cognitive restoration, positive emotional bonds and sense of identity. The study found that exist a direct positive association between psychological well-being and the extension of the green space, but the association was even stronger as biodiversity increased in the green space, independently from their area sizes.

The potential benefits of biodiversity to physical and mental health have been associated mainly with direct contact of people exposed to natural environments and to the presence of urban green spaces [24]. Figure 1 exhibits the places where studies have been reported world-wide about the association between mental health and green spaces. On the other hand, the urbanisation sprawl experienced by most of the countries world-wide prevent people from open and permanent contact to natural environments. This isolation could be related to an increased number of diseases associated with urban pollution, sedentary lifestyle, and the automobile traffic overflow [25]. Therefore, all the economic and technological advantages of living in urban settings, become trades-off that jeopardise human health by modifying the environmental conditions where people live and socialize. In reference [26], it is postulated that real progress in public health will only be possible from a more humane and ecological perspective. This approach should be rooted as two fundamental dimensions of public health, that is, capable of reducing social and health inequalities and at the same time promoting health-sustaining environments. In a classical clinical study [27], it was found that surgical patients recovered faster and required less use of pain-relief medication when they could see trees outside from their room windows, as compared to a control group that only could see the walls of neighbour buildings through the windows in the hospital rooms.

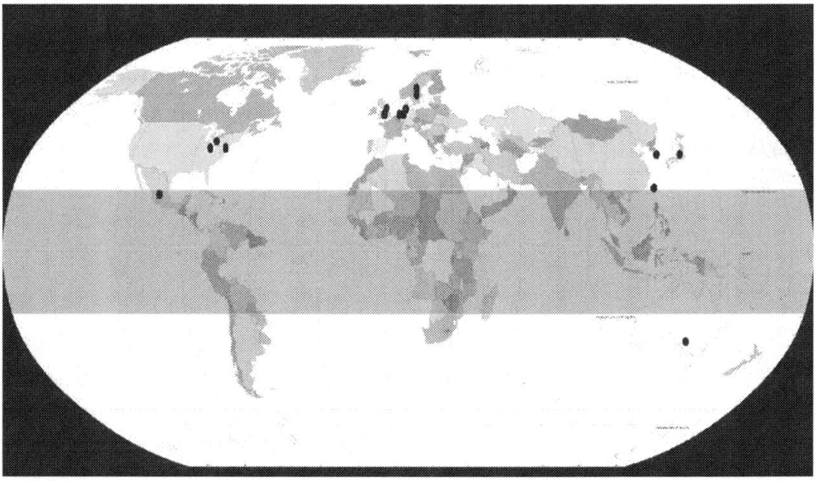

Figure 1. World mapping of the distribution of places were studies about the association of natural environments and green spaces with mental health have been conducted (according to references [2,30,82]).

In another study conducted in the Netherlands [28], it was found that people living in greener areas reported having less illness symptoms, and in general had better self-perception of their health status, including mental health. In such study, the separated effect of urban green spaces, agriculture space and natural environments were analysed, finding the strongest associations of the overall health status improvement with agricultural space living. According to the authors, this feature may reflect a Dutch condition not necessarily shared by other countries, where the green surface in agricultural areas is proportionally greater than in the other two settings. The results of this exploratory study suggested that the adults exposed to more green space (e.g. housekeepers and the elderly) report fewer symptoms, especially as the educational level increases. Recently, the same research team analysed the Dutch National Survey in General Practice to verify if the positive association found between green space exposure and good health status persisted after medical diagnoses [29]. The results indicated that not only the prevalence of 15 different group of diseases medically diagnosed were lower in residential areas with more green space (measured as the percentage of green space distributed around 1 km from the individuals' place of residence), but such an association was stronger for depression and anxiety.

For some authors, the studies linking the association of green spaces and biodiversity exposure with physical and mental health are still inconclusive, especially in urban settings [30]. However, other voices are claiming more conservation efforts, whether to enhance public health or improve aesthetics, despite any conflicting evidence [31].

3. Environmental and health components

According to reference [32], there are 5 characteristics of an area or place that influence individuals' health:

a. Physical features of the environment shared by all residents in a locality;
b. Availability of healthy environments at home, work, school and play;
c. Services provided (public or private) to support people in their daily lives;
d. Socio-cultural features of a neighbourhood; and
e. Reputation of an area.

The first three have to do with the physical infrastructure of the place, whereas the last two are more related with the collective functioning. These categories are not mutually-exclusive and could interact, which in turn will produce different health effects on people according to their particular biological, psychosocial and economic condition. From this perspective, the study of the health effects related to living in a particular place, need to switch the traditional epidemiological paradigm that blames the individual's behaviour as the cause (or causes) of many communicable and environmental diseases. The complexity of the contexts where the health-disease processes take place, turn the conduction of etiologic studies into searching efforts at multiple time-space levels, in order to avoid the constrictions imposed by the traditional epidemiology of proximal risk factors [33]. For example, it is a myth to think that population health is better in rural environments than in urban settings only because we assume rural people is less exposed to risk factors. Studies have demonstrated that despite the health benefits of contact with natural environments, the unfavourable socioeconomic conditions of many rural people could be as an adverse as to practically wipe out any potential benefit of natural exposure [34]. It is therefore important the inclusion of the context approach where specific risk factors take place in studying population health. Those factors that modify certain health condition in a population, act differently according to the level of organization and analysis [35].

The construction and functioning of the physical and social environment of a particular area may help ameliorate or affect the health of its residents both directly and indirectly. The presence of air pollution is an example of direct effect, when airborne pollutants affect the respiratory health of individuals; whereas food provision in good quality and quantity is an example of indirect effect, when malnutrition make individuals more susceptible to any form of infectious diseases. However, individuals not always can decide the best place for their health (with better environmental quality for instance), and often the selection is indirectly determined by social and economic pre-conditions of the individual related to his or her cultural and historic background [36].

The study of the role the physical environment plays in influencing human health is a key issue in public health. According to WHO reports, environmental factors are responsible for about 24% of the total global burden of diseases [37]. In Mexico, the National Health Program 2007-2012 indicates that about 35% of the total burden of diseases is attributable to environmental factors [38]. The increased rate of species extinction along with the degradation of more than 50% of the ecosystem services world-wide jeopardises life quality and the survival of humankind [1].

3.1. Mental health and biodiversity

Although the study of the effects of contact of nature on mental health is recent, the empiric evidence exists some time ago. Authors like Erik Erikson, Harold Searle, and Paul Shepard have explained about the destruction and exploitation of nature by the so-called Western Civilization along the settlement and development of new societies, which in turn made humans more vulnerable and dependent of the emerging conditions [39]. On the other hand, there are studies focusing on the mental health effects of contact with nature in vulnerable populations [40]. In a study conducted on 112 young adults [7], it was found that the exposed group to a natural environment while doing a hike reported less anger and better humour than the group that did the hike in just urban environment. In another study, patients that were exposed to fruit smell and natural scents, reported lower prevalence of depressive episodes [41]. Animal contact has also be an alternative support method for treating psychological disorders. In reference [42], found that patients with moderate depression interacting with dolphins reported lower depression prevalence after two weeks of treatment as compared to the control group.

There are three fundamental theories (developed in the 80's last century), which try to explain the positive effect on mental health of being in contact with nature:

1. **Biophilia**.- Represents an evolution-based theory defined as the innate emotional affinity of human beings to other living organisms and nature. This feature is rooted in the hereditary aspect of human essence [43]. It is hypothesized that this behaviour is determined by a programed genetic sequence along the course of human evolution which enables a positive response to natural environments in accordance with its own survival. This theory holds that even today human beings are attracted by these natural

environments as they are perceived with a sense of "belonging" (identity) and feel they act in a more efficient manner. Reference [44] considers this nature affinity to be bound deep inside human conscience, which emerges in a similar form as other psychic experiences such as myths, poetry and religion, with a vast and complex semiotic as well. This represents the fundamentals of the moral attitude of respect to any form of life and the value of biodiversity. Based on this perspective, a new concept was developed about the affinity towards diversity (ATD), defined as the individual predisposition to appreciate the variant dynamic interaction of human and nature in the everyday situations [45]. ATD has empirically explained that future-oriented individuals and with more socializing behaviours like altruism and cooperation, tend to high rate pro-social orientation that translate into pro-environmental behaviours. Interestingly, this attitude goes beyond passive acceptance or tolerance, but includes an emotional component that expresses the preference for nature, a sense of guilt for natural resource deterioration and discomfort for actions taken by individuals or companies affecting the environment [46].

2. **Attention-Restoration**.- This theory is based on the works of US psychologist W James at the end of the 19th century. According to this theory, in all individuals there are two areas of mental attention, a) direct attention, which is voluntary an intentional, i.e., one concentrates on aspects regarded as important for oneself. Other less important issues are classified as distractions and have to be blocked by the mind, which in turn produces mental fatigue (direct attention fatigue, or DAF); b) indirect attention (called fascination) which is involuntary and automatic, keeping concentration with low or no effort at all. This allows the brain to recover (or restore), before going back to direct attention [47]. Attention-restoration process takes place in the right side of the frontal cortex of the brain, which from an evolutionary standpoint, being alert and focused was necessary for survival. Natural environments provide the best conditions for restoration, as it allows staying away from daily routine, provide opportunity for fascination and pleasure, a sense of openness that invites the individual to explore, and the compatibility of the natural offering to one's own expectations. Moreover, just by observing a natural landscape may help restore the brain before moving to any direct attention [48].
3. **Psycho-physiologic stress recovery**.- This theory is based on the empiric results observed in the positive responses given by individuals exposed to natural environments [49]. According to this theory, the evolution-based ability of humans to recover from a dangerous situation was a natural selection factor that increased the probability for survival. Under stressful conditions, an individual react following a physiologic mechanism pattern known as the "fight or flight response" [50]. This reaction involves catecholamine secretion (including epinephrine) into the bloodstream, which causes muscular tension, rise blood pressure, accelerates pulse rate, constrict blood vessels and increase perspiration. Thus, an individual is prepared to respond adequately when facing a fatal situation, but can restore back to its original levels once the danger has disappeared or being controlled. Some studies have found that contact with nature causes people to lower their stress level, even at a short time after the

exposure has begun. The theory considers such a response due to a limbic-associated inherently reaction of the brain (a part even more ancient than the cortex), which enabled fit individuals to have greater chances of survival during the course of human evolution [51]. In a similar way as biophilia, genetic plays a crucial role in the development of this theory.

These three theories are still under development incorporating new findings of upcoming studies. Restorative theories (attention-restoration and the stress recovery) try to explain the mechanisms by which the brain may recover after a stressful episode or mental fatigue. The main difference between the two is that the former is a more voluntary mechanism that affects the cognitive process (brain cortex), and it is measured by psychological methods, whereas the latter is more an involuntary reaction involving primitive parts of the brain (limbic system), which is measured physiologically [52].

In summary, when there is a "disconnection" of the natural world where humans live and co-exist, many diverse psychological symptoms arise including anxiety, frustration and depression, which cannot be attributed only to intra-psychological or family driven issues. It has been observed that the contact with such natural world, by means of gardening practices, animal petting, green walk or green exercise, not only relief people from depressive symptoms, but increases human capacity to be healthier, strengthen self-esteem, promotes socializing and makes people happier [53]. Although the positive association between natural contact and mental health has been consistently reported, still remains a challenge determining "how close" this "green contact" should be most appropriate [29].

4. Depression

Depression is a frequent mental disorder that currently affects life quality not only of adults, but of younger people like teenagers and children world-wide [54]. It is characterized by an overall depressed mood, with a loss of interest and/or the inability to feel anymore pleasure for things or situations that formerly produced it, loss of self-confidence and a sense of uselessness [55].

Depression diagnostics is based mostly on self-reported symptoms of patients and on clinical observations, taking as standard criteria the Diagnostic and Statistical Manual of Mental Disorders (DSM-IV-TR[1]) of the American Psychiatric Association (APA). This diagnostic tool was designed to be used with populations in different clinical settings, and represents a necessary tool to collect and communicate statistical information for public health with higher precision [56]. DSM-IV-TR provides a rather descriptive nosology than etiological approach, because it relies more on severity patterns and symptoms duration than in the inferences about the causes of the patient's disorder. DSM-IV-TR uses a multi-axial classification for a complete and systematic assessment of the different mental disorders and medical illnesses, psychosocial and environmental problems and the level of activity. Depression belongs to Axis I clinical disorders as mood disorder, which in turn are classified in depressive disorder (unipolar depression), bipolar depression and two other

[1] New version (DSM-V) is expected to be ready by mid-2013 according to APA.

disorder based on etiologic causes (mood disorder caused by other diseases and mood disorder caused by drug use). All depressive disorders (e.g. major depressive disorder, dysthymia, and unspecified depressive disorder) can be distinguished from bipolar disorders because there is no history of previous maniac, mixed or hypo-maniac episodes. In general, unipolar depressive disorders are more prevalent than bipolar cases [57].

4.1. Risk factors

In most patients, depressive episodes occur due to a combination of genetic, biochemical and psychosocial factors. According to [58], factors associated with depression and anxiety in the elderly may be classified in those of biological, psychological and social origin. Among those of biological origin are concurrent chronic diseases, especially cardiovascular (high and low blood pressure), cerebrovascular and psychiatric; atherosclerosis, sleeping disorders, low activity level, obesity, hearing or vision impairment, alcohol consumption, tobacco and drug use, and in general with a poor health condition. Among the risk factors of psychological origin are personal traits such as neuroticism and the history of psychiatric disorders. Finally, social risk factors identified for old individuals are low level of socialization, small and scarce social networking, living alone (no partner or spouse), problems with partner or spouse, partner or spouse on depression, low social support, parental overprotection during infancy, stressful life events in infancy, constant victim of violence, aging, among others. In [59] were identified certain consistent risk factors that suggest at least in part, they are probably causally related to the development of a major depressive disorder, and are being female, having had stressful life events, adverse experiences during childhood (e.g. physical violence, parental absence, dysfunctional family, etc.) and certain personality traits. However, the list does not include genetic vulnerability that predisposes individuals to major depressive episodes, nor the severity of such symptoms in the wide variety of depression forms.

4.2. Epidemiology

Point prevalence of depression world-wide is 1.9% in men and 3.2% in women, while for a 1-year period is 5.8% and 9.5% respectively [60]. In USA, life-time major depression prevalence is estimated to be 10.4% in non-Hispanic whites and 8.0% in Mexican-Americans, but when depression is rather moderate and chronic (i.e. dysthymia), the order reversed probably due to the low socio-economic and education levels [61]. According to the Mexican National Assessment Performance Survey (ENED), major depression in Mexico has a global prevalence of 4.5%, having women more than double of men's prevalence (5.8% vs. 2.5%). It was observed that depression prevalence increased with age but decreased as school level of individuals raised [62]. It is noteworthy that ENED reported that major depression prevalence among women is the same in rural and urban settings, whereas in men, prevalence was higher in rural environments than in urban locations. In addition, no defined pattern could be observed in the distribution of major depression among the 32 Mexican states for men and women.

4.3. Seasonal Affective Disorder (SAD)

There is growing evidence that certain mental health problems develop only during autumn and winter season, remitting on warmer and sunny seasons [63]. In USA, between 4% and 6% of adults experience SAD, while 10% to 20% develop mild forms of the disease at the end of the fall season and beginning of winter [64]. Possible causes have been linked to ocular problems to process daylight and to a deficient melatonin secretion in patients that alters their sleep-wake circadian rhythms [65]. Other studies have shown that SAD is probably also associated with problems in serotonergic transmission, since patients under white light exposure treatment responded favourably [66]. Therefore, SAD could be a morbid condition affecting countries with longer winter seasons, even though the association not necessarily is entirely latitude-dependent, and other risk factors such as genetic susceptibility and socio-cultural context could also be playing important roles [67]. Recently, it has been argued the need to consider SAD as a well-defined psychology disorder, since DSM-IV-TR is still classifying it as a cyclic effect modifier in patients with mood disorders [68].

5. Problem statement

World Health Organisation establishes that it is not possible to improve health without including mental health, because it is a fundamental aspect for life quality [69]. If no action is taken, depression is estimated to be second in disability adjusted life years (DALYs) by 2020 world-wide, and will rank first in developed countries [70]. Recent calls for prevention action have set depression as a global priority, considering not only the burden of the disease in terms of treatment cost, but on the loss of productivity as well [71]. The implementation of preventive measures to treat any disease is always desirable over the usually costlier and bothersome curative methods [52]. However, it remains unknown what is the most effective strategy to reduce depression prevalence; it is still necessary to bear in mind that prevention is one of the first goals of public health. Although there is an increasing research production aiming at studying the association between biodiversity and mental health, it is unknown the existence of specific studies in low-to-medium income countries that focus on contextual determinants associated with depression.

Depression is one of the most important diseases among Mexican adults, being the second mental disorder reported in urban settings, just after alcohol consumption [72]. Some conditions of vulnerability were identified associated with major depressive episodes, such as aging, being women, having low educational levels, and living in socioeconomically deprived areas. There is no doubt about the association between the stressful urban way-of-life and depression in adulthood. However, in the Mexican National Assessment Performance Survey (ENED-2003), data showed the same prevalence of depression symptoms between urban and rural women, but was even higher in rural men than in city men dwellers. In addition, depressive symptoms prevalence distribution per political division (State) was different between men and women, with no clear geographical pattern [62]. For men, the States with the highest prevalence were Jalisco (5%), Veracruz (4.6%) and Tabasco (4.5%), whereas the last two in the list were Nuevo León (less than 1%) and Nayarit (less than 1%). In women, Jalisco was also high with (8.2%) just after Hidalgo (9.9%) and before Estado de México (8.1%). The Mexican States with the lowest prevalence of depressive symptoms in women were Campeche (2.9%) and Sonora (2.8%). In Table 2 is possible to see the results of the total prevalence of affective disorders (including depression) for each of 6 geographic zones identified in Mexico [73]. Of note are the lowest prevalence rates registered in the South-eastern states, where biodiversity and economic deprivation are high [74, 75]

Affective Disorder	Northwest	North	Central West	Central East	Southeast	Metropolitan Areas	TOTAL
Anytime	8.4 [1.6)	9.0 (1.1)	10.2 (1.5)	10.6 (1.6)	5.7 (1.5)	10.4 (0.9)	9.1 (0.6)
Last year	4.5 (0.9)	4.6 (0.7)	5.6 (1.0)	4.9 (0.6)	2.2 (0.6)	5.3 (0.9)	4.5 (0.3)
Last month	2.4 (0.4)	1.9 (0.3)	2.5 (0.7)	2.1 (0.6)	0.9 (0.3)	2.2 (0.5)	1.9 (0.2)

Table 2. Prevalence of affective disorders by geographic zone in Mexico according to reference [73]. Standard error values are between brackets.

It is important to remark that not only humans are under stressful conditions. Planet Earth as a whole is jeopardised on its basic functions due to alterations in its structure, composition and resilience. The UN Convention on Biological Diversity (UNCBD) estimates biodiversity loss currently is close to one thousand times the natural extinction rate, and it is possible to rise in the upcoming years: around 34,000 plant species and 5,200 animal species are in danger of extinction [76]. The Millennium Development Goals entails conservation efforts for biodiversity under its seventh proposal "Environmental Sustainability" [77]. In addition, the conservation and promotion of health-sustaining environments is one of the new challenges of public health intervention [26]. In cases like Mexico, a mega-diversity country, the efforts are more than justified since the benefits to improve population's mental health have been demonstrated.

6. Research evidence

In a recent systematic review, the results of 25 studies analysing the association between green spaces and overall health and well-being were compared, finding positive consistency between exposure and some mental health-related emotions [2]. In Table 3 can be observed the results of the study where the effects of the before-and-after exposure to natural environments were compared among individuals. Consistency of results was lower when the variables were physiologically measured. These meta-analytical findings provide high internal validity to the plausible association; however, the lack of context variability (whether physical or social environments), could limit the external validity of the results. This is what reference [78] calls *psychologistic fallacy,* where individual-level studies lack the inclusion of contextual variables that may explain the apparent variability observed.

OUTCOME	EFFECT SIZE	95% CI	No. STUDIES	RESULTS
Attention	0.23	(-0.30, 0.76)	3	No effect
Energy	0.76	(0.33, 1.22)	5	Improved
Anxiety	0.52	(0.25, 0.79)	6	Improved
Tranquillity	0.07	(-0.42, 0.55)	7	No effect
Anger	0.35	(0.07, 0.64)	6	Improved
Fatigue	0.76	(0.41, 1.11)	4	Improved
Sadness	0.66	(0.66, 1.16)	3	Improved
Diastolic BP	0.32	(-0.18, 0.82)	3	No effect
Cortisol	0.57	(-0.43, 1.57)	4	No effect

Table 3. Results of the effect size (*Hedges g*) of the studies that measured health status before and after exposure to natural environments. OUTCOME = Psychologic/Physiologic variable measured. EFFECT SIZE = Group measure (Hedges g). 95% CI = 95% Confidence Interval. RESULTS = Interpretation of statistical results

Most research on the aetiology of depression and its treatment, have focused on identifying individual risk factors [79]. From a public health perspective though, it is still desirable to keep efforts on preventing the occurrence of depression rather than only in improving diagnostics and treatment efficiency [52]. In a review including more than 30 randomised control trials, it was demonstrated that different preventive interventions can reduce the incidence of major depressive episodes by as much as 50% [80]

In another systematic review of 28 studies [81], the association between physical and social characteristics of the neighbourhoods and depression in adults was analysed. The study found evidence of the negative effects of economic deprivation and the protective effect as this economic condition improved. On the other hand, the association between physical environment and depression was less evident, probably due to the few studies that incorporated the physical dimension of the neighbourhoods. Therefore, socioeconomic characteristics of higher levels of aggregation (such as individuals' place of residence), have a demonstrated effect in the mental health and well-being of the exposed population, acting independently or as effect modifiers of individual risk factor (Figure 2), but this association is less clear with the physical attributes of the environment.

In an ecological study of the association of depressive symptoms prevalence and some biodiversity indicators (measured as non-aquatic animal and plant species richness and green areas) in Mexico, it was observed that at an aggregate-level of analysis, biodiversity was positive related to depressive symptoms [82]. In other words, the study suggests that as biodiversity increases (measures as all non-aquatic species richness) in a state, depressive symptoms increase as well. For this study, data analysed were obtained from different sources. The outcome set of depressive symptoms was taken from the Mexican National Health and Nutrition Survey, ENSANUT-2006 [83]. ENSANUT-2006 was a cross-sectional survey with a probabilistic, multistage, stratified and clustered sampling. The survey collected data from October 2005 through May 2006 on health and nutritional status of the Mexican population, health services quality, public health policy and programmes, and health expenditures of Mexican dwellers [84]. The survey's structure allows representative estimations to the national, state and local levels, for urban and rural areas defined according to the population size (rural settings with less than 2,500 inhabitants; urban settings from 2,500 up to 99,999 inhabitants; metropolitan areas from 100,000). Depressive symptoms in adults were defined as those of men and women aged 20 to 65 years old, who declared having at least 5 of the following symptoms during most of the day for a period of at least one-week (DSM-IV definition of major depressive episode establishes such symptoms over a period of two weeks, therefore, we kept the focus rather on depressive symptoms only): 1) depressed mood; 2) markedly diminished interest or pleasure in almost all activities; 3) significant changes in appetite or weight; 4) insomnia (or hypersomnia in some cases); 5) psychomotor agitation or retardation; 6) fatigue; 7) feelings of worthlessness; and 8) diminished ability to think, concentrate or make decisions [56].

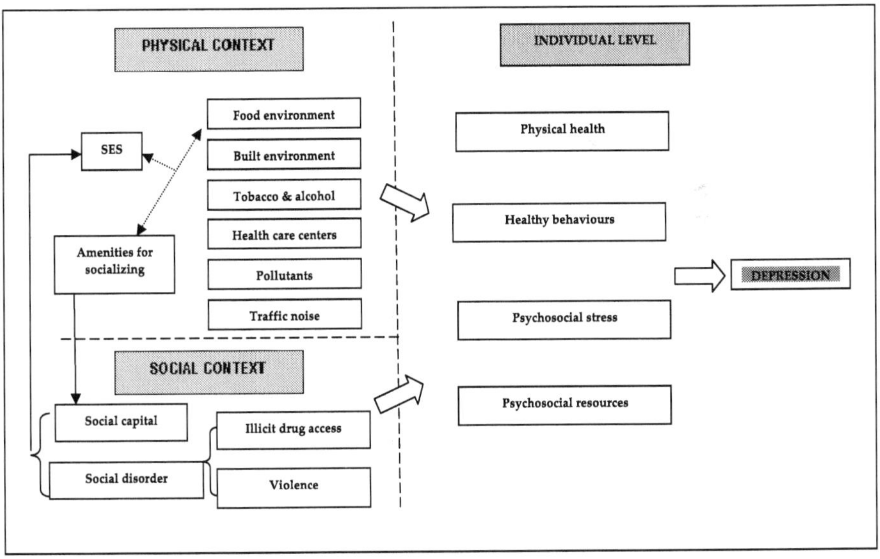

Figure 2. Neighbourhood contextual and individual risk factor model for depression in adult (modified from [81]).

The conditions were built from ENSANUT 2006 aggregating up to a state level the proportion of women, average age, and proportion of self-described as native-indigenous people. From the Mexican Compendium of Environmental Statistics 2008 [85], information was extracted per state for several biodiversity indicators such as animal (non-aquatic) and plant species richness, proportion of reforested land, proportion of natural protected areas, agricultural area and livestock grasslands. We based our ecological measure of biodiversity on the guidelines suggested by the Organisation for Economic Co-operation and Development for building environmental indicators [86].

Natural protected federal area proportions were determined according to the number of states included within its limits. The territory surface area for each state was used to calculate the proportion of green spaces occupied by re-forested, agriculture area-livestock grassland, and natural protected areas in all Mexican states. Animal and plant species richness were summed to account for the total species richness (non-aquatic biodiversity). In addition, economic disparity values were taken as Gini coefficients from the National Population Council [75], as well as the deprivation index data base. Both are measures of unfavourable socioeconomic conditions at group level, the former as an index of income-distribution inequality (the higher the value, the higher the inequality), whereas the latter measures the level of poverty based mainly on education and living conditions. Drug, tobacco and alcohol use data were obtained per state from the Mexican National Addictions Survey 2008 [87], whereas aggregate insecurity perception of individuals in every state was taken from the Mexican National Survey on Insecurity [88].

These unexpected findings are somehow in agreement with the results of a similar study in which a negative association of biodiversity with life expectancy at birth (LEB) was observed in Mexico [89]. Such eco-epidemiological study used 50 environmental indicators with information about demography, housing, poverty, water, soils, biodiversity, forestry resources, and residues were included in an exploratory factor analysis. Four factors were extracted: Population vulnerability/susceptibility, and biodiversity (FC1), urbanization, industrialization, and environmental sustainability (FC2), ecological resilience (FC3), and free-plague environments (FC4). Using ordinary least-squared regressions, it could be observed that whereas FC2, FC3, and FC4 were positively associated with life expectancy at birth, FC1 (biodiversity component) was negatively associated (Table 4). The results showed a South to North gradient inverse to the tendency with LEB. The author recommended including the physical environment as important macro-determinant when studying Mexican population health.

In another study conducted in USA, all-cause mortality in 47 largest USA cities was found to be higher in those having more green spaces [90]. They conclude that it is important the kind of contact that urban residents may have with their natural environment and the form of the green spaces as well, in order to expect the health benefits to the population, otherwise, the sprawling characteristics of USA cities may distort the positive association.

	Total population			Men			Women		
Variable	β	IC95%		β	IC95%		β	IC95%	
FC1	-0.71	-0.76	-0.64	-0.80	-0.88	-0.72	-0.62	-0.68	-0.56
FC2	0.14	0.07	0.21	0.14	0.06	0.22	0.13	0.07	0.19
FC3	0.07	0.00	0.14	0.09	0.01	0.17	0.06	-0.00	0.12
FC4	0.09	0.02	0.16	0.11	0.03	0.19	0.09	0.03	0.15
Adjusted r^2	0.9376			0.9344			0.9393		

FC1: population vulnerability/susceptibility and biodiversity.
FC2: urbanization, industrialization and environmental sustainability
FC3: ecologic resilience.
FC4: environments free of forest plagues.

Table 4. Impact of environmental factors on life expectancy at birth estimated with multiple linear regression models.

7. Conclusions

CE Winslow stated in 1920 that one of the goals of public health was prevention. Nowadays, the number of goals has increased as the population health becomes an emergent property of complex systems [91], but certainly prevention is still at the first place in the list. The challenge would be to find evidence-based effective preventive interventions [92]. Currently the relationship between biodiversity or green spaces and human health is not clear. The bulk of available evidence relating natural environments (with more biodiversity than built environments) and positive health outcomes is mainly based on data from regions with higher income and more development, which are not representative of heterogeneity of countries with less economic and human development. Studies from Latin American countries, Asia and Africa are urgently required to have a full understanding of the relationship, because there is evidence obtained in studies on other determinants of health suggesting a selection bias when data of countries with different levels of economic and human development are not included [93]. The limited evidence from developing countries, as Mexico, on biodiversity and depressive symptoms [82] and life expectancy at birth [89] is contrary to the findings in developed countries. Possible explanations to this difference include the high correlation between social determinants as income inequality, social capital, and level of democracy.

Despite of this methodological limitation to understand the causal relationships between biodiversity and depression, another plausible explanation can be related with latitudinal differences, because biodiversity decreases in regions distant from the tropics, thus, exposure to natural environments can exaggerate the positive effects. Some studies report an association between latitude and affective disorders [94-97]. An alternate explanation is related with the unit of analysis, because results of individual-level studies not necessarily are the same as those observed when analysing populations [98]. Favourable effects of biodiversity on health only have been observed in individual-level studies, whereas adverse effects have been reported in population-level studies [90]. These kinds of results are not surprising, and they are consequence of inherent limitations of science. Epidemiological and psychological studies are unable to detect the effects when low variability is present among the individuals or populations included in the studies since these approaches are based on the comparative methods.

In conclusion, we suggest that exposure to biodiversity can be good for health if the individuals are in built environments with adequate social conditions. These characteristics are frequent in Northern-European and North American countries. In contexts with higher biodiversity, the results can be ambiguous depending of the type of urbanisation [99]. As a consequence, more research in these regions is required because characteristics of the physical environment can be directly or indirectly correlated with social determinants. On the other hand, since different results are observed when studies are with individuals or populations, it is needed to include both approaches in multilevel studies. The inclusion of ecological concepts and methods will be useful to improve the quality of further studies on biodiversity and human health.

Author details

Hector Duarte Tagles
Department of Ecology and Environmental & Industrial Engineering Sonora State University (CESUES), Hermosillo, Sonora, México

Alvaro J. Idrovo
Center for Health Systems Research National Institute of Public Health, Cuernavaca, Morelos, Mexico

Public Health Department, School of Medicine, Industrial University of Santander (UIS), Bucaramanga, Santander, Colombia

8. References

[1] Millenium Ecosystem Assessment (2005). Ecosystems and Human Well-Being: Biodiversity Synthesis. World Resources Institute. Washington, USA.

[2] Bowler DE, Buyung-Ali LM, Knight TM, Pulin AS. (2010). A systematic review of evidence for the added benefits to health of exposure to natural environments. *BMC Public Health* (Open Access); 10: 456.

[3] Pretty J, Peacock J, Sellens M, Griffin M. (2005). The mental and physical health outcomes of green exercise. *Int J Environ Health Res*; 15(5): 319-37.

[4] Maas J, Verheij RA, Spreeuwenger P, Groenewegen PP. (2008). Physical activity as a possible mechanism behind the relationship between green space and health: A multilevel analysis. *BMC Public Health*; 8:206.

[5] Bauman AE. (2004). Updating the evidence that physical activity is good for health: an epidemiological review 2000-2003. *J Sci Med Sport*; 7: 6-19.
[6] Barton J & Pretty J. (2010). What is the best dose of nature and green exercise for improving mental health? A multi-study analysis. *Environ Sci Technol*; 44(10): 3947-55.
[7] Hartig T, Evans GW, Jamner LD, Davis DS, Garlin T. (2003). Tracking restoration in natural and urban field setting. *J Environ Psychol*; 32(2): 109-23.
[8] Seaman PJ, Jones R, Ellaway A. (2010). It's not just about the park, it's about integration too: why people choose to use or not use urban greenspaces. *J Behav Nutr Physical Activity*; 7:78.
[9] Bromet E, Andrade LH, Hwang I, *et al.* (2011). Cross-national epidemiology of DSM-IV major depressive episode. *BMC Medicine*; 9:90.
[10] Álvarez Alva R. (2002). Salud Pública y Medicina Preventiva. 3rd ed. Manual Moderno. México, 472 pp.
[11] Ives AR & Carpenter SR. (2007). Stability and diversity of ecosystems. *Science*; 317: 58-62.
[12] Tyler Miller G. (1994). Ecología y medio ambiente. Grupo Editorial Iberoamérica,S.A. de C.V. México, 867 pp.
[13] Magurran AE. (2010). Q&A: What is biodiversity? *BMC Biology*; 8:145.
[14] Purvis A & Hector A. (2000). Getting the measure of biodiversity. *Nature*; 405: 212-9.
[15] Chivian E & Berrnstein A. (2004). Embedded in nature: Human health and biodiversity. Editorial. *Environ Health Perspect*; 112(1): A12-3.
[16] CONABIO (2008). *Capital natural de México*. Vol. I: Conocimiento actual de la biodiversidad. Comisión Nacional para el Conocimiento y Uso de la Biodiversidad. México.
[17] Mittermeier R, Goettsch C, Robles-Gil P. (1997). Megadiversidad. Los países biológicamente más ricos del Mundo. CEMEX. México.
[18] Wilson EO. (1999). *The diversity of life*. USA: WW. Norton.
[19] Daily GC, Ehrlich PR. (1996). Global change and human susceptibility to disease. *Annu Rev Energy Environ*; 21: 125-44.
[20] Alves R, Rosa IMI (2007). Biodiversity, traditional medicine and public health: where do they meet?. *J Ethnobiology Ethnomedicine*; 3:14.
[21] Chivian E & Bernstein A. (2008). *Sustaining life: How human health depends on biodiversity*. Oxford University Press. USA
[22] Wilby A, Mitchel C, Blumenthal D, Daszak P, Friedman CS, Jutro P, Mazumder A, Prieur-Richard AH, Desprez-Loustau ML, Sharma M, Thomas MB, (2009). Biodiversity, food provision, and human health. Cap. 2. En: Sala OE, Meyerson LA, Parmesan C. (eds.) *Biodiversity change and human health. From ecosystem services to spread of disease*. Scientific Committee on Problems of the Environment (SCOPE). USA, 301 pp.
[23] Fuller RA, Irvine KN, Devine-Wright P, Warren PH, Gaston KJ. (2007). Psychological benefits of greenspace increase with biodiversity. *Biol Lett*; 3: 390-4.

[24] Dean J, van Dooren K, Weinstein P. (2011). Does biodiversity improve mental health in urban settings? *Medical Hypotheses*; 76: 877-80.
[25] Frumkin H (2002). Urban Sprawl and Public Health. *Public Health Reports*; 117: 201-17.
[26] McMichael AJ, Beaglehole R. (2000). The changing global context of public health. *Lancet*; 356: 495-9.
[27] Ulrich R (1984). View through a window may influence recovery from surgery. *Science*; 224(4647): 420-1.
[28] De Vries S, Verheij RA, Groenewegen PP, Spreeuwenberg P. (2003). Natural environments-healthy environments? An exploratory analysis of the relationship between greenspace and health. *Environment and Plannig A*; 35: 1717-31.
[29] Maas J, Verheij RA, de Vries S, Spreeuwenger P, Schellevis FG, Groenewegen PP. (2009). Morbidity is related to a green living environment. *J Epidemiol Community Health*; 63: 967-73.
[30] Lee ACK & Maheswaran R. (2011). The health benefits of urban green spaces: a review of the evidence. *J Public Health*; 33(2): 212-22.
[31] Schultz N. (2010). Country vs. city: Green spaces are better for you. *New Scientist*; 2785.
[32] Macintyre S, Ellaway A, Cummins S. (2002). Place effects on health: how can we conceptualise, operationalise and measure them?. *Soc Sci Med*; 55: 125-39.
[33] McMichael AJ. (1999). Prisoners of the proximate: loosening the constraints on epidemiology in an age of change. *Am J Epidemiol*; 149: 887-97.
[34] Schenker MB. (1996). Preventive medicine and health promotion are overdue in the agricultural workplace. *J Public Health Policy*; 17(3): 275-305.
[35] Diez Roux AV. (2008). La necesidad de un enfoque multinivel en epidemiología. *Región y Sociedad*; 20(2): 77-91.
[36] Verheij RA. (1996). Explaining urban-rural variations in health: a review of interactions between individual and environment. *Soc Sci Med*; 42: 923-35.
[37] WHO (2009). Fact sheets on environmental health. World Health Organization. Available: http://www.who.int/topics/environmental_health/en/. Accessed 2009 December 02.
[38] Secretaría de Salud (2007). *Programa Nacional de Salud 2007-2012. Por un México Sano: construyendo alianzas para una mejor salud*. México, 74 pp.
[39] Shepard P. (1982). *Nature and Madness*. Sierra Club, CA.
[40] Wells NM & Evans GW. (2003). Nearby Nature: A buffer of life stress among rural children. *Environ Behav*; 35(3): 311-30.
[41] Schiffman S. (1992). Aging and the sense of smell: Potential benefits of fragrance enhancement. In: Van Toller & G Dodd (eds). *Fragrance: The psychology and biology of perfume*. Elsevier. England, pp 51-66.
[42] Antonioli C & Reveley M. (2005). Randomised controlled trial of animal facilitated therapy with dolphins in the treatment of depression. *British Medical Journal*; 331(7527), 1231.

[43] Wilson EO. (1984). *Biophilia: The human bond with other species*. Cambridge: Harvard University Press.

[44] Kellert S & Wilson E (eds.) (1993). The biophilia hypothesis. Island Press. USA, pp. 334.

[45] Corral-Verdugo V et al. (2010). *Psychological approaches to sustainability*. Nova Science Publishers.

[46] Corral-Verdugo V, Bonnes M, Tapia-Fonllem C, Fraijo-Sing B, Frías-Armenta M, Carrus G. (2009). Correlates of pro-sustainability orientation: The affinity towards diversity. *J Environ Psychol*; 29: 34-43.

[47] Kaplan R & Kaplan S (1989). *The experience of nature: A psychological perspective*. New York: Cambridge University Press.

[48] Kaplan S (1995). The restorative effects of nature: Toward an integrative framework. *J Environ Psychol*; 15: 169-82.

[49] Ulrich R (1983). Aesthetic and effective response to natural environment. In: Altman I & Wolhwill JF (Eds.). *Behavior and the natural environment*. NY, Plenum Press, 85-125.

[50] McEwen BS, Stellar E (1993). Stress and the individual: mechanisms leading to disease. *Arch Int Med*; 153: 2093-101.

[51] Ulrich R, Simons RF, Losito E, Fiorito E, Miles MA and Zelson M. (1991). Stress recovery during exposure to natural and urban environments. *J Environ Psychol*; 11: 201-30.

[52] Bird W. (2007). *Natural thinking. Investigating the links between the natural environment, biodiversity and mental health*. The Royal Society for the Protection of Birds. UK, 116 pp.

[53] Chalquist C. (2009). A look at the ecotherapy research evidence. *Ecopsychology*; 1(2): 1-11.

[54] Benjet C, Borges G, Medina-Mora ME. (2008). DSM-IV personality disorders in Mexico: results from a general population survey. *Revista Brasileira Psiquiatria*; 30(3): 227-34.

[55] De la Garza Gutiérrez F. (2004). *Depresión, angustia y bipolaridad. Guía para pacientes y familiares*. Ed. Trillas, México, 233 pp.

[56] APA (1994). *Diagnostic and Statistical Manual of Mental Disorders (DSM-IV)*. American Psychiatric Association. 4th ed.

[57] Carney RM, Freedland KE. (2000). *Depression and medical illness*. Ch. 9. In: Berkman LF and Kawachi I (eds.). *Social Epidemiology*. Oxford University Press. USA, 191-212.

[58] Vink D, Aartsen MJ, Schoevers RA. (2008). Risk factors for anxiety and depression in the elderly: a review. *J Affective Disorders*; 106: 29-44.

[59] Fava M & Kendler K (2000). Major depressive disorder. *Neuron*; 28(2): 335-41.

[60] OMS (2001). Informe sobre la salud en el mundo. Salud mental: nuevos conocimientos, nuevas esperanzas. Organización Mundial de la Salud, Suiza. p. 29-30.

[61] Riolo SA, Nguyen TA, Greden JF, King CA. (2005). Prevalence of depression by race/ethnicity: findings from the National Health and Nutrition Examination Survey III. *Am J Public Health*; 95(6): 998-1000.

[62] Belló M, Puentes-Rosas E, Medina-Mora ME, Lozano R (2005). Prevalencia y diagnóstico de depresión en población adulta en México. *Salud Publica Mexico*; 47(suppl 1): S4-S11.

[63] Michalak EE, Wilkinson C, Dowrick C, Wilkinson G. (2001). Seasonal affective disorder: prevalence, detection and current treatment in North Wales. *Br J Psychiatry*; 179: 31-4.

[64] Cunningham A. (2011). Depression common with shorter days, less sunlight. USATODAY.com. Published on 2011 Jan 30.

[65] Wehr TA, Duncan WC Jr, Sher L, Aeschbach D, Schwartz PJ, Turner EH. (2001). A circadian signal of change of season in patients with seasonal affective disorder. *Arch Gen Psychiatry*; 58:1108–14.

[66] Virk G, Reeves G, Rosenthal NE, Sher L, Postolache TT. (2009). Short exposure to light treatment improves depression scores in patients with seasonal affective disorder: A brief report. *Int J Disabil Hum Dev*; 8(3): 283-6.

[67] Mersch PPA, Middendorp HM, Bouhuys AL, Beersma DGM, van der Hoofdakker RH. (1999). Seasonal affective disorder and latitude: a review of the literature. *J Affective Disorder*; 53(1): 35-48.

[68] Rosenthal NE. (2009). Issues for DSM-V: Seasonal affective disorder and seasonality. *Am J Psychiatry*; 166: 852-3.

[69] WHO (2005). *European Declaration on Mental Health*. World Health Organization.

[70] Murray CJ and Lopez AD. (1997). Alternative projections of mortality and disability by cause 1990-2020: Global burden of disease study. *Lancet*; 349: 1498-1504.

[71] Cuijpers P, Beekman ATF, Reynolds CF. (2012). Preventing Depression. A Global Priority. *JAMA*; 307(10): 1033-4.

[72] Medina-Mora ME, Borges G, Benjet C, Lara C, Berglund P. (2007). Psychiatric disorders in Mexico: lifetime prevalence in a nationally representative sample. *Br J Psychiatry*; 190(6): 521-8.

[73] Medina-Mora ME, Borges G, Lara Muñoz C, Benjet C, Blanco Jaimes J, Fleiz Bautista C, Villatoro Velázquez J, Rojas Guiot E, Zambrano Ruíz J, Casanova Rodas L, Aguilar-Gaxiola S. (2003). Prevalencia de trastornos mentales y uso de servicios: resultados de la Encuesta Nacional de Epidemiología Psiquiátrica en México. *Salud Mental*; 26(4): 1-16.

[74] SEMARNAT (2010). Compendio de Estadísticas Ambientales. Secretaría de Medio Ambiente y Recursos Naturales. México. Available: (http://www.semarnat.gob.mx). Accessed: 2012 April 02

[75] Tuirán-Gutiérrez R. (2005). La desigualdad en la distribución del ingreso monetario en México. *Consejo Nacional de Población* (CONAPO). México, 248 pp.

[76] UNCBD (2010). United Nations Convention on Biological Diversity. Available: http://www.cbd.int/2010/biodiversity/?tab=0,1,2 Accessed: 2012 April 02

[77] UN (2005). Millenium Development Goals. United Nations. Available: http://www.un.org/millenniumgoals/ Accesed : 2010 December 05.

[78] Diez-Roux AV. (1998). Bringing Context back into Epidemiology: Variables and Fallacies in Multilevel Analysis. *American Journal of Public Health* 88 (2): 216-222.

[79] Keller MB. (2003). Past, present and future directions for defining optimal treatment outcome in depression: remission and beyond. *JAMA*; 289: 3152-60.

[80] Muñoz RF, Cuijpers P, Smith F, Barrera AZ, Leykin Y. (2010). Prevention of major depression. *Annu Rev Clin Psychol*; 6: 181-212.

[81] Kim D. (2008). Blues from the neighborhood? Neighborhood characteristics and depression. *Epidemiol Rev*; 30: 101-17.

[82] Duarte Tagles H, Idrovo AJ (2012). Biodiversity and depressive symptoms in the tropics: a fsQCA of Mexican data. (Submitted)

[83] Olaiz-Fernández G, Rivera-Dommarco J, Shamah-Levy T, *et al.* (2006). Encuesta Nacional de Salud y Nutrición 2006. *Instituto Nacional de Salud Pública.* México.

[84] Rodríguez-Ramírez S, Mundo-Rosas V, Jiménez-Aguilar A, *et al.* (2009). Methodology for the analysis of dietary data from the Mexican National Health and Nutrition Survey 2006. *Salud Publica Mex*; 51(Suppl 4):S523-9.

[85] SEMARNAT (2008). *Compendio de Estadísticas Ambientales.* México: Secretaría de Medio Ambiente y Recursos Naturales.

[86] OECD (2003). *Environmental Indicators. Development, measurement and use.* France: Organisation for Economic Co-operation and Development.

[87] Rodríguez-Ajenjo C, Villatoro-Velázquez JA, Medina-Mora ME, *et al.* (2009). *Encuesta Nacional de Adicciones 2008.* México: Instituto Nacional de Salud Pública.

[88] De la Barreda L, Ruiz-Harrel R, Sayeg-Seade C, *et al.* (2005).*Tercera Encuesta Nacional Sobre Inseguridad 2005.* México: Instituto Nacional de Estadística y Geografía.

[89] Idrovo AJ. (2011). Physical environment and life expectancy at birth in Mexico: an eco-epidemiological study. *Cad Saude Publica*;27:1175-84.

[90] Richardson EA, Mitchell R, Hartig T, *et al.* (2012). Green cities and health: a question of scale? *J Epidemiol Community Health*; 66:160-5.

[91] Diex-Roux AV. (2007). Integrating social and biologic factors in health research: A systems view. *Ann Epidemiol*; 17: 569-74.

[92] Sackett DL, Rosenberg WM. (1995). The need for evidence-based medicine. *J R Soc Med*, 88: 620-4.

[93] Idrovo AJ, Ruiz-Rodríguez M, Manzano-Patiño AP (2010). Beyond the income inequality hypothesis and human health: a worldwide exploration. *Rev Saude Publica }*; 44: 695-702.

[94] Mersch PP, Middendorp HM, Bouhuys AL, Beersma DG, van den Hoofdakker RH (1999). Seasonal affective disorder and latitude: a review of the literature. *J Affect Disord*; 53: 35-48.

[95] Kegel M, Dam H, Ali F, Bjerregaard P (2009). The prevalence of seasonal affective disorder (SAD) in Greenland is related to latitude. *Nord J Psychiatry*; 63: 331-5.

[96] Levitt AJ, Boyle MH (2002). The impact of latitude on the prevalence of seasonal depression. *Can J Psychiatry*; 47: 361-7.

[97] Bloom DE, Canning D, Fink G (2008). Urbanization and the wealth of nations. *Science*;319:772-5.

[98] Rose G (1985). Sick individuals and sick populations. *Int J Epidemiol*; 14: 32-8.

[99] Rosen LN, Targum SD, Terman M, Bryant MJ, Hoffman H, Kasper SF, et al. (1990). Prevalence of seasonal affective disorder at four latitudes. *Psychiatry Res*; 31: 131-44.

Section 3

Agricultural Aspect

Chapter 10

Amaranthaceae as a Bioindicator of Neotropical Savannah Diversity

Suzane M. Fank-de-Carvalho, Sônia N. Báo and Maria Salete Marchioretto

Additional information is available at the end of the chapter

http://dx.doi.org/10.5772/48455

1. Introduction

Brazil is the first in a ranking of 17 countries in megadiversity of plants, having 17,630 endemic species among a total of 31,162 Angiosperm species [1], distributed in five Biomes. One of them is the Cerrado, which is recognized as a World Priority Hotspot for Conservation because it has around 4,400 endemic plants – almost 50% of the total number of species – and consists largely of savannah, woodland/savannah and dry forest ecosystems [2,3]. It is estimated that Brazil has over 60,000 plant species and, due to the climate and other environmental conditions, some tropical representatives of families which also occur in the temperate zone are very different in appearance [4].

The Cerrado Biome is a tropical ecosystem that occupies about 2 million km^2 (from 3-24° Lat. S and from 41-43° Long. W), located mainly on the central Brazilian Plateau, which has a hot, semi-humid and markedly seasonal climate, varying from a dry winter season (from April to September) to a rainy summer (from October to March) [5-7]. The variety of landscape – from tall savannah woodland to low open grassland with no woody plants - supports the richest flora among the world's savannahs (more than 7,000 native species of vascular plants) and a high degree of endemism [6,8]. This Biome is the most extensive savannah region in South America (the Neotropical Savannah) and it includes a mosaic of vegetation types, varying from a closed canopy forest ("cerradão") to areas with few grasses and more scrub and trees ("cerrado *sensu stricto*"), grassland with scattered scrub and few trees ("campo sujo") and grassland with little scrub and no trees ("campo limpo") [3,9]. Among the grassland areas there are some flat areas with rocky soil, called "campos rochosos", which are considered Cerrado areas because of their flora, especially when located in Chapada Diamantina (Bahia State), a transition area between Cerrado and Caatinga Biomes.

Although the Cerrado is considered a Hotspot for the conservation of global biodiversity, with plant species completely adapted to survive adverse conditions of soil and climate, only 30% of this Neotropical Savannah biodiversity is reasonably well known [8,10]. Coutinho [11] believes that the frequent occurrence of fire is one of the most important factors to determine this Biome´s vegetation, acting as a renewal element that selects structural and physiological characteristics. Nowadays it is believed that more than 40% of the original vegetation has already been converted into human-disturbed areas, due to the expansion of crops [12,13]. This process has accelerated the fragmentation of natural habitat, increasing the pressure on local biodiversity extinction and introducing exotic species, also amplifying soil erosion, water pollution and alterations in vegetation and hydrologic conditions [2,8,14].

The Amaranthaceae family is composed of 2,360 species and here will be listed those that occur in Brazil, emphasizing the Cerrado species and including information on endemism, endangerment and economic or potential use. We also provide a list of the most important bibliographical references for those who are interested in studying the species of this family. Some aspects of morphology, leaf anatomy and ultrastructure will be shown for six species found in the Neotropical Savannah core area (Chapada dos Veadeiros) and some of these aspects, as well as taxonomy and ecology, will be discussed in order to propose the use of this plant family as an indicator of the diversity in open areas of this Biome.

2. Methodology

2.1. List of the Brazilian Amaranthaceae species and those in RPPN Cara Preta

The Brazilian Amaranthaceae list (Table 1) was based on the research by the Brazilian taxonomists Marchioretto [15-21] and Siqueira [22-25] and on the most important taxonomic references to this Family both from the literature (Table 2) and Brazilian Herbaria (Table 3). All cited Herbaria are listed according to the Index Herbariorum [26,27].

The species to be detailed were collected in a Conservation Unit named Reserva Particular do Patrimônio Natural Cara Preta (RPPN Cara Preta), located in Alto Paraíso, Goiás State, Brazil. After obtaining authorization from the NGO Oca Brasil, random walks were done in order to locate, photograph and mark species with a Global Positioning System device, and to collect and make exsiccates for Herbaria deposits, from September 2006 until March 2009. Although some plant leaves were collected during the vegetative stage, these specimens were visited until flowering to identify them correctly. All exsiccates were deposited in Brazilian Herbaria as standard control material (prioritizing PACA, UnB and IBGE Herbaria) and these species are included in Table 1.

2.2. Leaf anatomy and ultrastructure

Completely expanded leaves, from 3rd to 5th node from the apex, of two to six specimens of each species were collected and sectioned. Part of the leaf medial region was fixed in ethanol, acetic acid and formaldehyde [28] for 24 hours and preserved in ethanol 70% until analysis to describe the anatomy and identify starch and crystal composition [28].

Some pieces of the leaf medial region were immediately submerged in a Karnovsky solution [29] of glutaraldehyde 2%, paraformaldehyde 2% and sucrose 3% in sodium cacodylate 0.05 M buffer for 12 to 24 hours and preserved in sodium cacodylate 0.05 M until processing for analysis under an electron microscope. For the latter analysis, these pieces were post-fixed in 2% osmium tetroxide and 1.6% potassium ferricyanide (1:1 v/v), followed by in-block staining with 0.5% uranyl acetate solution (overnight). These samples were then dehydrated in an acetone ascending series and slowly embedded in Spurr´s epoxy resin. Semi-thin and ultra-thin sections were obtained in ultramicrotome with glass and diamond knives. Semi-thin sections were stained with toluidine blue and analysed under the optical Zeiss Axiophot, and ultra-thin sections were analysed under the transmission electron microscope TEM JEOL JEM 1011.

3. Results and discussion

The Amaranthaceae family is composed of 2,360 species and 146 of them are found in Brazil (Table 1). Ninety-eight species within the family are found in the Cerrado and 73 spp. are endemic to Brazil, of which 13 are endemic to the Cerrado Biome (Table 1). Twenty Amaranthaceae species are exclusive to the Cerrado (Table 1).

At least 22 Amaranthaceae species are referred to as being used in folk medicine (Table 1). In Brazil, only two of these species are already used as commercial drugs, as capsules containing their powdered roots, with studies to support their medicinal activity: *Hebanthe eriantha* (Poir.) Pedersen and *Pfaffia glomerata* (Spreng.) Pedersen (Table 1), both known as "Brazilian-ginseng". However, there is neither registered success in isolating or synthesizing their components nor any economic studies about the viability of this kind of pharmaceutical procedure.

Although the species *Gomphrena macrocephala* St.-Hil. is not cited as medicinal (Table 1), the fructan content in its roots has been determined [30] because this species was considered synonymous with *G. officinalis* Mart. [31]. Later, it was determined that *G. officinalis* was synonymous with *G. arborescens* L.f. and not with *G. macrocephala* [22]. Studying *G. arborescens*, fructan was also determined as the principal carbohydrate in its subterranean system [32]. This species is used in popular medicine to heal respiratory diseases (asthma and bronchitis), to reduce fever and as a tonic [33-35]. An *in vivo* study (in cats) with the use of fructans isolated from *Arctium lappa* L. (Asteraceae) reported a cough-suppressing activity [36], and the presence of fructan in *G. arborescens* roots can partially justify the use of this species as a medicinal plant.

Most members of Brazilian Amaranthaceae are only known by taxonomists and 42 species are in danger of extinction according to Brazilian regional lists; 14 of them are recognized as endangered by the Brazilian Ministry of the Environment (MMA – "Ministério do Meio Ambiente") (Table 1). Most of the endangered species are classified according to the IUCN Red List of vulnerability categories, some even with the same criteria, and there is a wide range of research still to be done.

Species	**Bioma and level of endemism**	**Species Threat Level**	**Habit, popular name and species knowledge**
Achyranthes aspera L.	Cerrado		Herb; plant used as indigenous medicine in Ethiopia with chemistry study [37]
Achyranthes indica (L.) Mill.	Cerrado		Herb
Alternanthera adscendens Suess.	Cerrado exclusive		Shrub
Alternanthera albida (Moq.) Griseb.			Subshrub; C_4 photosynthesis physiology [38]
Alternanthera aquatica (D.Parodi) Chodat			Herb
Alternanthera bahiensis Pedersen	Cerrado, endemic to Brazil		Herb or subshrub
Alternanthera bettzichiana (Regel) G.Nicholson			Herb; popularly named "anador"; folk medicinal plant, used as analgesic and antipyretic [39]
Alternanthera brasiliana (L.) Kuntze	Cerrado, endemic to Brazil		Herb; called "perpétua-do-mato, periquito-gigante, penicilina" or Brazilian joyweed; folk medicinal plant, used as diuretic, digestive, depurative, bequic, astringent and antidiarrhoeal; ornamental plant; C_3 photosynthesis structure [40-42]
Alternanthera decurrens J. C. Siqueira	Brazilian Cerrado endemic (Januária - MG)	CR [43]	Subshrub
Alternanthera dendrotricha C.C.Towns.	Cerrado, endemic to Brazil		Shrub
Alternanthera flavida Suess.			Subshrub
Alternanthera hirtula (Mart.) R.E.Fr.		EN [44]	Herb
Alternanthera januariensi J. C. Siqueira	Brazilian Cerrado endemic (Januária - MG)	CR [43]	Subshrub
Alternanthera kurtzii Schinz			Herb
Alternanthera littoralis P.Beauv.			Herb; called "periquito-da-praia" [45]
Alternanthera malmeana R.E.Fr.		EN [44]	Herb
Alternanthera markgrafii Suess.	Brazilian Cerrado endemic (Serra de Grão Mogol - MG)		Herb

Alternanthera martii (Moq.) R.E. Fries	Cerrado, endemic to Brazil		Subshrub
Alternanthera micrantha R.E.Fr.	Endemic to Brazil	VU [44]	Herb; called "periquito-da-serra" [45]
Alternanthera minutiflora Suess.	Endemic to Brazil		Herb
Alternanthera multicaulis Kuntze	Endemic to Brazil		Herb
Alternanthera paronychioides A.St.-Hil.	Cerrado	VU [44,46]	Herb; called "periquito-roseta, periquito"; C_3-C_4 intermediary photosynthesis structure; C_4 photosynthesis physiology; ornamental plant [38,41,42,45,47]
Alternanthera philoxeroides (Mart.) Griseb.	Cerrado		Herb; called "perna-de-saracura, carrapicho-de-brejo" and alligatorweed [45]
Alternanthera pilosa Moq.			Herb
Alternanthera praelonga A.St.-Hil.		CR [44]	Herb
Alternanthera puberula D.Dietr.	Cerrado exclusive		Herb
Alternanthera pulchella Kunth			Herb; C_4 photosynthesis physiology [38]
Alternanthera pungens Kunth	Cerrado		Herb; called "erva-de-pinto"; folk medicinal plant, used to treat syphilis and skin diseases; C_4 photosynthesis physiology [38,39]
Alternanthera ramosissima (Mart.) Chodat	Cerrado		Herb
Alternanthera regelii (Seub.) Schinz	Cerrado exclusive, endemic to Brazil		Herb
Alternanthera reineckii Briq.	Cerrado	VU [44]	Herb; called "periquito-de-reineck" [45]
Alternanthera rufa (Mart.) D.Dietr.	Cerrado, endemic to Brazil		Herb
Alternanthera sessilis (L.) R.Br.	Cerrado	LC [48]	Herb
Alternanthera tenella Colla	Cerrado	VU [44]	Herb; called "apaga-fogo, carrapichinho, corrente, folha-de-papagaio, periquito, periquito-figueira, perpétua-do-mato, sempre-viva"and joyweed; folk medicinal plant, used as diuretic; this species is naturally infected by a potyvirus; C_3-C_4 photosynthesis physiology and structure [38,39,40,45,47,49,50]

Alternanthera tetramera R.E.Fr.			Herb
Amaranthus blitum L.			Herb; called "caruru"; folk medicinal plant, used to fight anemia; C_4 photosynthesis physiology [38,39,51]
Amaranthus caudatus L.	Cerrado		Herb; called "rabo-de-gato, cauda-de-raposa, disciplina-de-freira, rabo-de-raposa"; folk medicinal plant, used to treat pulmonary diseases and as emoliente; C_4 photosynthesis physiology; ornamental plant [33,38,39,41]
Amaranthus cruentus L.	Cerrado		Herb; called "caruru-vermelho, veludo, bredo-de-jardim, crista-de-galo"; folk medicinal plant, used as emollient and laxative; C_4 photosynthesis physiology [33,38,39,51]
Amaranthus deflexus L.			Herb; called "caruru-rasteiro"; C_4 photosynthesis physiology [38,51]
Amaranthus hybridus L.	Cerrado		Herb; called "caruru, bredo" and smooth pigweed; C_4 photosynthesis physiology [38,51]
Amaranthus muricatus (Moq.) Hieron.			Herb;, C_4 photosynthesis physiology [38]
Amaranthus retroflexus L.	Cerrado		Herb; C_4 photosynthesis physiology [38]
Amaranthus rosengurtii Hunz.		EN [44]	Herb
Amaranthus spinosus L.	Cerrado		Herb; called "caruru-bravo, caruru-de-espinho, bredo-de-espinho, caruru-de-porco"; folk medicinal plant, used to combat eczema and as emollient, laxative and antiblenorragic; C_4 photosynthesis physiology [33,38,39]
Amaranthus viridis L.	Cerrado		Herb; called "caruru-bravo, caruru-verdadeiro, cururu, caruru-de-soldado, caruru-de-folha-miúda, amaranto-verde"; folk medicinal plant, used as emollient and diuretic desobstruente; C_4 photosynthesis physiology [33,38-40,51]
Blutaparon portulacoides (A.St.-Hil.) Mears	Cerrado	VU [44]	Herb; called "capotiraguá"; folk medicinal plant, used to combat leukorrhea; C_4 photosynthesis physiology [38,39]
Blutaparon vermiculare (L.) Mears	Cerrado		Herb; C_4 photosynthesis physiology [38]

Celosia argentea L.	Cerrado		Herb; called "celosia-branca, celósia-plumosa, crista-de-galo, crista-de-galo-plumosa, suspiro, veludo-branco"; folk medicinal plant, used to combat diarrhea and as anthelmintic and astringent; ornamental plant [39,41,52]
Celosia corymbifera Didr.	Endemic to Brazil		Subshrub
Celosia grandifolia Moq.		EN [44]	Herb, subshrub; called "bredo-do-mato" [45]
Chamissoa acuminata Mart.		VU [44]	Subshrub; called "mofungo-rabudo" [45]
Chamissoa altissima (Jacq.) Kunth	Cerrado	VU [44]	Subshrub; called "mofungo-gigante" [45]
Chenopodium album L.	Cerrado		Herb
Chenopodium ambrosioides L.	Cerrado		Herb; called "erva-de-santa-maria, erva-santa, quenopódio" [40]
Chenopodium murale L.	Cerrado		Herb
Cyathula achyranthoides (Kunth) Moq.			Herb
Cyathula prostrata Blume	Cerrado		Herb
Froelichia humboldtiana (Roem. & Schult.) Seub.	Cerrado		Herb; C_4 photosynthesis physiology [38]
Froelichia interrupta (L.) Moq.			Herb; C_4 photosynthesis physiology [38]
Froelichia procera (Seub.) Pedersen	Cerrado		Herb; called "ervaço"; C_4 photosynthesis physiology [38,41]
Froelichia sericea (Roem. & Schult.) Moq.			Herb
Froelichia tomentosa (Mart.) Moq.	Cerrado		Herb; C_4 photosynthesis physiology [38]
Froelichiella grisea R.E. Fries	Braziliann Cerrado endemic (Chapada dos Veadeiros - GO)	VU [43]	Herb; C_3 photosynthesis structure [42]
Gomphrena agrestis Mart.	Cerrado, endemic to Brazil	EN [46]	Herb
Gomphrena arborescens L.f.	Cerrado exclusive		Herb, subshrub; called "perpétua, perpétua-do-campo, perpétua-do-mato, paratudo-do-campo, paratudo-erva, raiz-do-padre"; folk medicinal plant, used as tonic, to reduce fever and against respiratory deseases; potential use as ornamentalplant; roots are fructan-rich; C_4 photosynthesis physiology/structure [32,35,38,40,42,49,53,54]

Gomphrena basilanata Suess.	Endemic to Brazil		Subshrub; C_4 photosynthesis physiology [38]
Gomphrena celosoides Mart.	Cerrado		Subshrub; C_4 photosynthesis physiology [38]
Gomphrena centrota E.Holzh.	Endemic to Brazil	VU [43]	Subshrub; C_4 photosynthesis physiology [38]
Gomphrena chrestoides C.C.Towns.	Brazilian Cerrado endemic (Chapada Diamantina - BA)	VU [43]	Subshrub
Gomphrena claussenii Moq.	Cerrado, endemic to Brazil		Subshrub
Gomphrena debilis Mart.	Endemic to Brazil		Subshrub; C_4 photosynthesis physiology [38]
Gomphrena demissa Mart.	Cerrado, endemic to Brazil		Subshrub; folk medicinal plant, used to combat the flu; C_4 photosynthesis physiology [38,49]
Gomphrena desertorum Mart.	Cerrado, endemic to Brazil		Subshrub; C_4 photosynthesis physiology [38]
Gomphrena duriuscula Moq.	Endemic to Brazil	EN [43]	Subshrub; C_4 photosynthesis physiology [38]
Gomphrena elegans Mart.	Cerrado, endemic to Brazil	VU [46]	Subshrub
Gomphrena gardnerii Moq.	Cerrado, endemic to Brazil		Subshrub; C_4 photosynthesis physiology [38]
Gomphrena globosa L.	Cerrado		Subshrub; called "gonfrena, perpétua, perpétua-roxa, sempre-viva, suspiro, suspiro-roxo"; folk medicinal plant used to fight respiratory diseases; C_4 photosynthesis physiology; ornamental plant [38,40,45,55,56]
Gomphrena graminea Moq.	Cerrado	VU [44]	Subshrub; called "perpétua-gramínea"; C_4 photosynthesis physiology [38,45]
Gomphrena hatschbachiana Pedersen	Cerrado, endemic to Brazil	VU [43]	Subshrub
Gomphrena hermogenesii J.C. Siqueira	Brazilian Cerrado endemic (Chapada dos Veadeiros - GO)		Subshrub; C_3 photosynthesis physiology; C_4 photosynthesis structure [38,42]
Gomphrena hillii Suess.	Brazilian Cerrado endemic (Paraíso do Norte - TO)		Subshrub; C_4 photosynthesis physiology [38]
Gomphrena incana Mart.	Cerrado exclusive, endemic to Brazil		Subshrub; C_4 photosynthesis physiology [38]
Gomphrena lanigera Pohl *ex* Moq.	Cerrado exclusive, endemic to Brazil		Subshrub; C_4 photosynthesis physiology and structure [38,42]
Gomphrena leucocephala Mart.	Endemic to Brazil		Subshrub; C_4 photosynthesis physiology [38]
Gomphrena macrocephala A.St.-Hil.	Cerrado exclusive, endemic to Brazil		Subshrub; roots are fructan-rich; C_4 photosynthesis physiology [30,38]

Gomphrena marginata Seub.	Brazilian Cerrado endemic (Diamantina - MG)		Subshrub
Gomphrena matogrossensis Suess.	Cerrado exclusive, endemic to Brazil		Subshrub
Gomphrena microcephala Moq.	endemic to Brazil		Subshrub
Gomphrena mollis Mart.	Cerrado, endemic to Brazil		Subshrub; called "erva-mole, erva-rosa"; folk medicinal plant, used as tonic and carminative [39]
Gomphrena moquini Seub.	Brazilian Cerrado endemic (Serra do Cipó - MG)		Subshrub
Gomphrena nigricans Mart.	Cerrado, endemic to Brazil	VU [43]	Subshrub
Gomphrena paranensis R.E.Fr.	Cerrado exclusive, endemic to Brazil		Subshrub, C_4 photosynthesis physiology [38]
Gomphrena perennis L.		VU [44]	Subshrub; called "perpétua-sempreviva"; C_4 photosynthesis physiology [38,45]
Gomphrena pohlii Moq.	Cerrado exclusive		Subshrub; called "infalível, paratudo, paratudinho, paratudo-amarelinho"; roots are used in folk medicine against respiratory deseases; C_4 photosynthesis physiology and structure [38,39,42,49]
Gomphrena prostrata Mart.	Cerrado, endemic to Brazil		Subshrub; C_4 photosynthesis physiology and structure [38,42]
Gomphrena pulchella Mart.		EN [44]	Subshrub; C_4 photosynthesis physiology [38]
Gomphrena pulvinata Suess.	Endemic to Brazil		Subshrub; C_4 photosynthesis physiology [38]
Gomphrena regeliana Seub.	Cerrado exclusive, endemic to Brazil		Subshrub; C_4 photosynthesis physiology [38]
Gomphrena riparia Pedersen	Endemic to Brazil	CR [43]	Subshrub; C_4 photosynthesis physiology [38]
Gomphrena rudis Moq.	Cerrado exclusive, endemic to Brazil		Subshrub; C_4 photosynthesis physiology [38]
Gomphrena rupestris Nees	Cerrado, endemic to Brazil		Subshrub
Gomphrena scandens (R.E.Fr.) J.C.Siqueira	endemic to Brazil	VU [43]	Subshrub
Gomphrena scapigera Mart.	Cerrado, endemic to Brazil		Subshrub; C_4 photosynthesis physiology [38]
Gomphrena schlechtendaliana Mart.		EN [44]	Subshrub; called "perpétua-schlechtendal"; C_4 photosynthesis physiology [38,45]

Gomphrena sellowiana Mart.	Endemic to Brazil	VU [44]	Subshrub
Gomphrena serturneroides Suess.	Endemic to Brazil		Subshrub; C_4 photosynthesis physiology [38]
Gomphrena vaga Mart.	Cerrado, endemic to Brazil	VU [44]	Subshrub; called "thoronoé"; folk medicinal plant, used as analgesic [57]
Gomphrena virgata Mart.	Cerrado, endemic to Brazil		Subshrub; called "cangussú-branco, vergateza"; folk medicinal plant, antiletargic; C_4 photosynthesis physiology and structure [33,38,42]
Hebanthe eriantha (Poir.) Pedersen	Cerrado	EN [44], VU [58]	Subshrub, shrub; called "corango-açu, ginseng-brasileiro, picão-de-tropeiro,solidonia, suma"; folk medicinal plant, used to combat colic and enteritis; most of its chemical constituents are known and roots of this plant are already used by pharmaceutical companies [40,59]
Hebanthe grandiflora (Hook.) Borsch & Pedersen	Cerrado		Bush scandentia
Hebanthe occidentallis (R.E.Fr.) Borsch & Pedersen	Cerrado		Subshrub scandentia
Hebanthe pulverulenta Mart.	Cerrado	VU [58]	Subshrub scandentia; called "corango-veludo" [45]
Hebanthe reticulata (Seub.) Borsch & Pedersen			Subshrub, shrub scandentia
Hebanthe spicata Mart.			Shrub erect or scadentia
Herbstia brasiliana (Moq.) Sohmer		EX [46]	Subshrub
Iresine diffusa Humb. & Bonpl. *ex* Willd.	Cerrado		Subshrub; called "bredinho-difuso" [45]
Iresine poeppigiana Klotzsch			Subshrub
Lecosia formicarum Pedersen	Endemic to Brazil		Subshrub
Lecosia oppositifolia Pedersen	Endemic to Brazil	CR [43]	Herb or subshrub
Pedersenia argentata (Mart.) Holub			Herb
Pfaffia acutifolia (Moq.) O.Stützer	Cerrado		Herb or subshrub
Pfaffia aphylla Suess.	Brazilian Cerrado endemic (Gouveia - MG)		Subshrub
Pfaffia argyrea Pedersen	Cerrado exclusive, endemic to Brazil	VU [43]	Herb or subshrub
Pfaffia cipoana Marchior. *et al.*	Brazilian Cerrado endemic (Itambé do Mato Dentro - MG)		Subshrub

Pfaffia denudata (Moq.) Kuntze	Cerrado exclusive, endemic to Brazil		Herb, subshrub, shrub
Pfaffia elata R.E.Fr.	Cerrado exclusive, endemic to Brazil		Subshrub
Pfaffia glabrata Mart.	Cerrado exclusive		Herb, subshrub; called "corango-sempreviva" [45]
Pfaffia glomerata (Spreng.) Pedersen	Cerrado	VU [44]	Herb, subshrub; called "anador, canela-velha, ginseng-brasileiro, finseng, páfia, paratudo, corango-sempreviva"; folk medicinal plant, most of its chemical constituents are known and roots of this plant are already used by pharmaceutical companies; butanolic extract showed antihyperglycemic potential in vivo; C_3 photosynthesis physiology and structure [38,40,42,45,60]
Pfaffia gnaphaloides (L.f.) Mart.	Cerrado	VU [44]	Herb, subshrub, called "corango-de-seda", C3 photosynthesis physiology and structure [38,42,45]
Pfaffia hirtula Mart.	Cerrado exclusive, endemic to Brazil		Herb, subshrub
Pfaffia jubata Mart.	Cerrado, endemic to Brazil		Herb, subshrub; called "marcela-branca, marcela-do-campo, marcela-do-cerrado"and kytertenim; roots are used in folk medicine against intestinal problems [39,49]
Pfaffia minarum Pedersen	Cerrado exclusive, endemic to Brazil	VU [43]	Subshrub
Pfaffia rupestris Marchior. *et al.*	Brazilian Cerrado endemic (Rio Pardo de Minas - MG)		Subshrub
Pfaffia sarcophylla Pedersen	Brazilian Cerrado endemic (Niquelândia - GO)		Subshrub; nickel hyperaccumulator, it is one of the first species to recolonize the ground with high concentrations of total Ni in the soil (>1%) [61]
Pfaffia sericantha (Mart.) Pedersen	Cerrado, endemic to Brazil		
Pfaffia siqueiriana Marchior. & Miotto	Cerrado, endemic to Brazil		Subshrub
Pfaffia townsendii Pedersen	Cerrado, endemic to Brazil	VU [43]	Subshrub; C_3 photosynthesis physiology and structure [38,42]
Pfaffia tuberculosa Pedersen	Cerrado, endemic to Brazil		Herb, subshrub
Pfaffia tuberosa (Spreng.) Hicken	Cerrado		Herb, subshrub; called "corango-de-batata" [45]
Pfaffia velutina Mart.	Cerrado exclusive, endemic to Brazil		Subshrub

Pseudoplantago friesii Suess		PE [44]	Popular name is "caruru-açu" [45]
Quaternella confusa Pedersen	Cerrado exclusive, endemic to Brazil		Shrub
Quaternella ephedroides Pedersen	Cerrado, endemic to Brazil		Shrub
Quaternella glabratoides (Suess.) Pedersen	Endemic to Brazil	EN [44]	Subshrub; called "corangão" [45]
Xerosiphon angustiflorus (Mart.) Pedersen	Cerrado, endemic to Brazil		Subshrub
Xerosiphon aphyllus (Pohl ex Moq.) Pedersen	Cerrado, endemic to Brazil		Subshrub

Notes: The species threat level category is the same as that used in the IUCN Red List: CR (critically endangered), EN (endangered), EX (extinct), LC (Least Concern) and VU (vulnerable).

Table 1. Amaranthaceae species found in Brazil, identifying those endemics to Brazil and the ones found in the Neotropical Savannah (Cerrado), level of threat, habit, popular name (mostly in Portuguese) and some of the knowledge about the species.

3.1. Morphology, taxonomy challenge and species list of Brazilian Amaranthaceae

Taxonomy is the science that aims to identify and characterize species. It includes the study of the plant´s behaviour in nature and is based on plant morphology. The use of other data, such as anatomy studies, genetic characters, ecology and geographic pattern, aims to include and define affinities and parental relations among plant groups. Only by knowing the species is it possible for Botany to contribute to other scientific areas, including to the conservation of species *in situ*, not only of plants but also of animals.

It is not easy to correctly identify Brazilian Amaranthaceae species. Different species can be very alike in habit and vegetative morphology. The correct identification depends almost exclusively on some flower details, whose small dimensions make it especially difficult to work in the field, demanding a highly specialized work, only partially carried out for this family (15-22).

Brazilian species of this family are predominantly herbs, shrubs or climbing plants. They can be annual or perennial, with erect, prostrate, decumbent or scandent stem. In species from the Neotropical Savannah or from rocky fields, the underground organ is thickened and composed of roots and a xylopodium – a portion of the subterranean system which is responsible for the re-sprouting after a fire or other environmental stress [62]. The leaf arrangement can be opposite, alternate or with a basal aggregation of leaves. Leaves are exstipulate, glabrous or pubescent, with entire lamina and margins. Inflorescences can be cymoses, in spikes, in heads, corymboses or paniculates, axillary or axial. Flowers are bisexual or monoecious and small. The perianth is undifferentiated, actinomorphic, with five distinct or partially connated sepals. Flowers are associated with dry and papery bracts. Fruits are dry, usually a single-seeded achene or capsules with few seeds [15-22]. A short list of the most important Brazilian Herbaria to visit in order to study Amaranthaceae taxonomy is presented on Table 2 and the literature used to identify the species of this family is presented in Table 3.

Index	Herbarium Name	Institution and municipality
ALCB	Herbário da Universidade Federal da Bahia	UFBA/Campus de Ondina, Salvador, Bahia, Brazil
BHCB	Herbário da Universidade Federal de Minas Gerais	UFMG, Belo Horizonte, Minas Gerais, Brazil
BOTU	Herbário da Universidade Estadual Paulista	UNESP, Botucatu, São Paulo, Brazil
CEN	Herbário da EMBRAPA Recursos Genéticos e Biotecnologia	EMBRAPA/CENARGEN, Brasília, Distrito Federal, Brazil
CEPEC	Herbário do Centro de Pesquisas do Cacau	CEPEC, Itabuna, Bahia, Brazil
CESJ	Herbário da Universidade Federal de Juiz de Fora	UFJF, Juiz de Fora, Minas Gerais, Brazil
CPAP	Herbário do Centro de Pesquisas Agropecuárias do Pantanal	CPAP, Corumbá, Mato Grosso do Sul, Brazil
ESA	Herbário da Universidade de São Paulo	ESALQ/USP, Piracicaba, São Paulo, Brazil
GUA	Herbário Alberto Castellanos	FEEMA/INEA, Rio de Janeiro, Rio de Janeiro, Brazil
HTO	Herbário da Universidade Federal do Tocantins	UFTO, Porto Nacional, Tocantins, Brazil
HUEFS	Herbário da Universidade Estadual de Feira de Santana	UFES, Feira de Santana, Bahia, Brazil
IAC	Herbário do Instituto Agronômico de Campinas	IAC, Campinas, São Paulo, Brazil
IBGE	Herbário da Reserva Ecológica do IBGE	IBGE/RECOR, Brasília, Distrito Federal, Brazil
JPB	Herbário da Universidade Federal da Paraíba	UFPB, Cidade Universitária, João Pessoa, Paraíba, Brazil
MBM	Herbário do Museu Botânico Municipal	Prefeitura Municipal/SMA, Curitiba, Paraná, Brazil
PACA	Herbarium Anchieta	Instituto Anchietano de Pesquisas/UNISINOS, São Leopoldo, Rio Grande do Sul, Brazil
RB	Herbário do Jardim Botânico do Rio de Janeiro	JBRJ, Rio de Janeiro, Rio de Janeiro, Brazil
SP	Herbário do Instituto de Botânica	Secretaria de Meio Ambiente, São Paulo, São Paulo, Brazil
SPF	Herbário da Universidade de São Paulo	USP, São Paulo, São Paulo, Brazil
UB	Herbário da Universidade de Brasília	UnB, Brasília, Distrito Federal, Brazil
UEC	Herbário da Universidade Estadual de Campinas	UNICAMP, Campinas, São Paulo, Brazil
UFG	Herbário da Universidade Federal de Goiás	UFG, Goiânia, Goiás, Brazil
VIC	Herbário da Universidade Federal de Viçosa	UFV, Viçosa, Minas Gerais, Brazil

Notes: Herbaria are cited according to the Index Herbariorum and all of them have good collections of Amaranthaceae. The Institution/city where the Herbaria are located is also referred.

Table 2. List of the most important Herbaria references for researchers interested in studying the Brazilian Amaranthaceae

[15-21] Revisions of Brazilian *Froelichia, Froelichiella, Hebanthe* and *Pfaffia;* species list and phytogeography
[22-25] Revision of Brazilian *Gomphrena*; species list na phytogeography
[45] Amaranthaceae from Santa Catarina State, Brazil
[63] Restoring the *Hebanthe* genera
[64,65] Brazilian Amaranthaceae species and the Family in the World
[66] Revision of Amaranthaceae in the World
[67-71] Studies in South American Amaranthaceae
[72] Amaranthaceae in Flora Brasiliensis
[73] Studies of *Pfaffia* and *Alternanthera* genera
[74,75] Amaranthaceae in Central and South America
[51,52,56,76] Amaranthaceae from Rio Grande do Sul State, Brazil

Note: These references are ordereb by author and should be consulted in order to identify Brazilian species correctly.

Table 3. List of the most important bibliographical references for researchers interested in studying the Brazilian Amaranthaceae

The Reserva Particular do Patrimônio Natural (RPPN) Cara Preta, in Alto Paraíso, Goiás State, is a good representative of Neotropical Savannah vegetation, at about 1,500 meters of altitude and showing rocky slopes with Cerrado *sensu stricto* (Figure 1), grassland with scattered scrubs and few trees and grassland with few scrubby plants and no trees (Figure 2). The *Pfaffia* genus was restricted to a rocky slope and the other species were found in a level field of sandy soils, usually covered by Poaceae and Cyperaceae. It was very difficult to find all the species. It was only possible because of frequent visits to RPPN Cara Preta, using GPS to mark the local after finding any probable member of the family in order to be able to accompany them until the flowering stage. The area was monitored for one and a half year and only *Gomphrena hermogenesii* J.C. Siqueira and *Pfaffia townsendii* Pedersen (Figure 3) were localized, the first one always in vegetative stage. A key event to help finding all six species was a fire that burned out the vegetation in August of the year 2008: without the competition of the grasses, the Amaranthaceae species regrew and flowered rapidly, in order to spread their seeds before the grasses could fully recover (Figures 4-8).

Pfaffia townsendii is a shrub species with persistent aerial portions that flowers throughout the year (Figure 3). The herb *G. hermogenesii* is endemic to Chapada dos Veadeiros and also has permanent aerial portions (about 10-20 cm high), but it was commonly found in vegetative stage under the grass leaves; its flowering stage was stimulated by fire (Figure 4). *Froelichiella grisea* R.E.Fr. (Figure 5), *G. lanigera* Pohl. ex Moq. (Figure 6), *G. prostrata* Mart. (Figure 7) and *P. gnaphaloides* (L.f.) Mart. (Figure 8) species were recorded in the flowering stage at RPPN Cara Preta around 20 days after a fire that burned out all the vegetation in the area, which is evidence of the pirophytic behaviour of most Neotropical Savannah Amaranthaceae. Five of these species had never been recorded in this RPPN before and one of them was last recorded in 1966 (*F. grisea*), according to Herbaria data. Figures 1-8 are reproduced [77] with the authorization of the Biota Neotropica Editor, Dr. Carlos Joly.

Figure 1. (Figures 1-8) Photographs of the environment and of the studied species at Reserva Particular do Patrimônio Natural (RPPN) Cara Preta, Alto Paraíso, Goiás State, Brazil. **Fig. 1.** Rocky slope where were found the species *Pfaffia townsendii* Pedersen and *P. gnaphaloides* (L. f.) Mart. **Fig. 2.** Humid rocky grassland where were found the species *Froelichiella grisea* R.E.Fr., *Gomphrena hermogenesii* J.C. Siqueira, *G. lanigera* Pohl. ex Moq. and *G. prostrata* Mart. **Fig. 3.** *P. townsendii*. **Fig. 4.** *G. hermogenesii*. **Fig. 5.** *F. grisea*. **Fig. 6.** *G. lanigera*. **Fig. 7.** *G. prostrata*. **Fig. 8.** *P. gnaphaloides*.

Five species are herb to subshrub, and only *P. townsendii* is a shrub (Figure 3). Well-developed tuberous subterranean systems were found in *F. grisea* (Figure 5) and *G. hermogenesii*, while in *G. lanigera* and *P. gnaphaloides* the underground organ was less developed, also tuberous. *G. prostrata* and *P. townsendii* presented a well-developed and lignified underground organ. Leaves of *F. grisea* and *G. hermogenesii* are opposite and alternate in the other studied species. *F. grisea* and *G. lanigera* can present a basal aggregation of leaves. Leaves are always tomentose, with exception of the adaxial face of *F.*

grisea, which can be glabrous. Inflorescence is axial in all these species, spikes in *F. grisea* and *G. lanigera* and heads in the other studied species. Flowers are yellowish in *F. grisea, G. hermogenesii* and *G. lanigera,* with a tendency to turn red in the first and last one. In *G. prostrata, P. gnaphaloides* and *P.townsendii* flowers are white, turning beige in the last species. All the species have flowers associated with dry and papery bracts that persist alongside their dry fruits, usually a single-seeded achene, favouring anemocoric dispersion.

The fastest lifespan was observed in *G. lanigera,* which took around 20 days to regrowth and finish the flowering phase. In Figure 6, *G. lanigera* was about 20 days old and fruits were almost mature, indicating proximity to the seed dispersal phase. *Pfaffia townsendii* alone showed behaviour that was independent of fire, since even *G. hermogenesii* only flowered after being burned to the ground and regrowing from its xylopodium. The other four species were found after the occurrence of fire, all of them in the flowering stage.

In the Taxonomy and Morphology areas, studies of the genera Achyranthes, Alternanthera, Amaranthus, Blutaparon, Celosia, Chamissoa, Chenopodium, Cyathula, Iresine, Lecosia, Pedersenia, Pseudoplantago, Quaternella and Xerosiphon still need to be done, not only covering the revision of the Brazilian species, biogeography and morphological evolution, but also molecular biology to establish synonyms and to delimit variations among individuals of each species.

3.2. Leaf anatomy of Amaranthaceae species

Leaves of the six studied species have anatomical variation among the genera and are more similar between species of the same genus. Transverse sections show that *G. hermogenesii* (Figure 9), *G. lanigera* (Figure 10) and *G. prostrata* (Figure 11) have large nonglandular trichomes covering the single layered epidermis, dorsiventral mesophyll with upper palisade parenchyma and spongy parenchyma near the lower epidermis. All these three species are amphistomatic, and a complete well-developed parenchymatous sheath with thicker cell walls surrounds the vascular bundles (Kranz cells), in which starch accumulates. Calcium oxalate druses were found in the mesophyll. The leaf anatomy of the three *Gomphrena* spp. is compatible with the C_4 photosynthesis pathway.

Pfaffia gnaphaloides (Figure 12) and *P. townsendii* (Figure 13) have more undulating surfaces and a thinner leaf blade in relation to the *Gomphrena* species. Trichomes are also more frequent and thinner and the mesophyll is dorsiventral. The parenchymatous sheath has thinner walls than the neighbouring cells in *Pfaffia* species. Both species had elevated stomata on the lower epidermis and only *P. gnaphaloides* had few stomata on the upper epidermis. Starch was distributed in all mesophyll cells and calcium oxalate druses were rare. The anatomy of *Pfaffia* spp. leaves is compatible with C_3 photosynthesis metabolism.

Froelichiella grisea (Figure 14) has the only isobilateral mesophyll among the studied species, with palisade parenchyma near both upper and lower epidermis. Palisade cells are shorter near the lower epidermis. The parenchymatic vascular bundle is not conspicuous and organelles in these cells are positioned towards the outer cell walls, in the same way as they are found in the other mesophyll cells. Calcium oxalate druses were more common near the midrib, and the reaction to starch was similar to that of all the mesophyll cells. Its leaf anatomy is compatible with C_3 photosynthesis metabolism. Figures 9-14 [77] were reproduced with the authorization of the Biota Neotropica Editor, Dr. Carlos Joly.

Gomphrena trichomes are similar to the ones described for *G. arborescens* [32,54,78]. Although it is expected that stomata are reduced on the upper surface of land plants, the Cerrado *Gomphrena* species *G. arborescens, G. pohlii* and *G. virgata* have a similar number of stomata on both surfaces [78], subjecting them to a greater water loss, which is compensated by the well-developed subterranean systems that guarantee water supply during the lifespan of their leaves. The size and number of stomata on both leaf surfaces of *G. hermogenesii, G. lanigera* and *G. prostrata* is still to be verified, but simple observation indicates that it should be similar to the phenomena observed in the first cited species, since they also have a relatively well-developed subterranean system.

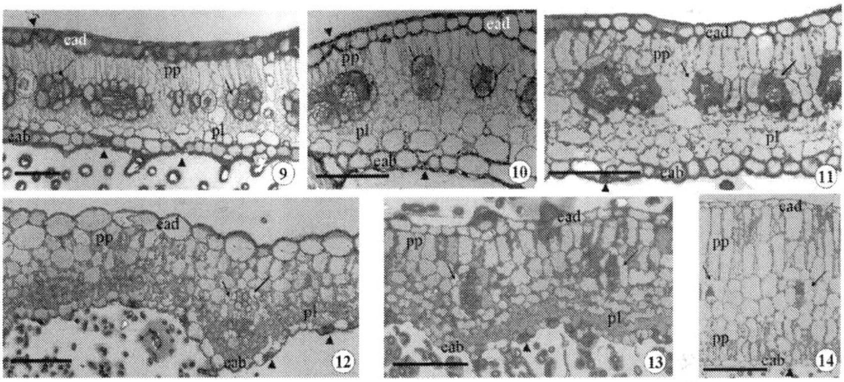

Figure 2. (Figures 9-14) Micrographies of the middle leaf transversal sections of the studied Amaranthaceae species. **Fig. 9.** *Gomphrena hermogenesii* - leaf blade thickness from medium to thick, dorsiventral mesophyll and complete parenchymatous bundle sheath, with thick cell walls and collateral vascular bundles. **Fig. 10.** *G. lanigera* - medium leaf blade, dorsiventral mesophyll and complete parenchymatous bundle sheath, with thick cell walls and collateral vascular bundles. **Fig. 11.** *G. prostrata* – thin to medium leaf blade, dorsiventral mesophyll and complete parenchymatous bundle sheath, with thick cell walls and collateral vascular bundles. **Fig. 12.** *Pfaffia gnaphaloides* – thin leaf blade, dorsiventral mesophyll, less defined parenchymatous bundle sheath and collateral vascular bundles. **Fig. 13.** *P. townsendii* – only hypostomatous leaf species, thin leaf blade, dorsiventral mesophyll, less defined parenchymatous bundle sheath and collateral vascular bundles. **Fig. 14.** *Froelichiella grisea* - thick leaf blade, isobilateral mesophyll with elongated palisade parenchyma under the adaxial epidermis, less defined bundle sheath and collateral vascular bundles. **Legend:** eab = abaxial epidermis; ead = adaxial epidermis; pl = spongy parenchyma; pp = palisade parenchyma; arrowhead = stoma; circle = druse; arrow = parenchymatous bundle sheath. Bar = 100 μm.

The leaf anatomy of the three *Gomphrena* spp. is similar to that of *G. arborescens* L.f. [54], *G. cespitosa, G. dispersa, G. nitida, G. sonorae* [79] and *G. conica, G. flaccida* [80] among others, most of them arranged in a *Gomphrena* atriplicoid-type of Kranz anatomy [81,82]. As expected, there was no significant variation in these species' leaf anatomy due to the life cycle stage, although older leaves collected during the vegetative stage have a thicker cuticle covering both epidermis surfaces, especially in *G. hermogenesii* species. The leaf anatomy observed in the two *Pfaffia* spp. is similar to that described in *P. jubata* [83], which also lacks the Kranz anatomy. There is no previous study about the anatomy of *F. grisea* leaves and its genus is monoespecif.

There are still a number of studies to be done in the field of anatomy and histology of Brazilian Amaranthaceae plants. Most of the medicinal species need to be analyzed and validated for their use as drugs, including anatomic description and an investigation of the secondary compounds of the used organs, by histology and by chromatography. Due to the difficulties in correctly identifying the species in the field, anatomical and morphological markers should be defined to guarantee these species' identity even during the vegetative stage. Besides that, anatomical studies can improve the taxonomy data and explain some morphological characters of this plant family, like the anatomical variations in the leaf that are connected to photosynthesis, or the secondary thickening and xylopodium development in underground organs, which is a character for the Cerrado species. The anatomy of few Brazilian Amaranthaceae species has been described, with the exception of some from the *Gomphrena* and *Pfaffia* genera [83-86].

3.3. Leaf ultrastructure of Amaranthaceae species

Leaves of the six studied species have less ultrastructural variation among the species of the same genus. *Froelichiella grisea* organelles are equally distributed among chlorenchyma tissues, usually near the cell walls. The chloroplasts of this species are always granal (Figure 15), even in the vascular cells, with large starch granules (usually one or two per organelle) in all tissues. Plastoglobuli are small and less numerous in mesophyll chloroplasts (Figure 15), but guard cell chloroplasts usually have just one large plastoglobulus and less conspicuous grana. Mitochondria and peroxisomes (Figure 15) were found in mesophyll and bundle sheath cells. Leaf ultrastructure is compatible with C_3 photosynthesis metabolism.

Mesophyll cell chloroplasts of *Gomphrena* species have conspicuous grana, rare starch granules and variable size of plastoglobuli: *G. hermogenesii* has larger ones in relation to *G. lanigera* and *G. prostrata*. Bundle sheath chloroplasts are completely devoid of grana or have few stacked thylakoids (Figure 16) in all studied *Gomphrena* species, but always have large starch granules and plastoglobuli. The larger the starch granules, the more deformed the chloroplasts' typical lens shape, as shown in *G. hermogenesii* (Figure 16). Mitochondria are usually numerous in bundle sheath cells and are always near chloroplasts, grouped next to the inner cell wall (towards the vascular bundle). Peroxisomes are rare, and a few were observed near chloroplasts in palisade and spongy parenchyma cells, but not in the bundle sheath cells. Phloem companion cells are mitochondria-rich in all *Gomphrena* species, as shown in *G. prostrata* (Figure 17). The presence of dimorphic chloroplasts, disposition of the organelles and the occurrence of Kranz syndrome seen in the leaf anatomy indicate that the C_4 photosynthesis pathway operates in the three studied *Gomphrena* spp.

Pfaffia species organelles are equally distributed among chlorenchyma tissues, usually near the cell walls. *Pfaffia* chloroplasts are granal even in the vascular cells, showing large starch granules and a similar size in all mesophyll cells, as can be observed in the palisade parenchyma of *P. townsendii* (Figure 18). Mitochondria and peroxisomes are common near chloroplasts (Figure 18). Phloem companion cells are mitochondria-rich and chloroplasts are smaller and granal, as in the other species of this study. Along with the aspects of *Pfaffia* anatomy described previously, their ultrastructure is compatible with the C_3 photosynthesis pathway.

Pfaffia gnaphaloides (Figure 19) and *G. hermogenesii* (Figure 20) leaves, collected during the flowering stage, were colonized by two distinct forms of microorganisms: (i) a smaller organism was found in the intercellular spaces (ics) of the spongy parenchyma (Figure 19); (ii) a larger and distinctly eukaryotic organism was found within distinct cells, with some morphological alterations suggesting an infectious process (Figures 19-20).

The external envelopae membranae system of the chloroplasts is disrupted in infected cells (Figure 20) and a size reduction was observed in the chloroplast plastoglobuli. All morphological characteristics observed in the intracellular microorganism suggest that it should be an obligate biotroph endophytic fungus belonging to the Ascomycete division (Figure 20). The invading fungus may be using the plastoglobuli lipids as its primary source of carbon and energy; the reduction of the plastoglobuli could also be due to its mobilization by the host plants in response to the stress caused by these biotic interactions. The complete identification of the fungus and its effect on the plants depends on its isolation from the environment/hosts and complementary studies.

The rare peroxisomes in *Gomphrena* spp. leaf cells and their presence among all chlorenchyma tissues of the *Pfaffia* spp. leaf cells is compatible with their possible photosynthesis metabolisms. Along with the presence of Kranz syndrome and dimorphic chloroplasts, the absence of peroxisome indicates that *Gomphrena* spp. perform photosynthesis via the C_4 pathway. In *Gomphrena* species, CO_2 concentration in the bundle sheath cells must be efficient, leading to a significant reduction in the oxygenase function of its RuBisCO enzyme. This leaves the species virtually free of the photorespiration process, aided by the large walls of the bundle sheath cells. Although a carbon isotope ratio study [38] indicates that *G. hermogenesii* is not a C_4 species, this species also has Kranz anatomy and ultrastructure compatible with C_4 metabolism, as do all the other studied *Gomphrena* spp. [42]. The distribution of its key photosynthetic enzymes will be carried out using immuno-cytochemistry, in our laboratory, in order to complete these data.

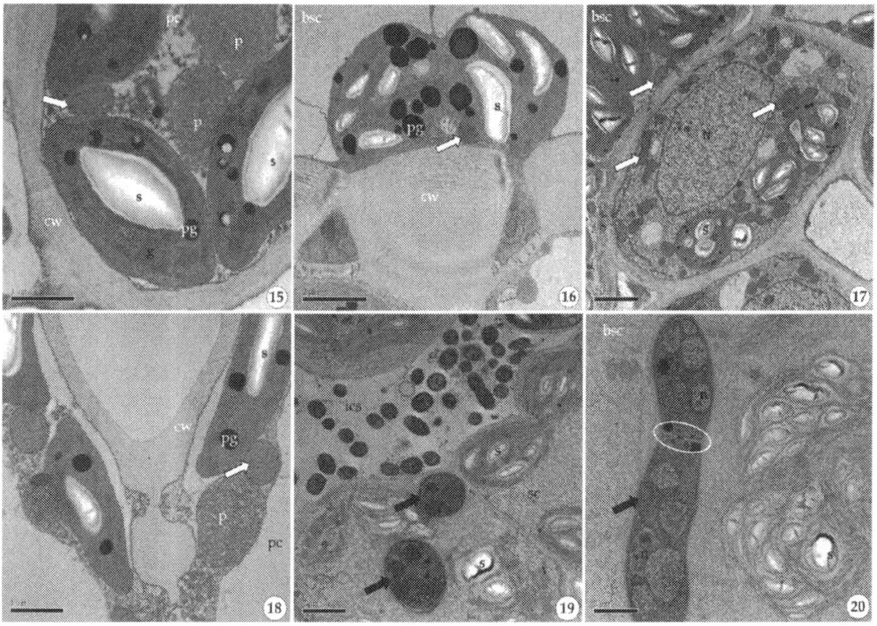

Figure 3. (Figures 15-20) Citological aspects of Amaranthaceae species as seen through a Transmission Electron Microscope. **Fig.15.** *Froelichiella grisea* palisade parenchyma cell. **Fig. 16.** *Gomphrena hermogenesii* bundle sheath cell. **Fig. 17.** *G. prostrata* phloem companion cell and bundle sheath cell on top. **Fig. 18.** *Pfaffia townsendii* palisade parenchyma cells.**Fig. 19.** *P. gnaphaloides* spongy parenchyma cells and invading microorganisms (black arrows). **Fig. 20.** *G. hermogenesii* bundle sheath cell and the invading Ascomycete fungus (black arrow) and the disrupted chloroplasts with smaller plastoglubuli. **Legend:** black arrow = invading organism; white arrow = mitochondria; ellipsis = septum with a simple pore; bsc = bundle sheath cell; cw = cell wall; ics = intercellular space; n = nucleus of the microorganism; N = nucleus of the plant species; p = peroxisome; pc = palisade parenchyma cell; pg = plastoglobulus in a chloroplast; s = starch granule in a chloroplast; sc = spongy parenchyma cell.

4. Conclusions and perspectives

This chapter presents data on Amaranthaceae species, with no pretension to explain the full potential of this plant family for scientific studies, but rather to provide a basic tool for those interested in amplifying studies on the species of this family. Based on our results, we are convinced of the importance of studying this family further, not only as a tool in the better preservation of endemic species, but also to explore its undoubted economic importance more fully. Basic research is still needed, with the aim of applying knowledge on these species to technological advances, especially in growing crops - since C_4 species have a faster metabolism and growing capacity, as observed in the species found in the RPPN Cara Preta - and to explore medicinal molecules of these plants. C_4 species are also important to balance CO_2 in the atmosphere because of their efficiency in the transformation of carbon into biomass; in Cerrado Amaranthaceae species, this storage is basically underground in their well-developed subterranean roots and xylopodium.

The number of medicinal plants among the Brazilian Amaranthaceae species may well be higher than already reported (Table 1), because Cerrado inhabitants are particularly interested in the highly developed subterranean systems of some medicinal species [34,49,58] which can be collected at any time of the year, even from species whose aerial portions are not persistent. Due to the morphological similarity among Amaranthaceae species in the Neotropical Savannah, their collectors can easily mistake one species for another during the vegetative stage, which confirms the need for further and more complete studies of the known medicinal and endangered species, at least.

Preparation of plant samples for transmission electron microscopy also proved to be useful in studying the morphology of fungi inside plant cells, as well as aspects of host-parasite interaction. This kind of study could be recommended for plants considered toxic to herbivores and to any medicinal plant consumed by humans, in order to give more information about the real source of poisoning or medicinal effect and for fine quality control. In both studied species (*G. hermogenesii* and *P. gnaphaloides*) the external macro aspects of the plants did not indicate the presence of the endophytic fungus.

RPPN Cara Preta is a small Private Conservation Unit (only 1.5% of the area of the Chapada dos Veadeiros National Park, a government-preserved area of 65,038 hectares). Both Conservation Units are separated only by a road, in Alto Paraíso municipality of Goiás State, Brazil. The latter site is registered by UNESCO as a natural protected Cerrado zone. RPPN Cara Preta has 245 species representing 47 family plants [75,86], which is 9.2% of the 2,661 plant species of Chapada dos Veadeiros [87], a good diversity of plants in relation to the occupied area. There are six Amaranthaceae species in the RPPN – 25% of the 27 species found in the National Park [87]. Considering that the RPPN Cara Preta Utilization Planning Report [86] indicated the presence of three endemic species, plus two Amaranthaceae species not reported initially [75], this Conservation Unit has 2% of endemic species – which is more than expected. According to [2,12], the Cerrado Biome is one of the priority hotspots for conservation because it has, among others, 4,400 endemic plants (1.5% of the Earth's 300,000 species). The Amaranthaceae family in RPPN Cara Preta can be considered a taxon indicator of the good diversity of the Neotropical Savannah. This taxon could be considered a plant diversity indicator in other works on flora in open areas of the Cerrado Biome. Because of the predominant habit (herbs and shrubs) and survival strategies, the presence of species from this family among the collected species clearly indicates a well performed collection effort.

There are a number of important factors indicating that this plant family deserves more studies for a greater understanding by researchers working in Brazil, and we recap them as follows: the Cerrado Biome holds 98 of the 146 Brazilian Amaranthaceae species (almost 70% of the total species) (Table 1); their pirophytic behavior and survival strategies (fast regrowth and seed dispersal before the complete recovery of grasses after fire) are coherent with the Biome's characteristics; their morphology shows exceptional adaptation to the seasonal climate and open areas (hairy aerial portions, partial or total loss of the aerial portions during the dry season, well-developed underground system with xylopodium, dry fruit dispersal by wind); their metabolism (evolution of C_4 and intermediary C_3-C_4 photosynthesis) may have importance for biomass conversion and CO_2 balance; and, finally, many of these plants are already used in medicines by Cerrado inhabitants and there may be much wider medicinal potential in other species of this family.

Author details

Suzane M. Fank-de-Carvalho*
National Council for Scientific and Technological Development - CNPq, Brasília, Distrito Federal, Brazil
Universidade de Brasília – UnB, Biological Institute, Electron Microscopy Laboratory, Brasília, Distrito Federal, Brazil

Sônia N. Báo
Universidade de Brasília – UnB, Biological Institute, Electron Microscopy Laboratory, Brasília, Distrito Federal, Brazil

Maria Salete Marchioretto
Instituto Anchietano de Pesquisas/UNISINOS, PACA Herbarium, São Leopoldo, Rio Grande do Sul, Brazil

Acknowledgement

We would like to thank CAPES, CNPq and FINEP for financial support; NGO Oca Brasil and Herbaria IBGE, UB and PACA for access authorization and research infrastructure; the aditional collectors for help in searching for and collecting the species at RPPN Cara Preta; and Susan Casement Moreira for the English review.

5. References

[1] Forzza, R.C.; Leitman, P.M.; Costa, A.F.; Carvalho Jr., A.A.: Peixoto, A.L.; Walter, B.M.T.; Bicudo, C.; Zappi, D.; Costa, D.P.; Lleras, E.; Martinelli, G.; Lima, H.C.; Prado, J.; Stehmann, J.R.; Baumgratz, J.F.A.; Pirani, J.R.; Sylvestre, L.; Maia, L.C., Lohmann, L.G.; Queiroz, L.P.; Silveira, M.; Coelho, M.N.; Mamede, M.C.; Bastos, M.N.C.; Morim, M.P.; Barbosa, M.R.; Menezes, M.; Hopkins, M.; Secco, R.; Cavalcanti, T.B. & Souza, V.C. (2010). Catálogo de Plantas e Fungos do Brasil – Vol. 1. Rio de Janeiro: Andrea Jakobsson Estúdio/Instituto de Pesquisa Jardim Botânico do Rio de Janeiro. 870 p.

[2] Myers, N.; Mittermeier, R.A.; Mittermeier, C.G.; Fonseca, G.A.B. & Kent, J. (2000). Biodiversity hotspots for conservation priorities. Nature, 403: 853-858.

* Corresponding Author

[3] Conservation International (2012). Cerrado. In: Biodiversity Hotspots. Available: http://www.biodiversityhotspots.org/xp/hotspots/cerrado/Pages/default.aspx Acessed 2012 Jan 31.

[4] Harley, R.M. & Giulietti, A.N. (2004). Wild flowers of the Chapada Diamantina – illustrated botanical walks in the mountains of NE Brazil. São Carlos: RIMA Ed. 344 p.

[5] Dias, B.F.S. (coord.). (1992). Alternativas de desenvolvimento dos cerrados: manejo e conservação dos recursos naturais renováveis. Brasília: FUNATURA/IBAMA. 97 p.

[6] Simon, M.F. & Proença, C. (2000). Phytogeographic patterns of Mimosa (Mimosoideae, Leguminosae) in the Cerrado biome of Brazil: an indicator genus of high-altitude centers of endemism? Biological Conservation, 96/3: 279-296.

[7] Eiten, G. (2001). Vegetação natural do Distrito Federal. Brasília: SEBRAE/UNB. 162 p.

[8] Klink, C.A. & Machado, R.B. (2005). A conservação do Cerrado brasileiro. Megadiversidade, 1/1: 147-155.

[9] Marinho-Filho, J.; Machado, R.B. & Henriques, R.P.B.H. (2010). Evolução do conhecimento e da conservação do Cerrado brasileiro. In: Diniz, I.R.; Marinho-Filho, J; Machado, R.B. & Cavalcanti, R.B. Orgs. Cerrado: conhecimento científico quantitativo como subsídio para ações de conservação. Brasília: Thesaurus Editora. Pp. 13-31.

[10] Paiva, P.H.V. (2000). A Reserva da Biosfera do Cerrado: fase II. In: Cavalcanti, T. B.; Walter, B. M. T. editors. Tópicos atuais em botânica - palestras convidadas do 51° Congresso Nacional de Botânica. Brasília: Sociedade Brasileira de Botânica/Embrapa-Cenargen. Pp. 332-334.

[11] Coutinho, L.M. (2006). O conceito de bioma. Acta Botanica Brasilica, 20/1: 13-23.

[12] Mittermeier, R.A.; Myers, N.; Thomsen, J.B.; Fonseca, G.A.B. & Olivieri, S. (1998). Biodiversity Hotspots and Major Tropical Wilderness Areas: Approaches to Setting Conservation Priorities. Conservation Biology, 12/3: 516-520.

[13] Ratter, J.A.; Ribeiro, J.F. & Bridgewater, S. (2000). Woody flora distribution of the Cerrado Biome: phytogeography and conservation priorities. In: Cavalcanti, T. B.; Walter, B. M. T. editors. Tópicos atuais em botânica - palestras convidadas do 51° Congresso Nacional de Botânica. Brasília: Sociedade Brasileira de Botânica/Embrapa-Cenargen. Pp. 340-342.

[14] Miranda, H. & Miranda, A.C. (2000). Queimadas e estoques de carbono no Cerrado. In: Moreira, A.G. & Schwartzman, S. editors. As Mudanças Climáticas e os Ecossistemas Brasileiros. Brasília: Ed. Foco. Pp. 75-81.

[15] Marchioretto, M.S.; Windisch, P.G. & Siqueira, J.C. (2002). Os gêneros Froelichia Moench e Froelichiella R.E. Fries (Amaranthaceae) no Brasil. Pesquisas-Botânica, 52: 7-46.

[16] Marchioretto, M.S.; Windisch, P.G. & Siqueira, J.C. (2004). Padrões de distribuição geográfica das espécies de Froelichia Moench e Froelichiella R.E. Fries Amaranthaceae) no Brasil. Iheringia Série Botânica, 59/2: 149-159.

[17] Marchioretto, M.S. (2008a). Os gêneros Hebanthe Mart. e Pfaffia Mart. (Amaranthaceae) no Brasil. Instituto de Biociências, Doutorado em Botânica, Universidade Federal do Rio Grande do Sul, Porto Alegre, Brazil. (Thesis). 255 p.

[18] Marchioretto, M.S; Azevedo, F.; Josende, M.V.F & Schnorr, D.M. (2008b). Biogeografia da família Amaranthaceae no Rio Grande do Sul. Pesquisas-Botânica, 59/2: 171-190.

[19] Marchioretto, M.S.; Miotto, S.T.S. & Siqueira, J.C. (2009). O gênero Hebanthe (Amaranthaceae) no Brasil. Rodriguésia, 60/4: 783-798.

[20] Marchioretto, M.S.; Miotto, S.T.S. & Siqueira, J.C. (2010). O gênero Pfaffia Mart. (Amaranthaceae) no Brasil. Hoehnea, 37/3: 461-511.

[21] Marchioretto, M.S.; Senna, L. & Siqueira, J.C.de. (2012). Amaranthaceae. In: Lista de Espécies da Flora do Brasil. Available: http://floradobrasil.jbrj.gov.br/2010/FB000042 Accessed 2012 Jan 09.

[22] Siqueira, J.C. (1992). O gênero Gomphrena L. (Amaranthaceae) no Brasil. Pesquisas-Botânica, 43: 5-197.

[23] Siqueira, J.C. (1995). Fitogeografia das Amaranthaceae Brasileiras. Pesquisas-Botânica, 45: 5-21.

[24] Siqueira, J.C. (2002). Amaranthaceae. In: Wanderley, M.G.L.; Shepherd, G. & Giulietti, A.M. editors. Flora Fanerogâmica do Estado de São Paulo. São Paulo: Ed. FAPESP-HUCITEC. Pp. 11-30.

[25] Siqueira, J.C. (2004). Amaranthaceae: padrões de distribuição geográfica e aspectos comparativos dos gêneros Africanos e Sulamericanos. Pesquisas-Botânica, 55: 177-185.

[26] Holmgren, P. K.; Holmgren, N.H. & Barnett, L.C. (1990). Index Herbariorum - Part I: The Herbaria of the World. New York, IAPT. 693 p.

[27] Thiers, B.M. (2012). Index Herbariorum: A Global Directory of Public Herbaria and Associated Staff. In: NYBG Herbarium. Available: http://sciweb.nybg.org/science2/IndexHerbariorum.asp Accessed 2012 Feb 24.

[28] Kraus, J.E. & Arduin, M. (1997). Manual básico de métodos em morfologia vegetal. Rio de Janeiro: EDUR. 198 p.

[29] Souza, W.de. (Ed.) (2010). Técnicas básicas de microscopia eletrônica aplicadas às ciências biológicas, 3.ed. Rio de Janeiro: Sociedade Brasileira de Microscopia. 357 p.

[30] Shiomi, N.; Odera, S.; Vieira, C.C.J. & Figueiredo-Ribeiro, R.C.L. (1996). Structure of fructan polymers from tuberous roots of Gomphrena macrocephala (Amaranthaceae) from the cerrado. New Phytologist, 133/4: 643-650.

[31] Figueiredo-Ribeiro, R.C.L.; Dietrich, S.M.C.; Carvalho, M.A.M.; Vieira, C.C.J.; Isejima, E.M.; Dias-Tagliacozzo, G.M.; Tertuliano, M.F. (1982). As múltiplas utilidades dos frutanos – reserva de carboidratos em plantas nativas do cerrado. Ciência Hoje, 14/84: 16-18.

[32] Fank-de-Carvalho, S.M. (2004). Contribuição ao conhecimento botânico de Gomphrena arborescens L.f. (Amaranthaceae) - estudos anatômicos e bioquímicos. Instituto de Ciências Biológicas, Mestrado em Botânica, Universidade de Brasília, Brasília, DF, Brazil. (Dissertation). 139 p.

[33] Corrêa, M.P. (1931). Dicionário das plantas úteis do Brasil e das exóticas cultivadas, Vol. II. Rio de Janeiro: Ministério da Agricultura. 707 p.

[34] Barros, M.G.A.E. (1982). Plantas medicinais – usos e tradições em Brasília – DF. Oréades, 8/14-15: 140-149

[35] Almeida, S.P. de; Proença, C.E.B.; Sano, S.M. & Ribeiro, J.F. (1998). Cerrado - espécies vegetais úteis. Planaltina: Embrapa. 464 p.

[36] Kardošová, A.; Ebringerová, A; Alföldi, J.; Nosál'ová, G.; Fraˇnová, S. & Hˇr´ıbalová, V. (2003). A biologically active fructan from the roots of Arctium lappa L., var. Herkules. International Journal of Biological Macromolecules, 33: 135–140.

[37] Kunert, O.; Haslinger, E.; Schmid, M.G.; Reiner, J.; Bucar, F.; Mulatu, E.; Abebe, D. & Debella, A. (2000). Three Saponins, a Steroid, and a Flavanol lycoside from Achyrantes aspera. Monatshefte für Chemie, 131/2: 195-204.

[38] Sage, R.F.; Sage, T.L.; Pearcy, R.W.; Borsch, T. (2007). The taxonomic distribution of C4 photosynthesis in Amaranthaceae sensu stricto. American Journal of Botany, 94/12: 1992-2003.

[39] Siqueira, J.C. (1987). Importância alimentícia e Medicinal das Amaranthaceae do Brasil. Acta Biologica Leopoldense, 1: 99-110.

[40] Lorenzi, H. & Matos, F. J. A. (2008). Plantas medicinais no Brasil - nativas e exóticas, 2.ed. Nova Odessa: Editora Plantarum. 576 p.

[41] Lorenzi, H. & Souza, H. M. (2008). Plantas ornamentais no Brasil - arbustivas, herbáceas e trepadeiras, 4.ed.. Nova Odessa: Editora Plantarum. 1120 p.

[42] Fank-de-Carvalho, S.M. (2011). Contribuição ao conhecimento da anatomia, micromorfologia e ultraestrutura foliar de Amaranthaceae do Cerrado. Instituto de Biologia, Doutorado em Biologia Celular e Estrutural, Universidade Estadual de Campinas, Campinas, SP, Brazil. (Thesis). 107 p.

[43] MMA (2008). Lista Nacional das Espécies da Flora Brasileira Ameaçadas de Extinção. In: Instrução Normativa no. 6 do Ministério do Meio Ambiente, de 23 de setembro de 2008. Availale: http://www.mma.gov.br/estruturas/179/_arquivos/179_05122008033615.pdf Accessed 2012 Apr 10.

[44] RS (2003). Lista Final das Espécies da Flora Ameaçadas – RS. In: Decreto estadual n 42.099, Estado do Rio Grande do Sul, publicado em 01/jan/2003. Available: http://www.fzb.rs.gov.br/downloads/flora_ameacada.pdf Acessed 2012 Apr 10.

[45] Smith, L.B. & Downs, R.J. (1972). Amarantháceas. In: Reitz, R. editor. Flora Ilustrada Catarinense. Itajaí: Herbário Barbosa Rodrigues. Pp. 1-110.

[46] SP (2004). Lista oficial das espécies da flora do Estado de São Paulo ameaçadas de extinção. In: Resolução SMA 48, Diário Oficial do Estado de São Paulo - Meio Ambiente, 22/set/2004. Available: http://www.ibot.sp.gov.br/pesquisa_cientifica/restauracao_ecologica/resolu%C3%A7%C3%A3o_%20sma48.pdf Accessed 2012 Apr 10.

[47] Rajendrudu, G.; Prasad, J.S.R. & Rama Das, V.S. (1986). C3-C4 species in Alternanthera (Amaranthaceae). Plant Physiology, 80: 409-414.

[48] IUCN (2011). The IUCN Red List of Threatened Species. In: International Union for Conservation of Nature and Natural Resources. Available: http://www.iucnredlist.org/apps/redlist/search Accessed 2012 Jan 21.

[49] Siqueira, J.C. (1988). Planta Medicinais – identificação e uso das espécies dos cerrados. São Paulo: Edições Loyola. 40 p.

[50] Almeida, A.M.R.; Fukushigue, C.Y.; Sartori, F.; Binneck, E.; Marin, S.R.R.; Inoue-Nagata, A.K.; Chagas, C.M.; Souto, E.R. & Mituti, T. (2007). Natural infection of Alternanthera tenella (Amaranthaceae) by a new potyvirus. Archives of Virology, 152: 2095–2099.

[51] Vasconcellos, J.M. DE O. (1985a). Amaranthaceae do Rio Grande do Sul, Brasil-II. Roessléria, 7/2: 107-137.

[52] Vasconcellos, J.M. DE O. (1985b). Amaranthaceae do Rio Grande do Sul, Brasil-III - Gêneros Celosia L. e Chamissoa H.B.K. Roessléria, 7/3: 165-182.

[53] Botsaris, A.S. (2007). Plants used traditionally to treat malaria in Brazil: the archives of Flora Medicinal. Journal of Ethnobiology and Ethnomedicine, 3: 1-8. Available: http://www.ethnobiomed.com/content/3/1/18 Accessed 2012 Jan 9.

[54] Fank-de-Carvalho, S.M. & Graciano-Ribeiro, D. (2005). Arquitetura, anatomia e histoquímica das folhas de Gomphrena arborescens L. f. (Amaranthaceae). Acta Botanica Brasilica, 19/2: 377-390.

[55] Vasconcellos, J. M. DE O. (1986a). Amaranthaceae do Rio Grande do Sul, Brasil V. Gêneros Pfaffia Mart. e Gomphrena Mart. Roessléria, 8/2: 75-127.

[56] Balbach, A. (1957). As plantas curam, 5.ed. São Paulo: Ed. Missionário. Pp. 299-300.

[57] Silva, I.M. (1983). A flora na vida do índio Karajá. Atas da Sociedade Botânica-secção RJ, 2/3: 21-32.

[58] ES (2005). Lista de Espécies Ameaçadas de Extinção no Espírito Santo. In: Decreto Nº 1.499-R, Diário Oficial do Estado do Espírito Santo, 14/jun/2005. Available: http://www.meioambiente.es.gov.br/web/flora.htm Accessed 2012 Apr 10.

[59] Siqueira, J.C. (1981). Utilização popular das plantas do cerrado. São Paulo: Editora Loyola. 60 p.

[60] Sanches, N.R.; Galletto, R.; Oliveira, C.E.; Bazotte, R.B. and Cortez, D.A.G. (2001). Avaliação do potencial anti-hiperglicemiante da Pfaffia glomerata (Spreng.) Pedersen (Amaranthaceae). Acta Scientiarum - Biological Sciences, 23/2: 613-617

[61] Reeves, R.D.; Baker, A.J.M.; Becquer, T.; Echevarria, G. & Miranda, Z.J.G. (2007). The flora and biogeochemistry of the ultramafic soils of Goiás state, Brazil. Plant and Soil, 293: 107–119.

[62] Appezzato-da-Gloria, B. (2003). Morfologia de sistemas subterrâneo - histórico e evolução do conhecimento no Brasil. Ribeirão Preto: A.S. Pinto Editor. 80 p.

[63] Borsch, T. & Pedersen, T.M. (1997). Restoring the Generic Rank of Hebanthe Martius (Amaranthaceae). Sendtnera, 4: 13-31.

[64] Martius, C.F.P. v. (1825). Beitrag Zur Kenntnis der natürlichen familie der Amaranthaceen. Bonn: Eduard Weber's Buchhandlung. 321 p.

[65] Martius, C.F.P.v. (1826). Amaranthaceae. In: Martius, C.F.P.v. editor. Nova genera et species plantarum – in itinere per Brasiliam, Vol.II. Munich, Typis C. Wolf. Pp. 1-64.

[66] Moquin-Tandon, A. (1849). Amaranthaceae. In: Candolle A. De editor. Prodromus Systematis Naturalis Regni Vegetabilis, Vol.13, part.2. Paris: Victoris Masson Ed. Pp. 419-423.

[67] Pedersen, T.M. (1967). Studies in South American Amaranthaceae. Darwiniana, 14/2-3: 430-462.

[68] Pedersen, T.M. (1976). Estudios sobre Amarantáceas sudamericanas – II. Darwiniana, 20/1-2: 269-303.

[69] Pedersen, T.M. (1990). Studies in South American Amaranthaceae - III (incluinding one amphi-Atlantic species) - Bull. Mus. Natl. Hist. Nat., B. Adansonia, Sér.4, 12/1: 69-97.

[70] Pedersen, T. M. (1997). Studies in South American Amaranthaceae – IV - Bull. Mus. Natl. Hist. Nat., B. Adansonia, Sér.3, 19/2: 217-251.

[71] Pedersen, T.M. (2000). Studies in South American Amaranthaceae – V. Bonplandia, 10/1-4: 83-112.

[72] Seubert, M. (1875). Amaranthaceae. In: Martius, C.F.P. Von, Endlicher & Urban editors.Flora Brasiliensis, Vol.V/part. 1. Munich, Typografia Regia. Pp.188-202.

[73] Stützer, O. (1935). Die Gattung Pfaffia mit einem Anhag neur Arten von Alternanthera. Feddes Repertorium Specierum Novarum Regni Vegetabilis, 88: 1-49.

[74] Suessenguth, K. (1934). Neue und kritische Amarathaceen aus Süd und Mittelamerika. Feddes Repertorium Specierum Novarum Regni Vegetabilis, 35: 298-337.

[75] Suessenguth, K. (1938). Amaranthaceen-Studien – I - Amaranthaceae aus Amerika, Asien, Australien. Feddes Repertorium Specierum Novarum Regni Vegetabilis, 44: 36-48.

[76] Vasconcellos, J. M. DE O. (1986b). Amaranthaceae do Rio Grande do Sul, Brasil – IV - Gêneros Pseudoplantago Suess., Iresine BR. e Blutaparon Rafin. Roessléria, 8/1: 03-15.

[77] Fank-de-Carvalho, S. M.; Marchioretto, M.S. & Báo, S.N. (2010a). Anatomia foliar, morfologia e aspectos ecológicos das espécies da família Amaranthaceae da Reserva Particular do Patrimônio Natural Cara Preta, em Alto Paraíso, Goiás, Brasil. Biota Neotropica, 10: 77-86. Available: http://www.biotaneotropica.org.br/v10n4/en/download?article+bn01310042010+abstract Accessed 2012 Apr 10.

[78] Fank-de-Carvalho, S.M.; Gomes, M.R.A.; Silva, P.I.T. & Báo, S.N. (2010b). Leaf surfaces of Gomphrena spp. (Amaranthaceae) from Cerrado biome. Biocell (Mendoza), 34/1: 23-35.

[79] Welkie, G.W. & Caldwell, M. (1970). Leaf anatomy of species in some dicotyledon families as related to the C_3 and C_4 pathways of carbon fixation. Canadian Journal of Botany, 48/12: 2135-2146.

[80] Carolin, R.C.; Jacobs, S.W.L. & Vesk, M. (1978). Kranz cells and mesophyll in the Chenopodiales. Austalian Journal of Botany, 26/5: 683-698.

[81] Kadereit, G.; Borsh, T.; Weising, K. & Freitag, H. (2003). Phylogeny of Amaranthaceae and Chenopodiaceae and the evolution of C_4 photosynthesis. International Journal of Plant Sciences, 164/6: 959-986.

[82] Sage, R.F. (2004). The Evolution of C_4 photosynthesis. New Phytologist, 161/2: 341-370.

[83] Estelita-Teixeira, M.E. & Handro, W. (1984). Leaf ultrastructure in species of Gomphrena and Pfaffia (Amaranthaceae). Canadian Journal of Botany, 62/4: 812-817.

[84] Gavilanes, M.L. (1999). Estudo anatômico do eixo vegetativo de plantas daninhas que ocorrem em Minas Gerais. 1. Anatomia foliar de Gomphrena celosioides Mart. (Amaranthaceae). Ciência e Agrotecnologia, 23/4: 881-898.

[85] Handro, W. (1964). Contribuição ao estudo da venação e anatomia foliar das amarantáceas dos cerrados. Anais da Academia Brasileira de Ciências, 36/4: 479-499.

[86] Handro, W. (1967). Contribuição ao estudo da venação e anatomia foliar das amarantáceas dos cerrados. II - Gênero Pfaffia. Anais da Academia Brasileira de Ciências, 39/3-4: 495-506.

[87] De Souza, A. (2004). Plano de Utilização RPPN Cara Preta. Nativa Proteção Ambiental, Alto Paraíso, Brazil. (Report). 327 p.

[88] Felfili, J.M., Rezend, A,V. & Silva-Júnior, M.D. (orgs). (2007). Biogeografia do Bioma Cerrado - vegetação e solos da Chapada dos Veadeiros. Brasília: Editora Universidade de Brasília/FINATEC. 256 p.

Chapter 11

The Ongoing Shift of Mediterranean Coastal Fish Assemblages and the Spread of Non-Indigenous Species

Stefanos Kalogirou, Ernesto Azzurro and Michel Bariche

Additional information is available at the end of the chapter

http://dx.doi.org/10.5772/50845

1. Introduction

Geological history of life on earth tells that continents have been isolated for long periods. It also reveals that collisions of land masses as well as lower sea levels allowed the spread of fauna and flora (Stachowicz and Tilman, 2005). In today's seas, marine communities are being altered and remodelled at an unprecedented rate, when compared to natural changes which occured over geological times. While many marine species populations are dwindling due to overfishing and habitat destruction (Jackson *et al.*, 2001), other species are invading new areas through anthropogenic vectors (Carlton 1985, Galil 2006, Galil et al. 2007). During the last centuries, human transport has increased the number of non-indigenous species (NIS) introductions. For example, half of the plant species of Hawaii are exotics (Sax *et al.*, 2002) as are about 20% of plants in California bay (Sax, 2002) and about 18% of fish species in the eastern Mediterranean Sea (Golani et al. 2002, Golani et al. 2006, EastMed 2010, Golani 2010).

Understanding invasion ecology requires a good knowledge of ecological processes in the systems under study, prior to invasion. Diversity, structure, and function of natural communities would give insights into fundamental ecological processes which could in turn give a better understanding of potential effects following the introduction of NIS.

From a societal perspective, species invasions might pose serious threats to human economic interests and health (Yang *et al.*, 1996; Sabrah *et al.*, 2006; Katikou *et al.*, 2009). Species invasions have also been considered to have negative impacts on native biodiversity (Reise *et al.*, 2006; Streftaris and Zenetos, 2006; Galil, 2007; Lasram and Mouillot, 2008; Zenetos *et al.*, 2009). Furthermore, invasions interacts with other disturbing factors to the marine ecosystem functioning such as habitat destruction, pollution and climate change

(Rilov and Crooks, 2009). Disturbance caused by habitat destruction may open up space for invaders but space can also be released by the invaders themselves. Consider the example given by Rilov and Galil (2009) where two non-indigenous siganids might have modified the competition between algae and mussels through intensive grazing, thus providing space for the non-indigenous mussel *Brachiodontes pharaonis*. Pollution can make environmental conditions less tolerable for native species, and perhaps provide opportunities for opportunists, among which non-indigenous species could be found (Occhipinti-Ambrogi and Savini, 2003; Wallentinus and Nyberg, 2007). Global warming is causing the shifts and poleward migrations of many taxa that are now extending their biogeographical range (Parmesan and Yohe, 2003; Perry *et al.*, 2005). This tendency is also observed in the Mediterranean Sea (Bianchi, 2007; Raitsos *et al.*, 2010). Some species, typically confined to the warmer parts of the Mediterranean, are currently colonizing the northern sectors. This phenomenon has been termed "meridionalization" (Azzurro, 2008).

The increase of water temperature is also allowing the success of tropical exotic species in the Mediterranean Sea, a phenomenon that has been called 'tropicalization' (Bianchi and Morri, 2003). Conditions facilitating invasions are usually related to the physical and biological attributes of the new colonized habitats. Biological impact studies include mostly those species of economic interests (e.g. fisheries) (Streftaris and Zenetos, 2006), human health (e.g. toxic species) (Yang *et al.*, 1996; Bentur *et al.*, 2008; Katikou *et al.*, 2009) and biodiversity (e.g. competition with indigenous species or habitat modifiers) (Golani, 1993a; Golani, 1994; Bariche *et al.*, 2004; Azzurro *et al.*, 2007a; Kalogirou *et al.*, 2007; Wallentinus and Nyberg, 2007; Bariche *et al.*, 2009). A lot of research has also focused on the factors controlling success or failure of invasive species by considering mechanisms of interactions between indigenous and NIS. There is no universal model explaining the mechanisms controlling the success or failure of an invading species (Stachowicz and Tilman, 2005). As far as the Mediterranean Sea is concerned, important mechanisms include competition for resources or space (Bariche *et al.*, 2004; Kalogirou *et al.*, 2007), top-down forces (Goldschimdt *et al.*, 1993), herbivory (Lundberg and Golani, 1995; Galil, 2007), and parasites (Diamant, 2010).

A widely cited theory in invasion ecology is about the relationship between diversity and invasibility of an ecosystem (i.e. more diverse communities should be more resistant to invasion) (Leppäkoski and Olenin, 2000). The mechanism suggests that as species richness increases the competition intensifies and less food resources remain available for new colonizers (MacArthur, 1955; Levine and D' Antonio, 1999). Less diverse ecosystems possessing fewer species and simpler food-web interactions would therefore provide available niches for the establishment of NIS. This hypothesis is known as the "biotic resistance hypothesis" (Levine and Adler, 2004). As an aid to understand this mechanism, both observational and experimental approaches have been applied with conflicting results (Levine and D' Antonio, 1999). Studies that employ both observational and experimental approaches show that high diverse systems does reduce invasion success (Stachowicz and Tilman, 2005). There is a long history of theoretical discussions about the relationship between species richness and productivity or stability of a system. Threats to global species diversity caused by human activities have raised concern on the consequences of species losses to the functioning of ecosystems. In ecology, this concern has received a lot of attention. During the last 20 years, experimental tests of the relationship between species richness and ecosystem processes such as productivity, stability and invasibility have increased rapidly (Stachowicz and Whitlatch, 1999).

Other theories go back to the work of Darwin. Darwin's "naturalization hypothesis" predicts that NIS are less prone to invade areas where closely related species are present. Those species would compete with their relatives and would encounter predators and pathogens. An opposing view is the "pre-adaptation" hypothesis predicting that NIS should succeed in areas where indigenous closely related species are present because they are more likely to share traits that pre-adapt them to their own environment. So far, these theories have been seldom tested on fish species and no clear pattern has emerged so far for these taxa (Ricciardi and Mottiar (2006). Ricciardi and Mottiar (2006) agreed with Moyle and Light (1996) that success is primarily determined by competitive interactions (e.g. "biotic resistance" hypothesis), propagulae pressure and environmental abiotic factors (i.e. the degree to which NIS physiological tolerances are compatible to local physical conditions). Rapid changes in environmental conditions, caused by human activities, have also been mentioned as to increase invasiveness (Occhipinti-Ambrogi and Savini, 2003). Habitats that lack predators are also suggested to be more prone to introductions of NIS (Moyle and Light, 1996). There is also a higher risk of further establishment of species in habitats that have already been invaded, referred as the "invasional meltdown" (Simberloff and Von Holle, 1999; Ricciardi, 2001). In a study from Great Lakes, Ricciardi (2001) found support for the "invasional meltdown" hypothesis by showing that positive interactions (mutualistic) among NIS are more common than negative (competitive). In further support of the "invasional meltdown" hypothesis, Ricciardi (2001) showed that exploitative interactions (e.g. predator-prey) among NIS are strongly asymmetrical to the benefit of one invading species at a negligible cost to another.

2. Current patterns of change of the Mediterranean biota

In the last century the Mediterranean Sea has been a receptacle of NIS, most of them arrived by mean of direct or indirect mediation of humans. Today, the Mediterranean Sea can be considered as one of the main hotspots of marine bio-invasions on earth (Quignard and Tomasini, 2000), and is by far the major recipient of NIS among European seas including macrophytes, invertebrates and fishes (Streftaris *et al.*, 2005; Zenetos *et al.*, 2010). The Mediterranean is unique because of its connection to the Indo-West Pacific realm via the Suez Canal (Fig. 1), allowing the so called Lessepsian migration (Por, 1978). The rate of this immigration has increased in recent decades and has ecological, social and economic impacts (Zenetos *et al.*, 2008; Bilecenoglu, 2010; Zenetos *et al.*, 2010). The Eastern Mediterranean basin is potentially more prone to introductions of subtropical and tropical NIS than the western basin. This has been mainly attributed to different physical and biological conditions between the two basins. It is to mention that the construction of the

Aswan Dam on the Nile River in 1966 reduced significantly the freshwater flood into the Mediterranean Sea. This led to an increased salinity of 2-3% along the Mediterranean coast of Egypt and to a reduction of the most important sources of nutrients in the eastern Mediterranean Sea (Galil, 2006). The damming of the Nile might have positively favoured the westward dispersion of Lessepsian NIS along the Northern African shores (Ben-Tuvia, 1973).

Figure 1. The Suez Canal

New terms have been recently created to describe current changes of the Mediterranean biodiversity. Due to the tropical nature of most of the exotic species that enter the Mediterranean, various authors have defined the process of entrance and spread of these organisms as 'tropicalization' (Bianchi and Morri, 2004; Bianchi, 2007). Another definition that has been used is "demediterranization"(Quignard and Tomasini, 2000) that put the emphasis on the process of biotic homogenization of the Mediterranean Sea. Instead Massuti et al. (2010) used the term 'meridianization' to indicate the increasing divergence (in terms of composition of the biological communities) between the Eastern and Western sectors of the Mediterranean, due to the continuous influx of Lessepsian and Atlantic biota. This latter term, 'meridianization' should not be confused with 'meridionalization', which instead would indicate the northward expansion of southern ('meridional') species towards the northern sectors of the basin (Azzurro, 2008). Several indigenous species such as *Sparisoma cretense* and *Thalassoma pavo* have been regarded as "meridional" (CIESM, 2008) since they have been recently found to reproduce and have established populations in the coldest part of the Mediterranean Sea (Ligurian Sea) (Guidetti *et al.*, 2002). Additionally, a reduction of temperate species followed the increase of tropical species in the Ligurian Sea (Bianchi and Morri, 2003). These trends of change in the biological diversity of the Mediterranean Sea are part of a general reshuffling of species, that is happening at the global level (Vitousek *et al.*, 1997), and climate warming would contribute to promote the shifts of species distribution, an evidence that is particularly clear among marine fishes (Perry *et al.*, 2005). As fish are particularly sensitive to changes in water temperatures, physiological processes directly alters behaviour, generating active movement and migratory patterns of these organisms (Roessig *et al.*, 2004). Other indirect effects of climate change, such as those related to the change of currents, could affect larval dispersal, retention and recruitment of marine organisms (Bianchi and Morri, 2004).

More than 700 fish species inhabit the Mediterranean Sea with a general decrease in number moving eastwards (Quignard and Tomasini, 2000; Lasram *et al.*, 2009) . Among these, at least 80 are non-indigenous of Indo-West Pacific and Red Sea origin (Cicek and Bilecenoglu, 2009; Bariche, 2010b; EastMed, 2010; Golani, 2010; Bariche, 2011b; Sakinan and Örek, 2011; Salameh, 2011; Bariche and Heemstra, 2012). The abundance of these non-indigenous species is not well documented. The list of non-indigenous fish species with quantitative information from the Mediterranean Sea can be found in Table 1.

Family	Species	Reference
Atherinidae	*Atherinomorus forskalii*	(Bariche *et al.*, 2007; Shakman and Kinzelbach, 2007)
Callionymidae	*Callionymus filamentosus*	(Gucu and Bingel, 1994; Kalogirou *et al.*, 2010)
Carangidae	*Alepes djedaba*	(Shakman and Kinzelbach, 2007)
Clupeidae	*Etrumeus teres*	(Bariche *et al.*, 2006; Bariche *et al.*, 2007; Carpentieri *et al.*, 2009)
Clupeidae	*Herklotsichthys punctatus*	(Bariche *et al.*, 2006; Bariche *et al.*, 2007; Carpentieri *et al.*, 2009)
Dussumieriidae	*Dussumieria elopsoides*	(Goren and Galil, 2005; Bariche *et al.*, 2007)
Fistulariidae	*Fistularia commersonii*	(Shakman and Kinzelbach, 2007; Carpentieri *et al.*, 2009; Kalogirou *et al.*, 2010; Kalogirou *et al.*, 2012b)
Hemiramphidae	*Hemiramphus far*	(Shakman and Kinzelbach, 2007; Carpentieri *et al.*, 2009)
Holocentridae	*Sargocentron rubrum*	(Carpentieri *et al.*, 2009)
Labridae	*Pteragogus pelycus*	(Kalogirou *et al.*, 2010)
Leiognathidae	*Leiognathus klunzingeri*	(Gucu and Bingel, 1994)
Monacanthidae	*Stephanolepis diaspros*	(Gucu and Bingel, 1994; Harmelin-Vivien *et al.*, 2005; Shakman and Kinzelbach, 2007; Carpentieri *et al.*, 2009; Kalogirou *et al.*, 2010; Kalogirou *et al.*, 2012b)

Mullidae	*Upeneus moluccensis*	(Gottlieb, 1960; Oren *et al.*, 1971; Gucu and Bingel, 1994; Golani and Ben-Tuvia, 1995; Sonin *et al.*, 1996; Goren and Galil, 2005; Harmelin-Vivien *et al.*, 2005; Carpentieri *et al.*, 2009; Kalogirou *et al.*, 2010)
Mullidae	*Upeneus pori*	(Gucu and Bingel, 1994; Golani and Ben-Tuvia, 1995; Goren and Galil, 2005; Shakman and Kinzelbach, 2007; Carpentieri *et al.*, 2009; Kalogirou *et al.*, 2010; Kalogirou *et al.*, 2012b)
Nemipteridae	*Nemipterus randalli*	(Carpentieri *et al.*, 2009)
Pempheridae	*Pempheris vanicolensis*	(Carpentieri *et al.*, 2009)
Scaridae	*Scarus ghobban*	(Bariche and Saad, 2008; Carpentieri *et al.*, 2009)
Scombridae	*Scomberomorus commerson*	(Shakman and Kinzelbach, 2007; Carpentieri *et al.*, 2009)
Siganidae	*Siganus luridus*	(Gucu and Bingel, 1994; Bariche *et al.*, 2004; Harmelin-Vivien *et al.*, 2005; Shakman and Kinzelbach, 2007; Carpentieri *et al.*, 2009; Kalogirou *et al.*, 2010; Kalogirou *et al.*, 2012b)
Siganidae	*Siganus rivulatus*	(George and Athanassiou, 1967; Bariche *et al.*, 2004; Bariche, 2005; Harmelin-Vivien *et al.*, 2005; Shakman and Kinzelbach, 2007; Carpentieri *et al.*, 2009; Kalogirou *et al.*, 2010; Kalogirou *et al.*, 2012b)
Sphyraenidae	*Sphyraena chrysotaenia*	(Golani and Ben-Tuvia, 1995; Carpentieri *et al.*, 2009; Kalogirou *et al.*, 2010; Kalogirou *et al.*, 2012a)
Sphyraenidae	*Sphyraena flavicauda*	(Kalogirou *et al.*, 2012b)
Synodontidae	*Saurida undosquamis*	(Oren *et al.*, 1971; Ben-Yami and Glaser, 1974; Gucu and Bingel, 1994; Golani and Ben-Tuvia, 1995; Galil and Zenetos, 2002; Goren and Galil, 2005; Harmelin-Vivien *et al.*, 2005; Shakman and Kinzelbach, 2007; Carpentieri *et al.*, 2009)
Pempheridae	*Pempheris vanicolensis*	(Harmelin-Vivien *et al.*, 2005)
Pomacentridae	*Sarogentrum rubrum*	(Harmelin-Vivien *et al.*, 2005)
Tetraodontidae	*Lagocephalus sceleratus*	(Carpentieri *et al.*, 2009; Kalogirou *et al.*, 2010; Aydin, 2011; Kalogirou *et al.*, 2012b)
Tetraodontidae	*Lagocephalus spadiceus*	(Carpentieri *et al.*, 2009)
Tetraodontidae	*Lagocephalus suezensis*	(Carpentieri *et al.*, 2009; Kalogirou *et al.*, 2010)

Table 1. List of the non-indigenous fish species of Indo-Pacific and Red Sea origin with references on quantitative information in abundance in the Mediterranean Sea

The arrival of these invaders raises plain concern on the ecological and economic impact that such migrants have but the available information is still scarce (Rilov and Galil, 2009) and there is an obvious lack of knowledge. It is at the same time obvious that the ecological effect of some species is significant (Kalogirou *et al.*, 2007; Bariche *et al.*, 2009). Competitive exclusion and displacement of native species are often potential expectations in ecological studies (Bariche *et al.*, 2004; Galil, 2007). A noteworthy example is the presence of the two Lessepsian herbivorous *Siganus rivulatus* and *S. luridus*. The impact of the two species on indigenous species has received some scientific attention and comments (Bariche *et al.*, 2004; Azzurro *et al.*, 2007b; Galil, 2007; Golani, 2010). Since their first record in the Mediterranean Sea, respectively 1924 and 1955, the two species have established significant populations in the eastern basin and have spread westwards as far as Sicily and Tunisia (Rilov and Galil, 2009). They constituted one third of the fish biomass over hard bottoms and their contribution to the guild of herbivorous fish species in shallow coastal areas reached 80% along the Levantine coast (Goren and Galil, 2001; Bariche *et al.*, 2004). The two siganids are commercially important in the eastern Mediterranean Sea. The native herbivorous *Sarpa salpa* (Sparidae) was relatively an abundant species and a possible competition with *S. rivulatus* was already highlighted along the coast of Lebanon (Gruvel, 1931; George and Athanassiou, 1967). Nowadays, records of *S. salpa* from Lebanon are very scarce and it has been suggested that the native species has been outcompeted by the Lessepsian invaders (Bariche *et al.*, 2004; Galil, 2007). In contradiction, Golani (2010) rejected this hypothesis and considered that Gruvel was unable to recognize *S. salpa* from another common sparid *Boops boops*. We consider that this assumption is not accurate as Gruvel described separately *B. boops* (as *Box vulgaris* C.V. = *Box boops* L.) and *S. salpa* (*Box salpa* Cuv.) and made clear reference to their local abundance and flesh palatability, showing that he knew well how to identify them (Gruvel, 1931). The author was assessing the fishery resources of the Levantine coast by mean of experimental trawling. The clear competitive superiority of siganids may be due to a greater adaptability to fluctuating environmental conditions and other biological advantages (Bariche *et al.*, 2004). Moreover, macrophytes are considered to be abundant along the eastern Mediterranean coast (Lipkin and Safriel, 1971; Lundberg and Golani, 1995). Lundberg & Golani (1995) compared the stomach contents of the siganids in relation to food availability in the source (Red Sea) and invaded (Mediterranean Sea) areas. They found a scarcity of food species underwater and abundance on vermetid reefs and beach rocks, which are situated at sea level and are thus rather inaccessible (Lundberg and Golani, 1995; Lundberg *et al.*, 2004). *Siganus rivulatus* and *S. luridus* were shown to be selective in the eastern Mediterranean, when macrophytes are diverse and abundant and will eat what is available when food is scarce (Bariche, 2006). Selectivity being more important in the Mediterranean Sea is probably due to a larger choice in food species (Golani, 1993b; Lundberg *et al.*, 2004). Nevertheless, the success of siganids also shows a larger trophic or eco-physiological flexibility in the Mediterranean Sea (Hassan *et al.*, 2003; Bariche, 2006). This also reveals the lack of available data regarding herbivorous Lessepsian siganids being better competitors than indigenous ones for food until it is proven that trophic resources constitute the most important limiting factor (Golani, 2010). Finally, negative consequences of fish invasions are not only restricted to native fish communities. The intensive grazing of macrophytes by the same siganids might have reduced the competition between algae and mussels, and thus released space for the establishment of a non-indigenous mussel, *Brachiodontes pharaonis*, on rocky shores along the coasts of the Levant (Rilov and Galil, 2009).

3. The case of the invasive pufferfish *Lagocephalus sceleratus*: Ecological consequences, economic impacts and risks for human health

Pufferfishes are marine fish species that are distributed in tropical and subtropical areas of the Atlantic, Indian and Pacific Ocean. Puffers include 121 species within the Tetraodontidae family among which nine (*T. flavimaculosus, L. sceleratus, L. spadiceus, L. suezensis, S. pachygaster, S. spengleri, T. spinosissimus, S. marmoratus, L. lagocephalus*) are found in the Mediterranean Sea. Some puffers contain the strongest paralytic toxin known to date, tetrodotoxin (Sabrah *et al.*, 2006). European legislation (854/2004/EC) states that toxic fish of the Tetraodontidae family should not enter the European markets. In a global perspective, occasional accidental poisonings have lead to numerous human deaths, the majority of which have been documented in southeastern Asia, including Malaysia, Taiwan, Hong Kong, and Korea (Kan *et al.*, 1987; Yang *et al.*, 1996).

Lagocephalus sceleratus received considerable public attention shortly after its first record in 2003 from the Gökova bay in the south-eastern coasts of the Aegean Sea due to the presence of significant amounts of tetrodotoxin (Akyol *et al.*, 2005). The distribution of *L. sceleratus* is currently limited to the eastern Mediterranean Sea but the species is showing a rapid spread westward.

Few dozens of tetrodotoxin poisoning cases occurred along the Levantine coast and in Cyprus (Bentur *et al.*, 2008). *Lagocephalus spadiceus*, which has been present in the Mediterranean for several decades, was rarely marketed but regularly consumed by fishermen in Lebanon and Syria without any noticeable concern. The sudden appearance of the highly toxic *L. sceleratus* had a serious impact on those fishermen, who used to eat *L. spadiceus* (Bariche *pers. obs.*). The large numbers of *L. sceleratus* that have been caught by coastal fishermen in the eastern Mediterranean has initiated major national efforts to alert fishermen and the public about the toxicity of this fish (Fig. 2).

These efforts have included setting up posters warning the public about the lethal effects if consumed, but also that small individuals could easily be misidentified with other small commercial edible species such as *Spicara smaris*, *Boops boops* and *Atherina hepsetus* (Kalogirou pers. obs.). Studies from the Mediterranean Sea showed that there is a significant positive correlation between toxicity levels and size of fish (Katikou *et al.*, 2009). According to the results of Katikou *et al.* (2009) individuals smaller than 16 cm in length do not possess toxicity levels that could be lethal. This reduces the risks in connection with misidentification since commercial *S. smaris*, *B. boops* and *A. hepsetus* rarely exceed this size.

Figure 2. A sampling from sandy bottoms on the coast of Rhodes Island in the southeastern Aegean Sea with a large number of Lagocephalus sceleratus individuals

Lagocephalus sceleratus has also been considered an economical pest by fishermen since it is affected the local fish markets in three ways; deterring customers from buying fish, introducing additional work to discard toxic fish and predating on local stocks of commercially important squids and octopuses (Fig. 3). Lebanese fishermen are considerably affected by the damage caused by *L. sceleratus* on their fishing gears and catches. In fact, the puffers damage considerably fishing nets and longlines with their strong fused teeth. This is evident from the numerous complaints done by Lebanese fishermen and the presence of fishing hooks and fishing nets fragments respectively in oral cavities and stomachs (Bariche, pers. obs.).

Lagocephalus sceleratus ranked among the 100 'worst' Invasive Alien Species (IAS) in the Mediterranean Sea with profound social and ecological impacts (Streftaris and Zenetos, 2006). Social impacts are obvious due to toxicity but the lack of quantitative data does not support ecological impacts. Despite the successful establishment in the eastern Mediterranean, little is known concerning the ecology of the fish (Kalogirou *et al.*, 2010; Kalogirou *et al.*, 2012b).

An invading species might sometimes go to a peak of density and then decline, a path often called boom and bust (Williamson and Fitter, 1996). This path followed the NIS bluespotted cornetfish *Fistularia commersonii* in Rhodes Island, SE Aegean Sea (Kalogirou, pers. obs.). When a NIS is established into waters where its preferred food is under-utilized by

Figure 3. Lagocephalus sceleratus individual from the coast of Rhodes Island in the south-eastern Aegean Sea

indigenous species the resulting population explosion is later brought into equilibrium with available resources. Even though this dynamic leads to the significant reduction of the invading species population size, only very few studies have reported subsequent extinction of the NIS. Competition, despite strong advocacy (Moulton, 1993), seems to be the least likely explanation for most of the examples. Decline and extinction from a build-up of enemies (predators and pathogens) and lack of sufficient resources is more likely to be important explanations in failure of invading animals to establish permanent populations (Williamson and Fitter, 1996).

4. Do we need new methodologies to monitor current changes of Mediterranean fish diversity?

Concern has been expressed to the lack of monitoring, coordination, and study in relation to the changing diversity of the Mediterranean Sea. As a matter of fact, exotic fishes spreading in the Mediterranean Sea are usually found by chance as specific procedures for their detection are lacking (Azzurro, 2010). Consequently, the extent of these changes may be underestimated as usually happens in several other marine systems (Witenberg and Cock, 2001). Increasing efforts are being devoted to the survey of marine habitats but one of the major obstacles to research remains the lack of data at large geographical scales. This would be important to perceive temporal and spatial trends and to fill important existing information gaps. New methodologies involving local communities have recently proved to be successful in discovering trends of change in Mediterranean fish diversity (Azzurro *et al.*, 2011). As a matter of fact, collaboration with local communities are increasingly used to approach the study of large scale changes in the natural world. As a matter of facts some countries, such as Australia and the USA (California; Hawaii) have already started monitoring projects which involve community-based actions for the detection of marine invasive species. People are basically asked to 'monitor' the marine environment around them, in the course of their daily activities and to provide reports of invasions and various tools and detection kits have been developed all around the world with the aim to widely disseminate information about potential invaders to target communities. In a pilot study called 'alien fish alert' fishermen and divers of the Sicily Strait were asked to provide reports of all "unusual occurrences" (Azzurro, 2010). Given the familiarity of fishermen with local species, no training on fish taxonomy was considered necessary and no black list was proposed, with the following slogan: *"there is no need of any expertise in identifying alien species – those familiar with our sea will immediately recognize a 'strange' fish that they have never seen before – it is such records that we are after!"*. An awareness campaign was realized by means of media promotion, posters and personal interactions. As a matter of fact, this organization was found to provide researchers with an excellent tool for early detection of newly established NIS. These activities are encouraged by the European Strategy of Invasive Alien Species for the up-building of public awareness and collection and distribution of information. In a parallel way, a set of posters with NIS fish photos were distributed among fishermen and at fish auctions along the coast of Lebanon and at port authorities and local fish markets of SE Aegean Sea, Greece. These posters showed only selected fish species displaying characteristic and recognizable family appearances, such as Apogonidae, Chaetodontidae, Scaridae etc. Additionally, posters in Lebanon also provided a phone number with a promise of payment when a new species is collected and delivered. Many fishermen showed a positive response to the advertisement and several first records were collected as such (Bariche, 2010a; 2011a; Bariche and Heemstra, 2012).

The collaboration with local fishery communities has several advantages when the species to monitor are fishes in respect to other groups of organisms. Fishery landings also provide quantitative data, samples and additional information. In addition, the identification of many fishes is relatively easy and this is an obvious advantage for their detection (Fig. 3). Therefore, members of local fishery communities, with broad geographical distributions and familiarity of natural environments could play a dynamic role for the early detection of environmental changes.

Another significant example of innovative ideas to monitor fish diversity changes in the Mediterranean Sea was "Local Ecological Knowledge" (LEK). In recent years, LEK has emerged as an alternative information source on species presence or qualitative and

Figure 4. Sargocentrum rubrum individuals caught by trammel nets in Lebanon

quantitative indices of species abundance (Rasalato *et al.*, 2010). Local Ecological Knowledge can be defined as the information that a group of people have about local ecosystems. We usually rely on knowledge gained by individuals over their lifetimes, and not on what information has been handed through generations. To extract data and information from individuals' memory, semi-structured or unstructured conversations between the researcher and the participant were used, a practice commonly called "oral history". In a recent study, Azzurro et al. (2011) provided evidence of a trend for thermophilic taxa to increase in the Central Mediterranean Sea on the basis of a set of interviews to local fishermen. The study was based on interviews to local fishermen and divers with more than ten years of experience. Species mentioned in each interview were used to build a presence-absence dataset that provided extremely coherent results about the northward expansion of families such as Carangidae and Sphyraenidae, whose expansion was only previously noted by occasional records in the scientific literature. These new methodologies give us the chance to get information that otherwise cannot be obtained from the efforts of single researchers. Hopefully in the next future their potential will be increasingly exploited for the monitoring and the understanding of the biodiversity changes in the Mediterranean Sea.

Author details

Stefanos Kalogirou
Hellenic Centre for Marine Research, Hydrobiological Station of Rhodes, Greece

Ernesto Azzurro
ISPRA, National Institute for Environmental Protection and Research, Livorno, Italy

Michel Bariche
Department of Biology, American University of Beirut, Beirut, Lebanon

5. References

Akyol, O., Unal, V., Ceyhan, T. & Bilecenoglu, M. (2005). First confirmed record of *Lagocephalus sceleratus* Gmelin, 1789 in the Mediterranean Sea. *Journal of fish biology* 66, 1183-1186.

Aydin, M. (2011). Growth, reproduction and diet of pufferfish (*Lagocephalus sceleratus* Gmelin, 1789) from Turkey's Mediterranean sea coast. *Turkish Journal of Fisheries and Aquatic Sciences* 11, 589-596.

Azzurro, E., ed. (2008). *The advance of thermophilic fishes in the Mediterranean sea: overview and methodological questions*. Monaco: *CIESM Workshop Monographs*

Azzurro, E. (2010). Unusual occurences of fish in the Mediterranean Sea: an insight into early detection. In *Fish invasions of the Mediterranean Sea: Change and Renewal* (Golani, D. & Appelbaum-Golani, B., eds.). Sofia-Moscow: Pensoft.

Azzurro, E., Fanelli, E., Mostarda, E., Catra, M. & Andaloro, F. (2007a). Resource partitioning among early colonizing *Siganus luridus* and native herbivorous fish in the Mediterranean: an integrated study based on gut-content analysis and stable isotope signatures. *Journal of the Marine Biological Association of the UK* 87, 991-998

Azzurro, E., Moschella, P. & Maynou, F. (2011). Tracking Signals of Change in Mediterranean Fish Diversity Based on Local Ecological Knowledge. *PLoS ONE* 6, e24885.

Azzurro, E., Pais, A., Consoli, P. & Andaloro, F. (2007b). Evaluating daynight changes in shallow Mediterranean rocky reef fish assemblages by visual census. *Marine Biology* 151, 2245-2253.

Bariche, M. (2005). Age and growth of Lessepsian rabbitfish from the eastern Mediterranean. *Journal of Applied Ichthyology* 21, 141-145.

Bariche, M. (2006). Diet of the Lessepsian fishes, *Siganus rivulatus* and *S. luridus* (Siganidae) in the eastern Mediterranean: A bibliographic analysis. *Cybium* 30, 41-49.

Bariche, M. (2010a). Champsodon vorax (Teleostei: Champsodontidae), a new alien fish in the Mediterranean. *Aqua, International Journal of Ichthyology* 16, 197-200.

Bariche, M. (2010b). *Champsodon vorax* (Teleostei: Champsodontidae), a new alien fish in the Mediterranean. *Aqua* 16, 197-200.

Bariche, M. (2011a). First record of the cube boxfish *Ostracion cubicus* (Ostraciidae) and additional records of *Champsodon vorax* (Champsodontidae) from the Mediterranean. *Aqua, International Journal of Ichthyology* 17, 181-184.

Bariche, M. (2011b). First record of the cube boxfish Ostracion cubicus (Ostraciidae) and additional records of Champsodon vorax (Champsodontidae) from the Mediterranean. *Aqua, International Journal of Ichthyology* 17, 181-184.

Bariche, M., Alwan, N., El-Assi, H. & Zurayk, R. (2009). Diet composition of the Lessepsian bluespotted cornetfish *Fistularia commersonii* in the eastern Mediterranean. *Journal of Applied Ichthyology* 24, 460-465.

Bariche, M., Alwan, N. & El-Fadel, M. (2006). Structure and biological characteristics of purse seine landings off the Lebanese coast (eastern Mediterranean). *Fisheries Research* 82, 246-252.

Bariche, M. & Heemstra, P. (2012). First record of the blackstrip grouper *Epinephelus fasciatus* (Teleostei: Serranidae) in the Mediterranean Sea. *Marine Biodiversity Records* 5.

Bariche, M., Letourneur, Y. & Harmelin-Vivien, M. (2004). Temporal fluctuations and settlement patterns of native and lessepsian herbivorous fishes on the Lebanese coast (Eastern Mediterranean). *Environmental Biology of Fishes* 70, 81-90.

Bariche, M. & Saad, M. (2008). Settlement of the lessepsian blue-barred parrotfish *Scarus ghobban* (Teleostei: Scaridae) in the eastern Mediterranean. *Marine Biodiversity Records* 1, e5.

Bariche, M., Sadek, R., Al-Zein, M. S. & El-Fadel, M. (2007). Diversity of juvenile fish assemblages in the pelagic waters of Lebanon (eastern Mediterranean). *Hydrobiologia* 580, 109-115.

Ben-Tuvia, A. (1973). Man-made changes in the eastern Mediterranean Sea and their effect on the fishery resources. *Marine Biology* 19, 197-203.

Ben-Yami, M. & Glaser, T. (1974). The invasion of *Saurida undosquamis* (Richardson) into te Levant basin - an example of biological effect on interoceanic canals. *Fishery Bulletin* 72, 359-373.

Bentur, Y., Ashkar, J., Lurie, Y., Levy, Y., Azzam, Z. S., Litmanovich, M., Golik, M., Gurevych, B., Golani, D. & Eisenman, A. (2008). Lessepsian migration and tetrodotoxin poisoning due to *Lagocephalus sceleratus* in the eastern Mediterranean. *Toxicon* 52, 964-968.

Bianchi, C. (2007). Biodiversity issues for the forthcoming tropical Mediterranean Sea. *Hydrobiologia* 580, 7-21.

Bianchi, C. N. & Morri, C. (2003). Global sea warming and "tropicalization" of the Mediterranean Sea: biogeographic and ecological aspects *Biogeographia* 24, 319-328.

Bianchi, C. N. & Morri, C. (2004). Climate change and biological response in Mediterranean Sea ecosystems - a need for broad-scale and long-term research. *Ocean Challenge* 13.

Bilecenoglu, M. (2010). Alien marine fishes of Turkey - an updated review. In *Fish Invasions of the Mediterranean Sea: Change and Renewal* (Golani, D. & Appelbaum-Golani, B., eds.), pp. 189-217. Sofia-Moscow: Pensoft Publishers.

Carpentieri, P., Lelli, S., Colloca, F., Mohanna, C., Bartolino, V., Moubayed, S. & Ardizzone, G. D. (2009). Incidence of lessepsian migrants on landings of the artisanal fishery of south Lebanon. *Marine Biodiversity Records* 2.

Cicek, E. & Bilecenoglu, M. (2009). A new alien fish in the Mediterranean Sea: Champsodon nudivittis (Actinopterygii: Perciformes: Champsodontidae). *Acta Ichtyological Et Piscatoria* 39, 67-69.

CIESM (2008). Climate warming and related changes in the Mediterranean marine biota. In *CIESM Workshop Monographs* (Briand, F., ed.), p. 152. Monaco.

Diamant, A. (2010). Red-Med immigration: a fish parasitology perspective, with special reference to Mxyxosporea. In *Fish Invasions of the Mediterranean Sea: Change and Renewal* (Golani, D. & Appelbaum-Golani, B., eds.), pp. 85-97. Sofia-Moscow: Pensoft Publishers.

EastMed (2010). Report of the sub-regional technical meeting on the lessepsian migration and its impact on eastern Mediterranean fishery. p. 25. GCP/INT/041/EC-GRE-ITA/TD-04.

Galil, B. & Zenetos, A. (2002). A sea change - Exotics in the eastern Mediterranean. In *Invasive Aquatic Species of Europe: Distribution, Impacts and Management* (Leppakoski, E., Olenin, S. & Gollasch, S., eds.), pp. 325-336: Kluwer Academic Publishers.

Galil, B. S. (2006). The marine caravan - the Suez Canal and the Erythrean invasion. In *Bridging divides* (Gollasch, S., Galil, B. S. & Cohen, A., eds.), pp. 207-300. Netherlands: Springer.

Galil, B. S. (2007). Loss or gain? Invasive aliens and biodiversity in the Mediterranean Sea. *Marine Pollution Bulletin* 55, 314-322.

George, C. J. & Athanassiou, V. A. (1967). A two year study on the fishes appearing in the seine fishery of the St. George bay, Lebanon *Ann. Mus. Civ. Stor. Nat. Genova* 76, 237-294.

Golani, D. (1993a). The biology of the Red Sea migrant, *Saurida undosquamis* in the Mediterranean and comparison with the indigenous confamilial *Synodus saurus* (Teleostei: Synodontidae). *Hydrobiologia* 271, 109-117.

Golani, D. (1993b). Trophic adaption of Red Sea fishes to the eastern Mediterranean environment - Review and new data. *Israel Journal of Zoology* 39, 391-402.

Golani, D. (1994). Niche separation between colonizing and indigenous goatfish (Mullidae) along the Mediterranean coast of Israel. *Journal of fish biology* 45, 503-513.

Golani, D. (2010). Colonization of the Mediterranean by Red Sea fishes via the Suez Canal - Lessepsian migration. In *Fish Invasions of the Mediterranean Sea: Change and Renewal* (Golani, D. & Appelbaum-Golani, B., eds.), pp. 145-188. Sofia-Moscow: Pensoft Publishers.

Golani, D. & Ben-Tuvia, A. (1995). Lessepsian migration and the Mediterranean fisheries of Israel. In *World Fisheries Congress* (Armentrout, N. B., ed.), pp. 279-289. New Delhi: Oxford & IBH Publishing Company.

Goldschimdt, T., Witte, F. & Wanink, J. (1993). Cascading effects of the Introduced Nile Perch on the Detrivorous/ Phytoplanktivorous Species in the Sublittoral Areas of Lake Victoria *Conservation Biology* 7, 686-700.

Goren, M. & Galil, B. S. (2001). Fish Biodiversity in the Vermetid Reef Of Shiqmona (Israel). *Marine Ecology* 22, 369-378.

Goren, M. & Galil, B. S. (2005). A review of changes in the fish assemblages of Levantine inland and marine ecosystems following the introduction of non-native fishes. *Journal of Applied Ichthyology* 21, 364-370.

Gottlieb, E. (1960). On the selection of *Upeneus moluccensis* and *Mullus barbatus* by trawl cod-ends in the Israeli fisheries. *General Fisheries Council for the Mediterranean* 8, 1-10 + VI.

Gruvel, A. (1931). Les Etats de Syria. Richness marines et fluviales. Exploitation actuelle, avenir. . In *Societe d'editions Geographiques, Marines et Coloniales*, pp. 72-134. Paris.

Gucu, A. C. & Bingel, F. (1994). Trawlable species assemblages on the continental shelf of the Northeastern Levan Sea (Mediterranean) with an emphasis on Lessepsian migration. *Acta Adriatica* 35, 83-100.

Guidetti, P., Bianchi, C. N., La Mesa, G., Modena, M., Morri, C., Sara, G. & Vacchi, M. (2002). Abundance and size structure of Thalassoma pavo (Pisces: Labridae) in the western Mediterranean Sea: variability at different spatial scales. *Journal of the Marine Biological Association of the United Kingdom* 82, 495-500.

Harmelin-Vivien, M. L., Bitar, G., Harmelin, J.-G. & Monestiez, P. (2005). The littoral fish community of the Lebanese rocky coast (eastern Mediterranean Sea) with emphasis on Red Sea immigrants. *Biological Invasions* 7, 625-637.

Hassan, M., Harmelin-Vivien, M. & Bonhomme, F. (2003). Lessepsian invasion without bottleneck: example of two rabbitfish species (Siganus rivulatus and Siganus luridus). *Journal of Experimental Marine Biology and Ecology* 291, 219-232.

Jackson, J. B. C., Kirby, M. X., Berger, W. H., Bjorndal, K. A., Botsford, L. W., Bourque, B. J., Bradbury, R. H., Cooke, R., Erlandson, J., Estes, J. A., Hughes, T. P., Kidwell, S., Lange, C. B., Lenihan, H. S., Pandolfi, J. M., Peterson, C. H., Steneck, R. S., Tegner, M. J. & Warner, R. R. (2001). Historical Overfishing and the Recent Collapse of Coastal Ecosystems. *Science* 293, 629-637.

Kalogirou, S., Corsini Foka, M., Sioulas, A., Wennhage, H. & Pihl, L. (2010). Diversity, structure and function of fish assemblages associated with *Posidonia oceanica* beds in an area of the eastern Mediterranean Sea and the role of non-indigenous species. *Journal of fish biology* 77, 2338-2357.

Kalogirou, S., Corsini, M., Kondilatos, G. & Wennhage, H. (2007). Diet of the invasive piscivorous fish *Fistularia commersonii* in a recently colonized area of eastern Mediterranean. *Biological Invasions* 9, 887-896.

Kalogirou, S., Mittermayer, F., Pihl, L. & Wennhage, H. (2012a). Feeding ecology of indigenous and non-indigenous fish species within the family Sphyraenidae. *Journal of fish biology* 80, 2528-2548.

Kalogirou, S., Wennhage, H. & Pihl, L. (2012b). Non-indigenous species in Mediterranean fish assemblages: Contrasting feeding guilds of *Posidonia oceanica* meadows and sandy habitats. *Estuarine, Coastal and Shelf Science* 96, 209-218.

Kan, S., Chan, M. K. & David, P. (1987). Nine fatal cases of Puffer fish poisoning in Sabah, Malaysia. *Medical Journal of Malaysia* 42, 199-200.

Katikou, P., Georgantelis, D., Sinouris, N., Petsi, A. & Fotaras, T. (2009). First report on toxicity assessment of the Lessepsian migrant pufferfish *Lagocephalus sceleratus* (Gmelin, 1789) from European waters (Aegean Sea, Greece). *Toxicon* 54, 50-55.

Lasram, F. B. R., Guilhaumon, F. & Mouillot, D. (2009). Fish diversity patterns in the Mediterranean Sea: deviations from a mid-domain model. *Marine Ecology Progress Series* 376, 253-267.

Lasram, F. B. R. & Mouillot, D. (2008). Increasing southern invasion enhances congruence between endemic and exotic Mediterranean fish fauna. *Biological Invasions*.

Leppäkoski, E. & Olenin, S. (2000). Non-native species and rates of spread: lessons from the brackish Baltic Sea. *Biological Invasions* 2, 151-163.

Levine, J. M. & Adler, P. B. (2004). A meta-analysis of biotic resistance to exotic plant invasions. *Ecology Letters* 7, 975-989.

Levine, J. M. & D' Antonio, C. M. (1999). Elton revisited: a review of evidence linking diversity and invasibility. *Oikos* 87, 15-26.

Lipkin, J. & Safriel, U. N. (1971). Intertidal zonation on rocky shores at Mikhmoret (Mediterranean, Israel). *J. Ecol.* 59, 1-30.

Lundberg, B. & Golani, D. (1995). Diet adaptations of Lessepsian migrant rabbitfishes, *Siganus luridus* and *S. rivulatus*, to the algal resources of the Mediterranean coast of Israel. *Marine Ecology* 16, 73-89.

Lundberg, B., Ogorek, R., Galil, B. S. & Goren, M. (2004). Dietary choices of siganid fish in Shiqmona reef, Israel. *Israel Journal of Zoology* 50, 39-53.

MacArthur, R. (1955). Fluctuations of Animal Populations and a Measure of Community Stability. *Ecology* 36, 533-536.

Massuti, E., Valls, M. & Ordines, F. (2010). Changes in the western Mediterranean ichthyofauna: signs of tropicalization and meridianization. In *Fish Invasions of the Mediterranean Sea: Change and Renewal* (Golani, D. & Appelbaum-Golani, B., eds.), pp. 293-312. Sofia-Moscow: Pensoft Publishers.

Moulton, M. P. (1993). The all-or-none pattern in introduced Hawaiian passeriforms: The role of competition sustained. *The American Naturalist* 141, 105-119.

Moyle, P. B. & Light, T. (1996). Biological invasions of fresh water: Empirical rules and assembly theory. *Biological Conservation* 78, 149-161.

Occhipinti-Ambrogi, A. & Savini, D. (2003). Biological invasions as a component of global change in stressed marine ecosystems. *Marine Pollution Bulletin* 46, 542-551.

Oren, O. H., Ben-Yami, M. & Zismann, L. (1971). Explorations of the possible deep-water trawling grounds in the Levant basin. *Proceedings of the General Fisheries Council for the Mediterranean, Studies and Reviews* 49, 41-71.

Parmesan, C. & Yohe, G. (2003). A globally coherent fingerprint of climate change impacts across natural systems. *Nature* 421, 37-42.

Perry, A. L., Low, P. J., Ellis, J. R. & Reynolds, J. D. (2005). Climate Change and Distribution Shifts in Marine Fishes. *Science* 308, 1912-1915.

Por, F. D. (1978). *Lessepsian migration: the influx of Red Sea biota into the Mediterranean by way of the Suez canal*. Berlin-Heidenberg: Springer.

Quignard, J. P. & Tomasini, J. A. (2000). Mediterranean fish biodiversity. *Biologia Marina Mediterranea* 7, 1-66.

Raitsos, D. E., Beaugrand, G., Georgopoulos, D., Zenetos, A., Pancucci-Papadopoulou, A. M., Theocharis, A. & Papathanassiou, E. (2010). Global climate change amplifies the entry of tropical species into the eastern Mediterranean Sea. *Limnology and Oceanograph* 55, 1478-1484.

Rasalato, E., Maginnity, V. & Brunnschweiler, J. M. (2010). Using local ecological knowledge to identify shark river habitats in Fiji (South Pacific). *Environmental Conservation* 37, 90-97.

Reise, K., Olenin, S. & Thieltges, D. (2006). Are aliens threatening European aquatic coastal ecosystems? *Helgoland Marine Research* 60, 77-83.

Ricciardi, A. (2001). Facilitative interactions among aquatic invaders: is an "invasional meltdown" occurring in the Great Lakes? *Canadian Journal of Fisheries and Aquatic Sciences* 58, 2513-2525.

Ricciardi, A. & Mottiar, M. (2006). Does Darwin's naturalization hypothesis explain fish invasions? *Biological Invasions* 8, 1403-1407.

Rilov, G. & Crooks, J. A. (2009). Marine Bioinvasions: Conservation Hazards and Vehicles for Ecological Understanding. In *Biological Invasions in Marine Ecosystems* (Rilov, G. & Crooks, J. A., eds.). Berlin Heidenberg: Springer-Verlag.

Rilov, G. & Galil, B. (2009). Marine bioinvasions in the Mediterranean Sea – History, distribution and ecology. In *Biological Invasions in Marine Ecosystems* (Rilov, G. & Crooks, J. A., eds.), pp. 549-575. Berlin Heidelberg: Springer.

Roessig, J. M., Woodley, C. M., Cech, J. J. & Hansen, L. J. (2004). Effects of global climate change on marine and estuarine fishes and fisheries. *Reviews in Fish Biology and Fisheries* 14, 251-275.

Sabrah, M. M., El-Ganainy, A. A. & Zaky, M. A. (2006). Biology and toxicity of the pufferfish *Lagocephalus sceleratus* (Gmelin, 1789) from the gulf of Suez *Egyptian Journal of Aquatic Research* 32, 283-297.

Sakinan, S. & Örek, Y. A. (2011). First record of Indo-Pacific Indian scad fish, *Decapterus russelli*, on the north-eastern Mediterranean coast of Turkey. *Marine Biodiversity Records* 4.

Salameh, P., Sonin, O., Edelis, D. & Golani, D. (2011). First record of the Red Sea orangeface butterflyfish *Chaetodon larvatus* Cuvier, 1831 in the Mediterranean. *Aquatic Invasions* 6, S53-S55.

Sax, D. F. (2002). Native and naturalized plant diversity are positively correlated in scrub communities of California and Chile. *Diversity and Distributions* 8, 193-210.

Sax, DovB F., Gaines, StevenB D. & Brown, JamesB H. (2002). Species Invasions Exceed Extinctions on Islands Worldwide: A Comparative Study of Plants and Birds. *The American Naturalist* 160, 766-783.

Shakman, E. A. & Kinzelbach, R. (2007). Distribution and characterization of lessepsian migrant fishes along the coast of Libya. *Acta Ichtyological Et Piscatoria* 37, 7-15.

Simberloff, D. & Von Holle, B. (1999). Positive Interactions of Nonindigenous Species: Invasional Meltdown? *Biological Invasions* 1, 21-32.

Sonin, O., Spanier, E. & Pisanty, S. (1996). Undersized fishes in the catch of the Israeli Mediterranean fisheries - are there differences between shallow and deeper water ? In *Preservation of our world in the wake of change* (Steinberger, ed.), pp. 449-454. Jerusalem: ISEEQS Publication.

Stachowicz, J. J. & Tilman, D. (2005). Species invasions and the relationships between species diversity, community saturation, and ecosystem functioning. In *Species Invasions: Insights into ecology, evolution and biogeography* (Sax, D. F., Stachowicz, J. J. & Gaines, S. D., eds.). Sunderland, Massachusetts: Sinauer Associates.

Stachowicz, J. J. & Whitlatch, R. B. (1999). Species Diversity and Invasion Resistance in a Marine Ecosystem. *Science* 286, 1577.

Streftaris, N. & Zenetos, A. (2006). Alien marine species in the Mediterranean - the 100 'worst invasives' and their impact. *Mediterranean Marine Science* 7, 87-118.

Streftaris, N., Zenetos, A. & Papathanassiou, E. (2005). Globalisation in marine ecosystems - The story of non indigenous marine species across European Seas. *Reviews in Oceanography and Marine Biology* 43, 419-454.

Vitousek, P. M., Mooney, H. A., Lubchenco, J. & Melillo, J. M. (1997). Human Domination of Earth's Ecosystems. *Science* 277, 494-499.

Wallentinus, I. & Nyberg, C. D. (2007). Introduced marine organisms as habitat modifiers. *Marine Pollution Bulletin* 55, 323-332.

Williamson, M. H. & Fitter, A. (1996). The characters of successful invaders. *Biological Conservation* 78, 163-170.

Witenberg, R. & Cock, M. J. W., eds. (2001). *Invasive alien species: a toolkit of best prevention and management practices*: GISP.

Yang, C. C., Liao, S. C. & Deng, J. F. (1996). Tetrodotoxin poisoning in Taiwan; an analysis of posion center data. *Veterinary and Human Toxicology* 38, 282-286.

Zenetos, A., Gofas, S., Verlaque, M., Cinar, M. E., Garcia Raso, J. E., Bianchi, C. N., Morri, C., Azzurro, E., Bilecenoglu, M., Froglia, C., Siokou I., Violanti, D., Sfriso, A., San Mart, G., Giangrande, A., Kata An, T., Ballesteros, E., Ramos-Espla, A., Mastrototaro, F., Oca A, O., Zingone, A., Gambi, M. C. & Streftaris, N. (2010). Alien species in the Mediterranean Sea by 2010. A contribution to the application of European Union's Marine Strategy Framework Directive (MSFD). Part I. Spatial distribution. *Mediterranean Marine Science* 11, 381-493.

Zenetos, A., Meric, E., Verlaque, M., Galli, P., Boudouresque, C.-F., Giangrande, A., Cinar, E. & Bilecenoglu, M. (2008). Additions to the annotated list of marine alien biota in the Mediterranean with special emphasis on Foraminifera and Parasites. *Mediterranean Marine Science* 9, 119-165.

Zenetos, A., Pancucci Papadopoulou, M. A., Zogaris, S., Papastergiadou, A., Vardakas, L., Aligizaki, K. & Economou, A. N. (2009). Aquatic alien species in Greece: tracking sources, patterns and effects on the ecosystem. *Journal of Biological Research-Thessaloniki* 12, 135-172.

Chapter 12

Impact of Land-Use and Climate on Biodiversity in an Agricultural Landscape

Andrzej Kędziora, Krzysztof Kujawa, Hanna Gołdyn, Jerzy Karg, Zdzisław Bernacki, Anna Kujawa, Stanisław Bałazy, Maria Oleszczuk, Mariusz Rybacki, Ewa Arczyńska-Chudy, Cezary Tkaczuk, Rafał Łęcki, Maria Szyszkiewicz-Golis, Piotr Pińskwar, Dariusz Sobczyk and Joanna Andrusiak

Additional information is available at the end of the chapter

http://dx.doi.org/10.5772/48653

1. Introduction

The term "biodiversity" was used for the first time by wildlife scientist and conservationist [1] in a lay book advocating nature conservation. The term was not adopted by more then decade. In 1980 use of the term by Thomas Lovejoy in the Foreword to the book "Conservation Biology" [2] credited with launching the field of conservation biology introduced the term to the scientific community. There are many definitions of biodiversity. One of them, formulated in Millennium Ecosystem Assessment [3] reads: " Biodiversity is the variability among living organisms from all sources, including terrestrial, marine, and other aquatic ecosystems and the ecological complexes of which they are part; this includes diversity within species, between species, and of ecosystems". **Biodiversity forms the foundation of the vast array of ecosystem services that critically contribute to human well-being**. Biodiversity is important in human-managed as well as natural ecosystems. Decisions humans make that influence biodiversity affect the well-being of themselves and others [3].

Biodiversity is the one of basic driving forces which influence and determine most of ecosystem services [3]. Interdisciplinary and long-term (50 years) researches carried out by Institute for Agricultural and Forest Environment were focused on recognition of factors, that brought about impoverishment of biodiversity in agricultural landscape and finding the ways to counteract these negative changes. During a few last centuries the rapid decrease of biodiversity has been observed all over the world [4]. Human activity brought to worsening the habitat ability to ensure the conditions for living rich plant, animal and fungi population. Conversion of more stable ecosystems like forests, meadows, and wetlands into less stable ones like arable land causes increase of threats to such fundamental processes like energy flow and cycling of matter in the environment [5]. Such factors like climate changes, simplification of landscape structure and changes in farming practices have resulted in worsening habitat conditions (water conditions, environment pollution, soil degradation etc.) leading in consequence to impoverishment of biodiversity. Agriculture is commonly considered to be one of the main threats to the biological diversity. Farmers try to eliminate all organisms which could diminish crops to increase agricultural production. In order to channel the solar energy and nutrients into products useful for man, farmers simplify plant cover structure both within cultivated fields (selection of genetically uniform and productive cultivars, elimination of weeds), and within agricultural landscape (eradication of hedges, patches of trees, mid-field wetlands and ponds, riparian vegetation strips). Increasing usage of fertilizers and pesticides caused environmental pollution that threats all organisms. Apart from this, the unfavorable climate changes were observed during the last century. Increase of air temperature and wind speed without clear increases of precipitation brought about worsening water conditions.

The agricultural landscape located in Wielkopolska (western Poland) around the Turew village (currently protected by law as landscape park) is a distinct example of such management which allows keeping intensive crop production without causing serious degradation of habitat and impoverishment of biodiversity. A network of diverse linear habitats (shelterbelts, hedgerows, roadside verges, tree alleys etc.) and small wood patches established 200 years ago, stretches of meadows, small mid-field water reservoirs, or wetlands provide good refuge for many organisms and unique facility for studying the effects of farming intensification at stable spatial arrangement of non-farmed habitats. In the Turew agricultural landscape near 850 species of vascular plants, 2600 insect species, 120 species of breeding birds, and about 700 species of macrofungi were detected, including many woodland species, as well as a variety of invertebrates including rare, threatened, protected by law and umbrella species [6]. As many, as 60-100% of animal taxa showed by regional lists of taxa live in this agricultural landscape.

The crucial factor for maintenance numerous and favorable habitats for various groups of animals is well developed structure of landscape. Guidelines for shaping of landscape structure towards ensuring maintenance end enhancement biodiversity as a crucial factor for sustainable development of agricultural landscape. For instance, such elements of landscape like shelterbelts, hedges, and strips of meadows, allows to survive animals (like bees), which population is reduced in the result of intensive use of pesticides. The old trees with a lot of hollows, existing within shelterbelts, are necessary for keeping rich population of many species of birds. In the chapter we summarized our knowledge on quantitative relationships between landscape features and diversity (species richness and abundance) of taxa mentioned above and to present the spatio-temporal pattern of species distribution and their abundance.

It is possible to reconcile very high level of intensive agriculture with protection and even enhancement of biodiversity. Creating a very mosaic landscape (biotic and abiotic component of landscape) is a toolkit for solving this problem. The integrity of biological and physical or chemical processes is a basic foundation of modern ecosystem or landscape ecological approaches. Recognition of this functional relationships leads to the conclusion that biodiversity cannot be successfully protected only by isolation from hostile surrounding, but its conservancy should rely on the active management of the landscape structures in a direction of their diversification. The very important guidelines for enhancement of sustainable system resistance and resilience to threats are diversification of its structure Generally speaking, the results of our study are in line with very recent findings, that due to the great diversity of climatic and physiographic conditions as well as customary regulatory and management of the landscape across Europe, makes it very much needed regional approach to biodiversity protection and management. For example, application of agri-environmental schemes, with the rules elaborated mostly in Western Europe (where agricultural landscape is in general extremely simplified, and where climate is mild, e.g. winters are warmer) to CEE countries seems to be inaccurate [7, 8].

2. Description of Study areas (Turew mosaic landscape and Kościan Plain)

The "Turew Mosaic Landscape" in this chapter is the part of the larger study area called "Kościan Plain" is located about 40 km south from Poznań – the capital of Wielkopolska region and its geographical coordinates are 16°45′ to 17°05′ E and 51°55′ to 52°05′ N (Figure 1). The Field Station of Institute is situated in the middle of the landscape near a small village called Turew. Therefore, the name Turew is used to identify the landscape. Wielkopolska region is known as the "bread basket". Agriculture is the dominant activity of the region.

Figure 1. Location of Turew Landscape

The study area "Kościan Plain" (area about 200 km^2) is located within radius of 5 – 10 km around Turew, in the area of the West Polish Lowland, which is a ground moraine created during the Baltic glaciation, that terminated about 10 000 year ago. Although the differences in altitude are small (from 75 m to 90 m a.s.l.) and the area consists of a rolling plain made up of slightly undulating ground moraine there are many drainage valleys. In general, light textured soils (Hapludalfs, Glossudalfs and less frequently met Udipsamments) with favorable water infiltration conditions are found in uplands. Deeper strata are poorly permeable and percolating water seeps to valleys and ditches or main drainage canal. In depressions Endoaquolls, poorly drained, collect water runoff and discharge water to surface drainage system [9].

The climate of the region is shaped by the conflicting air masses from the Atlantic, Eastern Europe and Asia (arctic 6%, polar martitime 59%, polar continental 28%, tropical 7%), which are modified by strong Arctic and Mediterranean influences. It results in a great changeability of weather conditions and the predominance of western winds brings strong oceanic influence that manifest in milder winters and cooler summers in comparison to the centre and east of Poland. Within Poland, this area is one of the warmest, with an annual mean temperature above 8°C (range from 6.9 to 8.5°C). Mean annual global radiation amounts 3700 MJ/m^2 and, mean annual net radiation equals to 1315 MJ/m^2.

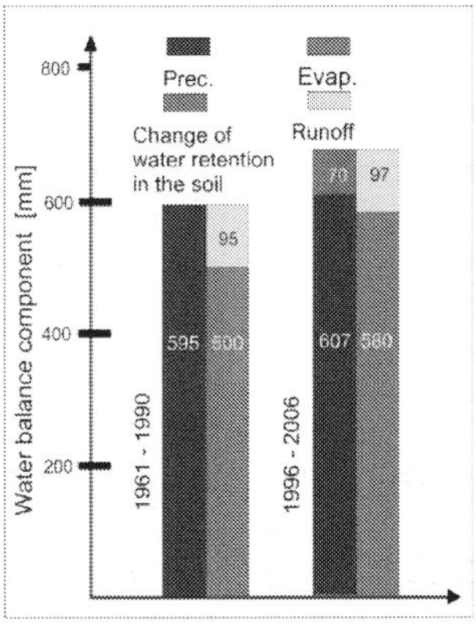

Figure 2. Water balance in Wielkopolska in 1961 – 1990 reference period and In 1996-2006 warm and dry period

Thermal conditions existing in the Turew landscape are favorable for cultivated plant growth. The mean plant growth season with temperatures above 5°C lasts from 21 March till 30 October. In reference 1961 – 1990 period (Figure 2) the mean annual precipitation is equal to 595 mm, of which 365 mm falls between April-September, and 230 mm in the period of October-March. Although the amount of precipitation in the spring-summer period is higher than in autumn and winter, a shortage of water occurs frequently in the plant growth season. This situation is aggravated by the dominance of light soils with poor water storing capacities. Average annual evapotranspiration amounts to 500 mm (485 mm in whole country) and water runoff is equal to 95 mm (212 mm in whole country). But in warm and dry period of 1996 – 2006 evapotranspiration was as high as 580 mm, which resulted in decreasing water retained in soil by 70 mm. It caused worsening of water conditions for plants.

In the land-use structure of whole catchment arable land makes 62.2%, forests and shelterbelts (mid-field rows of trees) cover 17.9%, meadows and pastures 12.5 %, water bodies 3.5 %, and villages, roads the rest of the area. There are no industrial facilities. The mean density of inhabitants equals to 55 individuals per 1 km^2.

- The natural forest complexes were replaced by woods planted by man or converted into cultivated fields that constitute up 70% of the total area, shelterbelts and small forests 16 % and grasslands 9 %.
- The structure of crops at the beginning of the 21 century was as follows: cereals (mainly, wheat and *Triticale*) including maize made 76.7%, vegetables 16%, potato, seed-rape, and sugar beets 6%.

The specific landscape of Turew neighborhood located in Wielkopolska was shaped during the twentieths of XIX century by general D. Chłapowski who farmed on 10 thousands hectares. He introduced essential changes in farming system as well as in the field today called landscape engineering. Conversion of the open and uniform agricultural landscape in mosaic which is rich in stable element like shelterbelts small mid-field water reservoirs was the results of his activity. Many wooded patches, shelterbelts, tree lines, clumps of trees were planted in the landscape. They were designed as a shelter for domestic animals and the measures against wind erosion. Since 1950s last century the investigation of agricultural landscape functioning has been carried out within that areas. About of one hundred km of linear and 10 hectares of new shelterbelts and woody patches were planted during the two last decades.

The "Turew Mosaic Landscape" (Figure 3A) is still abundant in various wooded patches located in upland parts of the landscape or along banks of the drainage water system as well as along other non-crop habitats, such as small water reservoirs, marshy habitats, bounds and so on. All together there are more than 800 shelterbelts forming the network in an area of 17200 ha. Cultivated fields make 70 % of the total area, forests and shelterbelts -16 % and grasslands - 9 %. The majority of farms are small. Only 27% are larger than 10 ha.

Comparative studies have been carried out in uniform agricultural landscape (Figure 3B) composed of large fields located about 10 km apart characterized by the same climatic conditions and similar soil types. This uniform landscape is almost entirely devoid of shelterbelts and the drainage system is mostly operating as an underground system. The cultivated fields are bigger than those in the mosaic landscape, and are ranging from 15 to 150 ha. A similar crop structure is appears in both mosaic and uniform landscape. The structure of the crops at the end of 1990s was as follows: cereals (68.1%), maize (9.6%), vegetables (16%), potatoes, seed-rape and sugar beets (6%). The substantial increase of cereal cultivations were observed over the period of 1984-2004. In 1985 cereals covered 48.1% of total arable land, in 1997 their contribution increased to 63.5% and in 2002 cereals were cultivated on 73.9% of arable land. Among the cereals wheat and triticale cultivations dominated.

(a)

(b)

Figure 3. A. Mosaic landscape. Phot. K. Kujawa; B. Uniform landscape. Phot. K. Kujawa

3. Methods

Due to a great variety of biological and ecological features of studied taxa, their distribution, abundance and diversity patterns were studied with the use of many methods differing strongly between the taxa. Diverse goals of former research projects carried out in the area described above involved various scales (approximately from hectares to tens of hundred square kilometers) and diverse habitats. Generally speaking, we refer our results mostly to three different objects:

a. landscape in broad sense (ca. 200 km^2, called in the chapter "Kościan Plain"), i.e. comprising all found habitat types, such as crop fields, meadows, semi-natural grasslands, woodlands, small wooded clumps (tree or shrub lines, belts or clumps), wetlands, roadside verges, water bodies and courses,

b. agricultural land in narrow sense (ca. 20 km^2, called "Turew Mosaic Landscape"), i.e. consisted of cultivated fields, meadows and grasslands, field margins (with all kinds of plant cover, roadside verges etc.), drainage ditches, small wooded patches, small water bodies and courses,
c. habitats (with area of tens hectares), called just wheat field, arable land, small wooded patches, drainage ditches, etc.

In other cases, described object (piece of landscape other than "Kościan Plain" and "Turew Mosaic Landscape") is described separately to distinguish it from the above listed. During the data analyzing, various statistical tests and procedures were applied, which are cited in appropriate place with commonly used names (t-test, Mann-Whitney test, Wilcoxon test, Analysis of Variance - ANOVA, Principal Component Analysis - PCA, General Linear Model - GLM, etc.). The significance level used for all tests was 0.05. For estimation total bird species richness, the Jack-Knife 2 estimator was used [10]. Arythmetic mean is given with standard deviation (SD).

3.1. Plants

Studies on spontaneous flora of the "Kościan Plain" have been carried out since 1975. First they were collected mainly in aquatic and marsh ecosystems and only occasionally apart from them. From 2000 they included the entire landscape and were taken with the method of mapping of all the species in the network of equal basic fields. Lists of the plant species of every investigated plot were analyzed. Material for the study on relationships between the floristic diversity of agricultural landscape and the diversity of its structural elements was collected on 37 square plots (1 km^2 each) [11]. Landscape diversity was assessed by number of elements of landscape (i.e. spatial units that differ in land use).

The most important measures used in floristic research include: species richness, floristic distinction (i.e. mean coefficient of species rarity), number of environmentally valuable species (recessive i.e. declining species, species included in red lists for Poland or Wielkopolska, protected species, and those rare regionally and locally), sociological groups, and geographic-historical groups. The status of species in the flora follows reports on Polish and regional floras. To diagnose the sociology of the taxa [12] elaboration was used. Names of alien geographic–historical groups follow [13,14]. Those are: archaeophytes (naturalized alien species introduced before ca. 1,500), naturalized neophytes (naturalized alien species introduced after ca. 1,500) and casual neophytes (casual alien species introduced after ca. 1,500 occurring sporadically or for a short time within the studied area). Among native species nonsynanthropic spontaneophytes (species which do not show permanent trends in occupying transformed anthropogenic habitats) and apophytes (indigenous species, permanently occupying strongly transformed anthropogenic habitats) are distinguished according to [15]. Species names follow [16].

Field inventory work on dendroflora was carried out during vegetative periods from 2001 to 2003. Prior to field work there were preparatory work on maps and aerial photos conducted. The inventory and evaluation encompassed all wooden plants assembles of open landscape, the area of which was smaller than 5 ha: shelterbelts, woodlots, avenues, tree lines and hedgerows. Desk work, which followed the field survey, was aimed at evaluation of wooden flora located in different landscape context among other by number of species and share of native and alien species [17].

The study of the flora and plant associations in the lake Zbęchy, peat bogs and Wyskoć Canal were carried first in 1976-1979. They were repeated in 2006-2007 [18 - 21]. Between these periods occasional studies were still performed. The commonly used Braun-Blanquet method was applied. The associations were distinguished and the taxonomy of syntaxa was adopted after [22]. The data for both investigated seasons were compared and valorized regarding to they threat and frequency of occurence according to [22] elaboration. The same method was used in the research on the flora of ponds that started in 1985 [23].

3.2. Insects

The basic apparatus (called "biocenometer" – Figure 4) used in the study of above-ground entomofauna in vegetation season is composed of cap in the form of a truncated pyramid with trapping surface 50x50 cm and height 70 cm. It is randomly barraged on the study area and trapped invertebrates are taken by suction apparatus, powered by a generator [24]. In all years (from 1970s till now) and habitats (crop fields, meadows, small wooded areas) the samples were collected with a frequency no less than three samples during one season (usually monthly, from May to October), in the series including 10 samples along a transect located in a given sample plot (habitat studied). Samples were separated by hand and the gained material was stored dried and identified to families (all insects) or genus and species (some groups of insects)[25].

Figure 4. Biocenometer. Phot. J. Karg.

The samples of insects wintering in litter and soil were taken from 1990s till now in shelterbelts and on adjacent crop fields. As many as 10 soil monoliths (without litter) with dimensions 10x10x10 cm were taken in each series three times per winter. Soil was manually separated. Obtained material was preserved in 75% alcohol and identified to the level of family or order (some larvae and pupae). Simultaneously, on the same dates and on the same places, litter samples from the shelterbelts were collected. Frames with dimensions 25x25 cm were used, and 20 samples were taken in each series. Further procedures were the same as in the case of soil samples.

The butterflies were studied mostly with the use of a transect method [26]. Eleven transects (with total length of 10.25 km) were located along the roads adjacent to various habitats (wooded areas, crop fields and meadows, young shelterbelts etc.) and the butterflies were usually counted and identified to species level in 5 m wide belt. The gathered data were used for the estimation of total or species abundance index estimation.

Since beginning of 1960s the observation of nocturnal butterflies were carried out using light trap method.

3.3. Spiders

Spider communities in winter cereals, sugar beet and alfalfa crops were studied in two types of agricultural landscape within the "Kościan Plain": mosaic one with the net of shelterbelts and the uniform one with lack of shelterbelts, where the majority of the terrain is covered by large sized crop fields. The results obtained from three fields in each landscape within three years were compared. To recognize an effect of wooded habitats on spider community occurring in crop fields, the spiders were collected in the distances of 10 m, 50 m and 100 m from shelterbelts (in case of mosaic landscape) or from a road (in case of uniform landscape). To assess the spider density, the samples were taken with biocenometer twice a year (in May and July). Foliage-dwelling spiders were taken by sweet-netting (50 sweeps were a sample), 2-3 times per season (from April to September).

3.4. Fishes

The results have been obtained by using an electrofishing method with IUP-12 device [27], used for catching fish in Wyskoć Canal. The most recent data were gathered in two 100 m long sections of this water course in 2007 and earlier – in 1997-1999 in 1000 m long section. The information on Zbęchy Lake was obtained from the user (Polish Angling Association, the Fishing Company in Osieczna).

3.5. Amphibians

Investigations of amphibians were carried out in the "Kościan Plain" mostly in the years 1995-2000, since March to October. Investigation covered mainly reproducing amphibian populations as well as water bodies in the spring because amphibians aggregate there for reproduction. The first step was cartographical analysis of the study area on the basis of topographic maps (1:10000) and aerial photographs. Then, location of ponds and all wet areas was checked in the field. As many as 150 water reservoirs were found. They were divided into 4 types, depending on impact of man activity: village ponds (located up to 100 m from buildings), reservoirs located on the meadows, in the fields and in the forests. In each water body the number of species and its abundance were studied. The presence of species was verified on the basis of direct observations of adults, juveniles or tadpoles. In some cases voices of males or unique characters of egg deposits were taken into account. Abundance of populations was estimated on the basis of direct observations of adults, counting of male voices during their high vocal activity (at night only) and numbers of egg clumps (for the brown frogs *Rana temporaria* and *Rana arvalis* only). The threats to amphibians (pollution, presence of fishes, ducks) were identified for each water body. Each reservoir was classified with respect to its usefulness to amphibian breeding (type of water plants, grade of shading, depth etc.). Moreover, in 32 reservoirs some chemical features of water (pH, ammonium, nitrates, and phosphates) were analyzed [28, 29].

3.6. Birds

At first, it is worth to mentioning that during the study species diversity in heterogenic habitat or landscape, serious methodological problem is defining the "representative" both study area and sampling effort. In case of birds, the representative agricultural landscape was defined as a mosaic of cultivated fields, meadows, pastures and small wooded area with the area not exceeding 4 ha and it was located in the "Turew Mosaic Landscape". To control biases related to sampling effort and to avoid formulation of biased conclusions, we applied here triple approach: a) description of species richness, b) comparisons to other comparable data in Poland, c) analysis of species richness indicators.

The distribution and population density of birds was studied with the aid of a variety of techniques, depending on the goal for a given research project.

a. Most data have been gathered with the use of a combined version of cartographic methods [30] , which was used in 1964-66, 1984, and since 1988 till now (with some time gaps). In all cases (plots and years) at least 6 counts/per plot were done (in most years – 9 counts) and number of breeding pairs was established on the basis of "paper territories" (clumps of at least 3 records of territorial males) or other observations of breeding behavior. The method was applied in all kinds of small wooded areas – clumps of trees, tree alleys and tree belts (in total of ca. 100 plots) as well as in few fragments of open landscape.
b. Transect method [31] was used for recognition of bird community patterns in an open landscape, among others for assessment of red fox impact on distribution and abundance of birds. The birds were censused on 200-300 m wide transects (several tens km), in the morning, twice a breeding season.
c. Point count methods [31] was applied when relationships between landscape structure and birds abundance and species richness had been studied in scale of tens of square kilometers (hundred points). Birds were counted twice a breeding season in the morning, up to 100 or 150 m from place of standing.

In next step, for understanding the effect of environmental factors on the bird communities in space and time, habitat structure was quantified with the aid of several variables in three scales: within the plot (tree belt, tree clump etc.), in close neighborhood (adjacent area) and in landscape around a plot (i.e. within a radius of 1500 m from given wooded patch).

a. Within the plot: tree, shrub, herb percentage cover, tree stem density, tree and shrub species diversity etc. [32, 33].
b. Area adjacent to studied plots : crop diversity, crop patch density etc. [33].
c. Landscape: woodiness index, percentage cover of crop fields, density of ecotones etc. [33].

In the next step the number of independent variable was reduced with the aid of PCA if needed, and the relationships between habitat or landscape structure were verified with the use of a various regression models.

3.7. Mammals

Small mammals were studied in 1999-2001 in two shelterbelts of different age located in crop fields in the "Turew Mosaic Landscape". The young one (16 m wide) was planted in autumn 1993 year and consisted of oaks, pines, birches, poplars, elms, maples, beeches and some other species in addition. The old one (36 m wide) was created with the shelterbelts' network in XIX century. The tree stand consisted of false acacia (with admixture of several oaks). Catch-Mark-Release (CMR) method was used. Traps were located in consistent lay out in both studied shelterbelts. 16 transects each 20 meters were set, 4 points including 2 traps were located in each transect. The traps were inspected every morning for 10 days within each season series (spring, summer, autumn).

In former years also other groups of mammal were studied. Carnivores were studied with trapping (small mustelids), recording of burrows and snow tracking. Night counting (with searchlight attached to a car) was also used. Belt assessment method with a line of beaters moving through selected area was used for estimation of hares. Roe deer and wild boar were counted by direct observation in late winter and early spring seasons [34].

3.8. Macrofungi

The study on macrofungi species diversity in the "Kościan Plain" was carried out basically with the use of three methods :

1. Route methods, enabling a basic (rough) recognition of species diversity in a large area.
2. Permanent plots, allowing for detailed description of fungal communities in various habitat types as well as for making comparison between the communities.
3. Transect method, which enables gathering the information from relatively large areas, comparing between them and detecting changes in species richness and diversity in a various ecosystems.

Additionally, in 2000-2003 study was carried out in 50 permanent plots (area of $400m^2$) located in managed forests, village parks, road tree alleys, tree belts and clumps of trees.

The route method has been used systematically since 1997 in various habitats. The rate of colonization of windbreak introduced in arable field is studied from 1998 [35], and preliminary study on the mycobiota of ephemeral habitats characteristic for a Polish lowland agricultural landscape (manure and straw piles) are realized from 2009. In result, in 1997-2010 numerous data on species diversity of a fragment of agricultural landscape have been collected.

3.9. Microfungi

Due to specificity of biological features of fungal pathogens, i.e. demanding special research techniques, the study was designed and carried out in other way than previously described taxa. The study of the fungal pathogens (infecting insects (Insecta), true spiders (Araneae) and mites (Acari)) was aimed at (1) recognition of their real resources and diversity in particular habitats, (2) – estimation of their effects on noxious and beneficial arthropod populations. They were conducted in German, France, and Poland during 10 vegetation seasons (1995-2004) in agroecosystems and neighboring forests. In agroecosystems three groups of elements were studied: annual crops (cereals and row plants), perennial crops (including meadows and pastures) and non-farmed clumps or strips of wild vegetation, including rushes and arborescent plants. This allowed for synthesis of the results for simplified (uniform) and diversified (mosaic) landscape types. The cases of epizootic appearance of particular pathogens' species were treated with special attention independently on the noxiousness or usefulness of their host species [36]. Before undertaking the extended studies in three countries, the methods were elaborated and tested in the "Kościan Plain" and the Wielkopolski National Park – both in the central Wielkopolska region (Poland). Both these areas have been included into the referred researches.

The sample plots were located along transects running across a chosen landscape fragments differing in plant cover, habitat structure or/and human impact. Material from 3-5 randomly scattered sample plots (2x2 m), where plant cover up to the height of 2.5 m and the litter including superficial layer of soil was carefully searched, served as basic samples [37]. The invertebrates were sampled two times during every growing season – in the turns of spring/summer and summer/autumn. Additional checking searches were made in other dates, but only in Poland. The applied methods give the best results in the recognition, sampling, and species isolation of these pathogens group, often allowing to discover spatial distribution of epizootic appearance and new species for science.

Apart from the study on pattern of occurrence, for some habitats and countries the investigations on the pathogens of the subcortical and wood-boring insects were carried out. Mortality causes of these insects – together with associated with them invertebrates in their feeding sites - have been investigated by the immediate stereo-microscopic searches of samples from invaded trees (logs and branches) and laboratory rearing of them, for periodical isolation of appearing pathogens [38, 39].

Occurrence and diversity of entomopathogenic fungi in the soil from different habitats of agricultural landscape, including the "Turew Mosaic Landscape" and closely situated Farm Karolew applying no-tillage system have been investigated in th period 1996 – 2006. In total

296 soil samples from 74 locations in Poland were collected in the years 1996-2006 mainly from agrocenoses (arable fields, meadows, pastures), semi-natural biotopes (shelterbelts, mid-field afforestations, balks) and forest biotopes. Fungi were isolated from soil samples by means of the "Galleria bait method" [40] and selective agar medium [41]. The infective potential of particular soil samples has been related to the mean colony forming units (CFU) (colony forming units) per 1 g of checked soil. Standard methods for χ^2 tests were used for comparison of their quantitative relation in soils samples from different habitats.

4. Characteristic of biodiversity

The methods used and goals of investigations conducted in the "Kościan Plain" differed strongly between the studied taxa or groups. That is, why some data included in this chapter should be regarded rather as the approximate information about "species pool" of organisms occurring there and as a general picture of wildlife taxonomical diversity and richness, but not in the term of detailed, complete datasets on species richness, concise across the taxa in scale of the landscape. Presumably, only the data on vascular plants, butterflies, amphibians and birds represent the groups for which we can evaluate or estimate the number of species for farmland (arable fields, meadows, field margins, small water bodies, drainage ditches) as well as for the whole landscape in broad sense. Other groups can be used for evaluation of biodiversity mostly in respect to farmland ("Turew Mosaic Landscape") or to selected habitats occurring in farmland (for example the fishes – for evaluation of small water bodies and water courses importance).

4.1. Plants

The total number of the spontaneously occurring vascular plant species, which have been noted until now within the "Kościan Plain" amounts to 848. It comprises just 54 % of the number of species in the Środkowa Wielkopolska Region. There are 773 naturalized species and 75 casual neophytes. Natives are 72 % and all alien species 28% of the total flora. Among the alien species the most important group are archaeophytes that present 39% of the total alien species number, whereas naturalized neophytes are 30% and casual neophytes 31% [11]. Among the naturalized neophytes there are 4 invasive species (*Heracleum sosnowskyi, Echinocystis lobata, Impatiens glandulifera, Reynourtia japonica*) from the ministerial list of alien species that especially threat the biodiversity in Poland. As many as 6 of the neophytes that are noted in the Landscape Park (*Echinocystis lobata, Elodea canadensis, Impatiens glandulifera, Prunus serotina, Reynourtia japonica, Robinia pseudoacacia*) are included in the list of 100 WORST in Europe [42]. The species typical for meadows show the highest share (22%) among the sociological groups of plants. The smallest is the group of thicket plants (5%). The share of other groups (aquatic, forest, xerothermic grassland, ruderal, segetal) is 11-14%.

Currently, the total landscape flora consists of 828 species due to the extinction of 20 species in the last decades. Nowadays 85 plant species of special care (rare, threatened, declining, protected) occur spontaneously in the "Kościan Plain". Among them, there is one species

(*Ostericum* palustre) protected in UE that is in the II Annex to the Habitats Directive, and two species (*Botrychium matricariifolium, Ostericum palustre*) listed in the "Polish Red Data Book of Plants" [43]. As many as 44 species from this group of special care plants are protected in Poland. Despite their exposure to extinction the other species from this group are not protected. Apart from them 12 valuable vascular plant species had been noted in the past, but they disappeared in the last decades.

The flora of fields that dominate in the landscape, consists of 224 plant species. It is 74% of the segetal flora of the Wielkopolska Region. Native species are 56% and alien species 44% of the total field flora. Archaeophytes that presents 65% of alien species are the biggest group among them. 21% are naturalized neophytes and 14% casual neophytes. In the group of naturalized neophytes there are often noted invasive species as *Amaranthus retroflexus, Conyza canadensis, Galinsoga parviflora* and *Veronica persica* and less frequently *Anthoxanthum aristatum, Bromus carinatus, Bidens frondosa* and *Galinsoga ciliata.* Among the sociological groups of plants the highest share show segetal species (34%) and from other sociological groups the biggest are meadow (20%) and ruderal species groups (19%). The share of remaining groups (aquatic, forest, xerothermic grassland, thicket) is 3-8%. 23 plant species of special care were noted in the fields. Among them there were species threatened in Poland and in the Wielkopolska Region (e.g. *Myosurus minimus, Valerianella locusta, Valerianella rimosa, Conium maculatum*), vulnerable archaeophytes (e.g. *Agrostemma githago, Anthriscus caucalis, Melandrium noctiflorum, Valerianella rimosa*) and only one species protected in Poland (*Ornithogalum umbellatum*) among them [11].

A studies conducted in the mid-1970s showed 131 plant communities in the "Kościan Plain", but actually 122 plant communities were confirmed. Number of aquatic associasions was reduced by 9 only in last four decades.

Forest communities represent less than 15% of the land cover and most of them are of anthropogenic origin. Pine monoculture stands are the most common forest communities in the area. Natural forest are represented by riparian (elm-oak) and alder stands – which covers narrow areas along water courses. Elm-ash riparian forests are noticed in the vicinity of the alder stands – characterized by lower level of groundwater.

As a result of thousands years of agricultural use of this area, dominating Middle European lowland oak-hornbeam forest habitats (*Galio-Carpinetum*) have been substituted with arable fields, whereas lowland ash-elm floodplain forest habitats (*Ficario-Ulmetum*) mostly with intensively cultivated meadows. Predominant deciduous forest habitats have been almost destroyed, and only small forest patches like manor park in Turew remained. As early as in the XIX century this area was among the most intensively used in this part of Europe. Arable fields occupied 65.2%, meadows – 13.5% , pastures – 3.4%, forests – 13.2%. In order to prevent the soil erosion, in 1820s the then holder D. Chłapowski introduced into fields a network of windbreaking shelterbelts which consisted mainly of *Robinia pseudoacacia* with some addition of *Quercus robur* and with hedgerows that consisted of *Crataegus monogyna*. Consequently, although most of them have been destroyed, the studied area is particularly rich in those species and shelterbelts and woodlots with *Robinia pseudoacacia* are distinctive features of the Park's landscape, distinguishing it from other parts of Poland. This species is not only predominating species or even the only woody species in many shelterbelts, but it was also recognized as the species which was most often found in all woody plant assembles surrounded by arable fields in the investigated area (found in 37% of all the studied tree lines, belts and clumps growing among crop fields – former Oak-Hornbeam habitat). The other frequently found species were: *Quercus robur, Pyrus pyraster* and *Acer platanoides*. In contrast, on meadows the most abundant species were *Salix alba* and *Alnus glutinosa* [17, 44].

Characteristic feature of "Turew Agricultural Area" is a network of tree or shrub lines or belts, which develop mostly along roads or ditches, and much less frequently constitute border line between adjoining crop fields. Tree and shrub lines constituted 80% of all woody vegetation clumps in this area [17]. However in most field margins, especially in the area consisted of very small (several hectares) pieces of arable land, there are no perennial (including woody) plant communities. Also thicket communities are very rare in the area. Usually, the anthropogenic habitats gives conditions for grassland communities which enriches biodiversity by creating favorable conditions for many rare plant species in the landscape.

Research of 50 small water reservoirs located in the Kościan Plain showed presence of 40 plant communities [23]. Those communities were often represented by pleustonic plants of Lemnetea class. Most of studied reservoirs with submerged vegetation formed hornwort phytocenoses - *Ceratophylletum demersi* while very rarely were observed stonewort communities (Characeae). Among the rush vegetation the *Phragmitetum australis* and *Typhetum latifoliae* were dominating (poor floristically). Sedge rushes - *Magnocaricion* were better formed.

In the investigated ecosystems There were noted 180 species of vascular plants, 3 of stoneworts and 1 moss were noted. The analysis of identified plant communities showed that 38 of them are of native origin. Communities of alien species that occurred in small mid-field ponds are formed by Canadian waterweed (*Elodea canadensis*) and sweet flag (*Acorus calamus*). The majority of plant communities in the mid-field ponds (29) were distinguished as native communities, which increase their areas as effect of the anthropogenic influence. Phytocoenoses in this category usually inhabit very fertile waters as mid-field ponds. Low diversity of plant species within phytocoenoses show the adverse changes in ecosystems as a result of human activities.

Mid-field ponds are marginal habitats in the agricultural landscape. Regardless of their small area they create conditions in which natural vegetation finds refuge, especially where the intensification of agricultural activities reduces ecological quality of landscape. Therefore, it is useful to enrich the landscape through creation of new and renovation of neglected water ponds.

One of the objects of investigations in the Turew landscape was the pond (1800 square meters) which was dug in the summer 1995 in natural, periodically flooded area. In this pond aquatic and marsh plant succession was observed in 1995-2005. During the first and second vegetation seasons the stonewort meadows covered 91% of bottoms area. In 1998 the stonewort domination finished. In that period there were not observed any phytoplankton blooms. After the third year of the pond existence the plants with floating leaves covered large area. The shade the bottom area was the reason of charophytes disappearance. In 2000 the communities of emerged plants dominated in the pond. Since 2003 the patches of hornwort are growing among the other submerged plants. It means the end of the early development stage of the pond. In the pond species rare in Poland and Wielkopolska Region occurred, for example: *Chara fragilis, Chara vulgaris, Ceratophyllum submersum* and *Teucrium scordium*.

The most valuable natural plant communities in the "Kościan Plain" are represented by aquatic (peat pits, mid-field ponds, lake and water courses) and meadow phytocenoses. Among the meadow communities the most important for biodiversity are marshy and swamp meadows endangered by drying and intensification of agricultural activities. The sedge (*Carex*), purple moorgrass (*Molinia caerulea*) and thistle (*Cirsium*) meadows were recognized as the most vulnerable to mentioned threats.

It is worth emphasizing that among 122 plant communities noticed in the whole "Kościan Plain", 54 are endangered in Wielkopolska. The most valuable communities are the 23 associations recognized in "V" category - at risk of extinction (rare or very rare communities or communities consisting of vulnerable species; those are also communities with simplified and poor species composition and with decreasing area of occurrence. Existence of these associations can be prolonged only by preserving actual conditions and reduction of anthropogenic pressure. In addition, 31 associations have been reported in high risk of withdraw - in the "I" category (communities of indeterminate threats because their distribution, dynamic tendency and systematics are weakly recognized). In the investigated area there have been identified 39 that are recognized as indicators of habitat listed in II Annex of the Habitats Directive.

4.2. Invertebrates

4.2.1. General characteristics of invertebrate diversity

„Turew Mosaic Landscape" favors to preserve high biodiversity level. According to data gathered by numerous authors, there were recorded ca. 3500 species of invertebrates, of that 2600 insects species since 1970s (Table 1), which represent a majority of Central European insect families. The most abundant are the species strictly related to agrocenoses (agrophags) and their natural enemies (predatory and parasitoid species). Some of the species are rare in Poland, many of them (carabid beatles, bumble bees, some butterflies) are protected by law or are listed in Red Lists [45]. Unfortunately, extremely strong differences in sampling efforts between the studied groups and qualitative character of much data do not enable to approximate total species richness.

Taxon/group	No of. species	Source	Taxon/group	No of. species	Source
Nematoda	40	[46]	Microlepidoptera	~150	Karg (unpubl.)
Enchytraeidae	16	[47]	Coleoptera	~500	Karg (unpubl.)
Lumbricidae	7	[48]	Hetero- and Homoptera	~200	Karg (unpubl.)
Acarina	146	[49]	Hymenoptera	~600	[52]
Araneae	224	[50]	- Apoidea	260	
			Thysanoptera	40	[53, 54]
Water insects	~190	[51]	Orthoptera	~30	Karg (unpubl.)
- Odonata	36		Dermaptera	>5	Karg (unpubl.)
- Heteroptera	41		Blattoptera	>2	Karg (unpubl.)
- Ephemeroptera	10		Diptera	~300	Karg (unpubl.)
- Coleoptera	>90		Others	~50	Karg (unpubl.)
Macrolepidoptera	~500	Karg (unpubl.)			
- butterflies	51	Sobczyk (unpubl.)			

Table 1. Number of species in invertebrate taxa.

4.2.2. Insects

The studies that have been carried out for above 40 years on above-ground insects of agrocenoses and small wooded patches, tree belts and lines, enable evaluation of their status and changes they are undergoing. The results of these investigations indisputably show the crucial role of semi-natural habitats in preserving insect diversity in an agricultural landscape. Biomass, abundance and diversity of above-ground insect communities in small wooded patches and in ecotone zone are on average significantly bigger than in open arable land. What is more, the abundance of predatory and parasitoid species, which play important biocenotic function, is in small wooded patches markedly higher, too.

Small wooded patches constitute refuge area as many species of insects can find there a favorable condition for wintering. The differences in density and biomass of insects between wooded area and crop fields are significant. In the first habitat and in ecotone zone (0.5 m wide) the values were several tens higher than in adjacent crop fields. More detailed information on distribution and abundance pattern of insects in relation to habitat and landscape structure are included in Chapter 5.2.

As many as 51 butterfly species were recorded in the "Kościan Plain". Most of them were found in two ecotones - between wooded area and crop fields (27 species) and between wooded area and meadows (41 species), where the butterfly abundance was 4-5-fold higher than in the previous ecotone. There were observed a few species listed in Red List: *Lycaena dispar, Colias myrmidone* and two other species worth to conserve regionally: *Heteropterus morpheus* and *Polyommatus amandus*. *Lycaena dispar*, hydrofilous species, inhabited meadows with drainage ditches, where larvae of the species could easily find their food - *Rumex hydrolapatum* Huds. Additionally, some observations of second generation of the species evidence for a favorable conditions, which characterize some habitats of the studied landscape. The population density of the species in wooded patches surrounded by crop fields amounted to 0.1 ind./km and in ecotones at wooded areas and meadows – 0.8 ind./km. The presence of dense network of shelterbelts, tree alleys, etc., favors relatively easy dispersion of the rare species from the breeding habitat across the landscape.

4.2.3. Spiders

In the "Kościan Plain" as many as 224 spider species (27% of Polish spider fauna) were observed and 17 species are listed in "The Red list of threatened and endangered animals" – most species with VU (vulnerable) status, one with EN (endangered) status. Of that, 204 species were found in the "Turew Mosaic Landscape" and 19 other species – exclusively in forest island constituted by a village park in Turew were noted [55,56].

In total, 21 spider families were noted, what constitutes 56 % of all spiders families of Poland. The first three spider families, which are most numerous in species, were in the same sequence as in the whole Polish spider fauna. Linyphiidae was the most species-rich spider family and it was represented by 63 species (32 % of the pool of species occurring in crop fields and shelterbelts), and the next ones were: Theridiidae – 19 species (30 %) and Lycosidae - 18 species (28%), respectively. The other frequently represented families were: Gnaphosidae -17 species, Salticidae -14, Araneidae -13, Thomisidae and Philodromidae – 9, Tetragnathidae – 7, Clubionidae and Dictynidae – 6 species. The rest of families was represented by 1-4 spider species.

4.3. Fish

Zbęchy Lake, the biggest basin in the "Kościan Plain" is inhabited by 11 species (bream *Abramis brama*, roach *Rutilus rutilus*, silvery crucian carp *Carassius gibelio*, common crucian carp *Carassius carassius*, tench *Tinca tinca*, carp *Cyprinus carpio*, ide *Leuciscus idus*, pike *Esox lucius*, perch *Perca fluviatilis*, pikeperch *Sander lucioperca*, eel *Anguilla anguilla*) from four families: *Cyprinidae, Anguillidae, Percidae* and *Esocidae* (by Polish Angling Association, the Fishing Company in Osieczna). Fish in the „Wyskoć Canal" are represented by six families (*Cyprinidae, Anguillidae, Percidae, Esocidae, Gasterosteidae, Gadidae* and *Cobitidae*) and 14 species (roach, rudd *Scardinus erythrophtalmus*, bream, gudgeon *Gobio gobio*, common crucian carp, silvery crucian carp, tench, three-spined stickleback *Gasterosteus aculeatus*, bleak *Alburnus alburnus*, perch, pike, eel, burbot *Lota lota*. Particularly valuable species, listed in Annex II of the Habitats Directive (Council Directive 92/43/EEC), is mud loach (*Misgurnus fossilis*).

4.4. Amphibians

The "Kościan Plain" is inhabited by 12 amphibian species. It is 67 % of all amphibian species living in Poland and 86 % of lowland amphibian species. The two species not recorded here, are rare (natterjack *Bufo calamita*) or very rare (agile frog *Rana dalmatina*) in Poland. The most common amphibians in the agricultural landscape belong also to the most common in Poland: water frogs (edible frog *Pelophylax esculentus* and pool frog *P. lessonae*) – inhabited

94% of all water bodies, brown frogs (common frog *Rana temporaria* and moor frog *R. arvalis*) - found in 89% of reservoirs. To the rarest species, inhabited less than 10 % reservoirs, belong marsh frog *Pelophylax ridibundus* and the European tree frog *Hyla arborea* [57, 28, 29]. One species (crested newt *Triturus cristatus*) is included in the Polish Red Book of Vertebrates (with the statuts NT - near threatened) and two other ones (crested newt and fire-bellied toad *Bombina bombina*) in the Polish Red List, with the status NT and DD (data deficient), respectively. The same two species are included in Appendix II of EU Habitat Directive. Crested newt and fire-bellied toad are not common in the study area – they were found in 20-30% of all investigated water reservoirs.

All types of water bodies and landscapes within the "Kościan Plain" were inhabited by similar number of species (11-12), but they differed in frequency and abundance of species. The highest number of species was noted in water bodies located in forests (mean 6.6 per reservoir) and in ponds surrounded by cultivated fields (6.2) and the lowest one – in reservoirs in meadow complexes (4.8). The species that occurred mostly in forest are: pool frog (70 % sites), moor frog (65%) and crested newt (35%). In the ponds in cultivated fields bred mainly (60-70% of sites) common spadefoot (*Pelobates fuscus*) and common frog. One species – green toad (*Bufo viridis*) – was found mostly (40%) in village ponds. The most abundant amphibian populations (> 500 individuals) were observed in forest (17% of all sites) and in meadow (8%) reservoirs.

4.5. Birds

4.5.1. Regional species pool

Total number of breeding species recorded in the "Kościan Plain" (in all habitats) amounted to *ca.* 120 [58]. Complete study on breeding bird community in the "Turew Mosaic Landscape" was carried out in 1991-1994 near Turew on the area of 1380ha (crop fields - 80 %, grasslands - 12 %, small wooded patches and lines - 5 %, roads - 2 %, water bodies and courses - 1 %) [59]. As many as 76 species were found to nest (with the population density of 140 pairs/100 ha) in that mosaic landscape. Small wooded patches were the most important as they covered only 5% of studied area, while constituted breeding habitat for as much as 88% of species (N=67) species and 54% of all breeding pairs. Although the studied community consisted mostly of species known as common and abundant in Poland, population density of several species which are considered to be threatened and/or rare in Europe was relatively high. It concerns e.g. ortolan bunting (*Emberiza hortulana*), corn bunting (*Emberiza calandra*) and red-backed shrike (*Lanius collurio*). The community included also some species from Polish Red List (DD – quail *Coturnix coturnix* and European turtle dove *Streptopelia turtur*) and from Appendix 1 of EU Bird Directive (white stork *Ciconia ciconia*, marsh harrier *Circus aeruginosus*, black woodpecker *Dryocopus martius*, barred warbler *Sylvia nisoria*, red-backed shrike, ortolan bunting, tawny pipit *Anthus campestris*).

4.5.2. Variability in landscape scale

Species number in the "Turew Mosaic Landscape" differed strongly between its fragments (with the area of 35-55ha). The minimal number of species (4) was observed in a plot without wooded areas, with distinctive predominance of crop fields (98.8%), the maximal one (33) was recorded in the mosaic-like area, rich in a variety of tree lines and patches. The population density was much less variable and ranged from 13 to 33 pairs/10 ha [59].

4.5.3 Variability of bird communities in small wooded patches and lines.

The analyses presented here were done for the data on 74 various tree lines (length < 3 km) and patches (area < 10 ha) studied in 1991-1994 [32], and for 68 patches (<4 ha) studied in 2005-2007 [60]. In 53 tree lines as many as 54 breeding species were observed. However, the use of species richness estimators for that dataset suggests that total number of species could be even higher – ca. 60 (ca. ¼ of breeding avifauna in Poland). The species number per single

plot (i.e. tree line) ranged between 1 and 28, with arithmetic mean amounting to 12±6.3 (SD). Total density amounted 1-134 pairs/km, on average 26±22 pairs/km. In small wooded patches as many as 56 species (3-23 per patch, on average 12.1±5.4) were observed. According the Jack-Knife2 species richness estimator one may expect more than 80 species (Figure 5) in all types of wooded patches, tree lines etc. Among dominating species (>5% of community) were finch (*Fringilla coelebs*), blackcap (*Sylvia atricapilla*), yellowhammer (*Emberiza citrinella*), great tit (*Parus major*), nightingale (*Luscinia megarhynchos*), blackbird (*Turdus merula*) and icterine warbler (*Hippolais icterina*). Total population of breeding birds amounted to 125 pairs/10ha.

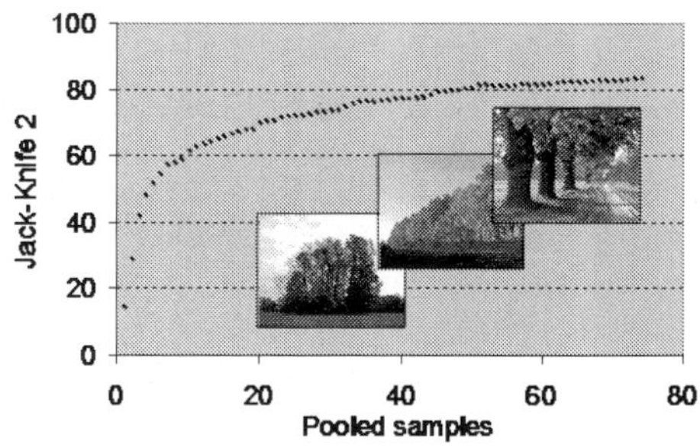

Figure 5. Species richness in wooded patches. Photo K. Kujawa.

4.6. Mammals

A list of big-sized mammal species in the "Kościan Plain" is based on field observations in the years 1998-2012 and references. Rodents were diagnosed on the basis of research carried in recent years by the CMR method, while the presence of other species was based on direct observation. Most of the recorded species are commonly encountered in this area commonly. There are no species for which the "Kościan Plain" would provide a key role in preserving their populations. However, the presence of protected species indicates the attractiveness of the area for those groups of animals. As many as 32 species (with the exception of bats), representing 14 families, were recorded in the area: mole *Talpa europaea* (partially protected), common shrew *Sorex araneus* (strictly protected), Eurasian water shrew *Neomys fodiens* (strictly protected), brown hare *Lepus capensis*, red squirrel *Sciurus vulgaris* (strictly protected), beaver *Castor fiber* (partially protected), water vole *Arvicola terrestris* (partially protected), red vole *Myodes glareolus*, common vole *Microtus arvalis*, field vole *Microtus agrestis*, house mouse *Mus musculus*, field mouse *Apodemus agrarius*, yellow-necked mouse *Apodemus flavicollis*, wood mouse *Apodemus sylvaticus* (partially protected), harvest mouse *Micromys minutus*, brown rat *Rattus norvegicus*, red fox *Vulpes vulpes*, raccoon dog *Nyctereutes procyonoides*, Eurasian lynx *Lynx lynx* (seen once), badger *Meles meles*, European otter *Lutra Lutra* (partially protected), pine marten *Martes martes*, beech marten *Martes foina*, least weasel *Mustela nivalis*, American mink *Neovison vison*, common raccoon *Procyon lotor*, wild boar *Sus strofa*, roe deer *Capreolus capreolus*, red deer *Cervus elaphus*, fallow deer *Dama dama*, Eurasian elk *Alces alces* (seen once), mountain sheep *Ovis ammon* [61, 34]. Two species – beaver and European otter – are listed in the annexes of the Habitats Directive. Traces of beaver were found mainly along of watercourses, e.g. Wyskoć Canal, but in melioration ditches, too. Fauna of bats requires additional research, but it is estimated that there are approximately 12 species of bats.

Mid-field shelterbelts are an important element of agricultural landscape for small mammals. Shelterbelts are mid-field refugees, food source and ecological corridors. They also join scattered elements of environment in mosaic landscape. Small mammals in the shelterbelts were studied e.g. by [62, 63]. Rodents can migrate for the longest distances along shelterbelts. In the shelterbelts, many species find the conditions for the reproduction and rearing, the overwintering or survival in the case of unfavorable weather conditions. Also wooded areas can be a food source for many species of mammals, often in the winter decisive for survival, or a supplemental food source. Seven species of small rodents (47% of Polish fauna) were recorded in the studied shelterbelts. In the old shelterbelt, yellow-necked mouse *Apodemus flavicollis* dominated markedly (55% of captured animals). Also numerous individuals of field mouse *Apodemus agrarius* (18%) and common vole *Microtus arvalis* (19%) were found. Additionally, individuals of wood mouse *Apodemus sylvaticus*, house mouse *Mus musculus*, harvest mouse *Micromys minutus* and bank vole *Myodes glareolus* were captured. In the young shelterbelts, small mammals were much more abundant but species structure was different than in the old ones. Field mouse dominated here (45%). Common vole was also numerous (32%). Other species (yellow-necked mouse, wood mouse, house mouse, bank vole) were much less frequent. According to classification by [64, 65], in which small mammals can be divided into "forest" species (yellow-necked mouse, bank vole), "field" species (house mouse, harvest mouse, common vole) and "intermediate" species (wood mouse, field mouse), our study indicates, that the old shelterbelt is an environment similar to the forest. In the young one, domination of "field" and "intermediate" species was observed. Small mammals were also studied in manorial park in Turew. Species typical for forest environment strongly dominated there (bank vole 54%, yellow-necked mouse 15%). Field mouse was also observed here (30%). Common vole was captured here only occasionally.

4.7 Fungi

4.7.1. Macrofungi

At first, it is worth to note that an agricultural landscape, till now, has been only rarely considered as the area, on which preservation of national species pool strongly depends. Only few studies on species diversity of Macromycetes (species, which can be observed with unarmed eye) have been carried out in agroecosystems and exceptionally in scale of landscape. The results of a study conducted since 1997 in the "Kościan Plain" indicate that mosaic-like agricultural landscape, rich in non-farmed habitats such as village parks, small wooded patches, menaged tree stands, is inhabited by a variety of fungal species, including rare and protected ones. In 1997-2011 as many as 687 species were found (Kujawa A., unpubl.), in that 99 from Ascomycota and 588 from Basidiomycota. According to substrate on which fungi grow, most frequent (55% of all species) were terricolous species (Figure 6A). Apart main type of substrates (soil, wood and litter), the fungi were found also on dung, dead fruitbodies, wood charcoal) and some others were parasites of herbs and insects. With respect to trophy, most species belongs to saprotrohic group (Figure 6B).

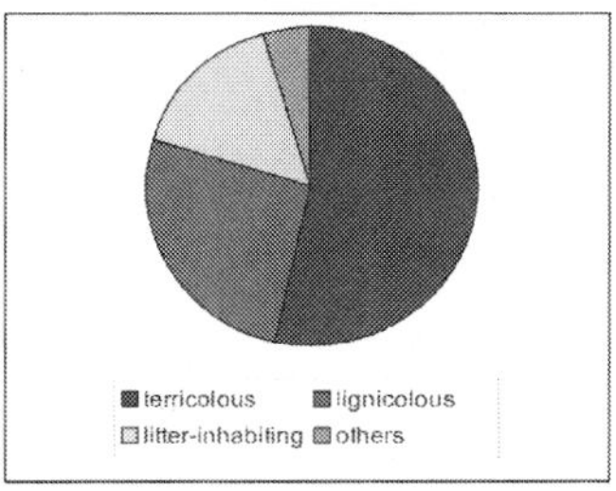

(a)

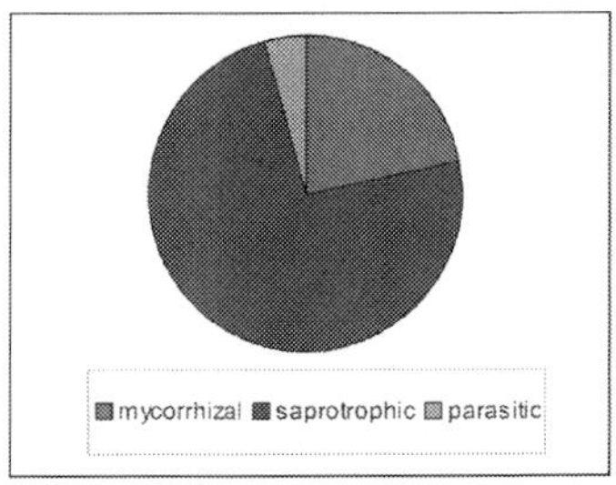

(b)

Figure 6. Share of species according to: a) substrate inhabited; b) their trophy.

Most of the studied elements of the „Kościan Plain" play a role of substitute habitats for some rare species and species protected by law. It is worth underlining, that 26 % of species observed in the "Kościan Plain" study area belong to the group of special concern species, i.e. these species which follow at least one of the below listed criteria:

a. Species protected by law [66] – 17 species, e.g. *Sarcocypha austriaca, Grifola frondosa, Sparassis crispa, Verpa conica, Fistulina hepatica, Geastrum striatum, G. berkeleyi, G. coronatum.*
b. Threatened species, listed in "Red list of the macrofungi in Poland" [67] – 98 species, e.g. *Coprinus bisporus, Crepidotus luteolus, Entoloma rhodocylix, Hygrocybe insipida, Inocybe calospora, Lepiota brunneoincarnata.*
c. Threatened species, listed in "European red list of the macrofungi" (Ing 1993) – 18 species, e.g. *Cordyceps capitata, Entoloma excentricum, Lepiota fuscovinacea, Mycenastrum corium, Perenniporia fraxinea*

d. Rare species, not protected by law and included in red lists, but observed in Poland only in 1-2 sites, in that the species found exclusively in the "Kościan Plain", according to checklist of Polish larger Basidiomycetes [68], checklist of Ascomycetes in Poland [69, 70] - 79 species, e.g. *Conocybe fuscimarginata, Entolona araneosum f. fulvostrigosum, E. cephalotrichum, E. incarnatofuscescens, Marasmius anomalus, Peziza ampliata,Xylaria oxyacanthae* .
e. Species found first time in Poland (in that, recorded exclusively in the "Kościan Plain") – 25 species, e.g. Desmazierella *Desmazierella piceicola, Entoloma parasiticum, Gammundia striatula,Hohenbuehelia cyphelliformis, Melanomphalia nigrescens, Pustularia patavina.*

4.7.2. Microfungi

The total number of the species of entomopathogenic fungi collected in the "Kościan Plain" amounted to 88. As many as 60 species were found in the "Turew Mosaic Landscape". The poorest resources of that group of organisms were regularly noted in both kinds of annual crops (cereal and row crops), where only 20-25 species occurred and the majority of them was connected with the roadside, balk or ruderal vegetation. However, even in such inhospitable conditions occur not rare epizootics in aphid colonies or adult anthomyid flies – mostly of the genus *Hylemya* - caused by *Zoophthora, Pandora* and *Entomophthora* species on big areas of cereals and rape seeds [71].

The main taxonomical units grouping the majority of obtained arthropod pathogenic species of fungi were the orders of Entomophthorales and Hypocreales. The greatest significance of the first order for the restriction of agrophagous and forest pest arthropods results from their spontaneous dispersal in dense host populations, ending often by almost total their mortality. Only few of representatives of the second order are able to form the perfect (ascomycetous) fructification form, but they produce as a rule abundant and strongly differentiated vegetative (conidial) sporulation forms (called anamorphs), allowing their overall dispersion and permanent restriction of arthropods. Their most common genera, as *Beauveria, Isaria, Lecanicillium, Metarhizium, Simplicillium* are polyphagous, non-selective with respect to a host species, permanently sustaining in soil conditions [72] and in "aeroplankton". Moreover, many of their strains have been applied as "active agents" in commercial biopesticides.

As many as eight entomopothogenic fungal species were found in soil of agrocenoses and semi-natural habitats [72]. Generally, three species of fungi: *B. bassiana, M. anisopliae* and *I. fumosorosea,* dominate in Polish soils. The dominance of particular species depended on habitat. In the soils from arable fields *M. anisopliae* dominated, with *I. fumosorosea* and *B. bassiana* as subdominants. *M. anisopliae, B. bassiana* and *I. fumosorosea* predominated in the soils from meadows and pastures, whereas *B. bassiana* in samples from forest soil and litter. It was found that infective potential and density of CFU of entomopathogenic fungi were significantly higher in soils from old or medium age shelterbelts than in adjacent large-area arable fields. Moreover, the soils from shelterbelts were characterized by more abundant fungal species composition (6 vs. 3 species respectively), being specific refuges for resources, diversity and persistence of entomopathogenic fungi in the agricultural landscape. Particularly valuable in this respect seem to be scrubby shelterbelts of rich species composition.

5. Impact of landscape structure and land-use

5.1. Plants

Floristic diversity of the whole agricultural landscape depends on landscape complexity. Results of the study on relationships between the floristic diversity of farmland and the diversity of its structural elements have shown a strong relationship between them [11]. The flora value increases with the number of spatial elements in the landscape (Figure 7).

Results of the study have proved that in a homogeneous landscape, composed of only arable fields and linear mid-field elements, the species number is half as high as in a landscape with numerous habitat islands. In fields, 224 plant species were noted. It was only 27% of the total landscape flora. The number of weed species in each field depended on method of cultivation and type and size of crop. The greatest species richness is characterized by small fields of farmers applying the extensive methods of cultivation.

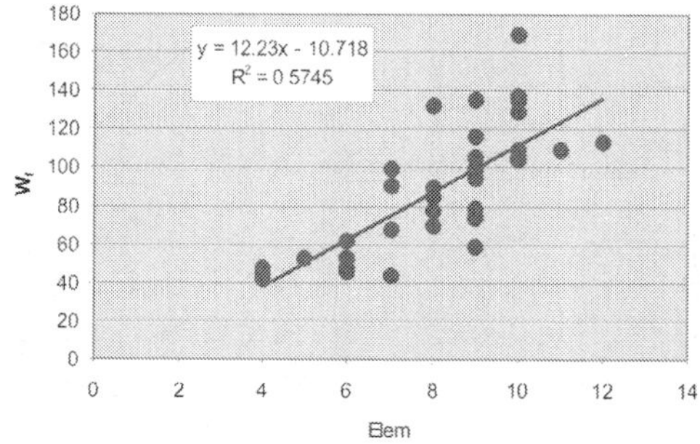

Figure 7. Relationship between the floristic value (*Wf) and the number of landscape elements (Elem) * Wf – the floristic value is the sum of the rarity coefficients of the naturalized plant species that are noted on the plots (1 km^2 each)

The studied landscape consists of various spatial elements with various land use. Evaluation of these elements allowed to point out those with the highest conservation value, which increase floristic diversity in agricultural landscape most significantly. Among them there are: water reservoirs and ditches, meadows, forests and manor parks. Nearly all the elements of the agricultural landscape, both patch-like and linear, are refuges for threatened and other environmentally valuable species.

The research on the flora carried on in the previous century was summarized by [6]. The study indicated that water reservoirs and meadows located in agricultural landscape are the habitats that provide refuge site for the greatest number of threatened species.

Lake differs from other landscape elements in plant cover, which is the most natural, and in numerous occurrences of species that are especially vulnerable to human impact. Rare species for the flora of Poland and Wielkopolska Region and taxa included in red lists of plants are noted more often in the water column and in the belt of rush of the lake than in other water reservoirs (e. g. *Hippuris vulgaris, Lathyrus palustris, Calamagrostis stricta, Carex disticha, Cladium mariscus, Dactylorhiza majalis, Juncus ranarius, Lotus tenuis, Nuphar lutea, Schoenoplectus tabernaemontani, Tetragonolobus maritimus, Teucrium scordium, Valeriana dioica*). They represent 32% of the total flora of the lake. The share of the native species group is very high (nowadays it is 95%) [11].

Species that are protected by law are the most numerous in peat pits. The flora of those biotopes is most similar to the flora of lakes in respect to natural value. Among threatened species, the following are worth mentioning are: *Achillea ptarmica, Calamagrostis stricta, Carex disticha, Cladium mariscus, Dactylorhiza majalis, D. incarnata, D. maculata, Hydrocotyle vulgaris, Lathyrus palustris, Menyanthes trifoliata, Nuphar lutea, Nymphaea alba, Pedicularis palustris, Schoenoplectus tabernaemontani, Teucrium scordium, Utricularia vulgaris, Viola palustris.* About 27% of peat pits total flora are valuable species.

High naturalness has distinguished also the flora of linear elements of landscape that are included in meliorative systems. Those are canals and ditches lying among meadows. Native species are 97% of this flora. Species typical for aquatic ecosystems and meadows have dominated in the sociological plant groups. The only records of *Batrachium fluitans, Sagittaria sagittifolia* and *Sparganium emersum* were found in canals and the only locality of *Alisma lanceolatum* in a ditch.

Ponds and midfield ditches are often the only aquatic ecosystems within rural areas. Thus the flora of them is especially valuable for biodiversity despite of their anthropogenic origin. This concerns particularly areas with big fields and intensive management. Those biotopes enrich poor agricultural landscape with a group of native species, also of the ones which are typical for natural ecosystems. Those are most of all aquatic, forest and ticket taxa. The flora of midfield ditches consists mainly of common species, but sometimes species from a vulnerable group can be observed. Those are e. g. protected by law *Batrachium trichophyllum, Centaurium pulchellum, Hedera helix* or sufficiently rare in Wielkopolska Region *Carex cuprina* and *Sagina nodosa*. Among all tree and shrub lines in the studied area, the dendroflora of the ones that were situated along ditches was the most natural and the share of native species was the highest [17].

Ponds are refuges for many aquatic and marsh species in agricultural landscape and therefore they are very important in maintenance of biodiversity. The only records of such species as *Potamogeton gramineus* and *Potamogeton trichoides* were observed in ponds. The flora of ponds is characterized by a higher share of alien species than the other aquatic biotopes. It is highest in case of naturalized neophytes [11].

The intensity of agriculture of the studied area rose during the last 30 years what resulted in the increase of the significance of agricultural nonpoint pollution leading to the eutrophication of ecosystems.

Thus changes in the flora of aquatic and marsh ecosystems (lake, peat pits, Wyskoć Canal) under increasing human impact were analyzed [20]. This was achieved by comparing the present flora of these habitats with their flora reported in the late 1970s. Vascular plant species richness during the last 30 years increased by 40, as 15 species disappeared and 55 new appeared. However, 6 moss and 4 stonewort species disappeared and only 3 new moss species were found (Table 2). The Wilcoxon test, taking into account changes in the dynamics of individual species, indicated statistical significant differences between the studied periods ($Z = 2.531$, $P = 0.011$). In the case of apophytes they were significant both in their number in both periods (sign test $Z = 6.019$, $P<0.001$) and their dynamics (Wilcoxon test $Z = 5.453$, $P<0.001$). Also the redundancy analysis (RDA) revealed significant differences in flora between the two study periods as the variables 'Present' and 'Past' were highly significant ($F = 7.7$; $P<0.0001$). In the case of apophytes they were significant both in their number in both periods (sign test $Z = 6.019$, $P<0.001$) and their dynamics (Wilcoxon test $Z = 5.453$, $P<0.001$).

Item	The number of plant species 1976-1980	2006-2007
Total flora	152	185
Vascular plants	129	169
Stoneworts	9	5
Mosses	14	11
Endangered from the red lists	25	18
New species of vascular plants	-	55
Disappeared vascular plants	-	15
Disappeared stoneworts	-	4
Disappeared mosses	-	6
New mosses	-	3
Nonsynanthropic native species	87	87
Apophytes (synanthropic native species)	40	73
Archaeophytes	0	2
Naturalized neophytes	2	6
Casual neophytes	0	1
Share of valuable species in the total flora (%)	51	37

Table 2. Transformations of the aquatic and marsh flora

Nearly all undetected species were rare in the Wielkopolska region. They are e.g. Batrachium trichophyllum, Carex diandra, Carex rostrata, Gentiana pneumonanthe, Hippuris vulgaris, Pedicularis palustris, Potamogeton friesii. However, complete disappearance of the water soldier (Stratiotes aloides) which was a very common species in peat pits in 1976-1980 was very spectacular. It is also interesting that hornwort (Ceratophyllum submersum) which was very rare in the Wielkopolska Region between 1976 and 1980 and only one record of it was known in the Kościan Plain (in a mid-field pond), widespread in the last 30 years. At present, patches of this species are often found in various types of water body.

Among the new species, the most common were apophytes (i.e. the synanthropic native species capable of occupying the habitats transformed by human activity). Among the species that disappeared from the flora, 84 % are nonsynanthropic spontaneophytes (i.e. the nonsynanthropic native species). An increase in the percentage share of apophytes (from 31% up to 43.2%) and of alien species (from 1.6% up to 5.3%) in the flora caused a decrease in the naturalness index (the percentage share of nonsynanthropic spontaneophytes in the flora) of aquatic and marsh flora from 67.4% to 51.5%.

In the group of new species there are 2 species of invasive neophytes. They are: *Bidens frondosa* and *Echinocystis lobata. Bidens frondosa* widespread already in the Kościan Plain landscape and outcompeted the native *Bidens* species, which completely disappeared from the flora of studied ecosystems. *Echinocystis lobata* has appeared in the natural vegetation patches in the last decade and its invasion in the next years is very probable.

In the group of species that disappeared there are only those characteristic for aquatic and meadow communities. New components of the flora are species from various sociological groups. Segetal and ruderal species, which mostly decrease the naturalness index, are 14% of the new species group.

14 moss species were found in aquatic and marsh ecosystems in 1976 - 1980. Their number decreased to 11 (Table 1). Two species protected by law disappeared (*Drepanocladus sendtneri* and *Leptodictyum humile*) and a new one appeared (*Amblystegium radicale*). Three calciphilous species vanished from the moss flora (*Drepanocladus sendtneri, Campylium polygamum, Campylium stellatum*). *Fontinalis antipyretica,* which in 1976-1980 was a very common component of the aquatic flora in the lake, currently belongs to the group of potentially endangered species. The frequency of occurrence of this species decreased by over 90%.

The flora of stoneworts (Characeae) decreased in number by 4 species (*Chara aculeolata, C. contraria, C. polyacantha, C. vulgaris*). All of them were included in the red list of stoneworts in Poland. *Chara polyacantha* was an especially valuable species because it is very rare in Polish flora and is protected by law.

A very significant decrease in the frequency of *Chara fragilis* and *Nitellopsis obtusa* (by 75% and 65%, respectively) was observed in the time interval of the investigations. These species, especially *Nitellopsis obtusa,* belonged to the most important components of submerged plant associations in the lake in 1976-1980.

Phytosociological studies on the vegetation of aquatic ecosystems were also carried out (Tab. 3). The total number of plant communities that were distinguished was 77. Half of them are mentioned in the list of threatened plant communities in Wielkopolska Region. The most valuable is the vegetation of peat pits – the highest number of species threatened, rare and vulnerable to human impact communities occurs in those biotopes. The highest phytocoenotical diversity characterized ponds and ditches lying among meadows. The smallest is the diversity of plant communities in midfield ditches.

Plant communities of the lake, peat pits and Wyskoć Canal were studied from the half of 1970, so it was possible to follow the transformation of them during the thirty-year period

Kind of ecosystem	Plant communities Number	R	V
Lake	22	4	12
peat pits	36	8	23
Canals	25	2	8
ditches among meadows	44	7	18
Ponds	48	7	19
ditches among fields	20	2	4
all ecosystems	77	15	38

Table 3. Phytocoenotic diversity of aquatic ecosystems (Explanations: R - rare, V – vulnerable)

(1976-2006). Nine communities of aquatic plants and one of bulrush community have perished in all the studied ecosystems. Only one of them is not endangered in the Wielkopolska Region. Almost all are very rarely met in the Kościan Plain landscape. During the thirty years the number of plant associations (mainly components of the rush belt), as well as the number of the species that constituted them, simultaneously increased.

In Lake Zbęchy 5 associations of submerged plants disappeared [18].The maximal depth of plant occurrence decreased from 3.6 m in 1976 to 2 m at present. This caused a reduction of total phytolittoral area by 50 percent. The area overgrown by submerged macrophytes decreased from 13 to 2 ha. Valuable plant associations like *Nitellopsidetum obtusae, Myriophylletum spicati, Najadetum marinae, Potametum lucentis, Cladietum marisci,* and *Scirpetum maritimi,* considered as endangered in the Wielkopolska Region, disappeared in the lake, or the area of their occurrence decreased.

At the same time cyanobacteria blooms which were not observed in the 1970s appeared. The appearance of blooms at present in the lake can be linked to the disappearance of submerged macrophytes, especially the extinction of *Nitellopsidetum obtusae* that dominated this area thirty years ago.

The most important change in Wyskoć Canal was the complete disappearance of patches of *Sagittario-Sparganietum emersi* association, which dominated in 1970s [19]. Three aquatic plants associations (*Charetum aculeolatae, Myriophylletum verticillati, Stratiotetum aloidis*), which nowadays are included in the list of endangered communities, disappeared from the peat pits [21].

The study on the flora of meadows has shown that they are very valuable for biodiversity of the whole agricultural landscape. They are distinguished by a high total number of species that is 45% of the total landscape flora, the highest native species diversity and the highest number of vulnerable species [6, 11]. However, during the last 30 years serious changes of the meadow flora were observed, particularly regarding marsh meadows, which are habitat of many endangered and protected plants. As a result of the changes, vegetation of meadows becomes homogeneous and the same everywhere. Rare plant species disappear and common grasses grow in their place. *Alchemilla monticola, Crepis praemorsa, Eleocharis quinqueflora, Eriophorum latifolium, Euphrasia rostkoviana, Pulicaria vulgaris, Viola stagnina* were among plant species that vanished during the last decades. Sites of many valuable species such as *Carex davalliana, Dactylorhiza incarnata, D. majalis, D. maculata, Dianthus superbus, Gentiana pneumonanthe, Parnassia palustris, Pedicularis palustris, Polygala amarella, Tetragonolobus maritimus* and *Triglochin maritimum* are threatened and they are constantly decreasing. As yet, *Ostericum palustre* occurs numerously and in many places, but it is also endangered because of a decrease of wet meadows area and intensification of meadow management.

Forests and manor parks are very important for biodiversity of agricultural landscape despite occupation of small surface. Their species richness is almost the same as in meadows (species noted in all the kinds of forests were 40% of the whole landscape flora). The number of vulnerable species is bigger than in forests only in meadows. The most interesting of those species is daisy leaf grape fern (*Botrychium matricariifolium).* It was found in habitats under big human pressure – economically used monocultures of oak and ash, where some typical practices linked with cultivation of trees were regularly carried out [73]. The species is very rare in flora of Poland and strictly protected. Apart from this species, in the group of vulnerable plants of forests and parks also such species as *Calamagrostis stricta, Campanula latifolia, Cucubalis baccifer, Gagea arvensis, G. minima, Leucoium vernum, Listera ovata, Lycopodium annotinum, Ophioglossum vulgatum, Platanthera bifolia, Teucrium scordium* etc. were noted.

Majority of forests on the explored area are of anthropogenic origin. Separate studies, carried out on the flora of 15 planted shelterbelts lying among fields, showed that 100 species of vascular plants occurred there – they were 30 tree and shrub and 70 herb species. 83% of all the species were native and 17% were alien species. It was more than in natural alder forest, where the share of alien species was only of 5%, but equal or less than in all the forests in the landscape, where the mean share of this species group was of 18% [11]. In the group of alien species there were archaeophytes (8% of all species noted in shelterbelts) and naturalized neophytes (9%). The share of alien species was bigger in the tree and shrub layer (27%) than in the herb layer (13%). The group of alien species in the herb layer in majority constituted from archaeophytes (10% of all species), while among tree and shrub alien species naturalized neophytes dominated (24%). Three of the naturalized alien species – common robinia (*Robinia pseudoaccacia*), black cherry (*Padus serotina*) and box elder (*Acer negundo*) – are mentioned as invasive species that endanger the biodiversity in Poland [42]. Common robinia and black cherry are included in the list of 100 worst alien species in Europe (DAISIE – 100 of the worst http://www.europe-aliens.org/ speciesTheWorst.do). The most recently found expansive and potentially dangerous alien woody taxon was *Amelanchier* sp. [17].

Out of 86 segetal weed species which occurred in the flora of all the landscape 76 species were observed on arable fields and only 9 in shelterbelts. In addition, only 5 of those noted in shelterbelts were met in the flora of fields. The other 4 species were very rare in the flora of the Turew landscape e.g. sharp-leaved fluellen (*Kickxia elatine*) which is a vulnerable archaeophyte in Poland [74] and is also published on the red list of plants of the Wielkopolska Region [75]. Shelterbelts were the only localities of this and several other species in the studied area. Therefore these anthropogenic biotopes may also constitute refuges of rare plant species. According to [76] segetal weeds have appeared in masses only in young shelterbelts. The vegetation of older and stabilized shelterbelts is peculiar and they are not a place from which weeds widespread on the fields.

As a result of investigation (2001-2003) of woody species in tree lines, avenues, shelterbelts, hedgerows and woodlands surrounded by arable fields or intensively used meadows in the central part of the Kościan Plain in 467 mature objects 86 species have been found [17]. The most often noted species was *Robinia pseudoacacia.* The other considerably often noticed species were: *Quercus robur, Pyrus pyraster* and *Acer platanoides.* In contrast, on meadows *Salix alba* and *Alnus glutinosa* most often occurred. The number of wooded species in each object varied from 1 to 22.

Generally, tree lines among arable fields were richer than the ones on meadows. On the other hand, the alien species were more common there and had higher share. Total number of alien and cultivated woody species in tree and shrub lines surrounded by arable fields was similar to the number of native species, whereas in those on meadows number of native species was higher. As a matter of fact, around arable fields there was no wooden vegetation patch (wood) which tree layer consisted of native species only. On the contrary, on meadows there were 29% such small woods.

5.1.1. Factors determining plant diversity

The key factor determining richness of plant species and diversity is the landscape structure. In the agricultural landscape all the elements that are different from arable fields, both patch-like and linear, are refuges for environmentally valuable species. Thus, the most serious threat to biodiversity is the loss of appropriate habitats, what results from increase of intensity of agriculture. The intensity of agriculture particularly increased in the 1970s and 1980s. The most serious threat was the increase in the nitrogen content in surface and groundwater. It resulted from increasing non-point pollutions of agriculture landscape and intensive nutrient leaching from dried up peatbogs. Both aquatic and marsh ecosystems are very important refuges for plant species and their communities which are the most vulnerable to human impact.

In recent years, probably as the result of temperature increase and eutrophication, increase of frequency of cyanobacterial toxin-producing blooms have been noticed in various water bodies [77]. Cyanobacteria are very expansive and their massive appearance, regardless if they produce harmful compounds or not, it is an unfavorable phenomenon. First of all, it is a sign of impaired balance of the ecosystem [78]. The effects of the cyanobacterial blooms are a large concentration of biomass, loss of biodiversity, reduction in biocenotic stability system, the presence of large numbers of heterotrophic bacteria (especially in the phase of cyanobacteria decay), inhibition of photosynthesis of planktonic algae which are associated with cyanobacteria as well as profound oxygen deficits in the lower layer of water column [79]. The most sensitive to cyanobacterial toxins are warm-blooded Vertebraters.

Another threat for aquatic ecosystems biodiversity is an incorrect fishery management. For example, large stock of grass carp (*Ctenopharyngodon idella*) results in adverse changes in the littoral of aquatic ecosystems. This species eats vegetation, leading to the elimination of spawning areas of fish and fry regrowth of the native valuable species. Small bodies of water rich in grass carp may be completely devoid of vegetation.

Common carp (*Cyprinus carpio*) at high densities can significantly influence on environmental conditions, accelerating the rate of eutrophication of natural waters and the same have a negative impact on native fish species. The foraging behavior of carp can also negatively affect the structure and dynamics of benthic macroinvertebrates community, as well as the physical condition of the substratum [80]. These disturbances may also intensify the algal blooms due to the re-suspension of sediments [81] and release of nutrients [82, 83]. Serious problem for biodiversity of aquatic environments and marshes results from overdrying of peatlands and changes in their management.

The transformations of vegetation in the Wyskoć Canal reflect in the eutrophication of waters and changes in water flow rate. Water management measures undertaken in the 1980s caused a lowering of both the river bed and water table of the ditch. As a consequence, there was observed drainage from surrounding meadows, what caused an accelerated peat mineralization and intensive leaching of nutrients. The observed transformations of reed and sedge communities in the Kościan Plain have partially natural and partially anthropogenic character, because the natural plant succession also accelerated due to the human pressure as well as due to climate changes, especially low precipitation. The increase in the number of identified plant associations is a result of human impact. However this process is jointed with withdrawal of plant species distinctive for waters with low nutrient concentrations and by expansion of plant associations that are indicators of high nutrient levels in waters.

The natural values of wet and marsh meadows that belong to the richest elements of the studied agricultural landscape are threatened not only by drainage but also by changes in the usage patterns. The extinction of these habitats results from the intensification of management (fertilization and sowing of common grass species) but on the other hand from the abandonment of meadows cultivation, which leads to a complete overgrowth of them by willow shrubs.

5.2. Insects (above-ground insects and butterflies)

As the importance of landscape (and habitat) for insect diversity was studied in the "Kościan Plain" in many various research projects, here we focus on a comparison between cultivated fields and small wooded patches (wide shelterbelts in that case). Such approach allows underlining the role of non-farmed habitats presence as a key factor for preserving high biodiversity level of an agricultural landscape. The results obtained during the last three years (2009–2011) show that the density, biomass, and diversity (expressed as number of above-ground insects families) in the shelterbelts are significantly bigger than in arable fields. Also in narrow ecotone (0–0.5 m from the shelterbelt) higher values of those parameters were noted. Thus, the results showing differences in density, biomass and diversity of entomofauna, depending on shelterbelt's age, which were reported earlier [84, 85] have been confirmed. The highest values of those parameters were observed in young (several years old) shelterbelts, in early stages of ecological succession. The lowest diversity (47–61 families) was observed in the several years old shelterbelt and the highest diversity (63–74 families) in a few years old one. In the oldest (over 100 years old) shelterbelt moderate values were noted (58–69 families). Similar pattern was found for density and biomass of insects. Also in the ecotone zone these differences were similar, but they completely disappeared in the arable areas distant 100 m from the shelterbelt, values of studied parameters are much lower than in shelterbelts and ecotones (Figure 8). That pattern have been constant for many years (several decades), while the proportions were changing and during recent years the mean individual weight and biomass of insects in the open fields (away from shelterbelts) increased.

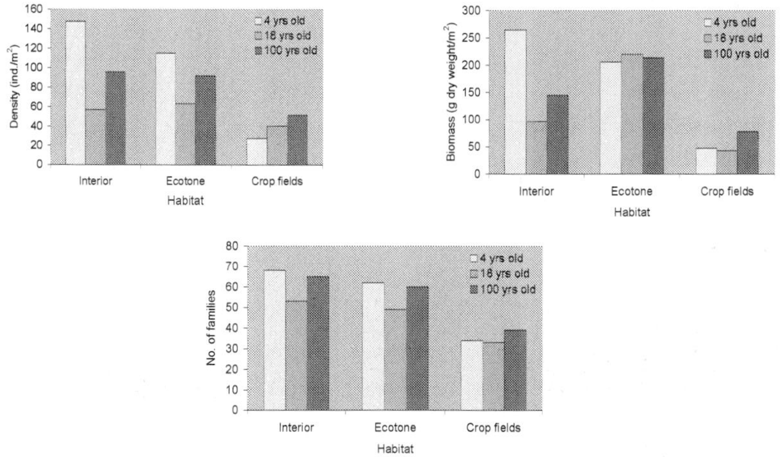

Figure 8. Relationships between insect communities (density, biomass, number of families) in vegetation season and age of shelterbelt with respect to three habitats: shelterbelt interior, ecotones between shelterbelt and cultivated fields and open fields.

The importance of non-farmed habitat for insect diversity was also studied in winter. As early as in fifth or sixth winter after introduction of the shelterbelts, they became an important place of wintering for numerous insect species. The diversity measured by the number of families reaches values (over 30 families) stable during next years of shelterbelt's growth (Figure 9). In the case of insect density, after rapid and strong increase of its value in 6th – 7th year of shelterbelt's growth (up to over 1300 ind./m^2), some decline and stabilization at the level about 200-1300 ind.m^2 was observed. Similar pattern of changes was noted for biomass. All studied parameters (density, biomass, number of families) reached the highest value in the oldest shelterbelt and the lowest – in the youngest one (Figure 10).

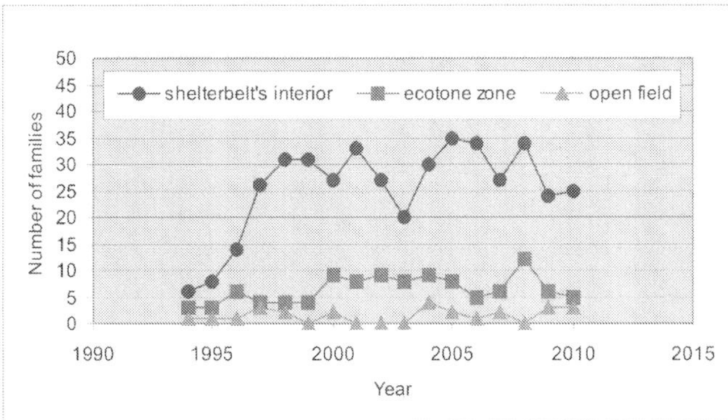

Figure 9. Number of insect families wintering in young shelterbelts and in adjacent crop fields.

Ongoing process of increasing share of cereals in crop structure creates favorable conditions for existence of species primarily living in grasslands, mainly steppes.

5.3. Spiders

Effect of landscape structure on spider assemblages is till now unclear and understanding it needs further investigations. However, gathered data also suggest positive effect of mosaic-like structure of landscape. Mean spider population density in cereal and sugar beet crops was similar across the landscape. In cereals it amounted to 7.5 ind./m^2 in mosaic landscape ("Turew Mosaic Landscape") and 7.3 ind./m^2 in homogeneous landscape (in the "Kościan Plain"). In

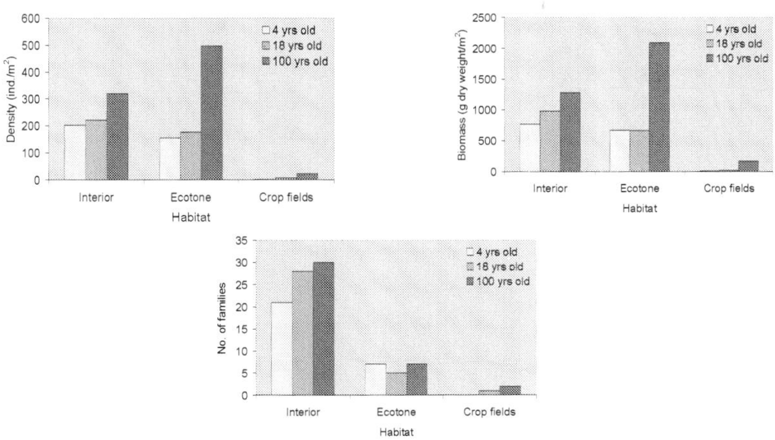

Figure 10. Relationships between insect communities (density, biomass, number of families) in winter and age of shelterbelt with respect to three habitats: shelterbelt interior, ecotone between shelterbelt and cultivated fields and open fields.

sugar beet crop it was equal to 3.1 and 2.9 ind./m^2, respectively. However, greater differences in alfalfa crop were noted. In homogeneous landscape spider density amounted 4.1 ind./m^2 and in the mosaic one - 10.8 ind. /m^2.

Spider species diversity in cereal and alfalfa crops was near twice higher in the mosaic landscape than in the homogeneous one. On the other hand, in sugar beet crop the same numbers of spider species were stated in both landscapes. In all studied crops in homogeneous landscape aeronautic spider species (small-in-size Linyphiidae) were more abundant than in the mosaic landscape. Foliage-dwelling spider assemblages structure in the studied crops were more diversified taxonomically in the mosaic landscape – higher numbers of spider families (cereals) or species (sugar beet and alfalfa) were noted. Moreover, the differences in share of particular spider families were stated. Non-web spiders were more abundant in these assemblages: ambush (families Thomisidae and Philodromidae) as well as actively hunting ones: Pisauridae, Salticidae, Mimetidae and Lycosidae.

In the cereal crops situated in homogeneous landscape the vast majority (75%) of the assemblage was composed of Araneidae – spiders hunting with orb webs, whereas in the mosaic landscape they constituted slightly more than 50% of the whole assemblage. In the last mentioned landscape type higher abundance of theridiid spiders was noted, which built three dimensional tangle webs. In the sugar beet crops the most abundant spider families were Theridiidae and Araneidae. The first one composed of over 40 % of assemblage in the homogeneous landscape, and share of Araneidae was similar (45 %) in the mosaic landscape. Among foliage-dwelling spiders in alfalfa crops linyphiid spiders (in majority Erigoninae) composed of 60% of assemblage in the homogeneous landscape and 25 % in the mosaic landscape. In the last one the next 25 % composed of Araneidae. The share of families Theridiidae and Tetragnathidae was similar in both landscapes.

The crop fields adjacency to the shelterbelts resulted in increasing of spider species diversity in above-ground layer of all studied crops. In the distance of 10 m from shelterbelts the highest numbers of spider species were stated and with the increasing distance the species numbers decreased until near twice lower in 100 m from shelterbelts.

5.4. Birds

The "Turew Mosaic Landscape" was in 1991-1994 inhabited by 76 breeding species [59] as mentioned in the chapter 4.5.1. The number may be evaluated as relatively high in Poland when agricultural areas are considered and the study area size is taken into account (Figure 11A). Moreover, regression analysis (GLZ) showed, that number of years had no significant effect on total number of species in this set of data. Thus, the Figure 11A evidences for high bird species richness in the "Turew Mosaic Landscape", presumably as a pure (not methodologically-biased) effect of habitat quality.

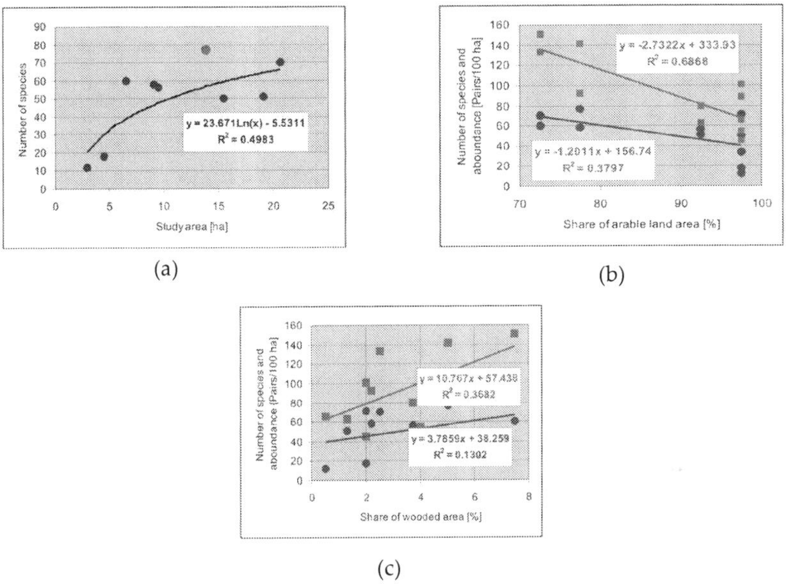

Figure 11. a) Species - area relationship for Polish farmlands [86]. Red dot – "Turew Mosaic Landscape"; b) Effect of cultivated fields share on breeding avifauna [86]. Red - species number, blue - density; c) Effect of wooded patches share on breeding avifauna [86]. Red - species number, blue - density

The data gathered and presented by [86] confirm clear relationships between landscape structure and bird richness and abundance. The values of both variables were negatively correlated with the share of arable land (Figure 11B) and positively correlated with the percentage of land covered by small wooded patches, lines etc. (Figure 11C). These two variables were weakly correlated ($r=-0.43$, $p=0.19$), so presumably both influenced the avifauna independently. It is worth to underline that species number of bird community in the "Turew Mosaic Landscape" is as high as observed in Podlasie, in eastern part of Poland [87], which is characterized by much less intensive farming practices than in Wielkopolska region and one could expect higher bird species richness just in E Poland. What is more, the species composition similarity between the "Turew Mosaic Landscape" and Podlasie was high (incidence-based Sörensen index = 82%, after [86], what indicates high efficiency of landscape structure (mostly wooded patches, lines etc. in that case) as a tool for mitigation of the impact of farming intensification on biota. Strong relationships between species richness and landscape structure is reflected also by the analyses performed for the data gathered in 25-55 ha subplots within the "Turew Mosaic Landscape" and in other places in Wielkopolska, which confirmed crucial importance of the saturation of the landscape with the wood patches and woodland-cropland ecotones [88].

The relationships between landscape structure and species richness (i.e. Jack-Knife 2) were studied with the use of the data from point count census presented by [89]. The analysis indicates similar overall species richness (65-80 species) in various landscape pieces with no respect to landscape structure (Figure 12), but the shape of estimated curves shows significant differences in species richness spatial distribution between the studied plots. In more simplified plots (arable fields – 95%) it is necessary to visit >25 points to observe 60 species, while in most diverse (arable fields – 52%) – only 15 points is needed to record 60 species It is also worth to underline that the landscape consisted of tree line, belts and patches (arable fields – 75%) had similar overall species richness to the area (arable fields – 52%) with relatively big forest complexes (Incidence- and abundance-based Sörensen index amounted to 76±2% and 94±3% S.E, respectively).

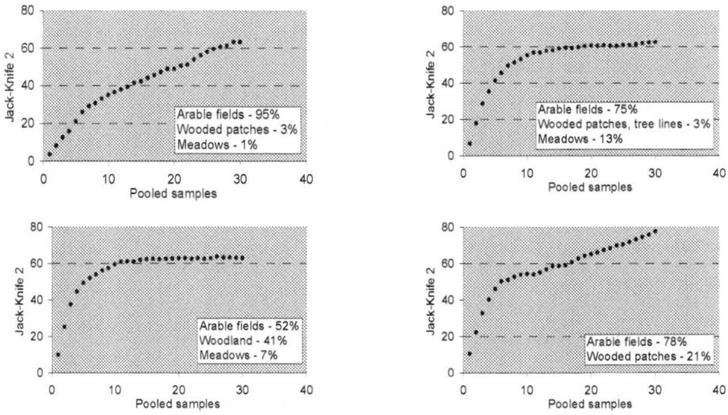

Figure 12. Bird species richness in various landscapes estimated with the use of Jack-knife 2.

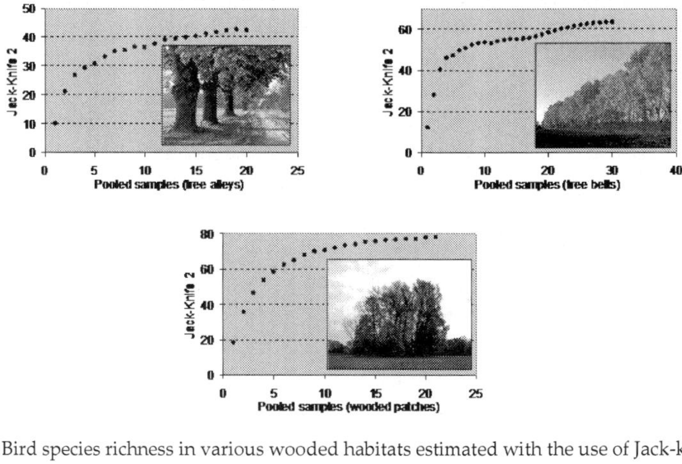

Figure 13. Bird species richness in various wooded habitats estimated with the use of Jack-knife 2. Photo K. Kujawa.

Due to a study focused on birds in small wooded patches, it could be possible to study the relationships between the bird communities and the variability in habitat structure (i.e. in scale of several hectares). The first issue was studied in small wooded patches, tree lines etc. Using the data presented by [32], the highest potential overall species richness (ca. 80 species) was estimated for wooded patches (Figure 13) and the value is close to the total species richness estimated for all kinds of patches (clumps, belts and lines) built of woody species.

Observed differences may be surely related to the differences in habitat structure between the three wooded habitat types as the analyses presented by [32] showed crucial importance of such habitat features as tree stand age, percentage cover of tree and shrub layer and the area (or length in case of line habitat), which all influenced species number positively. Indeed, the differences in bird species richness were in line with the differences in habitat structure between three habitat types [32].

However, the knowledge on relationships between individual habitat structure usually allows for explaining only a part of bird community variability in wooded areas. The analyses performed for the data on 66 islands located in the "Turew Mosaic Landscape" showed that taking into account landscape structure around studied habitats allows for understanding avifauna variability much better, especially for species strictly related to woodland, for which positive role for small wood island colonization played the presence of other wooded patches and tree lines in close adjacency [33]. Seemingly, these habitats significantly reduce the open landscape "resistance" for migrating woodland species and, as a result they enhance probability of colonization of given wood island by the birds.

5.4.1. *Effect of land-use changes on bird communities*

The quantitative investigations on birds in various mid-field wooded patches were conducted for the first time in 1964-1966 and were repeated in some plots in the years 1984, 1991-1994 [32], 1999-2002 [33] and 2005-2006 [89]. During the last 50 years, farming practices have been much intensified thus gathered data on birds enabled for testifying the effect of farming intensification on birds. Uniqueness of that research relies on dissecting the effect of landscape structure simplification and the effect of farming intensification. Commonly in Europe, landscape homogenization follows the intensification of agricultural techniques, but not in the "Turew Mosaic Landscape", where the spatial arrangement of non-farmed habitats is stable (even enriched with new tree lines and belts recently) due to establishing the Dezydery Chłapowski's Landscape Park, which successfully protects historically formed land mosaics (see "Study area").

First analysis of long-term changes in the breeding avifauna in the "Turew Mosaic Landscape" was done by [90]. The number of breeding species has increased from 44 in the 1960s to 51 in the 1990s (which can be at least partially explained by somewhat bigger sampling effort in the later period).

Total density of breeding birds remained unchanged in wooded patches (23–24 pairs/ha) with obvious exception of few areas where tree stand was cut. However, there was noticeable decrease in population density of some 'farmland specialists' like corn bunting and ortolan bunting. The analyses were then concluded that although the structure of bird community has changed only slightly in the study area, the population trend analysis suggests that agriculture intensification affects the avifauna [90].

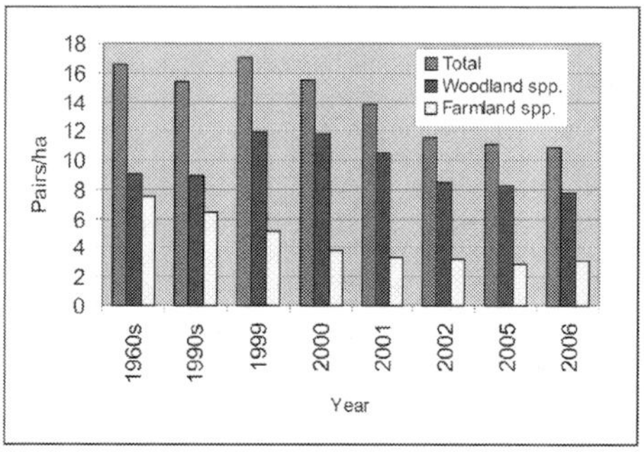

(a)

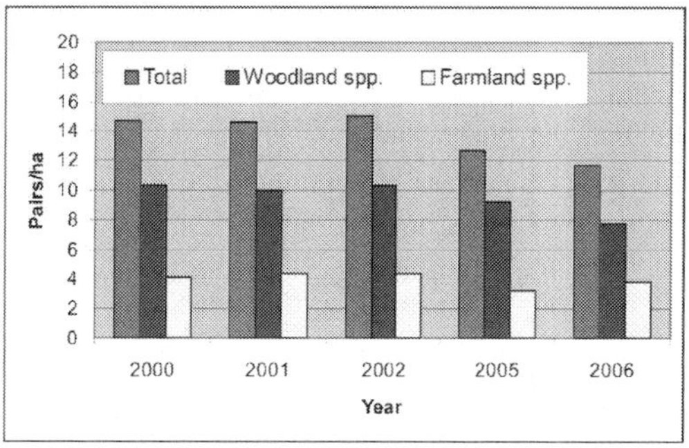

(b)

Figure 14. a) Bird population density in 6 wooded patches in the "Turew Mosaic Landscape" between 1964 and 2006 [90]; b) Bird population density in 55 wooded patches in the "Turew Mosaic Landscape" between 2000 and 2006 [90].

Recently, the changes that avifauna of small wooded patches is undergoing have been proved again [91]. The results of comparisons between the study periods (1960s, 1990s, 1999-2002 and 2005-2006) show large and significant different pattern of the changes for two main guilds of birds: woodland species (i.e. birds living in breeding season only in wooded areas, e.g. woodpeckers, robin *Erithacus rubecula*, nightingale, and spotted flycatchers *Muscicapa striata*) and farmland species (i.e. birds, which occur in wooded areas but using adjacent cultivated fields as feeding area or as place for building of nest, e.g. ortolan bunting, red-backed shrike, and goldfinch *Carduelis carduelis*). In the last 50 years the woodland species abundance and species number has been changing irregularly (increase in 1990s, some decline in XXI century), while the farmland species guild tends to decline permanently (Figure 14A). In consequence, the structure of the community has changed markedly. In 1960s the share of farmland species amounted to 45%, and it decreased in 2005-2006 to 27%.

Recently some decline in species richness has been observed too. Mean number of species per wooded patch decreased from 8.6±4.3 (SD) to 7.5±3.9 (SD) and bird abundance declined, too (Figure 14B). The differences between 2000-2002 and 2005-2006 were statistically significant for both guilds of species [91]. The observed pattern of changes shows that the intensification of agriculture is the main factor responsible for some bird impoverishment in small wooded patches in the "Turew Mosaic Landscape". Thus, it may be concluded, that mosaic of diverse habitat is necessary but not sufficient condition for successful protection of bird diversity in the agricultural landscape. To be as sufficient as possible, mosaics of diverse habitat has to accompany less intensive and more diversified (in terms of number of crops) farming.

5.4.2. *Factors influencing avian diversity*

The analyses and findings presented above show, that the bird communities in an agricultural landscape are influenced by a complex set of factors, which act in both habitat scale and landscape scale. The mechanisms involved are summarized below.

5.4.3. *Habitat scale*

- **Patch size.** The number of breeding species increases with the length and width of tree alleys and tree or hedge belts [32] and with the patch area [32, 33]. Birds are highly plastic species. Many species is able to breed in very small-sized patches. As many as 45 species were found to nest in wooded patches smaller than 1 ha [33].
- **Patch shape.** More complicated shape affects positively total species number as favors species preferring ecotonal and woodland edge species [33].
- **Patch vegetation structure.** More species breed in wooded patches with deciduous species, with old, diverse tree stands, with higher percentage cover of shrub layer, which means that more rich in species are wooded patches similar in structure to natural forest typical (but rarely occurring) for lowland regions in moderate climate [32, 33].
- **Diversity of crop fields.** This factor affected not only the number of species breeding in open area (more species in more fragmented and more diversified cultivated area) but also the birds occurring is small wooded patches (bird abundance and species number higher when the patch surrounded by diverse, fragmented crop fields) [33].

5.4.4. *Landscape scale:*

- **Landscape context (for wooded patches).** Although isolation of studied was on average small (distance from woodland amounted to maximum few kilometers, presence of dense network of tree line and belts) some bird species, especially those typical to woodland) were more abundant in wooded patches, which were located in a landscape rich in other wooded patches [33]. It means, that the knowledge on relationships between birds and inside habitat structure is not enough to understand the distribution of birds. It is necessary to take into account the landscape structure as well.
- **Landscape heterogeneity.** The analysis of species richness indicates that avian diversity in a mosaic-like agricultural landscape may be kept at same level as in a landscape with high

woodland dominance (in case of considering managed forest). If we take into account, that avifauna in our study included also rare and protected species (with abundant populations of e.g. Ortolan Bunting, Corn Bunting), the conclusion clearly arise that key way for preserving high bird species richness is making landscape as heterogenic as possible. Even though non-farmed elements cover only 5-7%, it is enough to preserve almost all species from the woodland generalists and woodland edge species. The observation can me generalized with respect to the issue of fragmentation. Although in ecology the habitat fragmentation is considered as negative phenomenon, in agricultural landscape that process has rather positive role for species diversity, when we discuss about small (<10 ha) forest patches. Concluding with the use of simple, model example, it is more profitable to keep e.g. five forest islands with the area 3 ha each, connected to each other by several kilometers of tree belts, than one "large" forest island with the area of 20 ha.

The studied area, especially the "Turew Mosaic Landscape", is characterized by most of features listed above. There are numerous, diverse non-farmed habitat patches bodies and wetlands. Moreover, most of wooded patches have old tree stand, they have diverse vegetation structure. Due to these features and mosaic-like structure, the "Turew Mosaic Landscape" constitutes a distinct example of an agricultural landscape, in which relatively rich avifauna occurs despite of high agricultural pressure [86]. Such agricultural landscape is favorable for at least 100 species, i.e. ca 40% of breeding species observed in Poland. It is striking number and we claim that the "Turew Mosaic Landscape" may play a role of a reference point for the studies on diversity of birds in lowland agricultural landscape in Central Europe.

However, recent publications on long-term changes in avifauna showed that even strongly stable amount and arrangement of non-farmed, semi-natural habitats is not enough to effectively preserve species richness, when farming practices become more and more intensive. Long-term decline in ecotonal species populations [89, 90] clearly show, that stable amount and structure of wooded patches, tree lines and tree belts does not mitigate the effects of farming intensification for all bird species guilds. Farming extensification seems to necessary to preserve total bird diversity.

5.5. Fungi

In chapter 4.7.1 it was mentioned that 26 % of species observed in the "Kościan Plain" study area belong to the group of special concern species (Species protected by law, threatened species listed in "Red list of the macrofungi in Poland" , threatened species listed in "European red list of the macrofungi", rare species).

Occurrence of these species in the study area provides great facilities in protection of fungal diversity in Poland (including agricultural area), due to proper landscape management and planning. Moreover , it may be useful for evaluation of response of given species to habitat alteration as well as for elaboration on nationwide strategy for protection of rare and threatened (according to IUCN's criteria) species of fungi. Proper landscape structure has significant influence on species diversity of fungi all over Poland. It allows for surviving of many woodland species under strong, unfavorable human pressure. Additionally, some rare

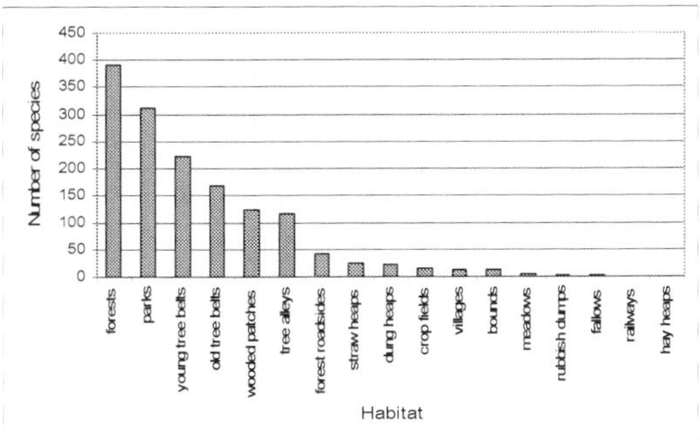

Figure 15. Number of fungi species in habitats located in the "Kościan Plain".

species dependent on traditional agriculture (growing in pastureland or organically fertilized fields) find appropriate niches in strongly diversified, mosaic agricultural landscape.

A key component for incorporating natural enemies, including fungi, in pest management in sustainable farming is the application of appropriate agricultural practices that improve conditions for these organisms in the agroecosystems and surrounding habitats. Agroecosystems suffer from mechanical disturbances due to necessary tillage regimes of various kinds, and these practices may negatively affect occurrence of entomopathogenic fungi in the soil [92, 93]. This was confirmed in [72] studies conducted in the years 2002-2004 in the area of Borek Wielkopolski directly adjacent to the "Kościan Plain".

6. Impact of climate change

Impact of climate change on biodiversity can be both direct and indirect. The direct influence lies in the fact that climate change will lead to changes in habitat conditions (mainly temperature and moisture) that can be favorable for some species and unfavorable for others. This influence is revealed by disappearing of existing species and by appearing of alien species, among which some can be invasive. The indirect impact is the fact that changes in habitat conditions are forcing farmers to change cropping patterns. For example, growth of thermophilic plants that require less rainfall (increase of cereals at the expense of root crop cultivation). This in turn influences the change of plant community and animals living in the agricultural landscape.

There is no doubts that climate change is not new in geological history of Earth, and organisms have adapted to most serious changes over evolutionary timescales. But the key question today is how will organisms respond to the current apparently rapid rate of anthropogenic climate change? [94].

Kędziora defined the process of climate changes that occurred in Poland as mediterraeanization. It's expressed by gradual increase of average air temperature with simultaneous absence of annual precipitation increase and move rainfalls to cold months. Similar trend for climate changes was observed in the neighboring state of Brandenburg. Such course of changes leads to deterioration of higrothermic conditions, especially during vegetative season.

It is obvious that impact of climate and land use changes is working synergistically. Therefore, the analysis of changes of land-use in the landscape during long period has been done in order to to distinguish the impact of climate from the impact of land use,. The study of a landscape structure which was conducted in the area of 16 km^2 in the vicinity of Turew, where the total area of meadows, pastures, and woody vegetation patches (the most important ecological elements in the agricultural landscape) was assessed at four time horizons: 1890, 1940, 1989, and 1996. Share of meadows, pastures and woody vegetation in total area of the landscape showed significant increase in the period 1890 – 1940 and remained slightly changed in the period 1940 - 1996 (Figure 15A). Although only mild increase of tree patches (from 64.5 to 69.5%) and decline of tree lines (from 35.2 to 30.5) and small decline of pastures was observed during the period 1940-1996 (Figure 15B), disadvantageous trend of changing their distribution occurred: disappearance of single trees and small strips of meadows distributed between fields [44].

Simplification of agricultural landscape may not only occur together with climate change, but it can also be a result of those changes. Deterioration of hydrothermal condition may cause disappearance of moist habitats, like swamps, marshes, and bogs. The presence of ecologically differentiated water reservoirs in landscape is also the key factor for maintenance of amphibian populations. Protection of such habitats in the agricultural landscape is especially difficult because this area is usually very (sometimes extremely) small and particularly vulnerable to influences. But the most important change, threatening biodiversity, is observed in the structure of the crop. Area of cereal crops increases rapidly at the expense of perennial and root crops. Until 1990s, [95] reported following structure: cereals 50%, row crops (including rapeseed) 20%, perennial fodder crops 10% and others 20%, in first decade of 21st century [96] noticed: cereals (including maize) 78%, legumes 16%, potato, rapeseed and sugar beets 6%.

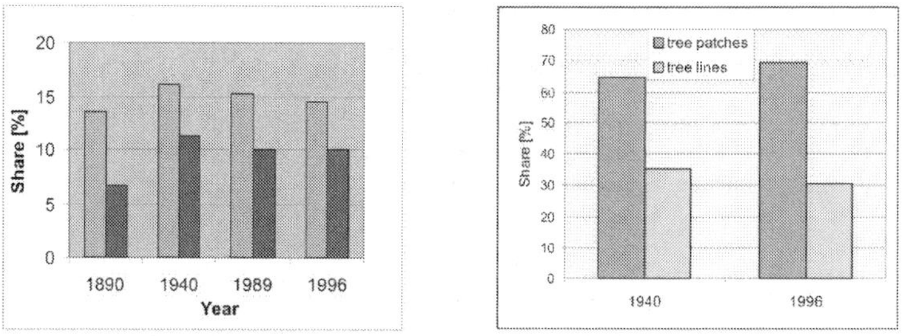

(a) (b)

Figure 16. a) Changes of share of meadows and pastures (light green) and woody vegetation (dark green) in total area of investigated landscape [44]; b) Changes of share of tree patches (light green) and tree lines (yellow) in total area of woody vegetation in investigate area [44].

In global scale temperature increase may cause serious changes in species distribution which will expand their ranges towards higher latitudes. Such processes were just observed in the case of fungi and lichens in UK, Netherland and Baltic See region. Herk [97] found, basing on a long-term monitoring, that many tropical and subtropical species of epiphytic and terrestrial lichen species are invading the Netherlands while 50% of the arctic-alpine and boreomontane species already shows a decline. In Baltic See basin, 39 distinguished species of Macromycetes extends range to the northern latitudes, for example from Central Europe to Scandinavia [98, 99]. Expected dramatic decline of forests and tundra biome towards desert and grasslands. This prediction was supported by [100]. Some results in literature, especially these made on climatic transects [101] showed that in many regions of the world we should expect increase of α diversity (species or highest taxonomic units number) as climate change will progress [102]. On the other hand due to dramatic speed of change we will lose mainly vulnerable species with narrow ecological niches [103, 104]. Such species are rare and characteristic for unique regions or habitats [105]. But then again common species having wide amplitude of distribution will move towards highest latitudes [106]. Increasing ranges of alien species might mean also the invasion of pests, both animals as fungi. Widely known is the case of Harlequin frog in mountain areas of tropical Central America (Monteverde (Costa Rica)). [107] showed that ongoing global warming process changes microclimate condition of Monteverde (Costa Rica) towards optimum for chytrids (warmer nights and increased daytime cloud) (*Batrachochytrium dendrobatidis)*. This fungi infects amphibians causing Chytridiomycosis - an infectious disease of amphibians responsible for extinctions of amphibian species in western North America, Central America, South America, eastern Australia, and Dominica and Montserrat in the Caribbean. Classic example is also horse chestnut leafminer (*Cameraria ohridella* - invasive species of uncertain origin, first observed in Macedonia in 1984 [108]. Pest invaded central and western Europe at an approximate rate of 60 km year. In Poland the species was discovered in 1998 for the first time in Lower Silesia [109]. The march of this insect, from south west to north east (according to the line of spring progress in Poland) suggests that its expansion is connected with climate changes. As far as plants are concerned changes in species distribution may lead to disturbances of balance between plants with different photosynthetic pathways. The system of interrelated factors influencing photosynthesis (CO_2, temperature, water availability) is very complicated and it is difficult to predict the direction of changes. For example, the increase of CO_2 concentration in the atmosphere will stimulate C3 plants [110-113] while in the case of C4 plants, biomass growth and competitive potency will be stimulated by temperature increase. Apart from that, C4 plants have better efficiency of water use (WUE) [114]. Therefore deterioration of humidity conditions caused by increase of temperature will cause better conditions for C4 plants than for C3 plants. Last observations suggest that despite of CO_2 elevation we may expect increasing share of C4 plants, in Central Europe. For example, between 372 species of exotic plants in Poland until 7.5% (28) consist C4 plants, while in native flora of Poland only 5 species represent C4 photosynthetic pathway. Some of them are aggressive invasive plants, (genus: *Setaria, Digitaria, Echinochloa, Eragrostis, Amaranthus,*). Many of them are common weeds, brought to the area before industrial era, with the crop plants, originated from the Middle East, but presently occurring of these species are not restricted to crop fields, some of them (*Echinochloa crus galli, Amarantus retroflexus*, different species from *Eragrostis* genus are aggressive invasive plants.

Changes in entomofauna of agricultural landscape, observed in last decades, are probably an effect of global climatic changes (since eighties). Increase of temperature without increase of precipitation, especially in summer is favorable for termophilous species. There are herbivorous species among them, so they are potential or real pests, mainly some of the beetles, bugs and butterflies. Several years ago, chemical protection from genus *Eulema,* strongly exceeding so-called threshold of harmfulness, started to be applied. Near this threshold there are tortoise bugs (*Eurygaster*), cutworms (*Agrotis*) and cereal ground beetle (*Zabrus tenebrioides*). The last species is an imposing beetle (family: Carabidae), feeding (larvae and adult forms) on cereals. High density of beetles may cause considerable loss, when they climb to ears and eat grain. In the last decade, number of Thysanoptera has also instantly grown. In 2002, in Wielkopolska region, they exceeded the harm level (mainly on oat). They were not chemically exterminated, and in last decades they were considered as potential pests Significant increase of Simulidae has also been observed. In the 70's and 80's years, only single individuals were found in large samples, containing several thousands of other Diptera, taken by motor net [25]. Increasing number of south-european (Mediterranean) species, especially moths (Sphingidae, Noctuidae) and Cicadidae, e.g. *Cicada orni* has been observed. For several years, the invasion of Asian ladybugs (*Harmonia axyridis*), is also observed.

The number of alien insect species will probably grow up. Some of them, with high invasive potential, are already present here, e.g. agrofagical western corn rootworm (*Diabrotica virgifera*), recorded in Poland since 2005. It is hoped that those species can not reproduce on a mass scale, because of natural, regulating biocenotic processes which efficiently act in an agricultural landscape. Such effect concerning horse chestnut leafminer (*Cameraria ohridella*) has been observed since the begin of the present century. Number of natural predators of this species grows from year to year and the presence of this pest decreases, what can be easily observed. On the contrary, 20-years study on insect fauna in Turew landscape showed graduating changes in its structure. During this period mean individual body mass of insects increased significantly (Figure 17A) as well as total biomass of insect community (Figure 17B). However biodiversity (expressed by number of families) significantly decreased much quickly in uniform landscape than in mosaic (Figure 18). This is probably an effect of the increase of the number of xerophillic and thermophillic insects on arable fields, including invasive large agrophagus, not connected with seminatural parts of agricultural landscape (*Zabrus, Agriotes, Eulema, and Eurygaster*) and predators (*Nabis*).

This phenomenon is a result of synergetic impact of increasing of cereal area and climate changes which leads to invasion of alien species having high body mass. Investigation showed that after 1990s a new species of butterflies have been observed. These species originated mainly from south and east Europe where climate is more dry and warm. It were thermofilic and xerophilic. For instance: *Ephesia fulminea, Syntomis phegea, Catephia alchymista, Arcthia villica* and more frequency inflying Mediterranean species (*Macroglossum stellatarum, Agrius convolvuli, and Phlogophora meticulosa*). Similar changes have been observed in other groups of insects (coleoptera, homoptera, hymenoptera, and heteroptera. Increase number of these species is especially evident in agrocenosis in uniform landscape, so diversity of insects living in cereals decreases slowly than in the mosaic landscape (Figure 18). Increasing share of maize in crop structure, which is the effect of climatic changes, brought about occurrence many efemeral habitats, like silage heap, which causes that some forest species migrates into agricultural landscape. For example: european rhinoceros beetle (*Oryctes nasicornis*)

Another change in insect community structure observed during 20 years was gradual vanishing of differences between communities in two type of landscape. In the eighties, mosaic landscape was characterized by higher biomass, higher individual body mass of insects and higher number of families. In first decade of 21st century those differences vanished and number of families in uniform landscape was even higher then in mosaic landscape. These data suggest declining role of landscape structure together with progress of climate change and simplification of crop structure. Long term study of bird assembles in Turew area confirms the results concerning insects. Since1964 up to 2006, in birds assembles in afforestations in gen. D. Chłapowski Landscape Park number of species declined of 20-25%, and this trend begun in the end of 20th century. Furthermore, permanent decrease of density of birds which use of agricultural fields during breeding period and typical forest species in 21st century (preceded by increase at the end of 20th century) has been observed. Comparison of the above results concerning birds rarely occurring in shelterbelts and hedges, typical for agricultural landscape with study results carried out 10 - 40 years ago leads to conclusion that farmland avifauna decreased approximately by 30 to 40%.

Data collected at the beginning of this century, together with those which were collected earlier indicate that the population density and species richness in the group of species living in afforestations, but using agricultural fields are decreasing, and this decrease albeit has been slowly is continued for many years. Moreover, recent data (from the years 2000-2006) indicate that the forest species begin to acquire negative changes. This is a new phenomenon, because so far the forest birds were at least stable, both in density as species richness, and in some periods showed even upward trends. It is hard to explain, but it should be emphasized that the phenomenon of declining of some species typical to forest, even common, like poor tit, robin and creepers, was noted also in the country scale [115].

In present climate change conditions amphibians are especially vulnerable animal group. The simplest measure of the scale decreasing amphibian population is change in number of water reservoirs as amphibian breeding sites through years. Probably the first time in Europe this method was used by Danish, which made the picture of changes in amphibian populations in years 1940-86 [116, 117]. In the Wielkopolska region in the end of 19th century there were 11 068 small water bodies (area less than 1 ha), but in 1960 only 2490 remained – decrease by 77% [118]. This process is ongoing. In the "Turew Landscape" between year 1960 and 1995 disappeared 55 out of 287 (19%) small reservoirs. The main causes of this process were direct impacts of man activity, such as drainage and filling 36 artificial reservoirs, but disappearance of remaining 19 reservoirs (35%) was caused by climatic changes – low precipitation and lowering of groundwater level [29].

Disappearance of these reservoirs together with biological degradation of many other caused decreasing of abundance of many amphibian populations (even up to 55-95 %), total disappearance of some other ones and drop of their diversity. The examples of such „climatic disasters" in Turew landscape are three large ponds situated in forest, far away from agricultural drainage system. All of them belonged in mid 1990s to the 10 most important amphibian breeding sites [28]. Pond 1 (Rabinek forest – 0,4 ha in 1988) – after lowering of water level number of water frogs droped from 1330 (in 1988) to 320 (in 1997)

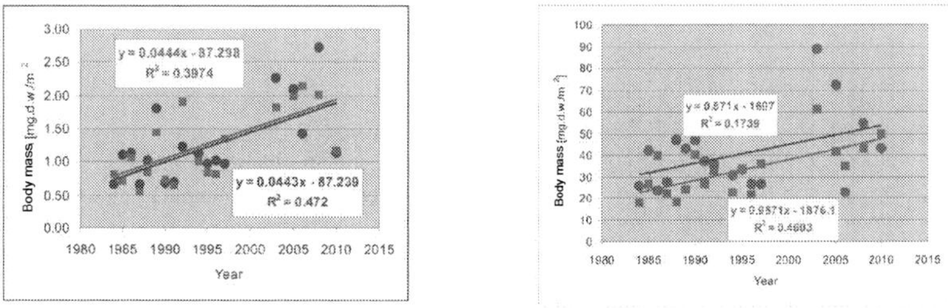

(a) (b)

Figure 17. a) Individual body mass of epigeic insects on grain crops in two types of landscape; b) Total biomass of epigeic insects on grain crops in two types of landscape.

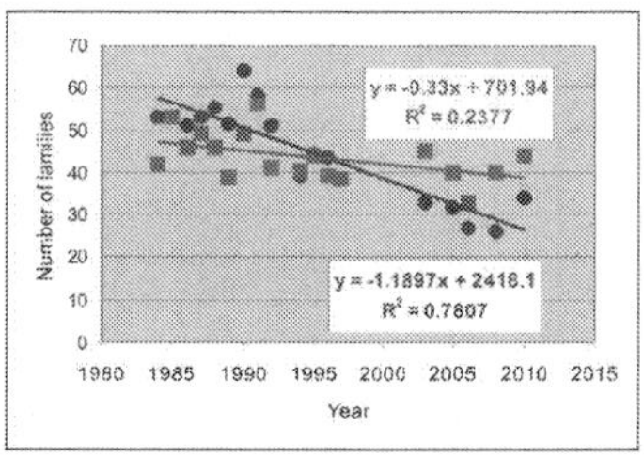

Figure 18. Diversity of epigeic insects (number of families) on grain crops in two types landscape.

individuals [51] at present there are ca. 50 frogs. Pond 2 and 3 (Błociszewo forest – 4 and 1,3 ha in 1995) – two biggest forest ponds in the Turew landscape inhabited by the largest populations (more than 100 individuals) of two amphibian species from Appendix II of the EU Habitat Directive: crested newt (*Triturus cristatus*), fire-bellied toad (*Bombina bombina*) and European tree frog (*Hyla arborea*). In the beginning of the 21st century these ponds were completely dried up and all amphibians disappeared.

In conclusion: long term observations evidenced that changes lead to uniformization of flora, fauna (and probably funga) of agricultural landscape. We cannot precisely distinguish changes caused by landscape (and crop) structure simplification and those caused by global change but there is no doubt that agricultural landscape diversity deteriorates. In recent 20 years it became obvious that we cannot stop decline of biodiversity only by use of protection of landscape heterogeneity, because the scale of changes in crop structure and that which are not directly connected with landscape, e.g. climate change or in the case of birds occurring in wintering areas is too high.

7. Conclusions

It seems obvious that human activity transformed nature in pursuit of safe food supplies, assertion of more comfortable housing conditions, exploitation resources and making transportation of people and goods easier and so on. The main change is transformation of stable ecosystems like forests, meadows, wetlands into unstable, mainly in farmlands. In order to obtain high yields, farmers must eliminate weeds, control of herbivores and pathogens, ensure that nutrients are easily accessible only for cultivated plants during their growth, increase mechanization efficiency. To increase production, farmers simplify plant cover structure both within cultivated fields (selection of genetically uniform cultivars and weeds elimination) and within agricultural landscape (elimination of hedges, stretches of meadows and wetlands, small mid-field ponds). Animal communities on cultivated fields are also impoverished (Karg and Ryszkowski 1996). Negative ecological effects of agriculture intensification are connected to:

- impoverishment of plant and animal communities
- decrease of humus resources
- decrease of capacities for water storage
- increase of pollution from non-point sources.

It must be clearly said that although farmers can moderate the intensity of these processes through proper selection of crops and tillage technologies, they are not able to eliminate them entirely. The higher control environmental threats efficiency evoked by agriculture could be achieved by structuring agricultural landscape with various non-productive components like hedges, shelterbelts, stretches of meadows, riparian vegetation, small ponds and so on. Therefore, any activity taken in order to maintain or increase landscape diversity is important not only for aesthetics and recreation reasons, but even more for environment protection, and by the same for the protection of living resources in the countryside. But it has been found that in a mosaic landscape composed of cultivated fields, rich in shelterbelts, stretches of meadows, small ponds and other semi-natural elements of landscape, appear animal communities richer and more diversified plant than in a uniform landscape that is composed only of large fields [34].

In order to maintain and even increase biodiversity we should act in two directions. First is the preservation and restoration of degraded habitats to ensure the existence of many species of plants and animals. Second, enriching crop structure in plants which less important from an economic point of view, but important as refuges for many animal species.

8. Guidelines

There is a growing body of ecological knowledge that management of agricultural landscape for its structural diversity becomes the important pillar of the sustainability of rural areas. Program of environmental protection in rural areas should aim not only at the introduction of environmental friendly technologies of cultivation within farm. They should also be concerned with challenge of how to increase the resistance or resilience of the whole landscape against threats. This could be achieved by stimulating natural processes underpinning the control of diffuse pollution and erosion, increase efficiency of water retention and biodiversity conservancy, which can not be controlled only at the farm level but have to be managed by increasing the landscape structure diversity.

The spatial net rangelands and shelterbelts in landscape of cultivated fields usually allows us to lower the concentration of harmful substances in environment to the level that is not dangerous for people's health and the normal course of natural processes. Understood this way, biogeochemical barriers constitute at the same time a kind of "shelters" for all living organisms. Thanks to this, they can influence, in a very positive way, the shaping and protection of the biodiversity in agroecosystems and agriculture landscape.

This does not mean that environmentally friendly technologies are not important. On the contrary, environment friendly technologies could mitigate negative effects of production but cycling of water or spreading pollution with water and wind operate in much larger scales than farm. It is also true for modification of microclimatic conditions or protection of biodiversity. The larger spatial scale e.g. watershed enables diversification of landscape structures which support higher stability and resistance of total landscape including individual farms to threats induced by production intensification.

The above considerations lead us to conclusion that activities aiming at optimisation of farm production and environment as well as biodiversity protection should be carried out in two different but mutually supportive directions. The first one involves actions within the cultivated areas. Their objective is to maintain possibly high level of the storing capacities of soil and to preserve or improve its physical, chemical and biological properties. They include agrotechnologies, which increase humus resources or counteract soil compaction, and rely on differentiated crop rotations. An important effect of humus resources augmentation would be improved water storage capacity, more intensive processes of ions sorption etc. Integrated methods of pest and pathogen control and proper dosing of mineral fertilisers adapted to crop requirements and to chemical properties of soil allow to diminish to same degree non-point pollution. The effectiveness of so directed activities, which could be called methods of integrated agriculture, depends on good agricultural knowledge.

The second component of the integration programme of farm production and nature protection is the management of landscape diversity. It consists in such differentiation of the rural landscape as to create various kinds of so -called biogeochemical barriers, which restrict dispersion of chemical compounds in the landscape, modify water cycling, improve microclimate conditions and ensure refuge sites for living organisms. In landscapes having mosaic structure higher doses of fertilisers can be applied than in homogenous ones which are composed of arable fields only (Ryszkowski 2002). This is a very important conclusion for the program of sustainable development of the countryside. Implementation of those ecological guidelines into the integrated agriculture policy will help to develop new environment friendly agro-technologies which, at the same time, enable intensive production balanced with ability of natural systems to absorb side effects of agriculture without being damaged.

Long term investigation carried out by Institute for Agricultural and Forest Environment showed **that increasing complexity of agricultural landscape mainly by introduction of non- productive elements like shelterbelts, strips of meadows, bushes and small midfield ponds** is one of the best tool for controlling water cycling and chemical pollution of surface and ground water in agricultural landscape.

Such findings open new frontiers for conservancy of living resources outside protected areas and open new prospects for reconciliation of agriculture with nature protection. Paradigms that are presently in force should be changed as follow:

Present paradigms	Future paradigms
Conservation	Management of protected ecosystems
Protection by isolation	Reconciliation with economic activities
Species and communities	Ecosystems
Inventory of species and recognition of their adaptations to environment	Understanding life supporting processes (energy flows, matter cycling and their control mechanisms, transmission of information)
Rare, endangered or protected species	Guilds or functional groups, key stone species
Homogeneity	Heterogeneity (diversity)

Table 4. Conclusions

Author details

Andrzej Kędziora, Krzysztof Kujawa, Hanna Gołdyn, Jerzy Karg, Zdzisław Bernacki, Anna Kujawa, Stanisław Bałazy, Maria Oleszczuk, Mariusz Rybacki, Ewa Arczyńska-Chudy, Rafał Łęcki, Maria Szyszkiewicz-Golis, Piotr Pińskwar, Dariusz Sobczyk and Joanna Andrusiak
Institute for Agricultural and Forest Environment, Polish Academy of Sciences, Poznań, Poland

Cezary Tkaczuk
Siedlce University of Natural Sciences and Humanities, Siedlce, Poland

Acknowledgement

Authors thank Mrs. M. Sc. Wenesa Synowiec for ensuring linguistic correctness of our elaboration.

9. References

[1] Dalesman R, F (1968) A Different Kind of Country. MacMillan Company, New York.
[2] Soulé ME, Wilcox BA (1980) Conservation Biology: An Evolutionary-Ecological Perspective. Sunderland, Massachusetts: Sinauer Associates Inc. 395 p.
[3] Millennium Ecosystem Assessment (2005) Washington DC: Island Press.
[4] Bourdeau P (2001) Biodiversity. In: Tolba MK, editor. Our Fragile World. Oxford: EOLSS Publishers. pp. 299-308.
[5] Kędziora A (2010) Landscape Management Practices for Maintenance and Enhancement of Ecosystem Services in a Countryside. Int. J. Ecohydrol. Hydrobiol. 10 (2-4), 133-152.
[6] Ryszkowski L, Karg J, Kujawa K, Gołdyn H, Arczyńska-Chudy E (2002) Influence of Landscape Mosaic Structure on Diversity of Wild Plant and Animal Communities in

Agricultural Landscapes of Poland. In: Ryszkowski L, editor. Landscape Ecology in Agroecosystems Management. Boca Raton: CRC Press. pp. 185-217.

[7] Tryjanowski P, Hartel P, Baldi A, Szymański P, Tobółka M, Goławski A, Konvicka M, Hromada M, Jerzak L, Kujawa K, Lenda M, Orłowski G, Panek M, Skórka P, Sparks TH, Tworek S, Wuczyński A, Żmihorski M (2011) Conservation of Farmland Birds Faces Different Challenges in Western and Central-Eastern Europe. Acta Orn. 46: 1-12.

[8] Ryszkowski L (2000) The Coming Change in the Environmental Protection Paradigm. In: Crabbe P, Holland A, Ryszkowski L, Westra L, editors. Implementing Ecological Integrity. NATO Science Series, Vol. 1. Dordrecht: Kluwer Academic Press. pp. 37-56.

[9] Marcinek J (1996) Soil of the Turew Agricultural landscape. In: Ryszkowski L, French NR, Kędziora A, editors. Dynamics of Agricultural Landscape. Poznań: PWRiL. pp. 19-27.

[10] Palmer MW (1991) Estimating Species Richness: The Second-order Jackknife Reconsidered. Ecology 72: 1512-1513.

[11] Gołdyn H (2009b) Effect of Landscape Structure on the Floristic Diversity of Farmland: a Case Study of the General Chłapowski Landscape Park [in Polish]. Rozprawy Naukowe 403. Poznań: Uniwersytet Przyrodniczy w Poznaniu. 197 p.

[12] Matuszkiewicz W (2001) Guide to Estimation of Plant Associations of Poland [in Polish]. Warszawa: PWN Sci. Publ. 537 p.

[13] Pysek P, Richardson DM, Rejmánek M, Webster GL, Williamson M, Kirschner J (2004) Alien Plants in Checklist and Floras: Towards Better Communication Between Taxonomists and Ecologist. Taxon 53: 131-143.

[14] Lambdon PW, Pysek P, Basnou C, Hejda M, Arianoutsou M, Essl F, Jarošík V, Pergl J, Winter M, Anastasiu P, Andriopoulos P, Bazos I, Brundu G, Celesti-Grapow L, Chassot P, Delipetrou P, Josefsson M, Kark S, Klotz S, Kokkoris Y, Kühn I, Marchante H, Perglová I, Pino J, Vila M, Zikos A, Roy D, Hulme P (2008) Alien flora of Europe: Species Diversity, Temporal Trends, Geographical Patterns and Research Needs. Preslia 80: 101–149.

[15] Jackowiak B (1990) Anthropogenic Changes of the Flora of Vascular Plant of Poznań [in Polish]. Poznań: Adam Mickiewicz University Press. 232 p.

[16] Mirek Z, Piękoś-Mirkowa H, Zając A, Zając M (2002) Flowering Plants and Pteridophytes of Poland. A Checklist. Kraków: W Szafer Institute of Botany, PAS. 442 p.

[17] Szyszkiewicz-Golis M (2004) Structure and Functional Evaluation of Midfield Afforestation in Agricultural Landscape of Greater Poland Region - Row Wyskoc Catchment's Case [in Polish]. PhD thesis. Poznań, Agricultural University. 188p.

[18] Gołdyn H, Arczyńska-Chudy E, Pińskwar P, Jezierska-Madziar M (2008a) Natural and Anthropogenic Transformation of Water and Marsh Vegetation in Lake Zbęchy (Wielkopolska Region). Ocean. Hydrob. Studies 37: 77-87.

[19] Gołdyn H (2009a) Changes in Plant Communities of a Drainage Ditch Over the Last 30 Years. EJPAU Seria Environmental Development. Avaiable: www.ejpau.media.pl/. Accessed 2009 Nov 09.

[20] Gołdyn H (2010) Changes in Plant Species Diversity of Aquatic Ecosystems in the Agricultural Landscape in West Poland in the Last 30 Years. Biodiversity Conserv. 19: 61-80.

[21] Gołdyn H, Arczyńska-Chudy E (2010) Changes of Plant Associations and Their Succession Processes Course in Peat Pits as an Effect of Peat Transformations and the

Intensification of Meadow Management. In: Szajdak LW, Karabanov AK, editors. Physical, Chemical and Biological Processes in Soils. Poznan: The Committee on Land Reclamation and Agricultural Environment Engineering, Polish Academy of Sciences. pp. 549-559.

[22] Brzeg A, Wojterska M (2001) Plant Communities of Wielkopolska, State of Knowledge and Threats [in Polish]. In: Wojterska M, editor. Plant Cover of Wielkopolska and South Pomerania Lakeland. Guide of Terrain Sessions of 52 PTB Congress. Poznań: Bogucki Wyd. Nauk. pp. 39-110.

[23] Arczyńska-Chudy E (2008) Small Mid-Field Ponds as Biogeochemical Barriers and Phytocenotic Diversity Richness [in Polish]. PhD thesies 199 p.

[24] Ryszkowski L, Karg J (1977) Variability in Biomass of Epigeic Insect in the Agricultural Landscape. Ecol. Pol. 25 (3): 501 517.

[25] Karg J (1980) Differentiation of Insect Biomass in Agricultural Landscape. Pol. Ecol. Stud. 6 (2): 317 328.

[26] Pollard E (1977) A Method for Assessing Changes in the Abundance of Butterflies. Biol. Cons. 12: 115-134.

[27] Świerzowski A (1973) Influence of Electricity on Fishes and Invertebrates [in Polish]. Roczniki Nauk Rolniczych 95: 141-149.

[28] Rybacki M, Berger L (1997) Amphibians of the General Dezydery Chłapowski Landscape Park [in Polish]. Biuletyn Parków Krajobrazowych Wielkopolski. 2: 22-40.

[29] Rybacki M, Berger L (2003) Contemporary Fauna of Amphibians in Wielkopolska Region in Relation to Disappearance of Their Breeding Sites [in Polish]. In: Banaszak J, editor. Stepowienie Wielkopolski Pół Wieku Później. Bydgoszcz: Wydawnictwo Akademii Bydgoskiej. pp 143-173.

[30] Tomiałojć L (1980) The Combined Version of the Mapping Method. In: Oelke H, editor. Vogelerfassung und Naturschutz. Proc. VI Intern. Conf. Bird Census Work. Göttingen. pp. 92-106.

[31] Bibby CJ, Burgess ND, Hill DA (1992) Bird Census Techniques. London: Academic Press. pp. 257.

[32] Kujawa K (1997) Relationships Between the Structure of Midfield Woods and Their Breeding Bird Communities. Acta Orn. 32: 175-184.

[33] Kujawa K (2006) Effect of Afforestation and Agricultural Landscape Structure on Breeding Bird Communities in Afforestations [in Polish]. Rozprawy Naukowe 381. Poznań: Uniwersytet Przyrodniczy w Poznaniu. 160 pp.

[34] Karg J, Ryszkowski L (1996) Animals in arable land. In: Ryszkowski L, French NR, Kędziora A, editors. Dynamics of an Agricultural Landscape. Poznan: PWRiL, pp. 138–172.

[35] Kujawa A, Kujawa K (2008) Effect of Young Midfield Shelterbelts Development on Species Richness of Macrofungi Communities and Their Functional Structure. Pol. J. Ecol. 56: 45-56.

[36] Bałazy S (2003) On Some Little Known Epizootics in Noxious and Beneficial Arthropod Populations Caused by Entomophthoralean Fungi. IOBC-WPRS Bull. 26,1: 63-68.

[37] Bałazy S (1997) Diversity of Entomopathogenic Fungi in Agricultural Landscape of Poland and France. In: Ryszkowski L, Wicherek S, editors. Ecological Management of Countryside in Poland and France. Poznań: RCAFE PAN. pp. 101-111.

[38] Bałazy S, Miętkiewski R, Tkaczuk C, Wegensteiner R, Wrzosek M (2008) Diversity of Acaropathogenic Fungi in Poland and Other European Countries. Exper. Appl. Acarol. 46,1-4: 53-70.

[39] Bałazy S (in press). 2012. Antagonistic Interactions Between Arthropods and Their Fungous Pathogens in Feeding Sites of Saproxylic Pest Insects [in Polish]. Poznań: Wydawnictwo Uniwersytetu Przyrodniczego.

[40] Zimmermann G (1986) Galleria Bait Method for Detection of Entomopathogenic Fungi in Soil. J. App. Entomol. 2: 213-215.

[41] Strasser H, Forrer A, Schinner F (1996) Development of Media for the Selective Isolation and Maintenance of Virulence of *Beauveria brongniartii* In: Jackson TA, Glare TR, editors. Microbial Control of Soil Dwelling Pests. Lincoln: AgResearch. pp. 125-130.

[42] DAISIE – 100 of the worst [Internet]. Available: http://www.europe-aliens.org/speciesTheWorst.do

[43] Kaźmierczakowa R, Zarzycki K (editors) (2001) Polish Red Book of Plants. Pteridophytes and Flowering Plants [in Polish]. Kraków: PAS, W. Szafer Institute of Botany, Institute of Nature Conservation. 664 p.

[44] Szyszkiewicz-Golis M., 2001, Land use changes and the structure of agricultural landscape in the Wielkopolska Region-western part of the Row Wyskoc catchment. The Problems of Landscape Ecology – volume X: Transformations of the natural environment of Poland and its functioning. Institute of Geography and Spatial Management of Jagiellonian University and Polish Association for Landscape Ecology, Krakow, 92-99 [In Polish]

[45] Ryszkowski L, Karg J, Kujawa K (1999) Protection and Management of Biodiversity in Agricultural Landscape [in Polish]. In: Ryszkowski L, Bałazy S, editors. Uwarunkowania Ochrony Różnorodności Biologicznej i Krajobrazowej. Zakład Badań Środowiska Rolniczego i Leśnego PAN. Poznań: pp. 59-80.

[46] Wasilewska L (1979) The Structure and Function of Soil Nematode Communities in Natural Ecosystems and Agrocenoses. Pol. Ecol. Stud. 5: 97-145.

[47] Kasprzak K, Ryl B (1978) The Influence of Agriculture on the Occurrence of Oligochaeta in Arable Soils [in Polish]. Wiadomości Ekologiczne. 24: 333-366.

[48] Ryl B (1984) Comparison of Communities of Earthworms (Lumbricidae) Occurring in Different Ecosystems of Agricultural Landscape. Ekol. Pol. 32: 155-165.

[49] Wolak M, Karg J (2002) Spiders Wintering in Mid-field Wooded Patches [in Polish]. In: Banaszak J, editor. Wyspy Środowiskowe. Bioróżnorodność i Próby Typologii: Wyd. Akademii Bydgoskiej. pp. 147–158.

[50] Kajak A, Oleszczuk M (2004) Effect of Shelterbelts on Adjoining Cultivated Fields: Patrolling Intensity of Carabid Beetles (Carabidae) and Spiders (Araneae). Pol. J. Ecol. 52 (2): 155–172.

[51] Mielewczyk S. Qualitative and Quantitative Changes in Occurrence of Ephemeroptera, Odonata, Heteroptera and Coleoptera in Young Mid-field Pond [in Polish]. In: Mat. Zjazdowe XVII Zjazdu Hydrobiologów Polskich. 1997 wrzesień 8-11; Poznań: 96.

[52] Banaszak J (1983) Ecology of Bees (Apoidea) of Agricultural Landscape. Pol. Ecol. Stud. 9: 421-505.
[53] Szeflińska D (1997) Thrips (Thysanoptera) of Grassland Habitats in the General Dezydery Chłapowski Landscape Park [in Polish]. Biuletyn Parków Krajobrazowych Wielkopolski. 2 (4): 72-77.
[54] Szeflińska D (2002) Thysanoptera of Cereals in Central Wielkopolska. Bulletin of the Polish Academy of Sciences. 50 (3): 177-181.
[55] Oleszczuk M (2010) Refugial Areas in Farmland as Habitats for Rarely Found and Threatened Species of Spiders (Araneae) in Poland [in Polish]. Chrońmy Przyr. Ojcz. 66: 361–374.
[56] Oleszczuk M (2011) Plant-dwelling Spiders (Araneae) of Selected Habitats in Manor Park in Turew [in Polish]. Biuletyn Parków Krajobrazowych Wielkopolski 19: 98-103.
[57] Berger L, Rybacki M (1999) Composition and Ecology of Water Frog Populations in Agricultural Landscape in Wielkopolska. Biological Bulletin of Poznań. Zoology 35(2): 103-111.
[58] Kujawa K (2000) Avifauna of the General Dezydery Chłapowski Landscape Park [in Polish]. Wielkopolskie Prace Ornitologiczne 9:89-121.
[59] Kujawa K (1996) The Influence of Agricultural Landscape Structure on Breeding Avifauna [in Polish]. PhD Thesis. University of Wrocław.
[60] Kujawa K, Łęcki R (2008) Does Red Fox *Vulpes vulpes* Affect Bird Species Richness and Abundance in an Agricultural Landscape? Acta Orn. 43:167-178.
[61] Ryszkowski L (1982) Structure and Function of the Mammal Community in an Agricultural Landscape. Acta Zool. Fenn. 169: 45–59.
[62] Kozakiewicz M (1993) Habitat Isolation and Ecological Barriers - the Effect on Small Mammal Populations and Communities. Acta Theriol. 38: 1–30.
[63] Kozakiewicz M, Jurasińska E (1989) The Role of Habitat Barriers in Woodlot Recolonization by Small Mammals. Holarctic Ecol. 12: 106–111.
[64] Hansson L (1987) Dispersal Routes of Small Mammals at an Abandoned Field in Central Sweden. Hol. Ecol. 10: 154-159.
[65] Yahner RH (1983) Small Mammals in Farmstead Shelterbelts: Habitat Correlates of Seasonal Abundance and Community Structure. J. Wildl. Manage. 47: 74–84.
[66] Rozporządzenie Ministra Środowiska z dnia 9 lipca 2004 r. w sprawie gatunków dziko występujących grzybów objętych ochroną. Dz.U. 04.168.1765 z dnia 28 lipca 2004 r.
[67] Wojewoda W, Ławrynowicz M (2006) Red List of the Macrofungi in Poland. In: Mirek Z, Zarzycki K, Wojewoda W, Szeląg Z, editors. Red List of Plants and Fungi in Poland. 3rd ed. Kraków: W. Szafer Institute of Botany, Polish Academy of Sciences. pp. 53-70.
[68] Wojewoda W (2003) Checklist of Polish Larger Basidiomycetes. In: Mirek Z, editor. Biodiversity of Poland. Vol. 7. Kraków: W. Szafer Institute of Botany, Polish Academy of Sciences. pp. 1-812.
[69] Chmiel MA (2006) Checklist of Polish Larger Ascomycetes. In: Mirek Z, editor. Biodiversity of Poland. Vol. 8. Kraków: W. Szafer Institute of Botany, Polish Academy of Sciences. pp. 1-152.

[70] Mułenko W, Majewski T, Ruszkiewicz-Michalska M (2008), editors. A Preliminary Checklist of Micromycetes in Poland. In: Mirek Z, editor. Biodiversity of Poland. Vol. 9. Kraków: W. Szafer Institute of Botany, Polish Academy of Sciences. pp. 1-752.

[71] Karg J, Bałazy S (2009). Effect of Landscape Structure on the Occurrence of Agrophagous Pests and Their Antagonists [in Polish]. Progress in Plant Protection 49, 3: 1015-1034.

[72] Tkaczuk C (2008) Occurrence and Infectious Potential of Entomopathogenic Fungi of Soils in Agrocenoses and Semi-natural Habitats in an Agricultural Landscape [in Polish]. Rozprawy naukowe 94. Siedlce: Wyd. Akademii Podlaskiej. 160 p.

[73] Kujawa A, Gołdyn H, Arczyńska-Chudy E (2005) A New Locality of Daisy Leaf Grape Fern (Botrychium matricariifolium) in Western Poland. Rocz. AR Pozn. 372, Bot-Stec. 8: 129-132.

[74] Zając M, Zając A, Tokarska-Guzik B (2009) Extinct and Endangered Archaeophytes and the Dynamics of their Diversity in Poland. Biodiv. Res. Conserv. 13: 17-24.

[75] Żukowski W, Jackowiak B (1995) List of Endangered and Threatened Vascular Plants in Western Pomerania and Wielkopolska (Great Poland). In: Żukowski W, Jackowiak B, editors. Endangered and Threatened Vascular Plants of Western Pomerania and Wielkopolska. Publications of the Department of Plant Taxonomy of the Adam Mickiewicz University in Poznań. Poznań: Bogucki Wyd. Nauk. pp. 9-96.

[76] Ratyńska H (2003) Plant Cover as a Result of Anthropogenic Changes in Landscape Exemplified by the River Główna Catchment Area (the middle Wielkopolska Province, Poland) [in Polish]. Bydgoszcz: Wyd. Akademii Bydgoskiej im. Kazimierza Wielkiego. 381 p.

[77] Pliński M (2009) Dangerous Water Blooms - Not Only a Hydrobiological Problem.XXI Congress of the Polish Hydrobiologist, Lublin, 9-12 Sep 2009. Session V- Ecotoxicology waters. p. 5.

[78] Mazur-Marzec H (2011) Toxic Blooms of Cyanobacteria in the Baltic Sea and Their Impact on Human Health. WWF Report Poland.

[79] Burchardt L, Pawlik-Skowrońska B (2005) Blue-Green Algal Blooms-Interspecific Competition and Environmental Threat. Wiad. Bot. 49: 39-49.

[80] Wilcox TP, Hornbach DJ (1991) Macrobenthic Community Response to Carp (Cyprinus carpio L.) Foraging. J. Freshwat. Ecol. 6: 170-183.

[81] Gehrke PC, Harris JH (1994) The Role of Fish in Cyanobacterial Blooms in Australia. Aust. J. Mar. Fresk. Res. 45: 905-15.

[82] Breukelaret AW, Lammens EH, Breteler K (1994) Effects of Benthivorous Bream (Aramis brama) and Carp (Cyprinus Carpio) on Sediment Re-Suspension and Concentrations of Nutrients and Chlorophyll a. Freshw. Biol. 32: 113-121.

[83] Williams AE, Moss W, Eaton J (2002) Fish Induced Macrophyte Loss in Shallow Lakes: Top-down and Bottom-up Processes in Mesocosm Experiments. Freshw. Biol. 47: 2216 – 2232.

[84] Karg J (2004) Importance of Midfield Shelterbelts for Over-wintering Entomofauna (Turew area, West Poland). Pol. J. Ecol. 52 (4): 421-431.

[85] Ryszkowski L, Karg J, Glura M (2009) Influence of Agricultural Landscape Structure on Diversity of Insect Communities. Pol. J. Ecol. 57 (4): 697 – 713.

[86] Tryjanowski P, Kuźniak S, Kujawa K, Jerzak L (2009) Ecology of Farmland Birds [in Polish]. Poznań: Bogucki Wyd. Nauk. 390 p.

[87] Pugacewicz E (2000) Breeding Avifauna of an Agricultural Landscape in Bielska Plain [in Polish]. Not. Ornitol. 41:1-28.

[88] Kujawa K, Tryjanowski P (2000) Relationships Between the Abundance of Breeding Birds in Western Poland and the Structure of Agricultural Landscape. Acta Zool. Hung. 46: 103-114.

[89] Kujawa K (1996a) The Influence of Landscape Pattern on Breeding Bird Communities in D. Chlapowski Landscape Park [in Polish]. Biuletyn Parków Krajobrazowych 3: 83-90.

[90] Kujawa K (2002) Population Density and Species Composition Changes for Breeding Bird Species in Farmland Woodlots in Western Poland Between 1964 and 1994. Agric. Ecosyst. Environ. 91: 261-271.

[91] Kujawa K (2008a) Long Term Changes (1964-2006) in Breeding Bird Community of Midfield Afforestations in the Dezydery Chłapowski Landscape Park (Wielkopolska) [in Polish]. In: Kaczmarek S, editor. Krajobraz i bioróżnorodność. Bydgoszcz: Wyd. Uczel. UKW. pp. 151-167.

[92] Sosa-Gomez DR, Delpin KE, Moscardi F, Farias JRB (2001) Natural Occurrence of the Entomopathogenic Fungi *Metarhizium, Beauveria* and *Paecilomyces* in Soybean Under Till and No-till Cultivation Systems. Neotrop. Entomol. 30,1: 407-410.

[93] Hummel RL, Walgenbach JF, Barbercheck ME, Kennedy GG, Hoyt GD, Arellano C (2002) Effects of Production Practices on Soil-Borne Entomopathogens in Western North Carolina Vegetable Systems. Environ. Entomol. 31: 84-91.

[94] Root TL, Price JT, Hall KR, Schneider SH, Rosenzweig C, Pounds JA (2003) Fingerprints of Global Warming on Wild Animals and Plants. Nature 421: 57–60.

[95] Ryszkowski L (1990) Agricultural Landscape of Turew Vicinity [in Polish]. In: Ryszkowski L, Marcinek J, Kędziora A, editors. Obieg Wody i Bariery Biogeochemiczne w Krajobrazie Rolniczym. Poznań: Wydawnictwo UAM. pp. 5-12.

[96] Ryszkowski L, Kędziora A (2007) Sustainability and Multifunctionality of Agricultural Landscapes. In: Plieninger T, editor. Zukunftsorientierte Nutzung Ländlicher Räume – Land Innovation. Materialien 15. Berlin-Brandenburgische Akademie der Wissenschaften, Berlin: pp. 5 23..

[97] Herk CM, Aptroot A, Dobber HF (2002) Long-term Monitoring in the Netherlands Suggests that Lichens Respond to Global Warming. Lichenologist 34: 1–15.

[98] Kreisel H (2006) Global Warming and Mycoflora in the Baltic Region. Acta Mycol. 41:79-94.

[99] Emanuel WR, Shugart HH, Stevenson MP (1985) Climate Change and the Broad Scale Distribution of Terrestial Ecosystem Complexes. Climatic Change 7: 29-43.

[100] Notaro M, Vavrus S, Liu Z (2007) Global Vegetation and Climate Change due to Future Increases in CO2 as Projected by a Fully Coupled Model with Dynamic Vegetation. J. Climate 20: 70-90.

[101] White P, Kerr JT (2006) Contrasting Spatial and Temporal Global Change Impacts on Butterfly Species Richness During the 20th Century. Ecography 29: 908 918.

[102] Kerbiriou C, Le Viol I, Jiguet F, Devictor V (2009) More Species, Fewer Specialists: 100 years of Changes in Community Composition in an Island Biogeographical Study. Diversity Distrib. 15: 641–648.

[103] Benning TL, LaPointe D, Atkinson CT, Vitousek PM (2002) Interactions of Climate Change with Biological Invasions and Land Use in the Hawaiian Islands: Modeling the Fate of Endemic Birds Using a Geographic Information System. Proc. Nat. Acad. Sc. J. 29: 14246–14249.

[104] Isaac JL (2009) Effects of Climate Change on Life History: Implications for Extinction Risk in Mammals. Endang. Species. Res. 7: 115–123.

[105] Malcolm JR, Liu C, Neilson RP, Hansen L, Hannah L (2006) Global Warming and Extinctions of Endemic Species from Biodiversity Hotspots. Conserv. Biol. 20: 538 548.

[106] Davey CM, Chamberlain DE, Newson SE, Noble DG, Johnston A (2011) Rise of the Generalists: Evidence for Climate Driven Homogenization in Avian Communities. Global Ecology and Biogeography 21: 568–578.

[107] Pounds JA, Bustamente MR, Coloma LA, Consuegra JA, Fogden MPL, et al. (2006) Widespread Amphibian Extinctions from Epidemic Disease Driven by Global Warming. Nature 439: 161–67.

[108] Deschka G, Dimic N (1986) Cameraria ohridella n. sp. aus Mazedonien, Jugoslawien (Lepidoptera, Lithocelletidae). Acta Ent. Jugosl. 22: 11-23.

[109] Łabanowski G, Soika G (1998) The Horse Chestnut Leafminer Infesting Chestnut in Poland. Ochrona Roślin. 42(12): 12.

[110] Gifford RM (1977) Growth Pattern, Carbon Dioxide Exchange and Dry Weight Distribution in Wheat Growing Under Differing Photosynthetic Environments. Aust. Jour. Plant Physiol. 4: 99-110.

[111] Sionit N, Morten DA, Strain BR, Helmers H (1981) Growth Response of Wheat to CO_2 Enrichment and Different Levels of Mineral Nutrition. Agr. Jour. 73: 1023-1027.

[112] Rogers HH, Thomas JF, Bingham GE (1983) Response of Agronomic and Forest Species to Elevated Atmospheric Carbon Dioxide. Science 220: 428 429.

[113] Kimball BA, Kobayashi K, Bindi M (2002) Responses of Agricultural Crops to Free-air CO2 Enrichment. Adv. Agron. 77: 293–368.

[114] Emmerich WE (2007) Ecosystem Water Use Efficiency in a Semiarid Shrubland and Grassland Community. Rangeland Ecol. Manage. 60: 464–470.

[115] Chylarecki P, Jawińska D, Kuczyński (2006) Monitoring of Common Breeding Birds - Report from Years 2003-2004. Warszawa: OTOP.

[116] Fog K (1988) Reinvestigation of 1300 Amphibian Localities recorded in the 1940s. Memoranda Soc. Fauna Flora Fennica 64: 134-135.

[117] Fog K (1997) A survey of the results of pond projects for rare Amphibians in Denmark. Memoranda Soc. Fauna Flora Fennica.73: 91-100.

[118] Stasiak P (1991) Disappearance of Small Mid-field Water Bodies in Wielkopolska Lowland in the Light of Cartographic Materials [in Polish]. M.Sc. Thesis. Adam Mickiewicz University, Poznań.

Section 4

Socio-Economic Impact

Chapter 13

Effects of Disturbance on Sandy Coastal Ecosystems of N-Adriatic Coasts (Italy)

Gabriella Buffa, Edy Fantinato and Leonardo Pizzo

Additional information is available at the end of the chapter

http://dx.doi.org/10.5772/47480

1. Introduction

Due to human driving forces, many terrestrial habitats are undergoing striking modifications, destruction and fragmentation at an increasing and historically unprecedented rate (Sala et al., 2000), drawing attention to ecosystems' resilience as a necessary condition for both biodiversity conservation and sustainable development (McLeod et al., 2005).

Among the most endangered and threatened ecosystems worldwide, there are seashore, coastal sand dunes and nearby wet infradunal downs which are facing escalating anthropogenic pressures (Defeo et al., 2009), chiefly from coastal development, direct human use, mainly associated with recreation, and sea level rise.

All coastal European Countries, and particularly those of the Mediterranean Basin (Curr et al., 2000; European Environment Agency [EEA], 1999), suffer from the loss and degradation of sand dune landscape which are leading to a dramatic biodiversity loss, caused by the alteration and disappearance of many habitats and the rarefaction and/or local extinction of the most typical and extremely specialized native species, sometimes replaced with alien species.

Coastal dune systems make up 20% of the area occupied by the world's coastal landscapes (van der Maarel, 2003) and contain diverse and productive habitats important for human settlements, development and local subsistence (Schlacher et al., 2008). According to data reported by the United Nations Conference on Environment and Development [UNCED] (1992), about half of the world's population lives within 60 km of the shoreline, and it is likely to rise to three quarters by the year 2020. Population increase, united with economic progress and development, and the growing demands for spare time opportunities represent the eventual drivers of escalating pressures on sandy beaches (Dugan & Hubbard, 2010). Particularly coast-bound tourism, which became a mass phenomenon after World

War II, is now considered the primary cause of degradation of coastal dunes (Acosta et al., 2000). Among European holidaymakers more than 60% prefer the coast (European Community [EC], 1998) and even more people use sandy beaches which attract the greatest percentage of tourists every year (Davenport & Davenport, 2006; Schlacher et al., 2007). Rapid growth of human populations on the coast is then expected to further influence beaches and coastal sandy ecosystems with effects on biodiversity, community composition and ecological function (Defeo et al., 2009; Dugan & Hubbard, 2010).

As for the Italian coastline, more than 3000 km are represented by sand dune systems, which maintained well preserved morphological, hydrological and naturalistic features until the nineteenth century (Garbari, 1984). From the twentieth century on, they have been suffering from a strong intensification of human activities that mainly include forestry, agriculture, fisheries and aquaculture, transport and tourism, with consequent urbanization, trampling and beach cleaning. During the last century, a loss of about 80% of dune systems has been calculated for the Mediterranean area as a result of increasing urbanization (EEA, 1995).

Similarly, from 1950 onward, in the N-Adriatic region large stretches of coastal seashores, foredunes and infradunal downs have been fragmented by housing and resort development, road construction and coastal armoring, and the remaining sites suffer from increasing erosion, reduction in sand supply, alteration of geomorphic processes and heavy human use in the form of trampling, mechanical grooming and berm building (Nordstrom et al., 2009).

Coastal dune systems are typical transitional ecosystems, linked both to marine environment and terrestrial river basins, usually extending, narrow and long, along the coastline (Acosta et al., 2007), where environmental factors deeply influence their size, shape and boundaries. Like other ecotones, they exhibit a sharp gradient both in biotic and environmental factors, mainly related to substrate coherence and salinity, wind, salt spray and wave regime, which differ with distance from the water and topographic sheltering (Acosta et al., 2007; Carranza et al., 2008; Lortie & Cushman, 2007; Nordstrom et al., 2009; Ranwell, 1972). This steep gradient makes them highly dynamic systems deeply influenced by environmental stressors and drivers (Barbour, 1992), but at the same time, as abiotic patterns change within a short distance, it is responsible for the high level of ecological diversity, environmental heterogeneity and for the coexistence of different communities within a relatively limited space (Frederiksen et al., 2006; Martínez et al. 2004; Wilson & Sykes, 1999). Moreover, coastal dune systems are inhabited by extremely specialized biotic assemblages, rarely shared with adjacent terrestrial ecosystems (Defeo et al., 2009; Kutiel et al., 1999; Schlacher et al., 2008).

Healthy dune ecosystems also provide unique ecological services, such as erosion and salt spray control, storm buffering, water filtration and purification, nutrients mineralisation and recycling, coastal fisheries, functional links between terrestrial and marine environments, provisioning of crucial habitats for endangered species such as birds, as well as cultural services like recreation and education (Barbier et al., 2008; Defeo et al., 2009; Schlacher et al., 2008).

Furthermore, in Europe, coastal dune habitats are listed in the CORINE biotope classification and some of them are regarded as priority habitats or habitats of community interest in Annex I of the EU Habitat Directive (CE 43/92), recognized as a cornerstone of Europe's nature conservation policy (Feola et al., 2011; Mücher et al., 2004). This status implies they deserve special conservation attention and, as asked by the most recent European legislation, a strategic approach to planning and management in order to achieve sustainable development. As the assessment of the state of ecosystems at a given time and place is at the basis of any planning process for the management and conservation of natural resources, the aim of this study was 1) to provide a comprehensive and up-to-date outline of N-Adriatic sandy coastal landscape; 2) to assess the impacts of anthropogenic disturbance on sandy coastal systems by integrating landscape-level and community-level approaches; 3) verify how habitat configuration is important in supporting landscapes', communities' and plant species' diversity and quality.

2. Study site

The investigation was undertaken on the N-Adriatic coast which represents the longest sandy coastal line in Italy. The 100 km long coastline encompassing the study sites correspond to the Venetian portion of this system, isolated by other areas of sandy coastal plain by the estuaries of large rivers: the Tagliamento river northwards and the Brenta-Adige-Po rivers system southwards (Figure 1).

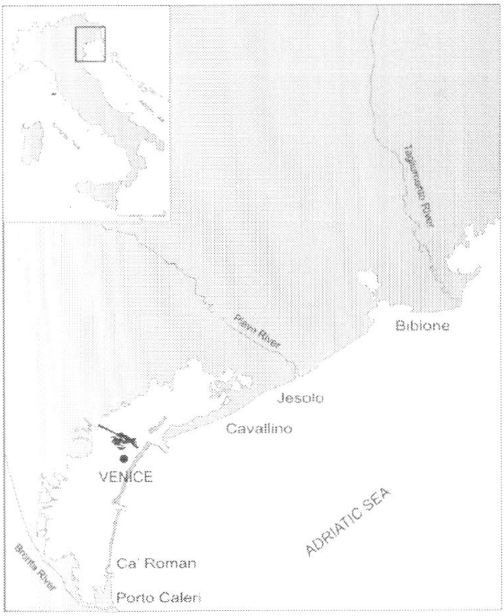

Figure 1. Map of the study site.

Sites consist of narrow, recent dunes (Holocene), that generally occupy a narrow strip along the seashore, bordered by river mouths and tidal inlets, mostly fixed by docks. The natural forces that shape and influence the dynamic of these sandy coastal systems are basically wind, waves and sea level fluctuations, climate and rivers run-off (Nordstrom et al., 2009). Sediments on the backshore and dunes are similar at all sites and are in the range of fine sand (Bezzi et al., 2009). Carbonate clearly rules the mineralogical composition of sands (especially in the northernmost area) due to the lithology of the catchment areas of corresponding rivers; southwards a slight magmatic component arises (Zunica, 1971). Dominant winds blow from the northeast and east (Bezzi et al., 2009). Annual average wave heights are lower than 0.50 m (Dal Cin & Simeoni, 1994); tides are semi-diurnal, with a spring range of about 1.0 m and a neap range of about 0.20 m (Polli, 1970); the combination of spring tides, winds and low atmospheric pressure can raise sea level up to 1.60 m.

Climate deserves a closer examination being one of the most characterizing aspects of the site. From a biogeographic point of view, the N-Adriatic seacoast can be included in the Eurosiberian region, Appennine-Balcanic province and Po-Valley subprovince. The mean annual temperature is about 13°C, with low winter (0.3°C) and high summer (17.7°C) values. Mean annual rainfall is 831.5 mm, with maximum precipitation (89.1 mm) in the spring-autumn season and a minimum (49.3 mm) in summer. Bioclimatic classification (Rivas-Martinez, 2008) shows a Temperate Oceanic type which allow reference to this area as the only sector of the Mediterranean Basin that does not belong to the Mediterranean climatic Region (Buffa et al., 2005; 2007; Sburlino et al., 2008).

Average annual values of both rainfall and temperature have been increasing since 1992 all along the Venetian coast: mean rainfall shows an increase from 688 mm in 1992 to the current 976 mm ($p<0.05$); mean temperature has grown from 13.04°C to 13.70°C ($p<0.10$). To evaluate a possible change in climate regime, we compared two time series monitored at the same site (weather station of Cavallino-Venice, located in the central part of the Venetian stretch), comprising the periods between 1961-1990 and 1992-2010 respectively. Although, the two mean annual precipitation do not show a significant difference (P_{61-90}=809.4 mm/P_{92-10}=845.5 mm; df=47; F=1.73; t=0.71), a seasonal rainfall redistribution has occurred with a shift from an oceanic regime to an equinoctial one that shows two maximum (in spring and autumn) and two minimum. Temperature has also changed: thermic regime remains similar, but with a general shift to higher temperature values; the mean annual temperature increased from 12.7°C to the current 13.5°C ($p<0.01$). Ombrothermic diagrams (Figure 2) allow us to highlight a change of climatic phase which shifted from Low Supratemperate upper subhumid climate to Upper Mesotemperate upper subhumid climate.

Besides the bioclimatic diagnosis, the variation of some bioclimatic indices (reported in Table 1) emphasises an increase of summer aridity, due to the simultaneous increase in temperature and decrease in precipitation, giving evidence to a process of "mediterraneisation" (Fernández-González et al., 2005). Moreover, the general warming lengthened the Period of Plant Activity (i.e., months with mean temperature >3.5°C, Rivas-Martinez, 2008) from 10 to 11 months.

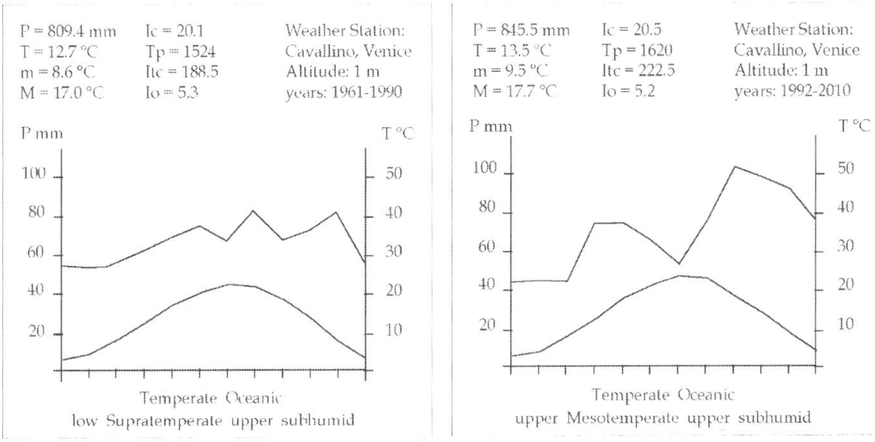

Figure 2. Bioclimatic classification of Cavallino weather station (central part of Venetian coastal stretch), for the period 1961-1990 (left) and 1992-2010 (right). For a detailed description of indices and general theoretical principles see Rivas-Martinez (2008).

variable	1961-1990	1992-2010
P mm (annual)	809.4	845.5
T °C (annual)	12.7	13.5
m °C (annual average of minimum values)	8.6	9.5
M °C (annual average of maximum values)	17.0	17.7
Tp °C (positive annual T, with Ti_{1-12} > 0ºC)	1524	1620
Ts °C (average T of summer three months)	651	686
Ps mm (average P of summer three months)	227.7	195.5
Ios3 (ombrothermic index (Pp/Tp) of summer three months)	3. 5	2.8

Table 1. Climatic parameters and bioclimatic indices. For a detailed description of indices and general theoretical principles see Rivas-Martinez (2008).

Many other studies identify the decade 1981–1990 as the onset of climate change in Europe, with a trend starting in the beginning of the 1970s (Werner et al., 2000). In the Veneto Region, this breakpoint is particularly evident for temperature and evapotranspiration: after the change point, in fact, temperatures show a significant increase (+1.5 and + 0.9 °C for yearly averages of maximum and minimum temperatures respectively relative to the previous phase) in all seasons and particularly clear in spring, summer and winter for maximum temperatures and in summer for the minimum ones (Chiaudani, 2008).

Human pressures are very high along the entire coastline. Until the 1950s, the Veneto coast was almost entirely fronted by dunes up to 10 m high (Bezzi & Fontolan, 2003; Pignatti, 2009), but few of them still survive and beaches generally suffer from the decrease of sediment supplies delivered by rivers, subsidence, and reduction in longshore sediment transport due to interruption by shore-perpendicular structures (Nordstrom et al., 2009).

To defend shorefront buildings and to provide space for leisure use, from 1950, large beach sectors have been protected, and still are, by different structures, such as groynes (shore-perpendicular constructions that catch sand moving alongshore), jetties and revetments, locally called "murazzi"; some of them, like in Pellestrina, were built by the Venetians in the 18th century and later rebuilt to larger dimensions (Nordstrom et al., 2009).

Most sites are managed by private corporations, through beach concessions by national government, and land behind beaches mostly developed as campsites, resorts, towns and villages; only very few sites (Ca' Roman, Porto Caleri and partly the "Laguna del Morto" near Eraclea) are still near-natural, undeveloped and underutilized. In 2011, beach summer tourism (from May to September) in the Veneto region numbered more than 25 million visitors. Summer beach tourism is one of the main resources of the region (Bezzi & Fontolan 2003) and facilities for accommodating people show an average density of about 76.3/100 sqKm. Private beaches management encompasses machine litter cleaning after major storm events and before the tourist season to clean and flatten the sand surface for tourist facilities and during the summer, plant litter and shell fragments are regularly removed and left outside the beaches (Nordstrom et al., 2009).

Therefore, the encroachment of human facilities has severely restricted the space available for natural landforms and vegetation, and environmental gradients have been largely truncated, fragmented or compressed. Scattered here and there, some stretches with high natural value still remain and host high levels of biodiversity. All these well preserved sites have been incorporated in Natura 2000 European network as Sites of Community Importance (SCI) and/or Special Protection Areas (SPAs) (Buffa & Lasen, 2010).

3. Landscape classification

3.1. Background

Modern ecology relies on the concept of ecosystem as a fundamental concept to consistent environmental policy making. As provided in Article 2 of the Convention on Biological Diversity (1992), ecosystem means a *"dynamic complex of plant, animal and micro-organism communities and their non-living environment interacting as a functional unit"*.

From a nature conservation point of view, main concerns should be focused on spatial extension of ecosystems and their quality, as well as on their adaptability and recovery potential. As different ecosystems types are not equally valuable or equally susceptible to human-induced environmental change, to ensure ecologically sound management aimed at sustainability and preservation of biodiversity, ecosystems need to be described, characterised and spatially located (Blasi et al., 2000; Rowe, 1996; Sims et al. 1996).

The increasing attention to biodiversity conservation and natural resources management has reawakened the interest in ecosystem classification and mapping, mostly focusing on flexible and vertical, namely hierarchical, methods (Acosta et al., 2005; Klijn & Udo de Haes 1994; Matson & Power, 1996; Zonneveld, 1995), which can help dealing with the complex and dynamic nature of ecosystems providing instruments to refer to any functional unit at any scale depending on the problem being addressed.

The basic idea is that different ecosystems are detectable as a function of their homogeneity, which depends on the scale of observation. The importance of scale is universally recognized as scale concerns all types of ecological data and it is a fundamental facet of ecological heterogeneity, whose interpretation depends on the level of observation established when studying an ecological system (Levin 1992; Rescia et al., 1997).

Hierarchical structuring of communities and ecosystems has been long recognized (Allen & Starr, 1982; O'Neill et al., 1986) and it basically means that, at a given scale, any biological system is composed of lower-level interacting components and, at the same time, is itself an element of a greater system (O'Neill et al., 1989). The recognition of spatially defined landscape units can thus be useful for stratifying landscapes into ecologically homogeneous units, whose patterns and functions, at each level, depend on both the potentiality of lower levels and the restraints imposed by higher levels (O'Neill et al., 1989). Ecosystem classification set up on the hierarchical concept can thus provide interconnected spatial units with different potential purposes, depending on the scale of the problems under investigation and the requisite precision of the results (Carranza et al., 2008).

Blasi et al. (2000; 2005) have recently proposed a hierarchical framework designed for describing and mapping Italian landscapes at different levels. It is a deductive and spatial explicit method based on the homogeneity of the physical environment aiming at defining land units with different vegetation potential (Blasi & Frondoni, 2011). According to the method, land attributes used for classifying landscapes are those widely recognized (Forman & Godron 1986; Zonneveld 1995) as ecologically relevant: climate, lithology, geomorphology, human activities, soil and vegetation, ordered from the most stable factors controlling processes occurring at larger ecological scales to more dynamic ones working at local levels, so reflecting their hierarchy both in time and space.

All these landscape attributes are integrated with the concept of potential natural vegetation (PNV). This concept is one of the most important concepts developed within plant ecology since it allows the provisioning of predictive models of plant communities dynamics (Biondi, 2011). The concept was firstly formalized by Tüxen (1956) and it can be defined as the plant community that would develop in a given habitat if all human influences would stop (Westhoff & van der Maarel, 1973); in other words, it delineates the spontaneous natural development of landscape within a homogeneous land unit. The concept of PNV correlates to that of vegetation series which is composed of dynamically linked plant communities developing into the same type of mature vegetation, i.e. the PNV (Biondi, 2011; Blasi et al., 2005).

The top-down, deductive approach supports the identification of homogeneous units which can be then classified according to the inner dynamic vegetation pattern.

From higher to lower level, Blasi's framework includes (Blasi et al., 2000; 2005):

- Land Regions (LR): represent the broadest level (scale >250.000); identified by means of macroclimatic features;
- Land Systems (LS): delimited according to significant lithological differences;

- Land Facets (LF): delimited through lithomorphological features (altitude, slope and aspect) and local bioclimatic types (rainfall and temperature regimes);
- Land/Environmental Units (LU): defined at a smaller scale (1:10.000/1:50.000), they are determined by a major vegetation series that evolves in one consequent PNV type.
- Land Elements (LE): correspond to the different plant communities.

Although successfully applied to different ecosystems (Acosta et al., 2003a; 2005; Blasi et al., 2004; 2005; Stanisci et al., 2005), the method appears less workable in sandy coastal landscape, where the identification of single land elements is quite impossible at the scale proposed. For this reason, according to the proposal of Carranza et al. (2008), in this study LU have been described through coastal dune geosigmeta which can be defined as a mosaic of adjoining vegetation series within a geomorphologic, biogeographical and bioclimatic unit (Biondi 1994; Blasi et al., 2005).

3.2. Application to N-Adriatic coastal dunes

Land Regions, Systems, Facets and Environmental Units of N-Adriatic coastal dunes were obtained by overlaying physical information digitalized as GIS data vector layers in ArcGIS 9.3 (Environmental Systems Research Institute [ESRI], 2008). As the entire studied area is included in the same macroclimate zone, starting thematic maps were a geological map (Bondesan et al., 2008), a geomorphological map (Bondesan & Meneghel, 2004) and a pedological map (Bini et al., 2002; Ragazzi et al., 2005; Ragazzi & Zamarchi, 2008). Bioclimatic characterization was set up according to Rivas-Martinez (2008) using data (1992-2010) from seven weather stations located all along the Venetian coastline.

Information concerning land elements distribution (potential coastal vegetation–zonation) was obtained using the phytosociological approach (Braun-Blanquet, 1964). Both up-to-date original data (41 surveys) and data from previous phytosociological studies (204 surveys) were used (Gamper, 2002; Gamper et al., 2008; Géhu et al., 1984; Poldini et al., 1999; Sburlino et al., 2008). Vegetation surveys (phytosociological relevés) were combined in a single matrix and classified by cluster analyses to be assigned to syntaxonomic taxa according to their floristic, structural and coenological features (Westhoff & van der Maarel, 1973).

3.3. Results

The entire study area belongs to the same Land Region and Land System (Table 2).

Rainfall and temperature regimes differentiated an upper Mesotemperate upper sub-humid climate northwards and a lower Supratemperate lower sub-humid climate, with a steppic variant, southwards allowing the identification of two different Land Facets (LF) which are separated by the Piave River that seems to act as a bioclimatic divide. Northern Land Facet (NLF) extends from the Tagliamento River to the Piave River; Southern Land Facet (SLF) from the Piave River to the Brenta-Adige-Po Rivers system in the south (see map in Figure 1).

Main land elements are summarized in Table 3 and grouped according to PNV classes.

LR	LS	LF	EU	LE
1. Temperate	1.1. Coastal sand dune	1.1.1. recent coastal dunes under upper Mesotemperate upper sub-humid climate	1.1.1.1. Coastal dune geosigmetum	1.1.1.1.1. Beach and mobile dunes
				1.1.1.1.2. Fixed dunes
				1.1.1.1.3. Dune slack transition to alluvial deposits
		1.1.2. recent coastal dunes under lower Supratemperate lower sub-humid climate, steppic variant	1.1.1.2. Coastal dune geosigmetum	1.1.1.2.1. Beach and mobile dunes
				1.1.1.2.2. Fixed dunes
				1.1.1.2.3. Dune slack transition to alluvial deposits

Table 2. Land Region (LR), System (LS), Facets (LF), Environmental Units (EU) and Elements (LE) of the Venetian coastal dune systems.

Each Environmental Unit is set up by three systems in contact with each other (Figure 3). The variation of structural types (namely, different life forms that dominate and characterize different plant communities) along the zonation is that typically found in sandy coastal systems worldwide (Biondi, 1999; Carboni et al., 2009; Frederiksen et al., 2006; Sykes & Wilson, 1991). The zonation develops moving inland from the sea edge along the steep environmental gradient, with the most pioneering annual communities on the beach and the woods in the inland sheltered zone.

Apart from the fixed zonation determined by their ecological requirements, N-Adriatic sandy coastal plant communities show a certain degree of uniqueness within the Mediterranean Basin (Pignatti, 1959; Géhu et al., 1984; Gamper et al., 2008; Sburlino et al., 2008), which is supported by various factors ranging from the present geographical and physical characteristics to the past climatic events that drove wide floristic migrations in Northern Italy. The synergy of all these features makes possible the presence of a wide range of species with different geographical distribution encompassing temperate, Mediterranean, western, eastern and mountain species (the latter mainly in the Northernmost part of the coast, carried downwards by torrential rivers like the Tagliamento and the Piave). This singular phytogeographic blend, also recognized for other north-eastern Italian ecosystems (Buffa & Villani, 2012), greatly increases the floristic value of this area, contributing to define plant communities and systems not found elsewhere (Lorenzoni, 1983; Buffa et al., 2007).

			Beaches and mobile dunes		Fixed dunes		Dune slacks transition	
		Land facets	1.1.1	1.1.2	1.1.1	1.1.2	1.1.1	1.1.2
		Land Elements	1.1.1.1.1	1.1.2.1.1	1.1.1.1.2	1.1.2.1.2	1.1.1.1.3	1.1.2.1.3
		PNV classes						
Annual and perennial grass vegetation		Salsolo-Cakiletum maritimae	*	*				
		Sporobolo-Agropyretum juncei	*	*				
		Echinophoro-Ammophiletum australis	*	*				
Annual and perennial grass, chamephytic, nanophanerophytic and phanerophytic vegetation	*Edapho-xerophilous series*	Tortulo-Scabiosetum (*e*)			*	*		
		Sileno coloratae-Vulpietum membranaceae			*	*		
		Teucrio-Chrysopogonetum grilli (*e*) (*1*)					(*)	
		Erico-Osyridetum albae (*e*)			*	*		
		Viburno-Phillyreetum (*e*)			*	*		
		Junipero-Hippophaetum (*e*)				*		
		Vincetoxico-Quercetum ilicis (*e*)					*	*
	Edapho-higrophilous series	Eriantho-Schoenetum nigricantis			*	*		
		Juncetalia maritimae communities			*	*		
		Phragmito-Magnocaricetea communities					*	*
		Erucastro-Schoenetum nigricantis (*e*)			*			
		Plantagini-Molinietum coeruleae (*e*)			*			
		Carici elongatae-Alnetum glutinosae					*	
Contact Land Systems		Alluvial deposits					*	*
		Lacustrine deposits with coastal lagoons					*	*
		Ancient dune					*	*

Table 3. Potential natural vegetation types that characterize coastal dune ecosystem zonation on the N-Adriatic coast. Asterisks point to the presence of the different plant communities; (*e*) indicates endemic plant communities; (*1*) *Teucrio-Chrysopogonetum grilli* dry grasslands are present only on ancient dunes located out of the surveyed area.

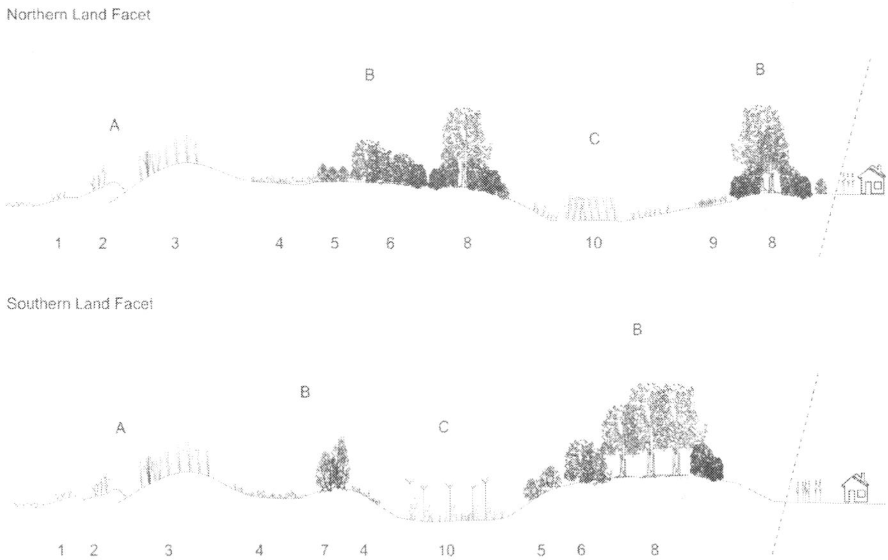

Figure 3. Simplified representation of the Potential Natural Vegetation (PNV) along the N-Adriatic coast. A=Beach and mobile dunes; B=Edapho-xerophilous series; C=Edapho-higrophilous series. 1=Salsolo-Cakiletum maritimae; 2=Sporobolo-Agropyretum juncei; 3=Echinophoro-Ammophiletum australis; 4=Tortulo-Scabiosetum; 5= Erico-Osyridetum albae; 6= Viburno-Phillyreetum angustifoliae; 7= Junipero-Hippophaetum fluviatilis; 8= Vincetoxico-Quercetum ilicis; 9=Teucrio-Chrysopogonetum grilli; 10=higrophilous mosaic.

The sandy coastal system begins with the pioneer, nitrophilous community dominated by annuals of the strandline zone (*Salsolo kali-Cakiletum maritimae* plant community); being exposed to wave inundation, salt spray and wind stress, the community is often patchy and fragmented. Beach land elements are species-poor communities, since few species can survive the stress and disturbance of sand mobility and salt spray (Nordstrom et al., 2009). Pioneer plants of *Salsolo-Cakiletum* are tolerant of salt spray and sand blasting and contribute to the formation of embryo dunes on the backshore while grasses form foredune ridges (Seabloom & Wiedemann, 1994). The *Cakile maritima* plant community is then followed by that of embryo dunes dominated by dune-forming plants such as *Elymus farctus* (*Sporobolo-Agropyretum juncei*). On mobile dunes (white dunes) the *Ammophila arenaria* community (*Echinophoro spinosae-Ammophiletum australis*) establishes; *Ammophila arenaria* is the dominant species and is responsible for stabilizing and building up the foredune by capturing blown sand and binding it together with its tough, fibrous rhizome system (Chapman, 1976). Beaches and mobile dunes soils are shallow, sandy, calcareous, mesic *Typic Xeropsamments* or *Typic Udipsamments* (Bini et al., 2002), depending on local topographical conditions, and have very low organic carbon content and scarce horizons differentiation.

Landward of the foredune crest, increased protection from physical stresses allows the development of woody shrubs in the seaward slopes of the dune and trees and upland species in the landward portions. The inner, more stable dunes host more developed soils: they are thicker and have better horizons differentiation with blocky structure, higher organic carbon content and higher Available Water Capacity (AWC). Depending on the soil moisture regime, they can be classified as *Typic Haploxerepts, Arenic* or *Typic Eutrudepts* (Bini et al., 2002). Consequently, vegetation evolves towards more structured forms, such as medium and high shrubs and ends in the forest dominated by holm-oak. Then, moving inland, mobile dunes systems are spatially replaced by the edapho-xerophilous series of fixed dunes, entirely composed of N-Adriatic endemic communities (Gamper et al., 2008). The series begins with a perennial dry microprairie (*Tortulo-Scabiosetum*) dominated by dwarf shrubs, perennial herbaceous species, mosses and lichens, that covers fixed dunes (grey dunes). On inland ancient dunes, it is replaced by an endemic dry grassland (*Teucrio capitati-Chrysopogonetum grylli*), whose structure is mainly characterized by perennial herbaceous species and lower dwarf shrubs cover. The coastal zonation ends in the dune slack transition with *Quercus ilex* woods (*Vincetoxico-Quercetum ilicis*). This community is currently present although scattered and fragmented as a consequence of agricultural claim, but it shows a good recovery potential under the canopy of senescent pine forests.

Intermediate communities (*Viburno-Phillyreetum angustifoliae* and *Erico-Osyridetum albae*) and the pseudo-macchia (*Junipero-Hippophaetum fluviatilis*) of the seaward side of the semi-consolidated dunes and fixed dunes, exposed to wind action, complete the zonation. Woody communities contribute to outline the biocenotic uniqueness of N-Adriatic coasts relative to Mediterranean coastal sand dune systems: particularly the widespread presence of *Juniperus communis* ssp. *communis* suggests links towards Atlantic coasts rather than Mediterranean ones (Gamper et al., 2008).

Interdunal depressions are colonized by wet grass communities (the edapho-hygrophilous series and the sub-halophilous mosaic of wet infradunal downs). The most interesting are the hygrophilous natural fens (*Erucastro-Schoenetum nigricantis*) and the semi-natural meadows *(Plantagini altissimae-Molinietum caeruleae)*, both endemic, growing on neutral to subalkaline soils enriched in organic matter. *Molinia caerulea* grasslands are a very rare semi-natural community, found where the water table is close to the surface, and its long term conservation needs constant agricultural management such as regular mowing. Managing slowdown determines littering, auto-manuring and development of common shrubs and woody vegetation. Soils of interdunal lowlands somehow resemble those of shallow dunes (coarse texture, subalkaline reaction, scarce horizons differentiation), but with an aquic soil moisture regime, reducing conditions and a slightly saline water-table, at least close to the coastline. Under these conditions, the *Schoenus nigricans* and *Erianthus ravennae* community *(Eriantho-Schoenetum nigricantis)* is the most widespread, although patchy and fragmented.

Inland wetland areas present counteracting aspects, since they are of interest both as regulators of hydrological conditions and for biodiversity conservation. They are characterized by soils with coarse-loamy texture, subalkaline reaction, organic matter accumulation, and a fresh water-table close to the surface (Bini et al., 2002). Plant communities change according to the water-table level and the nutrient content, from the hydro-hygrophilous cane brake (*Phragmition communis* communities) to fens and wet meadows (*Caricion davallianae, Magnocaricion elatae* and *Molinietalia caeruleae*). Further development would foresee the marshy willow shrub (*Frangulo-Salicetum cinereae)* and the black alder wood (*Carici elongatae-Alnetum glutinosae*), nevertheless still found only in a small patchy area southward.

Coastal dune vegetation has been often described as azonal (Acosta et al., 2003a; Buffa et al., 2007); this is mostly true for communities of beaches and mobile dunes which actually have been proved to have a wide geographical range (Biondi, 1999; Carranza et al., 2008). On the contrary, foredune systems, and particularly the edapho-xerophilous series, are delineated by floristic and coenological features which are more related to local conditions such as climate, morphology, lithology and history. In the study area, all these aspects give rise to an outstanding environment so explaining the high level of cenological endemism.

Compared to the xerophilous series, wetland systems are structurally simplified and host a lower number of plant species, usually with wide distributional range (Buffa et al., 2007), but their presence, though extremely patchy and scattered, contributes to increase the N-Adriatic landscape richness and diversity.

Although the two Land Facets show the same systems sequence (beach and mobile dunes-fixed dunes-dune slacks transition) they slightly differentiate in terms of number of land elements and landscape diversity. It is worth noting that the simplified landscape representation drawn in Figure 3 and the series described in Table 3 only represent the potential natural landscape of N-Adriatic sandy costal system but almost nowhere does it actually express from beach to dune slacks transition completely.

Particularly the most complex communities, i.e. woody communities, have almost disappeared and persist only scattered and patchy. The pseudo-macchia (*Junipero-Hippophaetum*), which only 30 years ago was widespread (Géhu et al., 1984), currently exists just in one site (Porto Caleri), thanks to the establishment of a protected area in 1991.

Despite anthropogenic pressure and changes in the coastal dune environment of the N-Adriatic coastline, 20 EU habitats of interest (EU 43/92) have been surveyed. Therefore, regardless of human pressures, the Venetian coastal ecosystem could be regarded as a biodiversity hotspot within the Mediterranean Basin and still conserves many valuable elements to be maintained and emphasized (Buffa & Lasen, 2010).

4. Assessing conservation status

4.1. Background

The increased rate of habitat change and natural resource utilization since the 1950s, and the consequent threats to biodiversity have led to increased concern for monitoring and protecting remaining natural areas.

N-Adriatic coastal dune system suffers from a severe and complex human utilization; meanwhile it holds high landscape, faunal and floral values. These characteristics make it an ideal site to test an analytical approach to conservation status assessment to provide management policy which takes into account ecological values, landscape complexity and driving processes.

In the late 1980', Franklin et al. (1981) recognized three primary attributes of ecosystems: composition, structure, and function, which frame and make up the biodiversity of an area. Composition concerns the identity and variety of elements and includes measures of species diversity; structure represents the physical pattern of a system, from habitat complexity to the landscape level; finally, function relates to ecological and evolutionary processes, such as gene flow, disturbances and nutrient cycling (Noss, 1990).

Recently, progress has been made in developing methods for monitoring compositional diversity and for assessing threats of individual species, mostly in support of the IUCN Red List (Akcakaya et al., 2000; IUCN, 2006). This approach has become so overriding, that ecosystems evaluation is commonly based on the proportion of the threatened or endemic species pool that they encompass (Bonn & Gaston, 2005). Much less advancement has been made in building up an adequate insight into structural and functional diversity and in developing sound methods for assessing threats to habitats and ecosystems. Consequently, structural simplification of ecosystems which leads to the disruption of fundamental ecological processes can remain unappreciated.

The well known complexity of biodiversity and its hierarchical organization make clear that composition, structure and function of ecosystems are interdependent, nested and bounded (Noss, 1990; Margules & Pressey, 2000). These two concepts lead to two main consequences: first, because of the complexity of biodiversity, conservation status assessment can be defined using surrogates or partial measures such as sub-sets of species, species assemblages and habitat types (Hermy & Cornelis, 2000; Margules & Pressey, 2000). In particular, plant communities and vegetation, owing to their specific nature, can be regarded as good indicators of overall biodiversity and specifically of ecosystem integrity of coastal dune ecosystems (Araújo et al., 2002; Carboni et al., 2009; Géhu & Biondi, 1994; Géhu & Géhu, 1980; Lopez & Fennessy, 2002). Second, hierarchy theory suggests that monitoring and assessment can not be limited to one single level of organization and that different levels of resolution are proper for different questions (Noss, 1990; Bonn & Gaston, 2005). Hence, a thorough analysis of the conservation status of landscapes should take into account multiple levels of biodiversity, from community level to the entire landscape, considering compositional and structural features as well. This can be performed choosing a consistent and multiscale set of key indicators, based on field data, embodying the entire complexity of the vegetation system (Carboni et al., 2009; Cingolani et al., 2010; Hermy & Cornelis, 2000).

4.2. Methods

According to a previous research (Buffa et al., 2005) bound to a small coastal area, the conservation status of N-Adriatic coastal landscape has been estimated at two different interconnected scales: at landscape and plant community level. The method has been slightly redefined following the recent proposal by Carboni et al. (2009) and Grunewald & Schubert (2007).

Assessment method grounds on the basic idea that the severe and stressing dunal environment is the dominant factor leading not only to the presence of highly adapted and specialized species and communities (which refers to the identity and variety of elements) but also to a typical and worldwide sea-inland spatial zonation of plant communities (i.e., the physical pattern of the system), which in absence of disturbance events tends to be fixed.

We limited our evaluation only to those PNV with a xeric soil moisture regime, namely "beaches and mobile dunes" and the edapho-xerophilous series of fixed dunes and dune slack transition (see Table 3); as for the edapho-higrophilous series, it was included in "natural surface" category.

4.2.1. Landscape level

Analysis were carried out on the basis of a digital map of the area (1:10.000) encompassing the major natural land cover types as well as artificial surfaces. Natural land elements belonging to coastal zonation were expressed in terms of EU habitats (following EU Habitat Directive).

The map was derived from panchromatic digital aerial ortho-photographs (dated 2010) with a resolution of 6000 x 5600 pixels, covering 1500 m wide stretch from the coastline inward. Land cover was manually interpreted on video, by means of a Geographic Information System (ArcGIS software 9.3), and field survey. The legend follows CORINE land cover expanded, where possible, to a fourth level of detail. If a CORINE land cover category embodied more than one EU habitat (i.e., 3.2.2.2 Termophilous shrubs), it has been accordingly split up.

GIS analysis tools were used to calculate some common landscape metrics (Table 4) regarding both composition and structure. Landscape metrics were figured out for the entire studied coastal stretch to define an overall conservation status; at the same time they allowed a comparative evaluation of the status of PNV land elements and landscape inside the two Land Facets recognized in the area.

Composition	Structure
Total LC type richness	Total number of patches
Natural LC type richness	Number of natural patches
Natural coastal LC type richness	Number of natural coastal patches
Total surface (ha)	Average natural patch size (ha)
Percent natural surface	Mean shape index for natural patches (MSI)
Percent natural coastal vegetation	
Percent urban surface	
Percent agricultural surface	
Shannon Diversity Index (H)	
Eveness (J)	

Table 4. Indices of landscape composition and structure used for assessment.

Landscape composition was evaluated in terms of richness and abundance of the main land cover categories. Particularly urban surface abundance, which causing physical changes strongly influence natural habitat, has been deepened as an indicator of human pressure to natural landscape (Margules & Pressey, 2000; McKinney, 2002). Finally, richness and abundance parameters of each single category have been utilized to compare Shannon diversity index (H) and evenness (J) of the two Land Facets.

Landscape structure was analyzed by means of statistics such as number of patches of each land cover category, mean patch size and their mean shape index. Patch shape and size are important structural features of the landscape mainly related to the concept of "edges", which have been recognized as functional components of the landscape (Cadenasso & Pickett, 2001; Forman, 1995), influencing fluxes of organisms, material, and energy between two adjacent habitats.

Patches spatial arrangement allow the quantitative measurement of heterogeneity of a landscape and the comparison of landscapes, while shape index evaluates landscape configuration in terms of the complexity of patch shape (McGarigal & Marks, 1995). Patton's shape index (Patton 1975) was chosen, among others, because it measures the complexity of patch shape compared to a standard shape (a circular standard) the same size, thus alleviating the size dependency problem. As natural communities of coastal ecosystems usually run stretched out parallel to the coastline, then high MSI values will testify for natural patterns, while less natural typologies or fragmented patches, with more isodiametric or round forms, will show lower MSI values (Carboni et al., 2009).

4.2.2. *Community level*

Number of categories, their proportions and diversity represent non-spatial system properties (Gustafson, 1998) mostly linked to compositional features, but composition being equal a landscape may exhibit many different patch arrangement, that is many different spatial configuration. When undisturbed, plant communities of sandy coastal ecosystems show a typical distribution pattern along the sea-inland gradient which can thus represent a reference model for evaluating actual spatial configurations.

Spatial configuration integrity was measured by means of richness of boundaries, n (Rescia et al., 1997) and the gamma connectivity, γ (Acosta et al., 2000; 2003b; Forman & Godron, 1986). According to Acosta et al. (2003b), the two indices were calculated through a one-dimensional approach which involves the projection of the plant communities found along sea-inland transects perpendicular to the coastline. Therefore, each transect presents a specific spatial sequence of plant community patches, which can be compared with the reference model.

The number of different types of boundaries (i.e. contacts between plant community types) along each transect represents n, the richness of boundaries, while gamma connectivity index refers to the position of a patch type in relation to other patch types and determines the boundary between patches and the links among them. Gamma connectivity index ranges from 0 (no links among patches) to 1 (every patch is linked to every possible patch), and higher values are normally considered an index of better environmental quality

(Forman, 1996). On the contrary, in coastal dune systems, where plant communities tend to have a strong linear distribution, the best structural quality reflects the natural and undisturbed sequence which has a low number of fixed links. Therefore, higher connectivity values are usually associated to disturbance events which destroy or modify the natural sequence. The two indices have been calculated along 30 transects arranged along the coastline at 3500 m interval, 10 in NLF and 20 in SLF.

Besides, for each plant communities we computed some indicators of species diversity and vegetation quality. We estimated the diversity of plant communities by calculating the diversity index and evenness, following the proposal by Grunewald & Schubert (2007) who adapted Shannon diversity and evenness index specifically for coastal dunes, H_{dune} and E_{dune}, incorporating the parameter "species density" (plant cover relative to the plot-size).

As aliens and ruderal species are predicted to increase as a result of increasing human disturbance (Richardson et al., 2000; Sax & Gaines, 2003), an effect in overall diversity may be expected. Recently, Carboni et al. (2009) proposed relating the number of species of a broad distribution type or of exotic origin, which are generally introduced and/or favoured by human disturbance, to the number of species of a chorological type characteristic for the examined area and strictly dependent on the studied region. As one of the main characteristics of N-Adriatic region is just the co-occurrence of many different chorological types, choosing the most typical was not possible. Therefore, to provide information about quality, or inversely, about the level of "anthropogenization" we calculated a natural diversity index (N) (Grunewald & Schubert, 2007). According to a common procedure in ecological and conservational studies (e.g., Grime, 2002; Martinez et al., 2004; Rodgers, 2002; Rodgers & Parker, 2003), the degree of natural diversity (N) was calculated by classifying species into typical native dune species and untypical dune species often associated with non-dune habitats and disturbed, nutrient-rich sites, including truly alien species as well. Being the ratio between diversity index H_{dune} computed with all species and that without "alien" species, the index N can be read as a sound evaluation of natural diversity; only a maximum value of one can be reached (if no species are excluded = complete natural diversity) (Grunewald & Schubert, 2007).

The significance of overall difference among communities in terms of diversity index, evenness and index N was assessed by one-way ANOVA on transformed data, in order to determine differences at the critical significance level $p<0.01$. One-way ANOVA on transformed data was also used to check differences in quality (H_{dune}, E_{dune}, N_{dune}) between the two coastal LF at the critical significance level $p=0.05$. Finally, we tested differences in the spatial patterns (connectivity and richness of boundaries) and quality (H_{dune}, E_{dune}, N_{dune}) of the two Land Facets comparing values through independent groups *t*-tests (Sokal & Rohlf, 1995).

4.3. Results

4.3.1. Landscape level

Land cover classification allowed the identification of 30 different CORINE Land Cover Types (Table 5), but as they could embody more than one habitats, total number of land categories summed up to 40.

Level 1	Level 2	Level 3	Level 4
1. Artificial surfaces	1.1. Urban fabric	1.1.1. Continuous urban fabric	
		1.1.2. Discontinuous urban fabric	
	1.2. Industrial commercial and transport units	1.2.1 Industrial or commercial units	
		1.2.2. Road and rail networks and associated lands	
		1.2.3. Port areas	
		1.2.4 Airports	
	1.3. Mine, dump and construction sites	1.3.2 Dump sites	
		1.3.3. Construction sites	
	1.4. Artificial non-agricultural vegetated areas	1.4.1 Green urban areas	
		1.4.2. Sport and leisure facilities	
2. Agricultural areas	2.1. Arable land	2.1.1. Non-irrigated arable land	
	2.2. Permanent crops	2.2.4 Other permanent crops	
	2.4. Heterogeneous agricultural areas	2.4.3. Land principally occupied by agriculture, with significant areas of natural vegetation	
3. Forests and semi-natural areas	3.1. Forests	3.1.1 Broad-leaved forests	3.1.1.1 Quercus ilex woods (9340)
			3.1.1.6 Higrophilous woods
		3.1.2. Coniferous forests	3.1.2.1. Pine woods (2270)
	3.2. Shrub and/or herbaceous vegetation association	3.2.2. Moors and heathland	3.2.2.2 Termophilous shrubs (2250/2160) (three different plant communities)
		3.2.4. Transitional woodland/shrub	
	3.3. Open spaces with little or no vegetation	3.3.1. Beaches, dunes, and sand plains	3.3.1.1. Open sand (1210)
			3.3.1.2. Partially vegetated dunes (2110)
			3.3.1.3. Densely vegetated dunes (2120)
			3.3.1.4. Moderately vegetated slacks (2130)
			3.3.1.5 Interdune annual grasslands (2230)
4. Wetlands	4.1. Inland wetlands	4.1.1. Inland marshes	
		4.1.2. Peatbogs	
	4.2 Coastal wetlands	4.2.1. Salt marshes	
		4.2.3. Intertidal flats	
5. Water bodies	5.1. Inland waters	5.1.1. Water courses	
		5.1.2. Water bodies	
	5.2. Marine waters	5.2.1 Coastal lagoons	

Table 5. CORINE land cover categories surveyed in the study area. For Level 4 correspondence with sandy coastal EU habitats is reported; numbers in brackets correspond to Natura 2000 codes. Note that some Natura 2000 habitats can be represented by more than one plant communities.

Total surveyed area resulted in about 15.800 ha; 30.3% is covered by urbanized surface (mainly represented by towns and villages, roads and tourist facilities) and 22.2% by agricultural areas (mostly arable lands); only 47.5% is included in natural or semi-natural categories, of which only 4.5% represented by dune systems (Figure 4 and Table 6).

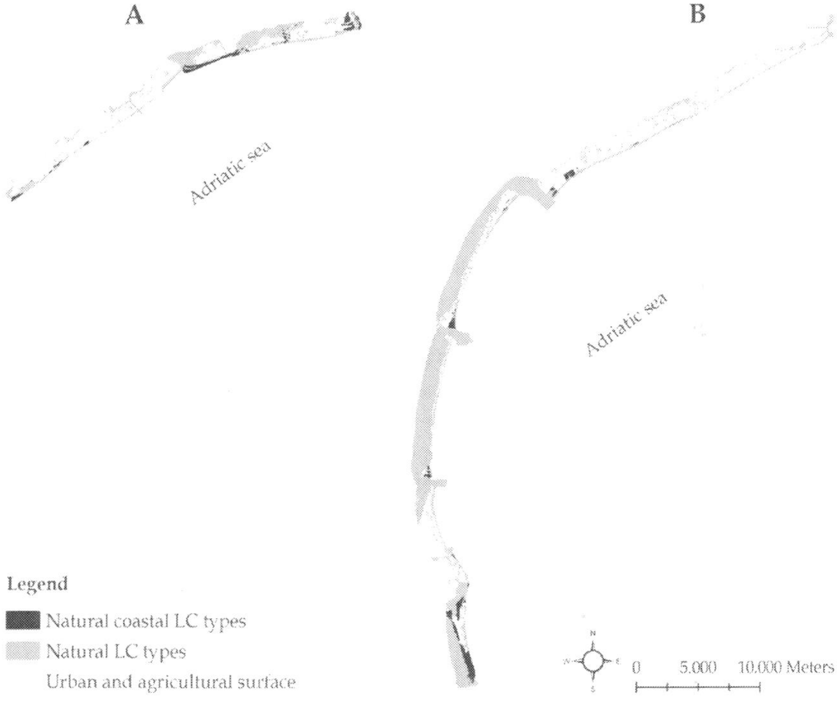

Figure 4. Land cover maps of the two Land Facets showing the main three categories as derived from the CORINE land cover map. A: Northern Land Facet, from Tagliamento River to Piave River; B: Southern Land Facet, from Piave River southwards to Brenta River.

Northern Land Facet (NLF) covers a surface of 4848 ha and extends from the Tagliamento River, in the north, to the Piave River; Southern Land Facet (SLF) covers a larger surface of 10942 ha, from the Piave River to the Brenta-Adige-Po Rivers system in the south.

As for composition, the two Land Facets resulted very similar. In NLF, the natural typologies were 25, 8 of which were coastal ones; the SLF showed 25 natural typologies as well, 9 of which of natural coastal LC types. Abundance of natural LC types discriminated between the two Land Facets: of the total surface surveyed, only 36% was included in natural or semi-natural categories in the Northern part, while in the Southern more than 50% of the total surface fell in this category. Interestingly enough, while urban surface is quite similar (32% in the north, 29% in the south), urbanization pattern differentiates; large and compact cities (such as Bibione, Caorle and Eraclea) predominate in the NLF, while in the SLF urban settlements permeate the landscape with some larger cities (like Jesolo) along with a continuous presence of houses, little villages and tourist facilities, whose spread is favoured by a more developed roads network. The situation is much impressive in the central part of SLF, corresponding to Venice barriers, where urbanization and widespread leisure facilities join a limited barriers width, leaving few sites for undisturbed natural vegetation. Therefore, although percentage of natural surfaces is lower in the Northern LF, they suffer from lower human pressure. Moreover, it is worth noting that the higher average size of natural patches in the SLF (Table 6) is mostly due to the presence of some categories such as coastal lagoons (i.e., particularly the Venice lagoon, in the central part) and salt marshes which usually cover broad areas.

variable	total	NLF	SLF
Composition			
Total LC type richness	40	36	37
Natural LC type richness	27	25	25
Natural coastal LC type richness	9	8	9
Total surface (ha)	15790.96	4848.79	10942.17
Percent natural surface	47.5	35.7	52.7
Percent natural coastal vegetation	4.5	5.8	3.9
Percent on natural surface		16.1	7.4
Percent urban surface	30.3	31.8	29.6
Percent agricultural surface	22.2	32.5	17.6
Shannon Diversity Index (H) (total surface)		2.52	2.34
Eveness (J) (total surface)		0.70	0.65
Shannon Diversity Index (H) (natural surface)		2.56	1.17
Eveness (J) (natural surface)		0.80	0.36
Structure			
Total number of patches	8100	2790	5310
Number of urban patches	4397	1652	2745
Number of agricultural patches	507	119	388
Number of natural patches	3196	1019	2177
Number of natural coastal patches	458	214	244
Average natural patch size (ha)	2.35	1.697	2.651
Average natural coastal patch size	1.54	1.303	1.743
Mean shape index for natural patches (MSI)	2.32	2.82	2.08
Mean shape index for natural coastal patches (MSI)	2.11	2.18	2.05

Table 6. Landscape composition and structure indices for total area and for each Land Facet examined (NLF=Northern Land Facet; SLF=Southern Land Facet).

Total Shannon diversity index, calculated taking into account the proportions of each land cover category, reflects the situation, with a slightly higher diversity northwards. The dominance of coastal lagoons and correlated salt marshes also affects evenness, which is slightly lower for the SLF than for the NLF. Shannon diversity index and evenness, calculated taking into account only the proportions of natural land cover category, still strengthen differences (Table 6) between the LF.

The analysis of the entire area resulted in the identification of a total of 8100 patches, 3654 of which were natural ones, 458 of which represented by coastal vegetation patches (Table 6). The NLF coastal landscape was much more structured and heterogeneous, especially considering the natural coastal patches. In this area number of patches is lower, but they sum to 16% of the total natural patches cover against only 7% of natural coastal patches in the SLF.

The higher heterogeneity reflects on the average size of natural coastal patches, which is little more than a hectare for the former and nearly two for the latter (*t*-test, $p<0.05$). Mean shape index does highlight significant differences in the shape of the natural coastal patches of the two areas (*t*-test, $p<0.01$), giving evidences, in the north, of a more natural landscape, where heterogeneity depicts typical features of coastal environments rather than a fragmented landscape, fruit of human disturbance.

Zooming inside the two land elements (beaches and mobile dunes and the edapho-xerophilous series of fixed dunes) helps to further clarify differences between Land Facets. Beaches and mobile dunes are much more abundant in the SLF, both in number of patches (123 against 35) and cover (25% against 15%), with a remarkable 10% covered by *Ammophila arenaria* community, which in NLF is almost replaced by an annual, nitrophilous community (*Sileno-Vulpietum*). Conversely, edapho-xerophilous series is much more widespread in the northern part and its importance is further underlined by the percentage cover of *Tortulo-Scabiosetum* community, one of the rarest and most vulnerable coastal plant association of the Veneto Region, whose abundance is nearly three time than in the south.

4.3.2. *Community level*

The coastal zonation comprises ten plant communities grouped in nine different EU habitats, as code 2250 embodies both *Erico-Osyridetum* and *Viburno-Phyllireetum* communities. They range from the pioneer beach communities to the holm-oak wood in the inland part.

Plant species richness typically increased following the sea-inland gradient and ranged between 19 species in 29 samples from the upper beach (the *Cakile maritima* community) to 84 species in 38 samples from the transition dune vegetation (*Tortulo-Scabiosetum* community), with a slight decrease in the stabilized dunes (in particular *Quercus ilex* wood).

As expected, H_{dune} index followed the same trend (Table 7), evidencing highly significant differences between communities of beaches and mobile dunes and those of the edapho-xerophilous series, with higher values that concentrated in the intermediate communities of the latter (ANOVA, $p<0.0001$, d.f. 195, $F=27.09$). On fixed dunes, plant succession leads to more mature and complex plant communities than on mobile dunes, and more species are characteristically present. Evenness also changed from young successional stages to more mature stages. In the species poor communities of mobile dunes, dominance by one or two species is usually high. Once plant succession has led to more complex plant communities with higher mean coverage, interspecific competition for limited resources becomes the dominant factor limiting dominance and favouring equidistribution of species. Index of natural diversity (N) showed an inverse trend and beaches and mobile dunes land elements were much more subject to invasion by human-favored ruderal or alien species. On the contrary, apart from *Tortulo-Scabiosetum*, intermediate and mature communities of the edapho-xerophilous series highlighted a very low degree of anthropogenization.

Quality of the vegetation	H dune	E dune	N dune
Salsolo-Cakiletum	0.583a	0.743a	0.702a
Sporobolo-Agropyretum	0.906b	0.779ab	0.673ab
Echinophoro-Ammophiletum	0.804abc	0.459c	0.619abc
Sileno-Vulpietum	0.91bcf	0.79ab	0.373g
Tortulo-Scabiosetum	1.540d	0.516cd	0.621abc
Erico-Osyridetum	1.252efg	0.575cdef	0.984de
Viburno-Phyllireetum	1.806deh	0.627ef	0.986def
Juniperum-Hippophaetum	1.519de	0.527cde	0.964d
Vincetoxico-Quercetum	1.426degh	0.402ce	0.931def

Table 7. Indices of overall quality of plant communities of coastal zonation (paired *t*-tests; where letters differ, values are significantly different at $p < 0.01$).

As for the analysis of the spatial arrangement of communities along the zonation no significant differences emerged between the two Land Facets neither in the spatial connectivity index, nor in richness of boundaries (Table 8).

variable	NLF	SLF	sig.
Spatial structure			
Richness of boundaries (n)	5.429	4.455	0.030
γ sea-inland	0.537	0.603	0.325

Table 8. Indices of spatial structure of the vegetation of the two Land Facets analysed (significance values refer to *t*-tests for independent groups; df = 28).

Richness of boundaries was hardly higher for the NLF as a consequence of the higher heterogeneity of this area, but the difference is only slightly significant. Moreover, the relatively high connectivity average values gave evidence for disturbance for both Land Facets, causing fragmentation of the communities and their repeated presence along the transects.

In Figure 5, some explanatory transects have been reported as examples. The comparison to reference model depicted in Figure 3 described two different disturbance patterns which confirmed the spatial and structural analysis. Transects of NLF mostly lacked first terms of zonation and the sequence started with open sand immediately followed by the edapho-xerophilous series and the edapho-higrophilous mosaic (both fresh and sub-halophilous communities where water-table was slightly saline). Conversely, in SLF edapho-xerophilous series is well conserved only in a few sites (see transect G); first terms of the sequence are much more widespread and sometimes well developed (see transect F), but in general their width is compressed by impending human settlements (see transect E).

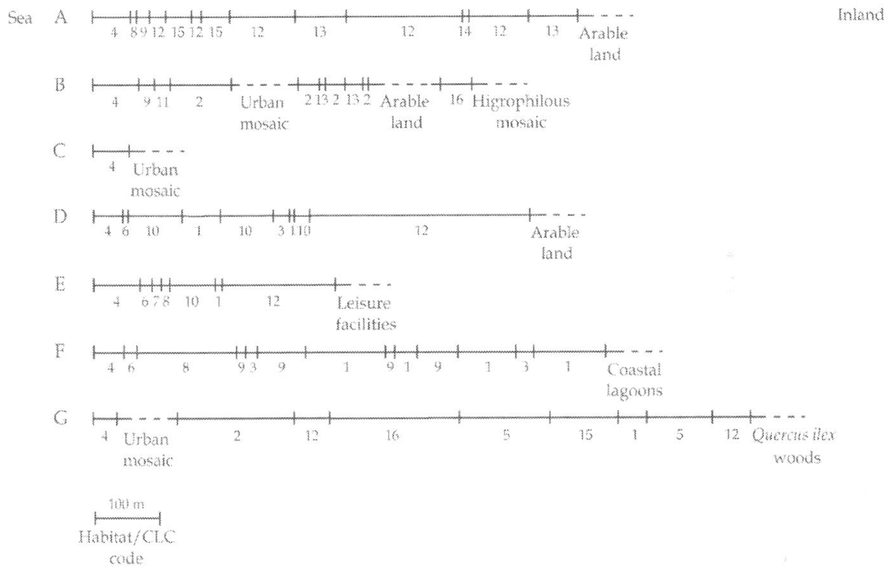

Figure 5. Schematic transects reporting sequence of natural and artificial land cover categories. Transects A and B are representative of NLF; in particular, transect A comes from the northernmost part corresponding to the Tagliamento mouth. Transects C-G come from SLF; F and G represent the sequence in Porto Caleri, at the southern boundary. 1= 3.1.1. Broad-leaved forests; 2=3.1.2. Coniferous forests; 3=3.2.4. Transitional woodland/shrub; 4=3.3.1. Beaches, dunes, and sand plains; 5=4.2.1. Salt marshes; 6=3.3.1.1. Open sand (1210); 7=3.3.1.2. Partially vegetated dunes (2110); 8=3.3.1.3. Densely vegetated dunes (2120); 9=3.3.1.4. Moderately vegetated slacks (2130); 10=3.3.1.5. Interdune annual grasslands (2230); 11=3.2.2.2. Termophilous shrubs (2250/2160) (three different plant communities); 12=3.1.2.1. Pine woods (2270); 13=4.1.1. Inland marshes; 14=4.1.2. Peatbogs; 15=3.1.1.6. Higrophilous woods; 16=3.1.1.1. Quercus ilex woods (9340).

Diversity and evenness indices, calculated taking into account only the proportions of coastal communities found in the two Land Facets, and particularly N_{dune}, the index of natural diversity, confirmed differences (Table 9). Apart from evenness, which never differentiated (E and E_{dune}), other indices were significantly higher for the dune system of NLF than for that of SLF indicating a richer and more diverse vegetation for the former. The better conservation status of the edapho-xerophilous series, which have been mostly eliminated in the southern area as a consequence of agricultural activities and leisure facilities development, is probably responsible for such higher diversity values.

The better conservation status in NLF has been also reinforced by the significant difference emerged for the N index. In this case, N represented the ratio between H calculated taking into account the cover of all communities and that without cover of "alien" communities. Values were significantly different and in the NLF, N index was nearly 1 (that is, almost no communities have been excluded), clearly highlighting how human pressures facilitate natural communities replacement by ruderal, not typical communities such as *Sileno-Vulpietum*, in the seaward slopes of the dunes, or by dynamic stages characterized by transitional woods and shrub communities, landwards.

variable	NLF	SLF	F	p
Shannon Diversity Index (H)	1.540	1.137	18.13	p<0,0001
Eveness (J)	0.582	0.550	n.s.	n.s.
H_{dune}	1.521	1.041	24.24	p<0,0001
E_{dune}	0.553	0.604	n.s.	n.s.
N	0.927	0.677	34.14	p<0,0001

Table 9. Differences in quality between NLF and SLF (one-way ANOVA; df=195).

5. Discussion and conclusion

Coastal areas are responsive systems affected by natural as well as anthropogenic pressures. Specifically, coastal sand dune dynamics, not only that linked to human disturbance, but even the natural cyclic dynamics (i.e. wave and tidal regime, sediment budget with the recurrence of regression and accretion phases), is associated with visible modifications in both plant communities and landscape, so that this sound relationship can be used as a monitoring tool in coastal areas (Araùjo et al., 2002). Particularly plant communities, that represent well-identifiable land elements with a relatively stable composition, structure and mutual relationships, all related to specific environmental conditions, can provide reliable monitoring activities (Loidi, 1994; Lomba et al., 2008).

At present, plant communities zonation along the N-Adriatic coast is complete only at few sites, mostly located at some distance from areas of urban development, where tourism is limited by legislation (Porto Caleri) or simply because of the difficulty in reaching them (i.e., Ca' Roman and the Tagliamento mouth). In almost all the rest of the Veneto coast, actual vegetation zonation is noticeably far from the potential one. Where disturbance events were very strong, some PNV communities completely disappeared and available space has been occupied by replacement ruderal communities.

The worst conservation status has been found on the central part of Venetian coastline, affected by urbanization, agriculture as well as heavy tourist presence, and here the natural landscape appears much more endangered in terms of trampling and alien invasions linked to human settlements and in some portion it has completely disappeared.

Factors causing the disturbance are of various types and act at multiple temporal and spatial scales, translating into effects that manifest themselves differently in space and time.

Moreover, it can be very difficult to distinguish actual sources of impact and to separate their individual effects (e.g., human trampling vs. erosion or embankment removing foredunes) (Defeo et al., 2009), or to find unaltered and natural beaches that could act as good control sites.

In Veneto region, most of the coastline is increasingly deprived of sand since groynes, jetties and revetments trap sediments that would otherwise supply beaches; activities such as land reclamation, urbanisation, afforestation and agricultural use further interfere with the sediment budget (Nordstrom, 2000). As a result, nearly 50% of the Venetian beaches are experiencing erosion (Bondesan & Meneghel, 2004), which mostly impacts beaches and embryonic dunes. Human's response to beach erosion and retreat has historically consisted in placing armouring structures (Charlier et al., 2005), that alter the natural hydrodynamic system of waves and currents, modifying sand transport rates, which in turn affect the erosion/accretion dynamics (Miles et al., 2001), possibly causing further deep habitat changes.

Particularly embryonic dunes are landfoms strongly related to beach dynamics and can thus be regarded as effective geo-indicators of coastal evolution. Absence of embryonic dunes is typical of coastal areas suffering from erosion phenomena which can be recognized through geomorphologic features, but as eroding beaches become narrower, the reduced surface directly reduce diversity of ecosystems, particularly in the upper intertidal zone (Dugan et al., 2008). Coastal erosion can thus lead to structural modifications in terms of denudation of some sites, thus truncating the coastal sequence completely removing the first terms (i.e., *Cakile maritima* and *Elymus farctus* communities) or drastically reducing their space thus causing communities merge together. Geomorphic events in mobile dunes even induce severe changes on inland areas as the defensive physical barrier provided by mobile dunes weakens. Disruptions of mobile dunes promote erosional gaps on the fixed dunes and a decline in vigour of *Ammophila arenaria*. As its resilience declines, marram is joined by more species, first by other specialised dune plants, then by less specialised grasses, drought-tolerant annuals such as those of *Sileno-Vulpietum*, a community rich in ruderal taxa.

Erosion is probably responsible for the small number of beaches and mobile dunes patches in NLF, but human pressures, mainly through trampling and beach grooming, can also promote the disruption of embryonic and foredunes, thus acting in synergy with erosion. As already pointed out by several studies (Brown & McLachlan, 2002; Carboni et al., 2009; Grunewald & Schubert, 2007; Kutiel et al., 1999; but see also Bonte & Hoffmann, 2005 for further references), trampling and other recreation-bound human activities, such as beach cleaning are among the most severe factors impacting sandy shores resulting in fragmentation, communities merging and/or replacement, alien invasion and in an overall lowering of diversity values. In SLF, where beaches are mainly in accretion phase, human disturbance is particularly intense and beach communities have almost completely disappeared from many sites, substituted by ruderal replacement communities.

Strong accumulation of mobile sand in interior areas has severe consequences also on plant communities of grey dunes, which in connection with trampling, causes the occurrence of plant species that are normally typical of embryonic dunes and a decrease in number of character species. Within the edapho-xerophilous series, grey dunes (*Tortulo-Scabiosetum*) show the highest amount of alien species, both exotic and human-favoured ruderal species. Particularly dwarf shrubs, which represent the most typical component, seem to be the least tolerant plants relative to trampling (Cole, 1995).

While pioneer stages mainly suffer from coastal erosion and tourism, at the other extreme, fixed dune communities are affected mostly by urbanization. As erosion and tourism pressures truncate the first elements of the zonation, towns and villages, coastal roads, pines plantations and agriculture truncate the last stages of the typical zonation. Human disturbance on intermediate shrub communities is chiefly represented by urban development, campsites, leisure facilities, roads and afforestation, while urban development and cultivated land have drastically reduced the area covered by holm-oak woods and at present well-preserved woods survive at only very few coastal sites. The effects of disturbance on coastal dune ecosystem vary according to the severity of the disturbance, but on fixed dunes and dune slacks transition disturbance usually drives to the complete disappearance of natural communities.

Therefore, the primary long-term threat coastal sandy ecosystems are facing is a "coastal squeeze" (Defeo et al., 2009; Schlacher et al., 2007), which causes sandy systems to be trapped between erosion on the sea side and human settlements inlands, thus leaving no space for natural sediment dynamics.

The two nested levels of analysis, considering higher organization level, the landscape level (with both composition and structure evaluation) as well as lower level (community level), such as species diversity and vegetation quality, provided a mutual description and evaluation of the naturalness of coastal landscape in the two coastal sectors studied and a sound assessment of their conservation status.

Coastal sand dune landscapes hold habitats of high economic, social and ecological value on a worldwide scale. At the same time, they are among the most threatened ecosystems on a national and European scale, facing escalating anthropogenic pressures. Sand dune habitat loss and degradation is leading to a remarkable biodiversity loss, which in turn can result in irreversible damage to ecosystem functions and ecological services. While some existing geo-physical models can be applied to predict climate-related changes for coastal areas (Zhang et al., 2004), no equivalents exist for the ecological effects of global change which could lead to a significant net loss of dune areas over the next century. Because of the scale of the problem, interdisciplinary and innovative approaches are required and the continued existence of sandy coastal areas as functional ecosystems is likely to depend on direct conservation efforts, which will have to progressively incorporate ecological aspects of coastal landscapes (Schlacher et al., 2007).

The most recent European legislation, in particular the Marine Strategy Framework Directive, which encompasses and reinforces other previous EU Directives such as WFD 2000/60 and Habitat Directive 92/43, calls for a strategic approach to coastal zone management providing sustainable development. The integrated and ecosystem-based approach used in this paper fulfils most EU requirements for policy making and by dealing with the problem both through a community-oriented and a landscape pattern-oriented approach it can provide a comprehensive framework for sustainable coastal management and development and for the improvement of projects or actions supporting biodiversity and ecosystem services. This approach could thus represent an innovative tool for the sustainable management assessment as it provides clear and easy applicable monitoring instruments allowing planners and stakeholders to evaluate the effectiveness of different action plans. Provided a solid classification, plant community types can be considered highly reliable indicators of environmental status in coastal areas. Moreover, the hierarchical landscape classification, coupled with the PNV concept, results in a reference model for environmental monitoring of anthropogenic pressures on coastal areas, providing interconnected spatial units which help dealing with the complex and dynamic nature of ecosystems.

Author details

Gabriella Buffa, Edy Fantinato and Leonardo Pizzo
DAIS - Dept. of Environmental Sciences, Informatics and Statistics, Ca' Foscari University, Italy

6. References

Acosta, A.; Blasi, C. & Stanisci, A. (2000). Spatial connectivity and boundary patterns in coastal dune vegetation in the Circeo National Park, Central Italy. *Journal of Vegetation Science*, Vol. 11, pp. 149-154.

Acosta, A.; Blasi, C.; Carranza, M.L.; Ricotta, C. & Stanisci, A. (2003b). Quantifying ecological mosaic connectivity and hemeroby with a new topoecological index. *Phytocoenologia*, Vol. 33, pp. 623–631.

Acosta, A.; Stanisci, A.; Ercole, S. & Blasi, C. (2003a). Sandy coastal landscape of the Lazio region (Central Italy). *Phytocoenologia*, Vol. 33, pp. 715-726.

Acosta, A.; Carranza, M. L. & Giancola, M. (2005). Landscape change and ecosystem classification in a municipal district of a small city (Isernia, Central Italy). *Environmental Monitoring and Assessment*, Vol. 108, pp. 323-335.

Acosta, A.; Ercole, S.; Stanisci, A.; De Patta Pillar, V. & Blasi, C. (2007). Coastal Vegetation Zonation and Dune Morphology in Some Mediterranean Ecosystems. *Journal of Coastal Research*, Vol. 23, pp. 1518–1524.

Akcakaya, H.R.; Ferson, S.; Burgman, M.A.; Keith, D.A.; Mace, G.M. & Todd, C. (2000). Making consistent IUCN classifications under uncertainty. *Conservation Biology*, Vol. 14, pp. 1001–1013.

Allen, T.F.H. & Starr, T.B. (1982). *Hierarchy: perspectives for ecological complexity*, University of Chicago Press, Chicago.

Araújo, R.; Honrado, J.; Granja, H.M.; De Pinho, S.N. & Caldas, F.B. (2002). Vegetation complexes of coastal sand dunes as an evaluation instrument of geomorphologic changes in the coastline, In: *Littoral 2002. The Changing Coast*. EUROCOAST/EUCC, pp. 337–339, Porto, Portugal.

Barbier, E.B.; Koch, E.W.; Silliman, B.R.; Hacker, S.D.; Wolanski, E.; Primavera, J.; Granek, E.F.; Polasky, S; Aswani, S.; Cramer, L.A.; Stoms, D.M.; Kennedy, C.J.; Bael, D.; Kappel, C.V.; Perillo, G.M.E.; Reed, D.J. (2008). Coastal Ecosystem–Based Management with Nonlinear Ecological Functions and Values. *Science*, Vol. 319, pp. 321-323.

Barbour, M.G. (1992). Life at the leading edge: The beach plant syndrome, In: *Coastal plant communities of Latin America*, Seeliger U. (Ed.), pp. 291–307, Academic, San Diego.

Bezzi, A. & Fontolan, G. (2003). Foredune classification and morphodynamic processes along the Veneto coast (N. Adriatic, Italy), In: *MEDCOAST '03*, Őzhan, E. (Ed.), pp 1425–1434, MEDCOAST Secretariat, Ankara.

Bezzi, A.; Fontolan, G.; Nordstrom, K.F.; Carrer, D. & Jackson, N.L. (2009). Beach nourishment and foredune restoration: practices and constraints along the Venetian shoreline, Italy. *Journal of Coastal Research*, Vol. 56, pp. 287–291.

Bini, C.; Buffa, G.; Gamper, U.; Sburlino, G. & Zilocchi, L. (2002). Soils and vegetation of coastal and wetland areas in Northern Adriatic (NE Italy). 7th Int. Meet. Soils with Mediterranean Type Climate (Selected Papers), *Options Méditerranéènnes*, Ser. A, Vol. 50, pp. 31-36.

Biondi, E. (1994). The phytosociological approach to landscape study. *Ann. Bot.*, Vol. 52, pp. 135–141.

Biondi E. (1999). Diversità fitocenotica degli ambienti costieri italiani, In: *Aspetti ecologici e naturalistici dei sistemi lagunari e costieri*, Bon, M.; Sburlino, G. & Zuccarello, V. (Eds.), pp. 39-105, Arsenale Editrice, Venezia.

Biondi, E. (2011). Phytosociology today: Methodological and conceptual evolution. *Plant Biosystems*, Vol. 145, no. 1, pp. 19-29.

Blasi, C.; Carranza, M.L.; Frondoni, R. & Rosati, L. (2000). Ecosystem classification and mapping: A proposal for Italian landscapes. *Applied Vegetation Science*, Vol. 3, pp. 233–242.

Blasi, C.; Filibeck, G.; Frondoni, R.; Rosati, L. & Smiraglia, D. (2004). The map of the vegetation series of Italy. *Fitosociologia*, Vol. 41, no. 1 Suppl. 1, pp. 21-25.

Blasi, C.; Capotorti, G.; Frondoni, R. (2005). Defining and mapping typological models at the landscape scale. *Plant Biosystems*, Vol. 139, pp. 155-163.

Blasi, C. & Frondoni, R. (2011). Modern perspectives for plant sociology: The case of ecological land classification and the ecoregions of Italy. *Plant Biosystems*, Vol. 145, pp. 30-37.

Bondesan, A. & Meneghel, M. (Eds.). (2004). *Geomorfologia della provincia di Venezia*, Esedra Editrice, Padova.

Bondesan, A.; Primon, S.; Bassan, V. & Vitturi, A. (Eds.). (2008). *Le unità geologiche della provincia di Venezia*, Provincia di Venezia, Università di Padova, Venezia.

Bonn, A. & Gaston, K.J. (2005). Capturing biodiversity: selecting high priority areas for conservation using different criteria. *Biodivers. Conserv.*, Vol. 14, pp. 1083–1100.

Bonte, D. & Hoffmann, M. (2005). Are coastal dune management actions for biodiversity restoration and conservation underpinned by internationally published scientific research?, In: *Dunes and Estuaries 2005*, Herrier, J.-L.; Mees, J.; Salman, A.; Seys, J.; Van Nieuwenhuyse H. & Dobbelaere I. (Eds.), pp. 165-178, Proceedings International

Conference on Nature Restoration Practices in European Coastal Habitats, Koksijde, Belgium, 19-23 September 2005.

Braun-Blanquet, J. (1964). *Pflanzensoziologie*. 3rd edn. Springer, Wien.

Brown, A.C. & McLachlan, A. (2002). Sandy shore ecosystems and the threats facing them: some predictions for the year 2025. *Environ. Conserv.*, Vol. 29, pp. 62-77.

Buffa, G.; Mion, D.; Gamper, U.; Ghirelli, L. & Sburlino, G. (2005). Valutazione della qualità e dello stato di conservazione degli ambienti litoranei: l'esempio del S.I.C. "Penisola del Cavallino: biotopi litoranei" (Venezia, NE-Italia). *Fitosociologia*, Vol. 42, pp. 3-13.

Buffa, G.; Filesi, L.; Gamper, U. & Sburlino, G. (2007). Qualità e grado di conservazione del paesaggio vegetale del litorale sabbioso del Veneto (Italia settentrionale). *Fitosociologia*, Vol. 44, pp. 49-58.

Buffa, G. & Lasen, C. (2010). *Atlante dei siti Natura 2000 del Veneto*, Regione del Veneto - Direzione Pianificazione Territoriale e Parchi, Venezia.

Buffa, G. & Villani, C. (2012). Are the ancient forests of the Eastern Po plain large enough for a long-term conservation oh herbaceous nemoral species? *Plant Biosystems*, in press.

Cadenasso, M.L. & Pickett, S.T.A. (2001). Effect of edge structure on the flux of species into forest interiors. *Conservation Biology*, Vol. 15, pp. 91-97.

Carboni, M.; Carranza, M.L. & Acosta, A. (2009). Assessing conservation status on coastal dunes: A multiscale approach. *Landscape and Urban Planning*, Vol. 91, pp. 17–25

Carranza, M.L.; Acosta, A.T.R.; Stanisci, A.; Pirone, G. & Ciaschetti, G. (2008). Ecosystem classification for EU habitat distribution assessment in sandy coastal environments: An application in central Italy. *Environ. Monit. Assess.*, Vol. 140, pp. 99–107.

Chapman, V. J. (1976). *Coastal Vegetation* (2nd Ed.), Pergamon Press, Oxford.

Charlier, R.H.; Chaineux, M.C.P. & Morcos, S. (2005). Panorama of the history of coastal protection. *Journal of Coastal Research*, Vol. 21, pp. 79–111.

Chiaudani, A. (2008). *Agroclimatologia statica e dinamica del Veneto. Analisi del periodo 1956-2004*, Tesi di Dottorato di Ricerca in Scienze delle Produzioni Vegetali, indirizzo Agronomia Ambientale, Università degli Studi di Padova, Padova.

Cingolani, A.M.; Vaieretti, M.V.; Gurvich, D.E.; Giorgis, M.A. & Cabido, M. (2010). Predicting alpha, beta and gamma plant diversity from physiognomic and physical indicators as a tool for ecosystem monitoring. *Biological Conservation*, Vol. 143, pp. 2570–2577.

Cole, D.N. (1995). Experimental trampling of vegetation. II. Predictors of resistance and resilience. *Journal of Applied Ecology*, Vol. 32, pp. 215-224.

Curr, R.H.F.; Koh, A.; Edwards, E.; Williams, A.T. & Daves, P. (2000). Assessing anthropogenic impact on Mediterranean sand dunes from aerial digital photography. *Journal of Coastal Conservation*, Vol. 6, pp. 15–22.

Dal Cin, R. & Simeoni, U. (1994). A model for determining the classification, vulnerability and risk in the Southern coastal zone of the Marche (Italy). *Journal of Coastal Research*, Vol. 10, pp. 18–29.

Davenport, J. & Davenport, J.L. (2006). The impact of tourism and personal leisure transport on coastal environments: a review. *Estuarine, Coastal and Shelf Science*, Vol. 67, pp. 280–292.

Defeo, O.; McLachlan, A.; Schoeman, D.S.; Schlacher, T.A.; Dugan, J.; Jones, A.; Lastra, M. & Scapini, F. (2009). Threats to sandy beach ecosystems: A review. *Estuarine, Coastal and Shelf Science*, Vol. 81, pp. 1–12.

Dugan, J.E. & Hubbard, D.M. (2010). Loss of Coastal Strand Habitat in Southern California: The Role of Beach Grooming. *Estuaries and Coasts*, Vol. 33, pp. 67–77.

Dugan, J.E.; Hubbard, D.M.; Rodil, I.; Revell, D.L. & Schroeter, S. (2008). Ecological effects of coastal armoring on sandy beaches. *Marine Ecology*, Vol. 29, pp. 160–170.

EEA (1995). *Europe's environment: the Dobříš assessment*, EEA, Copenhagen.

EEA (1999). *State and pressures of the marine and coastal Mediterranean environment*, E. Papathanassiou & G. P. Gabrielidis (Eds.), European Environment Agency, Environmental assessment series No. 5, Available from http://reports.eea.eu.int/ENVSERIES05/en/envissue05.pdf

E.C., DG XXIII (1998). *Fact and figures on the Europeans on holiday 1997-98*, Eurobarometer, no. 48, Brussels.

ESRI (2008). *ArcGIS 9.3*, Environmental Systems Research Institute, Redlands, California.

Feola, S.; Carranza, M.L.; Schaminée, J.H.J.; Janssen, J.A.M. & Acosta, A.T.R. (2011). EU habitats of interest: an insight into Atlantic and Mediterranean beach and foredunes. *Biodivers. Conserv.*, Vol. 20, pp. 1457–1468.

Fernández-González, F.; Loidi, J. & Moreno Sainz, J.C. (2005). Impacts on plant biodiversity, In: *Evaluación preliminar de los impactos en España por efecto del cambio climático*, Moreno J.M. (ed.), pp. 183-247, Ministerio de Medio Ambiente, Madrid.

Forman, R.T.T. (1995). *Land mosaics: the ecology of landscapes and regions*, Cambridge University Press, New York.

Forman, R.T.T. (1996). *Land Mosaic*, Cambridge University Press, Cambridge.

Forman, R. & Godron, M. (1986). *Landscape ecology*, Wiley & Sons, New York.

Franklin, J.F.; Cromack, K.; Denison, W.; McKee, A.; Maser, C.; Sedell, J.; Swanson, F. & Juday, G. (1981). *Ecological characteristics of old-growth Douglas-fir forests*, USDA Forest Service General Technical Report PNW-1, 18, Pacific Northwest Forest and Range Experiment Station, Portland, Oregon.

Frederiksen, L.; Kollmann, J.; Vestergaard, P. & Bruun, H.H. (2006). A multivariate approach to plant community distribution in the coastal dune zonation of NW Denmark. *Phytocoenologia*, Vol. 36, pp. 321-342.

Gamper, U. (2002). *Caratteristiche ecologiche della vegetazione a carattere mediterraneo presente sul litorale sedimentario nord-adriatico (Veneto–NE-Italia), con particolare riguardo alle problematiche di conservazione della biodiversità fitocenotica*, Tesi di Dottorato in "Biologia ed ecologia vegetale in ambiente mediterraneo", Università di Catania.

Gamper, U.; Filesi, L.; Buffa, G. & Sburlino, G. (2008). Diversità fitocenotica delle dune costiere nord-adriatiche 1 – le comunità fanerofitiche. *Fitosociologia*, Vol. 45, pp. 3-21.

Garbari, F. (1984). Aspetti della flora e della vegetazione delle nostre coste marine. *Agricoltura Ambiente*, Vol. 23, pp. 45-48.

Géhu, J.M. & Géhu, J. (1980). A methodology for the objective evaluation of natural environments. Coastal examples, *Third Seminary of Applied Phytosociology "Biological Evaluation of the Territory through Biocenotic Indices"*, Institute Européen d'Ecologie, Metz, pp. 75–94.

Géhu , J.M.; Scoppola, A.; Caniglia, G.; Marchiori, S. & Gehu-Franck, J. (1984). Les systèmes végétaux de la còte nord-adriatique italienne. Leur originalité à l'échelle européenne. *Doc. Phytosoc.*, N.S. Vol. 8, pp. 486-558.

Géhu, J.M. & Biondi, E. (1994). Anthropization of Mediterranean dunes, In: *Alterazioni Ambientali ed Effetti sulle Piante*, Ferrari, C.; Manes, F. & Biondi, E. (Eds.), pp. 160–176, Edagricole, Bologna.

Grime, J.P. (2002). *Plant Strategies, Vegetation Processes, and Ecosystem Properties*, Wiley & Sons, Chichester.

Grunewald, R. & Schubert, H. (2007). The definition of a new plant diversity index "H' dune" for assessing human damage on coastal dunes—Derived from the Shannon index of entropy H'. *Ecological Indicators*, Vol. 7, pp. 1–21.

Gustafson, E. J. (1998). Quantifying Landscape Spatial Pattern: What Is the State of the Art? *Ecosystems*, Vol. 1, no. 2, pp. 143-156.

Hermy, M. & Cornelis, J. (2000). Towards a monitoring method and a number of multifaceted and hierarchical biodiversity indicators for urban and suburban parks. *Landsc. Urban Plan.*, Vol. 49, pp. 149–162.

IUCN, (2006). *Guidelines for Using the IUCN Red List Categories and Criteria. Version 6.2.* Prepared by the Standards and Petitions Working Group of the IUCN SSC Biodiversity Assessments Sub-Committee in December 2006. Available from http://app.iucn.org/webfiles/doc/SSC/RedList/RedListGuidelines.pdf

Klijn, F., & Udo de Haes, H.A. (1994). A hierarchical approach to ecosystems and its implications for ecological land classification. *Landscape Ecology*, Vol. 9, pp. 89–104.

Kutiel, P.; Peled, Y. & Geffen, E. (2000). The effect of removing scrub cover on annual plants and small mammals in a coastal sand dune ecosystem. *Biological Conservation*, Vol. 94, pp. 235–242.

Levin, S.A. (1992). The problem of pattern and scale in ecology. *Ecology*, Vol. 73, pp. 1943-1967.

Loidi, J. (1994). Phytosociology applied to nature conservation and land management. In: Song, Y.; Dierschke, H. & Wang, X. (Eds.), Proceed. 36th IAVS Symp. in Shanghai, East China Normal Univ. Press.

Lomba, A.; Alves, P. & Honrado, J. (2008). Endemic Sand Dune Vegetation of the Northwest Iberian Peninsula: Diversity, Dynamics, and Significance for Bioindication and Monitoring of Coastal Landscapes. *Journal of Coastal Research*, Vol. 24, no. 213 (Supplement 20), pp. 113-121.

Lopez, R.D. & Fennessy, M.S. (2002). Testing the floristic quality assessment index as an indicator of wetland condition. *Ecol. Appl.*, Vol. 12, pp. 487–497.

Lorenzoni, G.G. (1983). Il paesaggio vegetale nord Adriatico. *Atti Mus. civ. St. nat. Trieste*, Vol. 35, pp. 1-34.

Lortie, C.J. & Cushman, J.H. (2007). Effects of a directional abiotic gradient on plant community dynamics and invasion in a coastal dune system. *Journal of Ecology*, Vol. 95, pp. 468–481.

Margules, C.R. & Pressey, R.L. (2000). Systematic conservation planning. *Nature*, Vol. 405, pp. 243–252.

Martínez, M.L.; Psuty, N.P. & Lubke, R.A. (2004). A perspective on coastal dunes, In: *Coastal dunes. Ecology and Conservation*, Martínez, M.L. & Psuty, N.P. (Eds.), pp. 3-10, Ecological Studies, Vol. 171, Springer , Berlin.

Matson, B.E., & Power, R.G. (1996). Developing an ecological land classification for the Fundy Model Forest, south-eastern New Brunswick, Canada. *Environmental Monitoring and Assessment*, Vol. 39, pp. 149–172.

McGarigal, K. & Marks, B.J. (1995). *FRAGSTATS: spatial analysis program for quantifying landscape structure.* USDA Forest Service General Technical Report PNW-GTR-351, USDA, Washington DC.

McKinney, M.L. (2002). Urbanization, biodiversity, and conservation. *BioScience*, Vol. 52, pp. 883–890.

McLeod, K.; Lubchenco, J.; Palumbi, S. & Rosenberg, A. (2005). *Scientific Consensus Statement on Marine Ecosystem-Based Management*, Communication Partnership for Science and the Sea, 2005; Available from http://compassonline.org/marinescience/solutions_ecosystem.asp

Miles, J.R.; Russell, P.E. & Huntley, D.A. (2001). Field measurements of sediment dynamics in front of a seawall. *Journal of Coastal Research*, Vol. 17, pp. 195–206.

Mücher, C.A.; Bunce, R.G.H.; Hennekens, S.M. & Shaminée, J.H.J. (2004). *Mapping European habitat to support the design and implementation of a pan-European network: the PEENHAB project*, Alterra report 952, Wageningen.

Nordstrom, K.F. (2000). *Beaches and Dunes on Developed Coasts*. Cambridge University Press, Cambridge, pp. 338.

Nordstrom, K. F.; Gamper, U.; Fontolan, G.; Bezzi, A. & Jackson, N.L. (2009). Characteristics of Coastal Dune Topography and Vegetation in Environments Recently Modified Using Beach Fill and Vegetation Plantings, Veneto, Italy. *Environmental Management*, Vol. 44, pp. 1121–1135.

Noss, R.F. 1990. Indicators for monitoring biodiversity: a hierarchical approach. *Conservation Biology*, Vol. 4, pp. 355–364.

O'Neill, R.V.; DeAngelis, D.L.; Allen, T.F.H. & Waide, J.B. (1986). A hierarchical concept of ecosystems. *Monographs in Population Biology*, 23, Princeton University Press, Princeton.

O'Neill, R.V.; Johnson, A. R. & King, A.W. (1989). A hierarchical framework for the analysis of scale. *Landscape Ecology*, 3: 193–205.

Patton, D.R. (1975). A diversity index for quantifying habitat edge. *Wildl. Soc. Bull.*, Vol. 3, pp. 171-173.

Pignatti, S. (1959). Ricerche sull'ecologia e sul popolamento delle dune del litorale di Venezia. Il popolamento vegetale. *Boll. Mus. Civ. Venezia*, Vol. 12, pp. 61-142.

Pignatti, S. (2009). Com'è triste Venezia, soltanto mezzo secolo dopo. *Parchi*, Vol. 58, pp. 59-70.

Poldini, L.; Vidali, M. & Fabiani, M.L. (1999). La vegetazione del litorale sedimentario del Friuli-Venezia Giulia (NE Italia) con riferimenti alla regione alto-Adriatica. *Studia Geobotanica*, Vol. 17, pp. 3-68.

Polli, S. (1970). *Tabelle di previsione delle maree per Trieste e l'Adriatico Settentrionale per l'anno 1971*. Istituto Talassografico Sperimentale Trieste, Trieste, pp. 20.

Ragazzi, F.; Vinci, I.; Garlato, A.; Giandon, P. & Mozzi P. (Eds.) (2005). *Carta dei suoli del bacino scolante in laguna di Venezia*. ARPAV Osservatorio Regionale Suolo, Castelfranco Veneto, pp. 399.

Ragazzi, F. & Zamarchi, P. (Eds.) (2008). *I suoli della provincia di Venezia*. Servizio Geologico e Difesa del Suolo della Provincia di Venezia, ARPAV Osservatorio Regionale Suolo, Castelfranco Veneto, pp. 268.

Ranwell, D. (1972). *Ecology of Salt Marshes and Sand Dunes*. Chapman and Hall, London, pp.258.

Rescia, A.J.; Schmitz, M.F.; Martín de Agar, P.; de Pablo, C.L. & Pineda, F.D. (1997). A fragmented landscape in northern Spain analyzed at different spatial scales: implications for management. *J. Veg. Sci.*, Vol. 8, pp. 343–352.

Richardson, D.M.; Pyšek, P.; Rejmánek, M.; Barbour, M.G.; Panetta, D.F. & West C.J. (2000). Naturalization and invasion of alien plants: concepts and definitions. *Diversity and Distributions*, Vol. 6, pp. 93-107.

Rivas-Martinez, S. (2008). Global bioclimatics (Clasificación bioclimatica de la Tierra). Version 1/12/2008. Available from http://www.globalbioclimatics.org

Rodgers, J.C. (2002). Effects of human disturbance on the dune vegetation of the Georgia Sea Islands. *Phys. Geogr.*, Vol. 23, no. 1, pp. 79–94.

Rodgers, J.C. & Parker, K.C. (2003). Distribution of alien plant species in relation to human disturbance on the Georgia Sea Islands. *Diversity and Distributions*, Vol. 9, no. 5, pp. 385–398.

Rowe, J.S. (1996). Land classification and ecosystem classification. *Environmental Monitoring and Assessment*, Vol. 39, pp. 11-20.

Sala, O.E.; Chapin, F.S. III; Armesto, J.J.; Berlow, E.; Bloomfield, J.; Dirzo, R.; Huber-Sanwald, E.; Huenneke, L.F.; Jackson, R.B.; Kinzig, A.; Leemans, R.; Lodge, D.M.; Mooney, H.A.; Oesterheld, M.; Poff, N.L.; Sykes, M.T.; Walker, B.H.; Walker, M. & Wall D.H. (2000). Global biodiversity scenarios for the year 2100. *Science*, Vol. 287, pp. 1770-1774.

Sax, D.F. & Gaines, S.D. (2003). Species diversity: from global decrease to local increase. *Trends in Ecology and Evolution*, Vol. 18, pp. 561-566.

Sburlino, G.; Buffa, G.; Filesi, L. & Gamper U. (2008). Phytocoenotic originality of the N-Adriatic coastal sand dunes (Northern Italy) in the European context: The Stipa veneta-rich communities. *Plant Biosystems*, Vol. 142, pp. 533-539.

Schlacher, T.A.; Dugan, J.; Schoeman, D.S.; Lastra, M.; Jones, A.; Scapini, F.; McLachlan, A. & Defeo, O. (2007). Sandy beaches at the brink. *Diversity and Distributions*, Vol. 13, pp. 556–560.

Schlacher, T.A.; Schoeman, D.S.; Dugan, J.; Lastra, M.; Jones, A.; Scapini, F. & McLachlan A. (2008). Sandy beach ecosystems: Key features, management challenges, climate change impacts, and sampling issues. *Marine Ecology*, Vol. 29, pp. 70–90.

Seabloom, E.W. & Wiedemann, A.M. (1994). Distribution and effects of Ammophila breviligulata Fern. (American beachgrass) on the foredunes of the Washington coast. *Journal of Coastal Research*, Vol. 10, pp. 178–188.

Sims, R.A.; Corns, I.G.W. & Klinka, K. (1996). Global to local: Ecological land classification - Introduction. *Environ. Monitor. Assessment*, Vol. 39, pp. 1-10.

Sokal, R.R. & Rohlf, F.J. (1995). *Biometry*.W.H. Freeman, New York.

Stanisci, A.; Feola, S. & Blasi, C. (2005). Map of vegetation series of Ponza island (central Italy). *Lazaroa*, Vol. 26, pp. 93-113.

Sykes, M.T. & Wilson, J.B. (1991). Vegetation of a coastal sand dune system in southern New Zealand. *J. Veg. Sci.*, Vol. 2, pp. 531–538.

Tüxen, R. (1956). Die heutige potentielle natürliche Vegetation als Gegenstand der Vegetationskartierung. *Angewandte Pflanzensoziologie* (Stolzenau), Vol. 13, pp. 5–42.

UNCED (1992). *United Nations Conference on Environment and Development, Agenda 21, Chapter 17: Protection of the Oceans, All Kinds of Seas, Including Enclosed and Semi-Enclosed Seas, and Coastal Areas and the Protection, Rational Use and Development of Their Living Resources.* United Nations Divison for Sustainable Development, New York, pp. 42.

van der Maarel, E. (2003). Some remarks on the functions of European coastal ecosystems. *Phytocoenologia*, Vol. 33, pp. 187-202.

Werner, P.C.; Gerstengarbe, F.W.; Friedrich, K. & Oesterle, H. (2000). Recent climate change in the North Atlantic/European Sector. *Int. Journal of Climatology*, Vol. 20, no. 5, pp. 463-471.

Westhoff, V. & van der Maarel, E. (1973). The Braun–Blanquet approach. In: *Ordination and classification of communities*, Whittaker, R.H. (Ed.), pp. 617–726, Handbook of vegetation science, Vol. 5, Junk, The Hague.

Wilson, B. & Sykes, M.T. (1999). Is zonation on coastal sand dunes determined primarily by sand burial or salt spray? A test in New Zealand dunes. *Ecol. Lett.*, Vol. 2, pp. 233–236.

Zhang, K.Q.; Douglas, B.C. & Leatherman, S.P. (2004). Global warming and coastal erosion. *Climate Change*, Vol. 64, pp. 41–58.

Zonneveld, I. S. (1995). *Land ecology: An introduction to landscape ecology as a base for land evaluation, land management and conservation*, SPB Academic Publishing, Amsterdam.

Zunica, M. (Ed.) (1971). *Evoluzione dei litorali dal Tagliamento all'Adige con particolare riguardo ai lidi della Laguna di Venezia (Relazione definitiva)*. Min. Lav. Pubbl. Com. St. Provv. Venezia, Padova.

Chapter 14

Creation of the New System of Management of Important Transition Zones in the Nature

Jing Tan, Tian Yan, Wang Shang-Wu
and Jie Feng

Additional information is available at the end of the chapter

http://dx.doi.org/10.577246057

1. Introduction

The construction of ecology management system in natural ecological reserve (reserve) has been a long-term and complex project led by Chinese government. Its fundamental purpose is to protect natural biodiversity, save endangered species, maintain ecology balance, explore the means and approach of using rationally natural resources and ultimately to promote the harmonious development between natural ecosystems and human socio-economy. International management and protection of reserve is a multi-level ecosystem management and protection including the core zone 、experimental zone 、buffer transition zone 、business district and living area. This idea has great significance to guide the management system construction of natural ecosystem in China.

Chinese western nature reserve system is well known for its characteristics such as large-scale, high-altitude, long-ecological transition zone and system vulnerability, which leading to the arduous task for government in natural ecological reserve protection and management, so the government usually attached more importance to the protection of the core and experimental zone, while relatively ignored the management and operation to transition zone. This situation doesn't meet the basic principles of natural ecological system protection and management of systematization, safety and integrity. Therefore, the most urgent priority is how to change the reserve excessive weak management present situation, save reserve biodiversity crisis, alleviate government management function pressure in future, which has important practical significance and long-term history meaning to ensure the natural ecological system security and integrity.

2. Section I The crisis and management system defects of the important transition zones of western nature ecology reserve

2.1. The crisis and pressure of the reserve transition zones

The transition zone is one of the important components of varieties of living organisms and the genetic variation patterns combination in the reserve biodiversity system. The 1997 statistics shows that the scale and amount of typical biology species transition zone in western region accounts for more than 80% of the nation totals [1], which consists of hundreds of thousands of hectares of transition zones, such as Sichuan Wo Long Nature Reserve and its transition zone with more than 20 hectares; Thousands of hectares of transition zone, such as the crossing transition zone between Sichuan MiYaLuo Natural Ecological Reserve System and Sichuan Cao PO Provincial Nature Ecology Reserve System ----- Mashan Village transition zone with 1110.2 hectares in Aba Prefecture. Sichuan Jin Fo Mashan reserve transition zone [2] with more than 900 hectares, etc.

However, these transition zones are facing double crisis in the long run, natural ecological security and system integrity crisis, the survival and development crisis from the transition zone community.

a. The Ecological Crisis Of The Reserve Transition Zone

The ecological crisis is mainly from three aspects which are as follows:

The first is the ecological crisis caused by Local residents' over-reliance on natural and ecological resources. Reserve transition zone residents are dependent on long-term eco-environmental resources for their livelihoods, thus the excessive deforestation and use leads to the fragmentation of biodiversity habitat, obvious reduction of biological species distribution, quick decreasing of herd numbers, species inbreeding and genetic heterozygosity falling; The excessive deforestation and use changed the forest composition and organic ecological relations among species, reduced the ability of resistance to diseases and pests for forest itself, increased the scale and frequency of forest diseases and pests occurrence, which made the transition zone natural forests appear many remnants, degradation even extinction. Such as recently decade, in China, 200 kinds of plants have extinct, 4,000 - 5,000 species of higher plants are facing threat, accounting for 15% -20% of total number of species. In China, the first batch rare and endangered plants published in the international convention have been up to 388 kinds of [3];

The second is ecology crisis resulted from the traditional hunting customs of community residents and the increasing demand for wild animals and plants consumer market. For a long time, because the transition zone has been non-control zone, which resulted in its acquisition and poaching behavior is always rampant. In addition to, lots of treasure herbs and edible fungi are collected, and the 20 kinds of national first level protection animals (giant panda, golden monkey, tibetan antelope and so on) are being endangered. Only in a smuggling raid inspection of 1996 in Sichuan, 146 copies of giant panda skin were seized by them [4].

The third is ecology crisis because of the community environment pollution in the transition zones. In this paper, the author's investigation indicates that the Sichuan Wang Lang national nature ecological reserve important transition zone ----- Ping Wu Guan Ba community and whose surrounding communities in Mian Yang City , and the Sichuan Miyaluo nature ecology reserve important transition zone----- Mashan in village in Aba Prefecture Li County and it's surrounding communities, face the waste and water pollution problems. And others data showed 44% of communities in the nature ecology reserve transition zones exist the garbage pollution diffusion phenomenon, 12% have the water pollution, 22% of the core area are suffering the direct threat [6] [7].

b. The Community Survival And Development Pressure Of The Reserve Transition Zone

Most of Western Reserve transition zones not only belong to minority areas which are representative of Tibetan, Qiang, Yi and so on, but also locate the poor communities in which the socio-economic conditions are relatively backward. Statistics show there are nearly 300 national standard poor counties surrounding the 926 national reserves [5]. According to the 85 investigations into the reserves and their surrounding communities, the population size is 14000 averagely in protection zone, while the transition zone is about 59000. In terms of population density, the population density of 85 protection areas and transition zones is 5.75 persons / km^2, which is 1.8 and 2.4 times compared with developed countries [8]. The survival and development pressure of transition zones is mainly from the following aspects:

The first is the pressure from cumulative increase of population. The long-term lenient family planning policy in these areas results in the obvious growth of population density and size for the three decades. Community residents have to live on subsidies relief and the original means of livelihood resulting from the inability to transfer and placement, when the government faces the pressure from community population cumulative increase.

The second is the dependence on ecological resources, excessive deforestation and hunting, which causes the area expansion of zoology forest, wasteland, mark place, and the frequent natural disasters such as regional climate, soil change and landslide. Therefore, some communities are facing the transfer pressure in the short run.

The third, Survival dilemma has resulted in the intensification of the contradictions between the conservation and community development. In some areas of the communities living in the alpine valley transition zone, farmers' own arable crop fails completely, due to water shortage, drought, soil fertility declination, decline in productivity as well as of human illegal logging and the destruction to wild animals; The lack of projects or technology, backwardness of transportation and educational and other factors make farmers be unable to save themselves; The difference of subsidies standard in different reserves transition zones cause the intensification of the contradictions among communities and governments.

2.2. The management system questions of the reserve transition zones

The geographical and scale characteristics of transition zones increase the protection difficulty of reserves. Statistics show that, by the end of 2008, China has established 2538 of different levels of nature ecology reserves, and the scale of them amounts to about 148.943 million hectares, accounting for 15.5 percent of the land area [6]. Moreover the area of protection transition zone is larger, which has far exceeded the scale and level of developed countries.

The investigations into nature ecology reserve in 1997 and 1999 , as well as "sustainable management policy research report of Chinese natural ecological conservation area" held by the Man and the Biosphere National Committee in 2000, pointed out that the majority of our reserves exist universally some problems such as poor management, irrational use of resources[8] and so on, in the ecosystems protection, promotion the ecological all-round development, scientific research, popular science and other functions. This requires we have to review from the management system of nature ecology protection.

a. The Management System Defects Of The Nature Ecology Reserve

Facing of the ecological crisis and community survival development pressure in the reserve, government should make efforts mainly from three aspects: First, the government need take leading management system, taking "rescuing protection mode, first program after construction, perfection gradually"[9] management measures to carry out mainly compulsory management and protection on the core areas; Second, the government should take system and legislation to restrict economy development of the community surrounding transition zone; Third, it's important to take community relief and subsidies measures to relieve the conflict between protection area and its surrounding communities development. For example, the protection of rare animals such as giant panda and Chinese alligator core area is a kind of typical crisis rescuing protection.

Under the guidance of "rescuing principle" imposed by the Government, the management system adopted of reserve in China is the government-led and top-down arrangement. Namely, it is the management system combining the comprehensive management with departmental management based on the category management and classification management. The so-called category management is that the area was divided into two levels including core area management and the experimental area management in line with degree of importance to protected objects; The so-called hierarchical management means that the management institution is composed of nation, province, city and county according to the representativeness, importance and crisis degree of protected objects; The so-called management combining the comprehensive management with departmental management, is according to the regulations provisions 1 , paragraph 2 1 in the law of 《People's Republic Of China Ecological Reserves Ordinance》 and the regulations paragraph 8 in the law of 《 Natural Ecological Reserve Ordinance 》: "the state environment protection administration departments shall be responsible for the comprehensive management of nature ecology reserves. Nearly ten administrative departments: State Forestry Administration, State Oceanic Administration, Ministry of Agriculture, Ministry of Construction, Ministry of Land Resources, Ministry of Water Resources and Chinese Academy of Sciences is in charge of the relevant conservation area in their respective responsibility range."

In theory, this management system arrangement is nearly comprehensive and thoughtful. In fact, it will lead to some problems such as overlapping management, cross-management, bull management and poor management in practice. First of all, a reserve can be managed simultaneously by 4 levels management agencies and multiple departments, which will lead to duties fuzzy and duplication and making responsibilities move down. But the corresponding resources and right don't move down, which leading to the separation of powers and responsibilities, management responsibility dislocation, management dislocation or management absence. As a result, this system causes high management costs and poor management effect because of coordination difficulty.

Second, category management generally just takes care of the core and experimental areas, ignoring the management and conservation of the whole natural protection system. Thus transition zone of protection system appears lots of management vacuum.

Third, hierarchical management、bull management and cross-management increase system construction costs and ineffective implementation. Different departments are in charge of the same protected area, thus they will easily make regulations from their own management or interests due to the dislocation of system or management standards introduced by them. That's to say, legislation and regulation has become the way of sharing resources and the means of vested interests distribution among government departments.

b. The Contradictions Between Ecology Protection And Community Development Intensified By The Mandatory And Exclusive Management Mechanism

The so-called "mandatory and exclusive management mechanism", established under the government-led management mode, has unfavorable effects on the protection activities. On the one hand, it take the core area as the mandatory protection and the scope of subsidies, these compensation policies objectively broke the balance among communities, which will sharpen the conflict among communities as well as the government; On the other hand, the current management system considers government as the management subject, while rules out other social members especially community residents, which will make the government locate the helpless and passive situation in the long run. This likely leads to the opposition sentiment between the reserve and communities, thus many farmers don't care the illegal behavior even develop and utilize maliciously the natural ecological resources, so management and protection effectiveness isn't satisfactory.

c. The Lack Of Scientific Partition And Systematic Concepts In Reserves

The natural ecology reserve refers to the ecology systems composed of representative, typical and integrated biological communities and the abiotic environment. In line with the biodiversity and systematic features, the protection areas is identified as the ecology system consisting of the core zone, experimental zone, buffer zone, transition zone, business district, tourist area, living service areas and other functional areas.

However, China's reserve only contains core area and experimental zone. According to the policy, only core area is the focus of compulsory protection and resource allocation. While the important ecology transition zone likely become vacuum area, because it has not been identified as the core area and the protected area. This kind of management thoughts isn't helpful for the development and balance of reserve ecology system, especially for the protection of important transition zone in the ecology system, which impacts directly on the integrity of the structure and function of ecology system.

Therefore, how to bring important transition zones into protection system? The author proposed the management system which introduces "Agreement Protection" mechanism based on project and brought farmers and social organizations into the legal management subjects, and they will participate in the protection with government together. This system can stimulate effectively the protection initiative and enthusiasm of community and social, ease the contradictions between the ecology protection and community development and functions pressure of government, ultimately, can resolve effectively the problem of subject vacancy of management system in western natural ecological reserve.

2.3. Creation the new management system of important transition zones in Chinese Western nature ecology reserves

a. The Basic Idea Of Creating Co-Management System Of The Reserve Transition Zone

Based on reserve transition zone crisis and the existing problems of reserve management system, firstly, the author believes that the reserve transition zone should be brought into the scope of protection and management to ensure the integrity of the ecology protection system; Secondly, community-based organizations and farmers in the transition zone also should be considered as the legal protection subject to make up for management system subject vacancy; Thirdly, on the condition of not changing ownership of state-owned ecological forest , let community-based organizations and farmers participate in the transition zone ecological forest management by the way of signing Agreement Protection based on projects. Simultaneously, in nature ecology Protection projections, entitle them some privileges such as information, participation, supervision and decision-making power to improve their host status and enhance responsibility consciousness in ecological protection; Create the new system (common management) of ecological protection in reserve transition zone (see Figure 1).

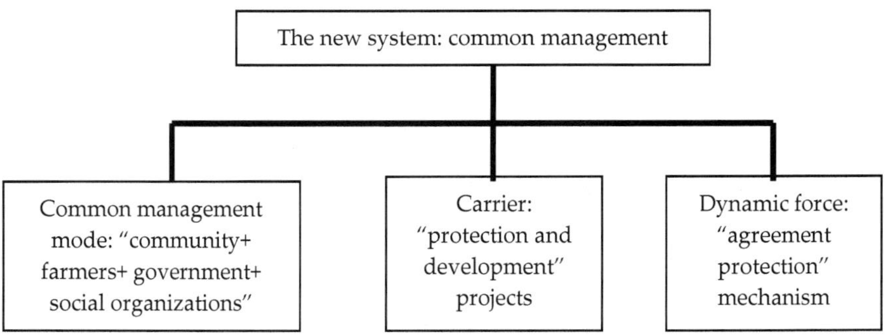

Figure 1. Common management system in natural protection reserve transition zone

b. Creation The New Mode Of Common Management: "Community+ Farmers+ Government+ Social Organizations"

The so-called common management mode: "community + farmers + government + social organizations" is that the government and community members consider the "Agreement Protection" mechanism as driving force to participate in jointly management such as the design, decision making, implementation, supervision and evaluation based on projects. The content of " Agreement Protection" agreed the common management rights, responsibilities and interests among "community and famers"(party a) 、 "government" (party b) and social organizations (third party), to form the interests joint through the resources integrity way, thus, the protection and construction tasks will be completed more easily. The focus of common management is target consistenc, interest-sharing, joint participation and risk-sharing. Compared with government-led mode, this mode has the following features:

1. Relying on community development projects

Common management mode regards a "Agreement Protection " projects as carrier, project target covers the transition zone protection target and community development goal, in details, including the international protection and development programs, the state development and poverty alleviation projects and new rural construction projects, especially GET (the global environment facility) aid projects (see Table 1).Projects aid agencies may be government and other social welfare organizations at home or abroad.

Projects type	Projects breakthrough point	Remarks
Economic development projects	Economic forest construction and its application; Ecological farming base construction; Ecological farm households tourism economic construction; Alpine ecological flowers construction; Special economic development projects	All kinds of fruit and vegetable planting base, medicinal plants base; Eco-bee breeding base; ecological poultry breeding base; Flower type ornamental and consumer base; Prickly ash and silkworm base, etc
Alternative Resource projects	Promotion and utilization of solar energy Digester construction and use Energy saving and environmental protection of life and production of construction Life and production construction of energy conservation and environmental protection	
Circulation resource industry construction projects	"Three links", human and livestock drinking water projects	"Three links" includes Water, electricity and road

Table 1. The coverage of community development projects

2. The main functions of "Agreement Protection" projects (see Table 2).

The functions of "Agreement Protection" projects	Specific tasks
The biodiversity management and protection of reserve transition zone	Eliminating of illegal logging and dredging; Eliminating illegal fishing; Eliminating foreign sabotage; Firewood cutting in accordance with provisions; Collecting the understory of grass and fruit, fungi, ferns and other non-wood products in accordance with regulations; Restrictive hunting wild boar and other destructive animals
Communities development	Infrastructure construction: human and livestock drinking project and new energy promotion, etc.; Rural economy construction: courtyard, fruit trees, rubber, walnut and other with local characteristics economic construction; Breeding and its processing; The new rural construction: electric light, TV ,phone, ecological toilet, rural cultural center construction, etc. The rural new practical technical training: planting and breeding industry, comprehensive prevention and control of disease, migrant workers' skills training, women's ability construction, etc.

Table 2. The main functions of "Agreement Protection" projects

3. The subjects' roles and responsibilities of co-management mode

The subjects of the common management mode include community-based organizations, farmers, government and social organizations. Government and social organizations are advocates and supporters of "Agreement Protection" projects. Community-based organizations and farmers are supporters and executors of "Agreement Protection" projects. First, the main responsibilities of the advocates and supporters: selection and determination of project, setting goals, drawing up jointly tasks with community organizations, providing project-related fund, materials and technical services; Second, community organizations and farmers considered as implementers participate in jointly the construction of project's implementation plan, management mechanism and management system with the advocates and supporters. Third, the advocates and supporters provide services such as guidance, service, monitoring and evaluation on the progress and performance of project of "Agreement Protection" projects (see Figure 2).

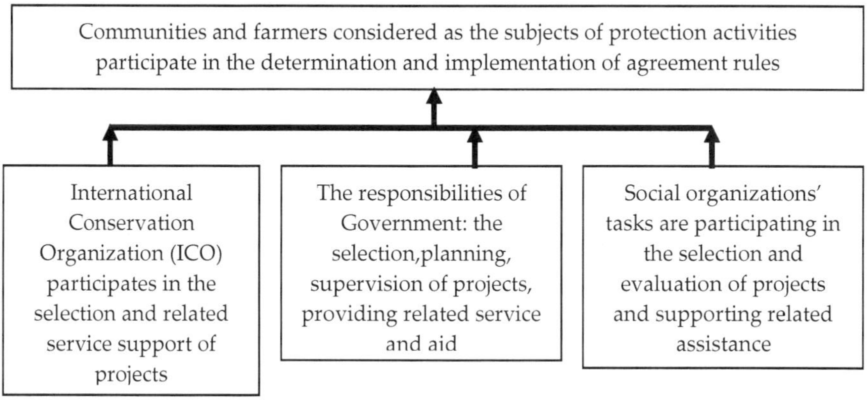

Figure 2. Subjects' responsibilities of co-management mode

c. Establishment the Related Mechanism Of Co-Management "Agreement Protection"

1. Establishment the "Agreement Protection" Mechanism Of Co-Management

The so-called "Agreement Protection" mechanism of co-management doesn't change the structure of forest rights, it make ecology forest management rights and protection rights delegated or transferred to the communities and any other organizations or individuals who are willing to assume the protection responsibility to ensure community's development and security and integrity of reserve ecosystem. This kind of mechanism sets responsibilities, rights and benefits of interested parties in the process of management and protection by means of agreement, forms the co-management mechanism based on community organizations and farmers, supplemented by Government and other organizations.

2. Establishing various forms of incentive and restraint mechanisms based on local conditions

In accordance with economic development regulations, under the condition of avoiding greater risk, to meet effectively personal interests is the direct driving force of promoting economic development. Therefore, the effective compensation and incentive way has positive significance to encourage communities and farmers to participate in the protecting action actively.

Compensation mechanism is a kind of coordinating manner for government and social organizations to support and relieve community livelihoods and development. So how to transfer the compensation way into the incentive and constraint functions is critical to encourage community and farmers to complete initiatively the management and protection task.

d. Establishment Of Multi-Level Education And Technical Training Mechanism

Education and technical training plays a very important role for the "Agreement Protection" mode. The persistence of farmers' protection consciousness, awareness and conservation actions on the transition zone usually can be reflected after education and training. For example, the provisions of contents and frequency of education publicity and the technical training in Mashan Village "Agreement Protection" project have achieved remarkable effect. Through education and technical training, farmers can really feel the importance of conservation actions on the ecologyl safety and community economic development, what's more, they will truly feel a range of democratic rights and master consciousness in the community ecology protection activities. Therefore, the establishment of education and technical training mechanism provide objectively the most fundamental and powerful guarantee for the continuous process of protection action. In summary, the root of management and protection issues in reserve transition zone is existing defects of management system. Through the establishment of "Agreement Protection" mechanisms and co-management mode, the management and protection of reserve transition zone was transferred to communities and farmers. This kind of new system is not only helpful to transfer resources and role of communities and other social organization, but also useful to relieve the pressure from protection function of government. So it can effectively make up for the poor management and the absence of subjects in the management system of nature ecology reserve. Therefore, it is the new breakthroughs for further reformation of natural protection management system.

3. Section II The "Agreement Protection" mechanism and pattern of Western conservation transition zone

3.1. Background

a. Overview of the "agreement to protection"

The "agreement to protect" research on Mashan Village of Li County, in Aba State, Sichuan Province(Hereinafter called as " agreement to protect "research) is raised by "Beijing Natural Landscape Protection Center"(Hereinafter referred to as "Landscape").It advocates and encourages local governments to transfer collective woodland protection, and lets the community and the villagers participate in the protection action. The " agreement to protect "based on various incentives such as ecological compensation is aimed to expand the scope of protection, ease the contradiction between protection and economic development, and finally promote community socio-economic development.

Based on the "Mutual agreement protection" signed by the Forestry Bureau of Li County and the Landscape, the Mashan Village is chosen as the pilot village of "agreement to protect". In December 2008, Forestry Bureau of Li County (Party A) and Mashan Village (Party B) signed the "agreement to protect" which continued from December 2008 to December 2010. According to the research requirements and conventions, we made mid-term evaluation in November 2009 and annual assessment in August 2010.

b. The nature and purpose of the study

Under the agreement, the implementations of the project are local government (Party A - Bureau of Forestry) and community residents (Party B - Horse Village community), and also invite independent experts to Party A, Party B and the third party (the "landscape" donors) group involved in the assessment. The study includes the implementation of the matters specified, the agreement's impact on community social and economic development ,and the agreement's effectiveness of community-based biodiversity protection etc..

There are three aspects of the study. Firstly, the study evaluated the effectiveness of the biodiversity conservation protection and socio-economic development; secondly, the study checked the process of the implementation, and provided an important reference for sustainable planning; thirdly, the study summarized the experiences and lessons of the protocol protected mode, timely detected problems and proposed recommendations for improvement, and finally provided technical support and enlightenment for the mode's demonstration and promotion.

c. The purpose of the agreement to protection

Ecology, society and economy form a complete organic system, and there are close internal relations and interactions among them. Therefore, the eco-environmental protection and economic and social development of the Mashan Village should promote each other and complement each other. The research on "Agreement to protect" is based on interactions of ecology, society and economy, and the decisions are made for the following reasons.

1. To regulate and restrain the villagers' behavior we should strengthen the protection awareness of the villagers, and let them recognize the vital and long-term interests they can get from the protection of the ecology; Only the villagers agree with the purpose of protection, will they take the appropriate measures and actions.
2. To improve the quality of the villagers' living environment by the protection of the natural environment and the using of the new energy, we can improve the living and health conditions of Mashan Village ,which will contribute to improving the quality of the living environment of villagers.
3. To provide financial compensation and maintain social stability. On the basis of the protection obligations and responsibilities on the villagers, we should provide financial compensation and spiritual encouragement to their protection behaviors. With a variety of compensation, we can resolve the contradiction between the protection and development, and fight for the win-win between community development and ecological protection.
4. 4) To protect and inherit Qiang culture. When it was designed, the agreement focused on the characteristics of the Qing culture and established cultural activity funds for purchasing necessary items, which provided important safeguard and support to the protection and promotion for national culture.

3.2. Natural and social status of Mashan Village

Observed from the perspective of the natural ecosystem's integrity and security, Mashan Village in Li County of Aba autonomous prefecture, Sichuan province, is a transition zone with great conservation value. Mashan Village is located in the arid valleys of the Hengduan

Mountainous and mountain forest Transitional Zone, the community is closed to the Wolong National Nature Ecological Reserve of Sichuan bordering on the Miyaluo Nature Ecological Reserve and Caopo Provincial Nature Ecological Reserve of Sichuan, which belongs to communities of Miyaluo Provincial Nature Ecological Reserve.

The topography in Mashan Village has created more abundant types of natural ecosystems, and gave birth to the rich resources of the species. Survey shows, there are 13 species of state emphatically protected plants and rare and endangered plants, 21 species of state-level protected wild animals and 18 species of state-level protected birds in nature reserve. The species of wild plants and animals under the first-grade state protection were *Fargesia spathacea Franch, Taxus chinensis, Rhinopithecus roxellanae, Ailuopoda melanolcuca, Panthera pardus, Panthera uncial, Moschus berezovskii, Moschus sifanicus, Budorcas taxicolor, Aqulia chrysaetos, Bonasa sewerzowi, Tetraophasis obscures, Lophophorus lhuysii* etc.

Mashan Village is a mixed village with Tibetan, Yi and Qiang where farmers rely on the natural environment, vegetation and animals, all of which are living resources of their lives. The nature ecological environment of community due to long-term deforestation, resulting in serious damage to vegetation, landslides, debris flows and other natural disasters, hillside planted almost reaped nothing. Over the past decade, local farmers mainly rely on government assistance. The farmers have not received ecological protection education and the technical training of the management and protection woodland, their unconscious awareness of participation in the protection of the ecological, turned a blind eye to community poaching, or even involved in illegal logging, all of which plunged the Miyaluo natural ecosystems security into a crisis.

How to change this situation, by whom or in what way to restore these damaged vegetation system? This section from the "agreement protection" mechanism and natural ecosystem co-management pattern of innovative practice, revealed that by optimizing the relations of production and a wider mobilization of social forces to participate in completion of the protection and management of natural ecosystems is the key to institutional innovation of the natural protection system.

3.3. "Agreement Protection" pattern and its characterisitics

"Agreement protection" is a new mechanism for biodiversity conservation. The concept is proposed by the Conservation International (Conservation International CI). Agreement protection is the protection right that aimed at separating ownership from managerial authority of protective area on the premise of unchanged ownership of woodland. And then participation management and protection of subject whose responsibilities, powers and interest are consolidated in the form of "agreement", and protection right and managerial authority are turned over to communities and farmers which promised to undertake management and the responsibility of protection, as well as any organization willing to undertake the responsibility of protection, and ultimately we achieve the goal of effective protection and development of natural ecosystems.

The author participated in the baseline survey and final research assessment of the project which called "Mashan Village in Xue town of Li County wild plants and animals agreement Protection projection" (hereinafter referred to "Protection projection") in Aba autonomous prefecture, Sichuan province in 2008 and 2010, considering that the "Protection projection "is a beneficial exploration of Nature Ecological Reserve transition zone joint management and protection system innovation, and very worthy of studying and summarizing.

a. "Agreement Protection" Pattern and Its Characteristics

Mashan Village's "Protection projection" operation pattern is based on the four subject containing community, farmers, government and society organizations which participate in operation pattern of joint management and protection (hereinafter referred to as "co-participatory pattern"). The pattern takes "Protection projection" as a support, the "agreement protection" as the ligament, agreed to communities with farmers (Party A) and Government (Party B) whose powers, responsibilities and interest in Protection projections. The basic features are as follows:

The first, the organization of the management of Protection projection is the common Management Committee including village committees and farmers representative of the Mashan Village, county-level Forestry Bureau and International Protection Organization; the second, the main role in the Protection projections: International Protection organizations are the sponsors and supporters of the project, county-level Forestry Bureau is the supporters and service providers of the project, communities and farmers are the specific actors of the protective action. The third, agreement protection content is developed by the members of the four parties and then was submitted to community farmers' general assembly for discussion to pass and practice. Fourth, "Protection" agreement clearly agreed responsibilities, rights and interest in order to make member's responsibilities and division clear. Under the framework of the agreement protection, county-level Forestry Bureau is responsible to assist the village committee of Mashan Village to build a conservation organizations and institutions including "patrolling management and protection group, "construction group of ecological forest and production forest", "oversight group of agreement Protection projection", "advocacy group of the national culture "and correlation protection institutions. International Protection Organizations and backbone of village committees are responsible for the coordination of Protection actions, and mobilize community residents' perceived protection and self-help management action and so on.

b. The Comparison of the "Agreement Protection" Pattern and the Government-led Pattern

1. Government-led pattern, in which government does as a dominant implementation of the protection and management of protected areas, and agreement protection pattern, in which the government, supporting and helping of social organizations, communities and farmers do as a dominant implementation of the protection and management of protected areas. Clearly defined communities and farmers are the main body of the natural ecological environment protection of community and the first responsible person status in the "agreement protection" pattern. The farmer should have right

relevant to the protection system to know, to speak, to participate, to make decision and management rights of democratic rights and civil rights, which completely changed the bystanders and saboteurs role of the community subject;

2. Agreement protection pattern changed the government-led functions. Government is no longer the only conservation managers in the reserve, but is important mentors, supporters and service providers of the ecological protection. Government provided adequate guidance, support and services to the protection action of communities and farmers, including construction of protection organizations, technical training, to guide management and provide appropriate compensation and activity funds. Moreover, through the guidance of the agreement Protection projection, the transition zone of protected areas is effectively integrated into the natural protection management system, which plays an important role to promote the integrity of the protected area system.
3. Agreement protection pattern introduce incentive and restraint mechanisms to encourage community farmers to join the protection action. Agreement Protection projection fully mobilize social resources ,establish various forms of incentives and constraints, assist Mashan Village community in establishing a stable monitoring and protection system, reward and punishment mechanisms implemented in the whole process of the protective action, not only have encouragement of material and spiritual dimensions but also incentives of technology, bonuses, and honors level. Moreover, it also provide with severe penalties to those who do not fulfill the agreement and not implement protection in accordance with the terms of the agreement. They can be used in the combination of incentive and restraint mechanisms, both simply and effectively.

In a word, "co-participatory protection pattern", on one hand, effectively gives the farmers a variety of democratic rights, increases their political status and enhances their sense of responsibility; on the other hand, objectively the effective means of rewards and punishments for the protective action which was smoothly carried out provide a mechanism guarantee and improve the farmers' enthusiasm, so this "agreement protection" pattern should receive good results.

3.4. "Agreement Protection "mechanism of Mashan Village

The "agreement protection" mechanism through helping the government to transfer the protection rights of woodland to the community organizations and farmers and by means of ecological compensation or a variety of other incentives promotes biodiversity protection and management of the Contract Area.

a. The Organizational Mechanisms and Decision-making Mechanism of Agreement Protection

Under the framework of agreement protection, Mashan Village firstly established organizational mechanisms and decision-making mechanism. In organization building, with the help of village committee, a organization carrier, it set up the "agreement protection" commission which is composed of governments, international protection organizations, village committees and farmers representative and strengthened the management of community protection organization. In the decision-making on major events, the village committee organizes villagers to discuss and make collective decisions, put the provisions of the agreement protection into system regulations of pacts agreed and implementation. Such as farmers can gather the fallen leaves and withered woods freely and fell the trees moderately in their own lands; through resolutions and announcements of the general assembly of farmers, require the implementation of strict state-owned forests and collective forest protection and management is required, and prohibited the community and non-community farmers into the protected areas for illegal logging.

b. According to Local Conditions, Establishing Classified Management Mechanism

There are a total area of 1110.2 hectares in the protection zone of Mashan Village, involving 502.8 hectares collective forests land and 607.4 hectares state-owned woodland. In view of woodland species and current situation, the community takes in the charge of classified management and protection mechanism of forest land. First, the community regards the collective forests and state-owned woodlands destroyed or degraded as a "reforestation land", through hands out saplings technical aid replants and protects; second, regards the collective forests and state-owned woodlands in good condition as "protection woodlands" for strict patrol and monitoring; third, the community farmers have deserted its own forest land, through compensates walnut saplings and technical guidance to plant and develop economic forest.

The local Forestry Bureau provide technical training and guidance for the three types of woodland, and international protection organizations and government should give appropriate subsidies for protecting behavior to ensure that replanted forest saplings of the survival rate could be more than 90%.

The classification management and protection mechanisms not only can achieve the goals of protective project, but also help communities to build economic self-help system to solve the contradiction between ecological protection and economic development and inspire the communities to actively participate in the awareness and responsibility of the protective actions.

c. Incentive and Restraint Mechanisms Ensure the "Protection projection" Carried Out Smoothly

Mashan Village "agreement protection" pattern runs in accordance with the principle of the "who builds, who benefits, who protects, who benefits". The incentive and restraint mechanisms are the key to ensure the protective action continued orderly development.

1. Incentives mechanisms. " Agreement protection " lay down the incentives methods: First, incentives methods such as free distribute walnut saplings and provide cultivation technology training and guidance to help farmers to develop economic forest and injected "blood transfusion and hematopoietic" mechanism for forest property rights of farmers to stimulate recovery and development of eco-forest; second,

encourage community organizations to carry out publicity education activities of conservation actions by way of bonuses and honors; third, methods such as on-site technical training and passing on management experience are used to promote ecological forest protection and forest management; fourth, reward hard-working people and punish lazy people by way of labor subsidies to strengthen the organization and institutionalization of the patrol work to ensure that regulatory work is normal and orderly development; fifth, encourage the community to carry out national characteristics cultural heritage and conservation activities by way of presenting cultural and promotional materials. Sixth, timely periodic summary assessment in accordance with the requirements of the "Agreement", give a reward and honorary title to what are effective protective behaviors and encourage farmers to develop the initiative, consciousness, honor sense, pride of the community eco-protective behaviors.

2. Constraint mechanisms. Mashan Village has developed the appropriate discipline and penalties focusing on the work of supervision and patrol team: Firstly, lay down the rules of the patrol mission and the punishment method of deforestation behavior to restrain the lazy violations. Secondly, the behavior such as those who fail to complete the task on time and timely report and process when have problem, take measures of withholding and temporary withholding subsidies or bonuses, and formulate the stipulation of reorganizing the deadline. Thirdly, appraisal protection motion regularly according to the agreement, the condition such as woodland management and protection and fulfill the guardianship duty are not in place which temporarily have not reward, pending completion of the rectification and then reissue. At last, make strict rules for collective protection activities about the number of participants (higher than 51% of the village) and the number of participation times (more than 4 times / year), and collective rewards and punishments are linked.
3. The distribution of benefits mechanism. In order to defend the collective economic sustainable development and the sustainability of community conservation actions, protocol protection provisions, the expected return of 20% of walnut grove set-aside for collective provident fund which used to guarantee the healthy and sustainable development of community protection organization in the provision of agreement protection. The expected return distribution mechanism, fully affirmed the value of the existence of collective organization, embodies the concept which the organization led farmers to implement protection actions. Such an interest distribution mechanism, increase farmers' hope and enthusiasm for expected income.

3.5. The effect evaluation of Mashan Village "Agreement Protection " pattern

5-7 August 2010, the authors carried out field research for Mashan Village "agreement protection" pattern. Before the survey they used a random number table, randomly selected 50 samples in the village 94 farmers of the General Assembly on the public, and at the same time took 30 samples as a standby; After three days of the household survey, observation, typical interviews and focusing discussion, we collected 54 valid questionnaires copies. Aiming at the discussion topics of research, the author selected 5 key indicators of data processing to form survey results.

a. "Agreement protection" pattern promote the transformation of the socio-economic status of Mashan Village

1. "Agreement protection" pattern significantly increase the protection awareness of farmers

In the investigation, the author selected two indicators to study the change of protection awareness for the farmers, awareness and value of acceptance. The awareness is measured and assessed by the Protection projection, project scope, protected plants, village regulation and so on (see Figure3). The other is described through protective object, community organization, community culture and organizational activities and so on.

The survey results show that the farmers' protection awareness changes significantly. Implementation of the agreement Protection projection made communities and farmers show the master attitude to protecting responsibility and protective behaviors. Farmers have a clear understanding of the practical significance and long-term significance of the ecological protection behavior for themselves and the environment, farmers are willing to participate in patrol work, and they reached a consensus on the validity of agreement protection at 4 aspects.

The First, the natural disasters such as landslides, mudslides can be decreased by protecting forest land.

The Second, the forest land protection is our responsibility. We should do better on our own beneficial protection actions;

The third, the "agreement protection" pattern play a role in improving the natural ecological environment of the community. After two years of hard-working, community and ecological protection zone should be able to see more wild rare animals, and promote the economic development of tourism;

The fourth, the construction of Walnut economic forest is indeed a good project can benefit for the future generations.

2. The "agreement protection" pattern promoted the construction of community organizations system

Protection organizations is the most foundational and most important basic-level organization which mobilizes community farmers' positivity of protection action, organize publicity education, consolidate community the continuity of protection action, organizes publicity education, consolidates the sustainability of community protection action, and improve the protective effect. Because of the imperfect natural resource management system, as all rural communities throughout the country in the past, Mashan Village did not have protective organizations which allow community to participate. "Agreement protection" pattern promoted the system construction of community organization (see table 3).

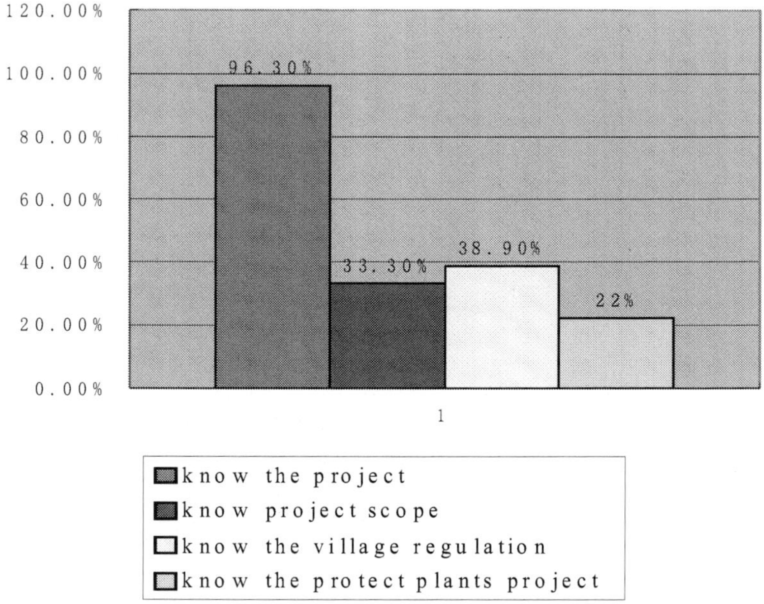

Note: Combined with questionnaire and interview survey comprehensive data, ① 96.3% of the farmers know the "agreement protection" project and its content; ② 54.5% of the farmers think that the woodland should be protected by themselves, 34% of the farmers believe that the woodland should be protected by the government and farmers together, 11.5% is not clear. ③ 61.8% of the farmers think that once found the actions such as deforestation and poaching will resolutely put a stop; Data derived from the obtained first-hand information by the authors' investigation

Figure 3. The awareness for studying protection consciousness

Method of data sources	Indicators: the construction of Organizations and institution
Interviews, Observations, Questionnaires,	The village regulation increased from 3 to 14, including the entire contents of community ecological protection and civilized behavior. Established 4 kinds of protective organizations: "the management and protection group of patrolling mountain", "ecological forest and economic forest construction group", "supervision groups of agreement Protection projection", "advocacy group of national culture". Put forward 4 systems and a require, that is "cultural propaganda system", "management system of ecological forest and economic forest construction", "supervision system", "supervision groups rules and regulations", and "a claim to the management and protection group of patrolling mountain" .

Note: Data derived from the obtained first-hand information by the authors' investigation

Table 3. Organizations and institutions building

The formulation of the organizations and systems effectively not only provide an important safeguard for participating, regulating and supervising protective actions, but also promise the way of making use of natural resources and management oversight, set up a woodland management and protection of technical training platform to improve the protection consciousness, awareness and capacity of farmers.

3. "Mashan Village" pattern improved the protection ability of farmers

The survey shows that the behaviors such as unauthorized felling bamboo, illegal hunting, forest destruction, digging herbs randomly, picking up mushrooms freely have not happened in experimental communities in 2 years; community farmers are able to moderate harvesting wood according to the provisions of the agreement protection and dependence on forest resources significantly reduced. The "agreement protection" pattern enhance the protection abilities of the community farmers (see table 4).

Method of data sources	Index: the change of the protection abilities
Interviews, Observations, Questionnaires, On-site assessments	The behaviors of random cutting scarcely appear during the term of agreement protection. 15% of the households use their own planting trees as part of live energy resources. More than 95% of the households use electricity to cook food and boil water, even a small number of villagers use solar power. Forest fires did not happen in the village during the term of agreement protection

Note: Data derived from the obtained first-hand information by the authors' investigation

Table 4. The protecting abilities of farmers

4. The "agreement protection" pattern promoted economic conditions diversification of farmers

Under the joint efforts of governments and international protection organizations ,"agreement protection " takes planting the economical trees as the breakthrough point to help the community farmers learned a variety of techniques such as woodland planting, management, disease prevention and patrolling to help farmers improve the value of knowledge of ecological forestry and the management level. In this paper, the author selected 4 indexes to research the changes of the economic development, that is the disposable income per household, living consumption expenditure per household, number of walnut trees per household and number of livestock per household(table 5).

Method of data sources	Index: the economic changes condition
Observations, Interviews, Questionnaires, On-site assessments	100 walnut trees planted per family. Farmers attach great importance to the walnut forest expected return, after two years, and every single walnut tree can increase the income of ￥200 -￥500. The average annual income of walnut tree per household can reach about ￥30 000 per household. After Walnut trees mature, the average annual income will also increase more than 10%. Approximately 85% of the rural households are no more engaged in livestock breeding, which caused the annual income will be reduced about ￥500-￥1000 per household. Reduce farming income has little effect on livelihoods while plays a significant role in protecting ecology. The structure of households' income which changes from "Logging and hunting forest and products income + original planted income + the original aquaculture income" two years ago into "conversion of cropland to forest subsidies + post-disaster reconstruction subsidies + wage income + the original pepper and other forest income". The household disposable income reached￥15,000-￥35,000 per year right now. The average household expenditure is mainly education and living expenses, and other expenditure performance was not significant

Note: Data derived from the obtained first-hand information by the authors' investigation

Table 5. The economic development changes of the farmers

The survey shows that the farmers' income structure has changed from aquaculture and migrant workers into the mainly farming gradually after the implementation of the "agreement protection " pattern.

b. The "agreement protection" pattern promoted the protection and management of natural ecosystems

1. Promoted the integrity of natural ecosystems. Transferred the transition zone of protected areas over to community to manage which can make up for the lack of traditional protected areas division and management system and also promote the returning to nature and integrity of protected areas ecosystems.
2. Fill the subject vacancy in the law. Integrating the community farmers into the category of management' main body is the breakthrough point of the management system. This not only solved the problem of government's functions pressure and shortage of human resources, but also more important thing is that the transition zone transformed from the vacuum state to the protection state. Farmers effectively have the democratic status related to protection, which have far-reaching social and political significance.
3. The management functions of government have changed. Government is no longer the protection functions that are monopolized, but is important advocates, supporters and service providers in the natural ecological protection. So they also look forward to the new mechanism to give a guarantee on the allocation of resources.

4. Conclusions

Mashan Village is a transition zone of protection areas which only have nearly 100 families and the transition zone area of 1, 000 hectares. The practical effect of the village in the implementation of the "agreement protection" pattern and mechanism is significant. First, it has expanded the area of the nature ecological reserves to alleviate the contradiction between the ecological protection, community production and economic development; second, it stimulated the protective awareness and consciousness of community organizations and farmers individuals, mobilized the ability of ecological protection and promoted the diversification of community protection organizations and personal relationships, enhanced the cohesion of the community farmers, responsibility and collective honor sense; Third, community and government are propelled to participate protection action together, objectively enhanced the positive interaction development of community ecology, economy, social; fourth, "agreement protection" pattern completely changed a single government-led management model. Similar to the major transition zone of topography and biodiversity of Mashan Village there are still have a lot in western china. Mashan Village "agreement protection" pattern provided creating the management system of nature ecological reserves major transition zone of western parts with strong evidence and basis.

Author details

Tan Jing, Tian Yan, Wang Shang-Wu
Sichuan Agriculture University, Sichuan, China

Feng Jie
Forestry Bureau Li County, Sichuan, China

5. References

[1] Wang Zhi, Jiang Ming-kang. Comparison of Chinese nature reserve classification with IUCN protected area categories. Rural Eco-environment,2004,(4)

[2] Xue Qiao, Fan Hua. Try commenting on the relationship that Conservation Steward Program with Forestry Ecological and environmental protection of Li county.Journal of Sichuan Forestry Science and Technology.2009.Vo1.30, (3)

[3] Liu Jian-Guo, M. Linderman. Ecological Degradation of nature reserves: The case of Wolong giant panda Nature Reserve. Science, 2001 (4)

[4] Su Yang. The research and practice of harmonized development of the nature reserves and their ambient communities in West China. Sustainable development in China,2003(4)

[5] Chen Hong-Mei, Research on the Legal System of Nature Reserve's Management in China[D]. HoHai university.2006.(2)

[6] Li Yong-hua. Thoughts of Function and Intensive Management of Nature Protection Zone. Ecological Economy,2001,3,30~32

[7] Zhao Jun. Defects and Improvements of Public Participation s Principle in Chinese Environment Law[J]. Environmental Science and Technology,2005,28(2).
[8] Xia Shao-min, Yan Xian-wei. Management system of China's nature reserve. Journal of Zhejiang Forestry College,2009,26(1)
[9] Han Nian-Yong. A policy study on sustainable management for China nature reserves. Journal of Natural Resources,2000,3,32~37

Chapter 15

Implication of Alien Species Introduction to Loss of Fish Biodiversity and Livelihoods on Issyk-Kul Lake in Kyrgyzstan

Heimo Mikkola

Additional information is available at the end of the chapter

http://dx.doi.org/10.5772/47183

1. Introduction

Lake Issyk-Kul in Kyrgyzstan is the second largest mountain lake in the world (after Lake Titicaca in South America). It is situated in a basin surrounded by high mountains. While its water level is at 1608 m altitude, the mountain ranges of Kungei Ala-Too in the north reach 4711 meters, and those of Terskei Ala-Too in the south 5216 m. These mountains represent the major part of the Issyk-Kul catchment of 22,080 km^2 and provide most of the water to the lake.

Issyk-Kul Lake is 180 km long and 60 km wide, its surface area is 6240 km^2, and the shoreline 670 km (Figure 1.). The mean depth is 280 m, and the maximum 702 m, making it the fifth deepest lake in the world. The area covered by a depth of 0-100 m represents 38% of the total area. This is the major production zone of the lake [1].

Native fish stocks in this high-altitude saline lake have been subjected to predatory pressure from large number of introduced alien fish species. Previous papers and fishers are convinced that these predators are the most destructive to fish biodiversity in the lake, but this study wants to raise also other reasons which could explain at least part of the loss of fish stocks. The rapid growth in human activities with the development of tourism industry; irrigation; water eutrophication and pollution, and climate change impacts are alternative factors this presentation focuses on. This chapter also reviews the fish stocks and fishery management measures to increase the fish yields at the Lake.

Measures taken in order to protect the decreased fish stocks and endemic fish species include a Moratorium for Artisanal and Commercial Fish catching for a period of 10 years (2003-2013). Despite of the Moratorium at least 500 people continue their activity as illegal fishermen. Impacts of illegal and over-fishing are evaluated as anthropogenic activities. It is also noted that the disintegration of the Soviet Union had profound economic and social effects on many of the newly independent transition economies, like Kyrgyzstan.

Figure 1. Lake Issyk-Kul is large like a sea. Opposite shore is often not visible. Photo: Azat Alamanov

2. Description of study region

2.1. Physical and chemical environment

Although located at a high altitude, Issyk-Kul Lake never freezes over. The water temperature does not fall below the temperature of maximum density (2.75 °C at the mineral concentration of 6‰) except in shallow Rybachinsky and Tyup bays. The climate is continental with a short hot summer and a cold long winter. In summer the surface water temperature in the central part reaches 18°C, in winter it is seldom above 4°C [2]. The temperature may drop by 12-14°C down to 50 m depth and a further 1.5-2°C to the depth of 100 m. The water layer at 100-200 m depth maintains almost a stable temperature with changes kept within 0.1-0.3°C [3].

The chemical composition of the lake is as follows in mg/l^{-1}: Calcium – 121, Magnesium – 287, Sodium+Potassium – 1544, Chloride – 1596, Bicarbonate+Carbon trioxide – 318, Sulfate – 2102. Total cations: 1952, total anions 4016 [4]. So its mineral content is chloride/sulfate/sodium/magnesium based. With the drop in water level also comes a certain increase in salinity. Data from 1932 shows that the salinity measured 5.8‰, and by 1984 it had increased to 6.0‰. Over this period water level dropped by 2.5 m. Current measurements show that the salinity between October 2008 and November 2009 varied in Bosteri between 2 and 9‰ and the average was 5.9‰; which indicates that it is going down (Mikkola, unpublished). Since 1986, the decline in water level has stopped and the lake level has started to rise again. Low salinity (less than 20% that of seawater) indicates that in historical terms Issyk-Kul has only relatively recently become a closed lake. Hydrologists have suggested that, deep underground, the lake water filters into the Chui River. It looks as though the river Chui never "found" its way to the lake as the river bends a mere 4 km of the lake to the west, disappearing into the desert of Kazakhstan. During the very high water cycles the lake's water may overflow to the river through a natural depression – the Kutemaldu channel.

Currently the pH ranges between 7.7 and 7.9. The waters of Issyk-Kul are rich in oxygen, as a result of aeration and movement of lake waters. First of all, water is well oxygenated because it is regularly mixed by strong winds. During the warm period of the year, the surface water moves from the central part of the lake towards the shores and it is replaced by deeper cool water. In the middle of the lake the water is stratified down to 5-10 m whereas near the shores the thermal discontinuity is at a depth of 20-30 m owing to the warm water inflow. Apart from the central upwelling there is also lateral upwelling that is, caused by the wind driving surface water from the shore to the open parts of the lake [5]. Two major currents, driven by two wind streams, can almost always be observed: one follows the northern shore in a westerly direction, and the other flows east along the southern shore. The transparency of Issyk-Kul waters approaches that of seawater, and in the open part of the lake Secchi disc readings range from 30-47 m, but are reduced even down to 50 cm at the mouths of the inflowing rivers.

2.2. Biological features

Lake Issyk-Kul flora contains emerged macrophytes, like *Phragmites australis, Typha latifolia* and *Scirpus tabermaemontani* until the depth of 1.5 m. Submerged macrophytes like *Potamogeton pectinatus, Myriophyllum spicatum* and *Najas marina* and attached algae can go down to 30-40 m. Mean annual macrophyte production is about 277 g/m^2 [6]. *Characeae* are the most common macrophytes, representing 96% of the total annual macrophyte production, and are present in almost all plant associations. Four species of *Chara* grow in shallow water, and three benthic species exist further down. Dense growth of *Charophyte* green algae extends to 40 m depth. Issyk-Kul Lake water is rich in phytoplankton, with 299 identified species. Blue-green algae (*Cyanophyceae*) dominate, but their standing crop is low [7]. Phytoplankton production is at the level of 49 g/m^3 [6].

Zooplankton includes 117 species and is dominated by *rotifers* (84%), followed by *cladocerans* (9%) and *copepods* (7%). Zooplankton production is 91g/m^3 [6]. Zooplankton and phytoplankton distribution in the lake is uneven, with bays and shallows being richer than open water. *Arctodiaptomus salinus* is present in all parts of the lake and over the year it may represent 75-95% of the total number of zooplankton and 95-99% of the biomass. This species migrates during the night into the surface water where its concentration reaches up to 35,000 ind/m^3 [8], thus representing an important food source for all plankton-eating fish like Issyk-Kul Dace *Leuciscus bergi.*

Zoobenthos comprises 224 species. Most benthos occurs between the shoreline and 40 m depth, which comprises the *Charophyte* zone. According to [6] the mean annual production of zoobenthos is 10 g/m^3. It has been calculated that the average biomass of zoobenthos in the gulfs with open zones is 93.6 kg/ha [9]. *Chironomids*, mollusks, *gammarids* and *Mysis* comprise 75-80 per cent of the total. In the deeper zone beyond the zone of *Characeae* and down to about 70 m, the biomass is 2.5-3.5 g/m^2 and is dominated by *chironomids* and the mollusk *Radix auricularia*. Three *Mysis* species introduced into Issyk-Kul from Lake Balkhash, Kazakhstan, in 1965-1968 are now permanently established in shallows, mostly in 1.5-1.8 m depth, but reaching down to 10 m. Their mean biomass in such waters has been measured to range from 1.5 to 2.5 g/m^2 [10].

2.3. Distribution and abundance of fish fauna

The original fish fauna comprised twelve indigenous species and two subspecies particular to this lake (Table 1). The long historical and geographical isolation of the lake favoured the formation of endemic forms. This fauna is a typical example of the local Central Asian fish complex originating from Central Asian Mountain fauna (a term used by Berg, 1949), which is characterized by the presence of the *loaches* and *cyprinids*, with a small addition of *leuciscins* of Siberian origin. In the native fish fauna of the lake there were no predators although large Naked Osmans *Gymnodiptychus dybowskii* are said to feed partly on small fish [11].

Strictly endemic fish Schmidt's Dace *Leuciscus schmidti* is present throughout the shallow littoral zone but goes during winter down to 35-40 m. It appears in two forms, a common fast-growing lake form and a slow-growing bay form. The fast-growing form reaches 31 cm, a weight of 650 g, and age of 11 years. It spawns on stony beds at depths of 0.5-10 m between the end of March (water temperature 5°C) and mid-May. Fecundity is 6,000-65,000 eggs per year. It feeds largely on *Characeae*, but also on mollusks. The slow-growing form is present throughout the shallows. It reaches a length of 23 cm, a weight of 220 g, and a maximum age of 13 years. It has a similar fecundity and it spawns on the same substrate as the other form, but later.

Issyk-Kul Dace was the dominant fish until 1997, when Schmidt's Dace became for the first time the most numerous commercial fish in the lake. Issyk-Kul Dace inhabits the whole littoral zone, but is more pelagic than Schmidt's Dace. During the winter it is found down to depths 120-150 m, and reaches a maximum body-length of 17.5 cm and weight of 60 g. It spawns in shallow waters at depths between 1-8 m, and feeds mostly on plankton. During recent years the number and distribution of this species have sharply declined.

There are two endemic species distributed in mountain waters of Middle and Central Asia. The Scaly Osman *Diptychus maculatus* inhabits high-mountain streams, but enters also into Lake Issyk-Kul. It can grow 50 cm long and weighs up to 1 kg. It feeds on vegetation and invertebrates. The fish spawns in the spring and summer. It has a dwarf form, which lives mainly in the incoming small rivers, and may not live in the lake. It does not exceed 25 cm in length and weighs less than 200 g.

The Naked Osman is found in mountain rivers and lakes (Figure 2.). It also appears in two forms: one inhabits in rivers and the other in lakes. Lake living fish appear to have two ecological morphs: a winter lake morph and a summer migratory morph which spawns in rivers with a sandy bottom. Forms and eco-morphs would indicate that taxonomic studies are needed. The winter morph spawns from February to April and its fecundity is 13,000-14,500 per year. The summer morph is smaller, has a lower fecundity of 5,500-12,000, and spawns from April until September [12,13]. Both morphs are omnivorous. In the lake it feeds mostly on mollusks over muddy and loamy bottoms at 15-30 m deep. The largest Naked Osmans in the lake attain the age of 20 years and can grow up to 60 cm long and 3 kg of weight. It was once important commercial fish in the Issyk-Kul Lake, but there are indications that it is close to extinction [14].

Figure 2. The first Naked Osman captured alive 2009 in the UNDP/GEF Project. Photo:Azat Alamanov

Scientific name	Common name	Indigenous	Introduced
Onchorhynchus mykiss	Rainbow Trout		+
Salmo ischchan	Sevan Trout		+
Coregonus lavaretus	Common Whitefish		+
Coregonus widegreni	Valaam Whitefish		+
Coregonus autumnalis	Baikal Omul		+
Leuciscus schmidti	Schmidt's Dace	e	
Leuciscus bergi	Issyk-Kul Dace	e	
Phoxinus issykkulensis	Issyk-Kul Minnov	e	
Tinca tinca	Tench		+
Gobio gobio latus	Issyk-Kul Gudgeon	e	
Schizothorax pseudoaksaiensis issykkuli	Issyk-Kul Marinka	e	
Diptychus maculatus	Scaly Osman	e	
Gymnodiptychus dybowskii	Naked Osman	e	
Alburnoides taeniatus	Striped Bystranka		+
Abramis brama orientalis	Oriental Bream		+
Cyprinus carpio	Common Carp	o	
Ctenopharyngodon idella	Grass Carp		+
Hypophtalmichtys molitrix	Silver Carp		+
Carassius auratus auratus	Goldfish		+
Pseudoraspora parva	Stone Moroko		+
Capoeta capoeta capoeta	Transcaucasian Barb		+?
Triplophysa stoliczkai	Tibetan Stone Loach	e	
Triplophysa stoliczkai elegans	Tyanschan Loach	e	
Triplophysa dorsalis	Grey Loach	e	
Triplophysa strauchii strauchii	Spotted Thicklip Loach	e	
Triplophysa labiata	Plain Thicklip Loach		+
Triplophysa ulacholicus, including T.u. dorsaloides	Issyk-Kul Naked Loach	e	
Sander lucioperca	Pike-perch		+
Micropercops cinctus	Eleotris or Odontobutid		+
Glyptosternum reticulatum	Turkestan Catfish	e	
Aspius aspius	Asp		+?
Coregonus albula	Vendace (Ryapushka)		+?
Coregonus peled	Peled		+?

e= Indigenous, += Introduced, o= not known if indigenous, and +?=not known if the introduction failed

Table 1. List of fish species in the Issyk-Kul Lake [15].

Issyk-Kul Lake and incoming rivers have five indigenous and one alien loach species which are common in littoral underwater meadows, but are also found down to 100 m depth. They feed on benthos, plankton and eggs of other fish [13]. They have never been recorded by name in the catch of the lake except maybe in the "others" component. Subspecies would urgently require taxonomic revision, especially *Triplophysa ulacholicus versus Triplophysa u. dorsaloides* which are here synonymized.

The Issyk-Kul Gudgeon *Gobio gobio latus* spawns in June-July in shallows and feeds on benthos, detritus and fish eggs. It is preyed upon by Spotted Thicklip Loach *Triplophysa s. strauchii*, Sevan Trout *Salmo ischchan* and Pike-perch *Sander lucioperca* [13]. Again this fish has no commercial value and falls into "others" category in fish statistics.

Issyk-Kul Minnov *Phoxinus issykkulensis* is one of the strictly endemic fish species of Issyk-Kul Lake, but unfortunately there is no data on biology or abundance as it has never been important in commercial fishery.

Common Carp *Cyprinus carpio* is a widespread freshwater fish which has been introduced from Asia to every part of the world and it is included in the list of the world's worst invasive species. Many people in Kyrgyzstan, however, see Common Carp as indigenous calling it 'wild form' of Common Carp (Sazan). Most likely it was also introduced into the lake, but probably during the ancient times. It was known to be cultured in Kyrgyzstan at least since 1852 [15]. If accepting the 'wild origin' then the Issyk-Kul populations can be considered vulnerable to extinction.

Issyk-Kul Marinka *Schizothorax pseudoaksaiensis issykkuli* is an endemic species, which reaches 70 cm and a weight of 8 kg, and spawns from May until mid-July on rocky substratum in shallows near aquatic plants (Figure 3.). Its fecundity is 25,000 per year. It is omnivorous. Between 1985 and 1989 it formed 6% of the fish catch but after 1992 it disappeared completely (Table 2).

3. Background and historical overview

3.1. Introductions of alien fish species

At least 19 species have been introduced to the lake by humans, either on purpose or accidentally. Introduction of Vendace *Coregonus albula* and Peled *Coregonus peled* failed, and also survival of Asp *Aspius aspius* and Transcaucasian Barb *Capoeta c. capoeta* is doubtful as these species have not been reported recently. Formerly, the small Issyk-Kul Dace was the major item in fish catches, where it represented about 90% of total biomass. It was, however, considered to have a low value and this led to a proposal to introduce new fish species into the lake [16]. The introduction of the Sevan Trout from Armenia was recommended and, in 1930, 755,000 fertilised eggs were released, followed in 1936 by a further 800,000. Until 1964 Sevan Trout remained rare in the lake due to the shortage of suitable spawning grounds (Figure 4.). At its best, 1976 and1979, 51,6 and 53,8 tonnes of Sevan Trout were captured from the Issyk-Kul Lake. This was mainly due to state owned hatcheries which released 79 million fry into the lake during the 1970s [17].

Figure 3. One of the few Issyk-Kul Marinka captured alive during the study 2008-2011. Photo: Azat Alamanov

Fish species	1955-1964	1965-1969	1970-1974	1975-1979	1980-1984	1985-1989	1990-1994	1995-1999	2000-2003
Schmidt's Dace	482	225	544	496	292	241	223	105	44
Issyk-Kul Dace	9586	10741	9147	5736	1123	1064	790	94	12
Issyk-Kul Marinka	263	39	16	3	34	138	1	-	-
Osmans*	114	10	13	17	10	10	19	-	-
Common Carp	75	85	29	5	7	32	22	-	-
Whitefish	-	-	1	35	106	248	163	57	11
Sevan Trout	0.3	30	123	457	244	206	91	29	1
Pike-perch	-	287	1364	895	340	320	227	25	1
Others	46	51	51	15	15	48	98	74	12
Total	**10566**	**11468**	**11288**	**7659**	**2171**	**2307**	**1634**	**384**	**81**

Original data from Fisheries Department/Mairam Sarieva. Note that the data is in centners of Soviet Union. One centner is 100 kg not one ton as so often misquoted in previous publications. *Osmans = Scaly- and Naked Osman

Table 2. Fish catch from the Issyk-Kul Lake in five year averages from 1955-2003.

Figure 4. Sevan Trout is a colourful fish which grows well in the Issyk-Kul Lake. Photo: Azat Alamanov

Following its introduction, Sevan Trout became an active predator of other fish in the lake and developed several special features. Its growth rate was 4 to 6 times that in Sevan Lake in Armenia. In Issyk-Kul it grows to a bigger size. It matures earlier, and its fecundity has increased five-fold to 3,300-17,300 eggs per fish. The limiting factors for this species in Issyk-Kul are food resources and habitat for reproduction.

In the 1950s, there were further introductions of fish species in order to establish diverse stocks of piscivorous fish, introduce species feeding on phytoplankton and aquatic *macrophytes*, and increase the number of benthos- and plankton-feeding fish [18]. Pike-perch, Tench *Tinca tinca* and Oriental Bream *Abramis brama orientalis* were introduced in 1954-1956 [19]. They became established predominantly in the eastern part of the lake but started soon spreading all over the lake. The introduction of Grass Carp *Ctenopharyngodon idella* and Silver Carp *Hypophtalmichtys molitrix*, and with them inadvertent introductions of Goldfish *Carassius a. auratus*, Stone Moroko *Pseudoraspora parva* and Eleotris *Micropercops cinctus*, were successful but caused a disaster to 'wild carps'. Grass and Silver Carp brought infectious *ascites* of carps into the lake, and the numbers of 'wild carps' started to decrease due to disease.

In the early 1970s, a decision was taken to transform the lake into a trout-whitefish water body at the expense of the local Issyk-Kul Dace population. However, Whitefish *Coregonus lavaretus* is mainly a plankton and benthos feeder, but large individuals can occasionally take other fish. Common Whitefish was introduced from Lake Sevan, Valaam Whitefish or Ludoga *Coregonus widegreni* from Lake Ladoga, and from Lake Baikal came Arctic Cisco or Baikal Omul *Coregonus autumnalis* and Peled [20]. Eggs of Whitefish from Sevan Lake were transferred to the Ton hatchery from which four-day-old fry were released into Lake Issyk-Kul. During 1966-1973 87 million fry were released. In 1974, the first 500 kg of Whitefish were harvested from Tyup Bay.

There were also proposals to replace the Issyk-Kul Dace with the Peled and Vendace (Ryapushka), more nutritious food fish species. However, Peled and Vendace soon disappeared from the lake most likely due to lack of suitable reproduction conditions. Baikal Omul was still observed in the lake as of late 1970s. After that only Whitefish established itself as commercial fish and the highest catch recorded was 35.3 tonnes in 1989. After that the catch started to go down mainly due to reproduction problems and hatchery failure (Table 4).

Most harmful introduction took place accidentally from the cage culture of Rainbow Trout *Onchorhynchus mykiss* (Figure 5.). Since 1980s a lot of small and some large fish started to escape from the culture operations, and now Rainbow Trout is very common all over the lake, but especially near eight cage culture farms. Rainbow Trout moves to fish diet at the size of 35-40 cm [21]. It is not clear if Rainbow Trout would be able reproduce in some incoming rivers, as nowadays the lake seems to have all aged and sized Rainbow Trout.

3.2. Introduction of alien food species for fish

The fish food base was successfully enriched by the introduction of *mysids*, which became targeted by the introduced *coregonids*. In the Issyk-Kul Lake the introduced Whitefish benefitted most of mollusks and *mysids*. Their growth rate was faster than that of the original stock in Lake Sevan, and they also started maturing at an earlier age. However, there was a high mortality of *coregonid* fry which found insufficient food in Issyk-Kul Lake and were heavily preyed on by the endemic fish. Also, it was believed that the higher salinity of Issyk-Kul as compared to Sevan and Baikal could have had a negative impact on fry [20]. The decision was made to stock advanced fry and fingerlings, of which millions were stocked in the following years (at least 10 million from 1977to 1988).

3.3. Recreational and commercial fishing activities

Amateur fishing in the lake started in the 1870s. At first it was unorganized and no statistics were collected. During the 1890s fish catch ranged between 17 – 105 tonnes [2]. At that time, and for many years after, the major commercial fish were Naked Osman, Common Carp, Issyk-Kul Marinka, Issyk-Kul Dace and Schmidt's Dace. According to available information in 1941-1945 harvests of Issyk-Kul Marinka, Naked Osman and Common Carp reached 61 tonnes per year. During the same period Issyk-Kul Dace catch varied from 551 to 900 tonnes per year. It is important to note that during this period lake had already one alien predator, Sevan Trout, but catch of these had not started. It is also interesting to note that 'native' species did not start to go down after the introduction of Sevan Trout [11].

More detailed catch statistics are available from 1955 until the moratorium in 2003 (Table 2). As Table clearly illustrates, fish landings shrunk sharply after the Issyk-Kul Dace population collapsed. Increased fishing activity could be one explanation, but by the end of the Soviet period the state fishing industry was at its peak and encompassed only 300 fishers, 122 boats and 8640 nets [11]. This number of fishers is not much for that size of lake, and over fishing hardly explains alone the decline of fish catches. However, by 1990 Issyk-Kul landings were barely twenty percent of the levels recorded a quarter of a century earlier.

Introduction of one more predator, Pike-perch, into the lake 1954-56 had no immediate and dramatic impact on both lake landings and the species composition of landings. It seems that the production of the lake went down clearly only 1975 onwards, as in the transition from one feeding level (plankton and benthos feeders) to another (predatory fish); the lost feed coefficient against the productivity is approximately ten times. The Pike-perch production reached its peak in 1974, and then the production of Issyk-Kul Dace started to go down very fast as shown in Table 2. The role of Sevan Trout is not as clear as it achieved the highest population only in 1979 after Issyk-Kul Dace had already diminished to one third from the starting level.

Whitefish is often listed as predatory fish but its possible bad habits are not showing clearly in this material. Whitefish achieved the peak population in 1989 when all other species had started to go down very rapidly. Interesting is that the peak populations of Osmans occurred already 1961-62, when all predatory species recorded zero catch. Issyk-Kul Marinka had the peak population in 1988. Total disappearance of Osmans from the catch data happened 1988 and that of Issyk-Kul Marinka 1993. Disappearance of Common Carp took place during the same year (1993) as that of Issyk-Kul Marinka. The predators alone cannot explain these losses, and it is difficult to see any direct relation with fishing either.

Issyk-Kul Dace made 94 per cent of the catch in 1965-69 and less than 15 per cent in 2000-2003, same time the total catch of the lake went down from 1147 to 8 tonnes. Issyk-Kul Marinka, Naked Osman and Sazan Carp are those commercial and indigenous species which are most seriously endangered.

3.4. Fishing as livelihood

The contribution of fishing to annual average income of Issyk-Kul district families is from 5 to 10 per cent and only for some small groups up to 30%. The monthly income of fishermen is not more than 40 USD and that of women processing the fish 54 USD. Women's income is little higher than men's [22]. Although income from fish is small, it allows the families to have cash on daily basis and facilitates implementation of other cash requiring activities (purchase of seeds, forage for animal-breeding etc.). For more details see [15].

3.5. Water level

In historical terms the water level of Issyk-Kul Lake has obviously fluctuated. Some changes are gradual, others sudden and disastrous since they were caused by earthquakes and torrents of water rush from mountains. Large ancient city has been located at depths of 5 to 10 meters near the north coast of the lake, but it was destroyed maybe some 2500 years ago by one of the many local floods which are known to occur every 500 to 700 years [23]. Between years 600 and 1200 AD Issyk-Kul shoreline was again some 500 m lower and after that in the fifteenth century the water level of the lake was more than 10 m higher than it is today.

On Issyk-Kul basin118 rivers and streams flow toward the lake, but only 42% of them actually drain into it and 25% have discharges less than 1 m^3 s^{-1}. Only 9% of these are rivers with catchment areas of more than 300 km^2. Rivers are fed predominantly by melt water from glaciers and snow above 3300 m. The river system reflects also the distribution of rainfall in the basin with low precipitation in the west, where the river system is poorly developed. In the east, where the precipitation is heavier, the hydro-network is denser and the rivers fuller. The greatest volume of flow comes through rivers on the basin's eastern side. Water from most of the rivers has been completely diverted for irrigation before it enters the lake. Therefore, bays in the northern and western coast suffer from increased mineralization. The rivers supply the lake with 3720 million m^3 of water per year [24], but they are not the only water supply the lake is receiving. The annual surface water discharge, precipitation and groundwater discharge to the lake are 21, 29, and 33cm respectively; the evaporation from the lake is 82cm. For more details see [7].

Until 1985 water level in Issyk-Kul was falling. Between 1876 and 1972 the decrease was 9 m [25]. During 1960-1979 when the fish catches started to decrease clearly the total decline of water level was 140 cm, at an average rate of seven cm annually. That loss of water level has been one important factor affecting the fish stocks and fisheries.

3.6. Irrigation

Uptake of water for irrigation is one of the factors seen to be responsible for the present changes in the water level. Irrigation also hinders the river spawning of many species, as it prevents small rivers and streams from reaching the lake. Irrigation has led to drying and silting up of spawning grounds and death of the fry themselves as they are poured out with the river water to irrigated fields. During 1960-1979 irreversible uptakes of water from rivers for irrigation reduced the volume of river water entering the lake by an estimated 23 per cent [6]. While in 1930 there were 50,000 ha of irrigated area in the Lake Issyk-Kul catchment, by 1980 the irrigated area reached 154,000 ha [26]. However, even without this irrigation loss, the lake would still have declined at the rate of five cm annually between 1960 and 1979 [27]. This would indicate that climatic factors have also been involved in the fall of the water level.

4. Material and methods

This chapter is based on my field work and data collection in Kyrgyzstan when working in the UNDP/GEF Project: "Strengthening policy and regulatory framework for mainstreaming biodiversity into fisheries sector" as International Fishery Policy Adviser 2008 to 2009 and in the FAO Project GCP/KYR/003/FIN: "Support to Fishery and Aquaculture Management in the Kyrgyz Republic" 2009-2010 as Technical Advisor. A lot of generally unknown 'grey literature' in Russian has been translated and used in this text. More information about interviews and field experimentation is documented in [15].

5. Results and discussion

5.1. Species introduced

It is obvious that Sevan Trout and Pike-perch introductions can be blamed for the reduction in catch. There is clear positive correlation between Sevan Trout and Pike-perch and Schmidt's Dace catch (Table 3). Interestingly Sevan Trout has correlation also with Pike-perch catch. There is positive correlation between prey species, like between Marinka and Osmans as well as between Issyk-Kul and Schmidt's Dace and Sazan Carp, but only Whitefish has strong negative correlation with Issyk-Kul Dace.

Species	Pike-perch	Sevan Trout	White-fish	Marinka	Osmans	Issyk-Kul Dace	Schmidt's Dace	Sazan Carp
Pike-perch	1	,535	-,056	-,176	,088	,268	,559	-,012
Sevan Trout	,535	1	,311	,013	,036	-,136	,339	-,193
White-fish	-,056	,311	1	,143	-,316	-,574	-,209	-,149
Marinka	-,176	,013	,143	1	,492	,116	,124	-,288
Osmans	,088	,036	-,316	,492	1	,480	,424	,261
Issyk-Kul Dace	,268	-,136	-,574	,116	,480	1	,622	,468
Schmidt's Dace	,559	,339	-.209	,124	,424	,622	1	,065
Sazan Carp	-,012	-,193	-,149	,288	,261	,468	, 065	1

Red marked numbers correlation is significant at the 0.01 level (2-tailed). Green marked have correlation at the 0.05 level.

Table 3. Correlations of main fish species in the catch between 1955 and 2003.

But correlation does not necessarily mean causation. This far the introduction of predatory fish species has been seen as the major if not the only reason why native fish stocks collapsed [11,15,28]. This same conclusion was also made in Africa, where introduction of the Nile Perch *Lates niloticus* was believed to have caused the greatest vertebrate mass extinction in recorded history [29,30]. Approximately 150 different species of *Haplocromis* chicklids became extinct in recent times in Lake Victoria. Now, however, reevaluated data shows that Nile Perch did not really succeed until, and after, its prey (the *haplochromines*) had disappeared. The increased eutrophication of the lake and oxygen problems may explain more the diversity changes than the single species predation or fisheries exploitation [31,32].

This does not mean that one should support the alien introductions, and precautionary principles are necessary. Precautionary principle states that one has to expect that new introduced species, although in closures or in the cage, tend to escape for one reason or another into nature [33]. Any new voluntary alien introduction should be understood with that background and the rule should be clear: that new human introduced alien species are not allowed to enter into the country. Still it remains inevitable that some invasive species will arrive without any help of humans.

As shown in Table 4 there are seven clear reasons which could explain the loss of fish stocks and biodiversity and another five reasons which have had at least some negative impact. The reviews and field work highlights that all these twelve negative factors have been present more or less at the same time, so it is not possible to single out any one of them as the most important. Surely they have rather caused the loss of fish stocks and biodiversity together. Some of these factors have been listed already before, like over fishing, disintegration of the Soviet Union, irrigation and water level. Over fishing of Issyk-Kul Dace stocks by the Soviet fleet based at Issyk-Kul Lake was presented in [11]. The disintegration of the Soviet Union had profound economic and social effects, especially in the fisheries sector of the newly independent transition economies [28]. Nowhere were these production shortfalls bigger than in Kyrgyzstan where Lake Issyk-Kul fish landings were in 2003 less than 7 per cent of the catch level recorded in 1989. The major consequence for the fisheries sector was the spectra of uncertainty which included the uncertainty of, how the sector was to be managed, how access to water bodies was to be regulated, how to maintain the backward and forwards supply chains which underpinned pond aquaculture, and livelihoods – as the Soviet guarantee of job security was rescinded. Many experts and professional fishers left the sector to find employment in other sectors in Kyrgyzstan or abroad. Intensive irrigation led to reduced water levels in the lake and more importantly heavy water abstraction caused drying of many incoming streams that the endemic fish species previously used for feeding and/or spawning [34,35]. It has been shown that biological productivity of Lake Issyk-Kul decreased from 1973 to 1981 when the water level was declining at a rate of 7-10 cm per year [36].

5.2. Hatchery failure

Reproduction of many alien and endemic fish species was severely constrained by the limited number of suitable spawning rivers. As a consequence, the state established hatcheries on the Ton (1964) and Karakol (1969) rivers – with the brief to capture spawning fish, extract the eggs, raise the fry-fingerlings produced, and thence restock the lake. According to [7], the minimum Sevan Trout return in landings is given as 2% of releases; that means that at least 750,000 fry, each of 1 g weight, must be produced and released annually. Assuming an egg mortality of 50%, hatcheries should produce 1.5 million eggs per year. Ton hatchery produced 9 million fry annually in 1989-91. After the breakdown of the former Soviet Union the state hatchery production went down sharply. Over the period 2004-8 Ton Hatchery continued to restock the lake with Sevan Trout at much reduced rate, 446,000 fingerlings annually. Nowadays (2010) Ton Hatchery is able to release some 900,000 fry with 40% egg survival. No endemic fish fry have been produced despite of the capacity, but Rainbow Trout fingerlings have been produced on a contractual basis for the cage farmers.

Estimated IMPACT	**Strong negative** impact	**Some negative** impacts	**Not visible** impact	**Some positive** impacts
Introduction of alien fish species	Yes			
Introduction of alien food species				Yes
Over fishing	Yes			
Illegal fishing			Yes	
Disintegration of the Soviet Union 1991		Yes		
Cage culture	Yes			
Moratorium	Yes			
Hatchery failure	Yes			
Tourism		Yes		
Water level		Yes		
Irrigation	Yes			
Water pollution		Yes		
Climate change		Yes		
Radioactive leakage			Yes	
Military activities			Yes	
Mining activities	Yes			

Table 4. Impact evaluation of different natural and anthropogenic factors on fish stocks and biodiversity

During 1966-1973 over 12 million Whitefish fry were released annually from the state-owned Karakol Hatchery, but 1977-1988 fingerling production went down to 1 million per year. After privatization Karakol Hatchery has been able to produce below 2.5 million Whitefish fry annually, explaining why the collapsed Whitefish stocks are not recovering, as obviously very little or no natural reproduction takes place in the lake.

These drastic declines in restocking have undoubtedly been one contributor to the decrease in recorded fish landings at Issyk-Kul Lake.

5.3. Cage farming of fish

The cage farming of Rainbow Trout started in 1988 by Alfa Laval Avose, but was not economically viable due to the high cost of feed. Obviously large number of fingerlings escaped into the lake, when the storm was turning the experimental cages around. In 1989-

90 the company was able to produce 20 tonnes of Rainbow Trout. After the collapse of the USSR there was no development of this activity before 2006 when Ecos International commenced cage farming activities at Issyk-Kul Lake. Since that time exponential growth in trout culture has taken place.

Nowadays the existing eight cage farms and their 26 cages (as in April, 2009) are producing well over 300 tonnes of Rainbow Trout per year [37]. This production is causing pollution in the form of medicaments used for the treatment and prevention of diseases and pathogenic bacteria and parasites. By authorizing lake-based cage culture of Rainbow Trout, the authorities are allowing inevitable eutrophication. Extra nitrogen and phosphorus from unused feed will add to the primary production of algae and lower oxygen level. The second problem is the excess feed which sinks to the bottom of the lake through the net cages. At the bottom, sinking feed and faeces and urine of fish will cause the formation of hydrogen sulfide gases harming the other users and fauna of the lake. The worst, however, is the unwanted new continuous introduction of that predatory fish to the lake, because especially large specimen can and will escape the cages. After that they move around the lake and eat a lot of endemic fish species. According to fishermen (personal interviews in 2009) Rainbow Trout is the main predator in the lake, even more predacious than Pike-perch, because it comes to prey in shallow waters near the shoreline while Pike-perch often remains in the deeper waters (Figure 5).

5.4. Moratorium and illegal fishing

In order to protect the decreased fish stocks in the lake, the President of the Kyrgyz Republic declared a Moratorium for Artisanal and Commercial Fish catching for a period of 10 years (2003-2013). The need for total ban was stated to be illegal and over-fishing which was seen as the only reason to loss of fish resources and endemic species. But the moratorium can become an effective measure of restoration of fish resources in the lake only if mechanism of implementation and realization (fish inspections etc.) is developed as well. Otherwise the moratorium will not work. Despite of the total ban at least 500 people continue their activity as illegal fishermen. On average they are catching 5 to 20 kg per night, but every fourth night is stormy making artisanal fishing with small boats impossible. If fishing 100 nights per year, they are catching between 250 and 1000 tonnes per year. Should this fairly conservative estimation be true, the lake is fished at level of 0.4 - 1.6 kg/ha. So this hardly can be seen as over fishing in a lake where theoretical production capacity is estimated at 4.5 kg/ha [34]. This of course by assuming that the fish stocks have in the last ten years recovered from 2003 level after the total ban of commercial fishing.

5.5. Management and conservation possibilities

It is far too easy to blame over-fishing that some species became nearly extinct and that fewer fish are caught. More important problem is the absence of any fisheries management and lack of controlled protection of fish resources. Removal of the fishing ban is necessary, since it cannot be controlled and monitored. Exploring co-management arrangements is a

Figure 5. This kind of 7 kg Rainbow Trout eats a lot of small indigenous fish species. Photo: Azat Alamanov

better option than command and control as the resources are not available for such policy measure [15]. If more than 500 people are continuing fishing despite of the moratorium, the policy needs to be evaluated for better stewardship outcomes. Actually it is far more important to continue to fish large predatory fish, if having any concern of the survival of native non-predatory species.

However, commercial fishing needs to be reconsidered after the moratorium, in 2014, as recreational and food fishery may be far more sustainable. Due to the growing importance of Lake Issyk-Kul for recreation, fishery management might go in the direction of producing valuable sport and recreational fish to satisfy the demand of the tourists and visitors. Such recreational fishing will basically target large predatory species- Rainbow and Sevan Trout and Pike-perch, which are the favourite species for sport fishing. Recreational fishing of large fish will promote Issyk-Kul Lake as more attractive for tourists. It will also help fishery managers to shift proportions of predator and prey fish species and diminish the negative effect of alien predators towards vulnerable stocks of endemic fish species.

Rare indigenous species stocks will not improve without artificial propagation in local hatcheries. This production of fry has started through UNDP/GEF Project, but stocking the lake with small indigenous is not viable before considerable harvesting of large predators. The number of Sevan Trout is easy to regulate, as it mostly depends on the stocking rate of fingerlings into the lake and these are reproduced artificially in the hatcheries. Rainbow Trout and Pike-perch are more difficult to remove if they are able to multiply in the lake (Figure 6.). Improved stocks of small endemics could take care of these predators by eating their eggs and small fry, like small fish did in Lake Victoria by preying on eggs and fry of the Nile Perch [32].

5.6. Water pollution

Widespread mining operations are causing disruption of soils, terrain and water tables but more serious water pollution comes from illegal dumping or storing of toxic chemicals currently in use at Kumtor gold mine, in Tian Shan Mountains. It is the largest gold mine, as well as a major government revenue source, which routinely ignores national environmental legislation. Kumtor mine reportedly uses up to 10 tonnes of cyanide per day in its mining operations, and number of chemical constituents is released into the environment [38]. By sure this is affecting fish populations downstream and the health of local people using the contaminated water or fish (Table 5).

One of the worst regional environmental disasters in recent history occurred on 20 May 1998, when a truck hauling toxic chemical crashed just upstream from the mouth of the Barkuum River, which empties into Lake Issyk-Kul. As a result, 1762 kg of sodium cyanide, a chemical used in the processing of gold ore at Kumtor, were dumped into basin waters [6].

Lack of both adequate infrastructure and financial means to support public utilities (let alone any resort or tourism industry) has made it impossible to improve wastewater treatment plants. This in turn leads to further pollution and unwise use of lake waters. The gradual increase in settlements and industries around the lake has led to an increase in pollution. Although most enterprises have wastewater treatment facilities they are not efficient and some effluents still reach the lake.

Figure 6. Artificial nests are used to remove the eggs of Pike-perch from the lake to reduce the numbers of that alien predatory fish. Photo: Azat Alamanov

Agriculture, through the use of fertilizers and pesticides, also contributes to the lake pollution, but level of fertilizer application on crop fields is known to be moderate. However, the Issyk-Kul area produces 12% of total national cereal crops and over 40% of potato crops. Of the total area of orchards nationwide, 20% are in Issyk-Kul. Numbers of domestic animals in the catchment area is very high: Cattle 163,500; sheep 1,944,400; swine 32,700; poultry 623,400 and horses 48,500 [39]. Dairy product processing covers 50% of the national dairy product supply. Animal breeding is growing, with average annual sheep and cattle surplus at 5-6%. Grazing land is overloaded by 1.5 times its capacity (Figure 7). Grazing practices have changed so that all livestock owned by small proprietors are now grazing near the lake as the farmers have no transport nor money to drive their animals upland to outlaying pastures. That could cause social conflict (grazing on beaches and resort areas) and eutrophication of the lake but luckily people are commonly collecting the manure for fire or fertilizing. While the large volume of 1738 km^3 of water in the lake may have at present considerable diluting capability and with the good water mixing is also able to quickly oxygenate organic matter inputs to the lake, sheltered shallows are subject to eutrophication. As the shallows are also important spawning and feeding areas for a number of fish, such eutrophication may affect especially those cold-water fish species which require pristine waters, like Whitefish.

Figure 7. These camels are the only memory of the Silk Road at the Issyk-Kul Lake coast of which is heavily overgrazed by the domestic animals. Photo: Azat Alamanov

Eutrophication caused by birds is not often considered as a problem, but in the Issyk-Kul Lake the amount of migratory birds is such that it will affect the lake. Anywhere from 44,000 to 68,500 birds belonging to 30 to 35 species winter on the lake, and even more birds use it as stopover and feeding place during spring and autumn migration [40].

5.7. Climate change implications

Within the Issyk-Kul basin there are 834 glaciers of various sizes ranging from less than 0.1 to 11 km^2. For example, a typical Issyk-Kul glacier Karabatkak has shown in long-term study between 1957 and 1997 that ice loss exceeded snow mass gain by 18 m.

This thinning of ice is due to climate change, summers have been 0.6 degrees Celsius warmer, although the annual average temperature has risen only 0.2 degrees Celsius. Based on this it was calculated that before 2005 overall glaciation area near the lake will go down 32% on the northern slopes and even 77% on south-facing slopes [41].

The continuing retreat of glaciers in the Issyl-Kul catchment, the melt water from which is one of the major contributors to the lake, seems to be going in parallel with the declining lake water level [42]. Without glacial runoff overall drop in lake's water level would have been much greater.

The Kyrgyz Republic is within a high seismic activity area, and Issyk-Kul is a tectonic lake, and the lake bottom is believed to have numerous warm water springs. These explain partly why the lake never freezes over, except in the shallow Rybachinsky and Tyup bays. The water stays warmer than the air for seven months per year [39].

Hot springs at the lake and on the bottom change water quality and may facilitate winter spawning of some introduced species like Pike-perch. During the test fishing in early April 2009 we found after opening a 40 cm Pike-perch that it had preyed another (12 cm), and even that small prey had eaten a few juvenile Pike-perch, not more than 5 cm long. It was estimated that these 3rd level victims of cannibalism must have been born early February. This kind of winter spawning is not known before but could obviously take place due to the hot springs.

5.8. Other human activities

There are recent reports on the radioactive waste contamination in Central Asia [43] showing that the situation is critical especially in Kyrgyzstan, with 36 tailing sites and 25 uranium dumps in the country. Kadzhi-Say, the country's largest tailing site (containing 150,000 m^3 of radioactive waste), is located barely 1.5 km from Lake Issyk-Kul. Yet although some information is available on the impact of radioactivity on humans, it is not well studied or understood what direct impact the current radioactivity levels have on the aquatic biodiversity in Kyrgyzstan. The monitoring of water bodies for radioactivity is not done consistently and to date, no assessment has been made of the uranium contamination of fish populations of indigenous species and its consequences for fish stocks in Kyrgyzstan. It is not clear, if and how much radioactive waste has already gone into the lake or still goes from incorrectly closed tailing sites and uranium dumps.

During the Soviet period, the USSR Navy operated an extensive facility at the lake's eastern end, where submarine technology was evaluated. Also Navy tested torpedoes built in Tashkent. Not known if torpedoes exploded in these experiments. If so this must have killed a lot of fish. In 2008 Kyrgyz newspapers reported that Russian Navy is planning to establish a new naval testing facility around the Karabulan peninsula on the lake. This may affect the fish stocks in the future depending on the tests undertaken.

During the Soviet era, the lake became a popular vacation resort, with numerous sanatoria, boarding houses and vacation homes along its northern shore. These fell on hard times after the break-up of the USSR, but from 2005 onwards hotel complexes are being refurbished and simple private bed- and –breakfast pensions are being established for a new generation of health and leisure visitors (Table 5).

Tourism has become one of the most dynamically developing sectors of economy of the Kyrgyz Republic. The number of arrivals of foreign tourists is expected to exceed 2 million persons per year. International tourists are primarily from Kazakhstan, Russia, and

Uzbekistan. If half of these tourists will visit the lake, there is need to accommodate an additional 1 million persons per year at the lake in hotels using natural beaches for recreation. Nowadays the lake has 343 tourist enterprises, including cafes and restaurants.

Years	Population in '000	Visitors in hotels	Visitors at homes	Total in '000
2006	430	198	296	924
2007	433	199	245	877
2008	435	194	349	978
2009	438	169	318	925
2010	441	181	227	849
2011	445	185	231	861

Table 5. Permanent population and annual visitors in the Lake Issyk-Kul area [44].

In addition, large hospitals have been built to use medicinal mud and hot springs along the coasts for medicinal purposes. Regulations exist for water system supply and to use fully purified sewerage systems. Recycled waste water could be used for irrigation. Unfortunately, some entrepreneurs have forgotten these regulations, and continue to pollute the lake as no corruption free control exists.

Asian Development Bank study [45] has concluded that available water and sanitation and waste disposal infrastructure in the Issyk-Kul area is decrepit, dysfunctional, poorly managed and negatively impacts the surrounding environment. The planned tourist influx equivalent to four times the resident population applies excessive pressure on the existing infrastructure, which results in the pollution of the lake. The proposed Issyk-Kul Sustainable Development Project initiated by the Asian Development bank (ADB) would address the environmental and institutional issues around the Lake Issyk-Kul. The Japanese International Cooperation Agency will also develop the sewerage system and sewage treatment plant in Cholpon-Ata through parallel financing with the ADB.

6. Concluding remarks

Issyk-Kul Lake is the second largest high-altitude lake in the world providing recreational and small-scale fishing activities as well as cage culture of Rainbow Trout in the Kyrgyz Republic. The original fish fauna comprised twelve indigenous species and two subspecies particular for this lake. At least 19 species have been introduced to the lake by humans, either on purpose or accidentally. The populations of several indigenous fish are seriously threatened, because many of the introduced fish species are potential predators. Issyk-Kul Marinka, Naked Osman and Sazan Carp are those commercial and indigenous species which are most seriously endangered. In 1986 a total ban was declared for catching Naked Osman, but it did not lead to positive results, indicating that anthropogenic activities were not the only reasons for the suffering of the endemic fish species.

Fishers and most of the previous papers are convinced that the predatory fish species have been the most destructive to biodiversity. Addressing the introductions, the basic rule should be that new human introduced alien species are not allowed to enter into the lake. At least any further fish introductions into the lake should be carefully evaluated to prevent unwanted changes in fish stocks. Issyk-Kul, as an oligotrophic lake of low productivity, has a low carrying capacity for fish; hence it will never become a water body which would sustain high levels of fish stocks.

Dissolution of the Soviet Union explains to some extent the collapse of the fisheries sector (including the hatchery operations) in Kyrgyzstan, but maybe not the loss of biodiversity. Rapid growth in human activities with the development of tourism industry; irrigation; water eutrophication and pollution, and climate change impacts seem to be important root causes for loss of fish stocks and biodiversity degradation.

Uptake of water for irrigation is one of the factors seen to be responsible for the present changes in the water level as water from most of the rivers has been completely diverted for irrigation before reaching the lake. Irrigation hinders also the river reproduction of many species as it prevents spawning fish from entering the rivers or fry to return to the lake. During 1960-1979 when fish catches started to decrease the total decline of the water level was 140 cm, at an average rate of seven cm annually. That loss of water level has been one important factor affecting the fish stocks and fisheries.

There are recent reports on the radioactive waste contamination in Kyrgyzstan, where the country's largest uranium tailing site is located barely 1.5 km from Lake Issyk-Kul. It is not clear, if and how much radioactive waste has already gone into the lake or still goes in from incorrectly closed tailing sites and uranium dumps. Maybe even more serious water pollution comes from illegal dumping or storing of toxic chemicals currently in use at a gold mine in Tian Shan Mountains. It is the largest gold mine, as well as a major government revenue source, which routinely ignores national environmental legislation. This mine uses daily up to 10 tonnes of cyanide in its operations, and many of toxic chemicals are released into the environment. This is surely affecting fish populations downstream and the health of local people using the contaminated water or fish for drink and food.

Existing water and sanitation and waste disposal infrastructure in the Issyk-Kul area is decrepit, dysfunctional, poorly managed and has negative impacts on the environment. The planned tourist influx equivalent to four times the resident population will apply excessive pressure on the existing infrastructure, which will result in further pollution of the lake.

Important problem is the total absence of any fisheries management and lack of controlled protection of fish stocks and diversity. Fishing ban is not helpful as it cannot be controlled and monitored. Exploring co-management arrangements is a better option than command and control as the resources are not available for such policy measure. Rare and endangered indigenous fish species will not increase without artificial propagation in local hatcheries. Stocking the lake with small indigenous species is not viable, if not first harvesting the large predators. Pike-perch would just eat small indigenous species and grow bigger and spawn more.

Over-fishing of introduced species, like Rainbow Trout and Pike-perch, could be a good thing. As popular food fish and recreational catch, they could be severely over fished. This could lead to population reduction, and several populations of endemic fish species should soon show signs of increasing numbers. So the authors should allow the local fishing communities capture large introduced fish species as much as they can rather than restricting them through moratorium.

Any new development initiatives must be consultative and participatory in order to be more consistent with local habits and cultural values. Inherited customs provide an important element for the development of locally based resource management system. Consequently, allocation and sustainable management of natural resources is one of the key issues for the local population, whose daily cash economy is directly dependent on availability of fish resources.

Before being able to define the best management ways, one needs further research in taxonomy and fish biology. The knowledge of fish stock parameters is essential for the determination of appropriate fisheries management and definition of sustainable fish yield. Impact of mining toxins and radioactive waste is important to study and to know for control measures. Water pollution is a continuous risk for this important lake, and has to be halted in the future. The Kyrgyz public must be engaged in the future through environmental education in conservation and preservation of natural and cultural riches of the Issyk-Kul area.

Author details

Heimo Mikkola
University of Eastern Finland, Kuopio Campus, Kuopio, Finland

Acknowledgement

My deepest gratitude goes to Mrs. Burul Nazarmatova who translated all grey Russian literature into English and collected a lot of relevant Soviet time data from the internet. Mrs. Mairam Sarieva from the Fisheries Department of the Kyrgyz Republic assisted me in collecting the fish catch statistics of the Issyk-Kul Lake. Some unpublished data from fish stocks was also given by Dr Muchtar Alpiev from the Kyrgyz Academy of Science. Last but not least I want to give my special thanks to Dr Ahmed Khan, Memorial University of Newfoundland, Canada, who's useful and critical suggestions improved this chapter a lot.

7. References

[1] Kodyaev G.W (1973) Morphometric characteristics of Lake Issyk-Kul. Izv. Vsesoyuz Geogr. Obschestva 105(4): 362-365 (In Russian).
[2] Berg L.S (1930) The present state of fisheries in Issyk-Kul. Issykkulskaya Ekspeditsia 1928 g. 1: 1-48 (In Russian).
[3] Molchanov I.V (1946) Lake Issyk-Kul. Gidrometeoizdat, Leningrad. 53pp (In Russian).
[4] Kadyrov G.K (1986) Hydrochemistry of Lake Issyk-Kul and its basin. Ilim, Frunze. 212pp (In Russian).
[5] Shabunin G.D (1981) Thermodynamic processes in Lake Issyk-Kul. Biol. Osnovy Ryb. Khoz. Vodoemov Srednei Azii i Kazakhstana: 394-396. Ilim, Frunze (In Russian).

[6] Baetov R (2003) Management of the Issyk-Kul basin. Mimeo 28/08/2003, 22 p.
[7] Savvaitova K.A, Petr T (1999) Fish and Fisheries in Lake Issyk-Kul (Tien Shan), River Chu and Pamir Lakes. FAO Fisheries Technical Paper 385, Rome.
[8] Foliyan L.A (1981) Zooplankton of Lake Issyk-Kul. Biol. Osnovy Ryb. Khoz.Vodoemov Srednei Azii i Kazakhstana: 389-390. Ilim, Frunze (In Russian).
[9] Pavlova M.V (1964) Zoobenthos of Issyk-Kul Lake Bays and their use by fishes. Ilim, Frunze. 95pp (In Russian).
[10] Ivanova L.M (1986) Distribution of mysids in Lake Issyk-Kul. Biol. Osnovy Ryb. Khoz.. Vodoemov Srednei Azii i Kazakhstana: 65-66. Ilim, Ashkhabat (In Russian).
[11] Konurbaev A.O, Kustareva L.A, Alpiev M.N, Kabataev D.T, Konurbaev E.S (2005) Conditions of Issyk-Kul Lake Ichtiofauna, fishery and its management. Mimeo, 24 pp. Bishkek.
[12] Berg L.S (1949) Freshwater fishes. AN SSSR. Moscow and Leningrad. 2: 478-925 (In Russian).
[13] Turdakov F.A (1963) Fishes of Kirgizia. Ilim, Frunze. 283pp (In Russian).
[14] Mikkola H (2010) Introductions causing serious problems to endemic fish. Suomen Luonto 69(4): 15-16 (In Finnish).
[15] Alamanov A, Mikkola H (2011) Is Biodiversity Friendly Fisheries Management Possible on Issyk-Kul Lake in the Kyrgyz Republic? Ambio 40: 479-495.
[16] Berg L.S (1929) Fisheries in Issyk-Kul. Izv. Inst. Opytnoi Agronomii (Moskva) 7:179-181 (In Russian).
[17] Borisov S.I (1981) Induced reproduction of trout in Issyk-Kul. Biol. Osnovy Ryb. Khoz. Vodoemov Srednei Azii i Kazakhstana : 40-42. Ilim, Frunze (In Russian).
[18] Turdakov F.A (1954) On the development of fisheries in Lake Issyk-Kul. Trudy Inst. Zool. Parazitol. Kirgiz. Fil. AN SSSR 2:31-38. Frunze (In Russian).
[19] Turdakov F.A (1961) Reconstruction of fisheries in Lake Issyk-Kul. Vestnik AN SSSR 2: 54-56 (In Russian).
[20] Nikitin A.A (1976) Introduction and induced reproduction of whitefish in Kirgizia.. Ilim, Frunze. 122pp (In Russian).
[21] Svärdson G, Nilsson N-A, Dahlström H, Tuunainen P (1968) Fish, management of fishing waters and fish farming. 302 pp. Kirjayhtymä, Helsinki (in Finnish).
[22] UNDP (2007) Access of men and women to the natural resources of Kyrgyzstan, 83 p., Bishkek
[23] Lukashov N (2007) Remains of ancient civilization discovered on the bottom of a lake. Novosti 27/12/2007 Available: http://en.rian/analysis/2007/1227/94372640-print.html. Accessed 2008 Jun 11.
[24] Zabirov R.D, Korotaev V.N (1978) Characteristics and morphometry of the lake: In: Ozero Issyk-Kul: 3-48. Ilim, Frunze (In Russian).
[25] Shnitnikov A.V (1979) Lake Issyk-Kul: nature, conservation and the prospects for lake utilization. Ilim, Frunze. 84pp (In Russian).
[26] Leontiev S.A, Sevastyanov D.V (1988) Problem of managing Issyk-Kul water level. Vodnye Res. 1: 121-128 (In Russian).
[27] Sevastyanov D.V, Smirnova N.P (1986) Lake Issyk-Kul and trends in its natural development. Izd. Nauka, Leningrad. 246pp (In Russian).

[28] Thorpe A, Van Anrooy R, Niyazov B.N, Sarieva M.K, Valbo-Jørgensen J, Mena Millar A (2009) The collapse of the fisheries sector in Kyrgyzstan: an analysis of its roots and it prospects for revival. Communist and Post- Communist Studies 42:141-163.

[29] Ball D (2004) Lake Victoria: The problems and options. 1-2. Available: http://aquarticles.com/articles/management/Ball_2Victoria.html. Accessed 2010 Oct 21.

[30] Chege N (1995) Lake Victoria: A sick giant. People & the Planet, 1-4. Available: http://cichlid-forum.com/articles/lake _victoria_sick.php. Accessed 2010 Nov 21.

[31] Kolding J, van Zwieten P, Mkumbo O, Silsbe G, Hecky R (2008) Are the Lake Victoria Fisheries Threatened by Exploitation or Eutrophication? Towards an Ecosystem-based Approach to Management. Pp. 309-355 in Bianchi, G. & Skjoldal, H.R. eds.: The Ecosystem Approach to Fisheries. CAB International.

[32] Kolding J, van Zwieten P, Mkumbo O, Hecky R, Silsbe G (2012) When prey becomes predators: eutrophication and the transition of Lake Victoria's fish community from a cichlid to a Nile Perch dominated state. Paper for the World Fisheries Congress in Edinburg 2012: G4: Impact of marine or aquatic acidification and eutrophication on fish and fisheries.

[33] FAO (1996) Precautionary approach to capture fisheries and species introductions. FAO Technical Guidelines for Responsible Fisheries. No. 2:1-54, FAO, Rome.

[34] Alpiev M.N (2006) Internal water ecosystems. In Third National Report on Saving of Biodiversity in the Kyrgyz Republic. Bishkek (In Russian).

[35] Mikkola H (2009) GCP/KYR/003/FIN " Support to fishery and aquaculture management in the Kyrgyz Republic". FAN FAO Aquaculture Newsletter 43: 6-7.

[36] Konurbaev A.O (1981) Decrease in biological productivity of Lake Issyk-Kul due to the declining water level. Biol. Osnovy Ryb. Khoz. Vodoemov Srednei Azii i Kazakhstana: 100-103. Ilim, Frunze (In Russian).

[37] Alamanov A, Mikkola H (eds.) (2009) Workshop report on Lake Issyk-Kul biodiversity friendly fisheries management regime proposal and fisheries co-management. Ecocentre, Cholpon- Ata, Kyrgyzstan, 10-12 September 2009. GCP/KYR/003/FIN FAO Field Document 1/2009, 92 p.

[38] Trilling D (2012) Kyrgyzstan: Gold Mine Could Exacerbate Central Asian Water Woes-Report. Eurasia Net's Weekly 31/01/2012 Available: http://www.eurasianet.org/node/64928. Accessed 2012 Feb 7.

[39] International Lake Environment Committee (1999) Lake Issyk-Kool. 9p. Available: http://www.ilec.or.jp/database/asi/asi-55.html. Accessed 03 Feb 2012.

[40] Van der Ven J (2002) Looking at Birds in Kyrgyz Republic, Central Asia. 180 p. Rarity, Bishkek.

[41] Dikikh A.O (2000) Glaciations in the Issyk-Kul Basin; its role as a flow source. Nature and People of Kyrgyzstan. Special Edition, Bishkek.

[42] Artemev V.V, Klinge R.K, Selivanov A.O (1992) History of formation of water resources and fluctuations in water level of Lake Issyk-Kul in late Pleistocene and Holocene. Vodnye Res. 4: 71-78 (In Russian).

[43] UNDP (2009) Uranium Tailing in Central Asia. Final Framework Document, 114 p. Bishkek.

[44] National Statistics Committee (2011) Permanent population and Tourism of the Issyk-Kul Oblast. Brief Statistical Handbook, Bishkek.

[45] Asian Development Bank (2009) Environmental Impact Assessment of Issyk-Kul Sustainable Development Project, Kyrgyz Republic. ADB TA No. 7228 KGZ, 2:1-120.

Chapter 16

Efforts to Combat Wild Animals Trafficking in Brazil

Guilherme Fernando Gomes Destro, Tatiana Lucena Pimentel, Raquel Monti Sabaini, Roberto Cabral Borges and Raquel Barreto

Additional information is available at the end of the chapter

http://dx.doi.org/10.5772/48351

1. Introduction

Wildlife trafficking, including the flora, fauna and their products and byproducts, is considered the third largest illegal activity in the world, after weapons and drugs trafficking. Considering only the wild animals trafficking in Brazil, it is estimated that about 38 million specimens are captured from nature annually and approximately four million of those are sold. Based on the data of animals seized and their prices, it is suggested that this Country deals with about two billion and five hundred million dollars a year [1].

The wildlife trafficking networks, like any other criminal network, have great flexibility and adaptability and join with other categories or activities (legal or illegal), such as drugs, weapons, alcohol, and precious stones. Their products are often sent from the same regions and have similar practices such as forgery, bribery of officials, tax evasion, fraudulent customs declarations, among many others [1].

In some cases, the criminals are infiltrated in public agencies to entice public officials and, in case of problems in the target Country, they can move with ease to other destination. Moreover, people involved can be easily replaced by others more efficient, reliable and qualified for the activity. This great power of mobility and changeability is one of the major problems to map the criminal networks and their local of action [2].

Although modern techniques has been used, around the world, to help the enforcement in the combat of illegal wildlife trade [3,4,5], the trafficking structure still presents features in common with the set of network information, because it requires equipment that enables the continuous exchange of information on routes, on the most quoted animals at the black market, on new forms of fraud and on corruption pathways. The new technologies are more and more used to increase the possibility of success on criminal operations, either through the use of cell phones, computers to defraud documentation, or internet sales, among others [2].

According to the report of the National Network to Combating Wild Animal Trafficking [1], there are four methods that encourage the illegal trade in wild animal: (a) animals for zoos and private collectors, (b) for scientific use/ biopiracy, (c) for pet shops and (d) for products and byproducts.

However, it is known that identifying the site of capture isn't an easy task, because the locals where the animals are confiscated usually differs from where they were captured. Furthermore, the capture and the sale of wild animals and their byproducts are not concentrated in one only place and do not always follow the same destiny: the movement is intense, with many destinations. After being captured, the animals commonly pass through small and medium traffickers who make the connection with Brazilian and international large dealers, however, the animals can also be sold by internet, pet shops and illegal fairs [2].

Although the trafficking consequences are numerous, it is possible to group them into three main branches: (a) Sanitary, since illegal animals are sold without any sanitary control and can transmit serious diseases, including unknown ones, onto domestic breeding and people [1,2,6]; (b) economic/social, as the trafficking moves incalculable amounts of financial resources without bringing income to the public coffers [1]; and (c) Ecological, since the capture from nature done without discretion accelerate the process of extinction of species, causes damage to ecological interactions and loss of the genetic heritage. Moreover, the trafficking can also bring ecological damage arising from the introduction of exotic specimens, that, although acquired as pets, are being abandoned by their owners in various natural areas [1].

Illegal wildlife trafficking is an extremely lucrative crime with serious consequences yet relatively low penalties and few prosecutions [3]. Besides all the complicating factors inherent to the trafficking, the researchers of this subject are facing yet the lack of organized and systematized data and information [2]. In addition, the studies on trafficking and its impacts on biota are also scarce [7], what makes the task of systematization even more complex.

Thus, through this work, we presented a national view of the control and combat actions towards the wild animals trafficking in Brazil through the existing information at the corporative systems managed by the Brazilian Institute of Environment and Renewable Natural Resources - IBAMA. As specific goals, we aimed to:

- Historically evaluate the gradual development of the Brazilian environmental enforcement related to fauna;
- Map the Brazilian States where there are greater efforts against wild animals trafficking, as well as the most confiscated species;
- Evaluate the major forms of admission and destination of the wild animals present at the Rehabilitation Centers;
- List the main perspectives and recommendations of actions to combat wild animals trafficking in Brazil.

2. The enforcement for conservation of wild animals in Brazil

For the preparation of this paper we used, primarily, historical information present in four information systems (Table 1), all managed by IBAMA. This information was compiled, systematized and analyzed together with literature data.

System	Name	Objective
SICAFI	Recording, Levying and Enforcement System	Responsible for recording data and information relating to environmental enforcement activities performed by IBAMA and partners institutions
SISPASS	Recording Passeriform Amateur Breeders System	Responsible for the control of the activity of Amateur and Commercial Passeriform[1] Breeders
SISFAUNA	Fauna Management System	Responsible for the management of wild animals in captivity, including the emissions of permits, stock control, domestic trade, licenses issued and carried out transactions
SISCITES	System for the importation and exportation of specimens, biological stuff, native and exotic wildlife products and byproducts	Controls the importation and exportation of species listed in the CITES[2] appendices

Table 1. Information systems related to wildlife and managed by IBAMA.

In Figure 1 we summarized data from the Wild Animals Rehabilitation Centers – CETAS, on all the confiscated wild animals placed there during eight years. The CETAS are responsible for receiving, identifying, marking, selecting, evaluating, recovering, rehabilitating and placing wild animals. Furthermore, they are important allies to the actions for the repression of trafficking because they provide relevant information about confiscated wild animals or from voluntary delivery.

As recommended by the Brazilian Environmental Policy and showed in Figure 1, the State supervision related to illegal wildlife, under the responsibility of the Environmental Military Police, has steadily increased in number and efficiency, thanks to ongoing efforts to decentralize responsibilities in the Country. Thus, IBAMA has been able to focus on major crimes, with significant results through the dissuasion of his actions. It is important to inform that in Brazil, the fines are applied per animal. So, due to that, Minas Gerais (which had the highest participation of environment military police) has the largest number of fines (Figure 2), but it doesn't reflect the absolute value. This happens because of the type of inspection that fights against the final receptors of wild animal traffic.

[1] IBAMA's Normative Instruction No 15/2010

[2] CITES: Convention on International Trade in Endangered Species of Wild Fauna and Flora (www.cites.org). The Brazilian CITES Management and Enforcement Authorities are represented by IBAMA.

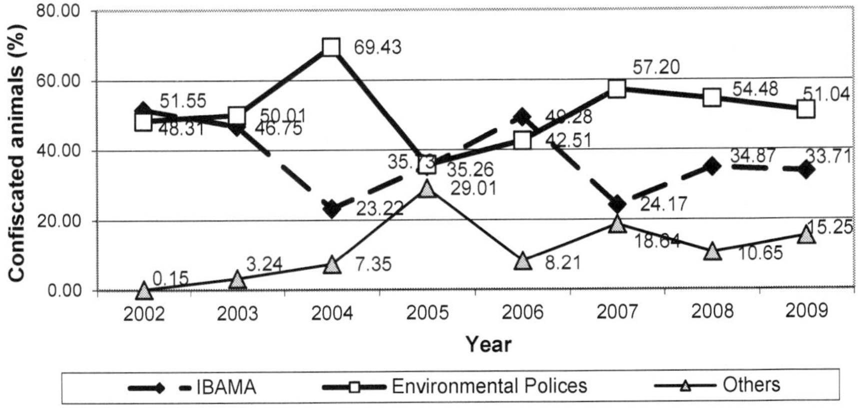

Figure 1. Number of confiscated wild animals received by CETAS

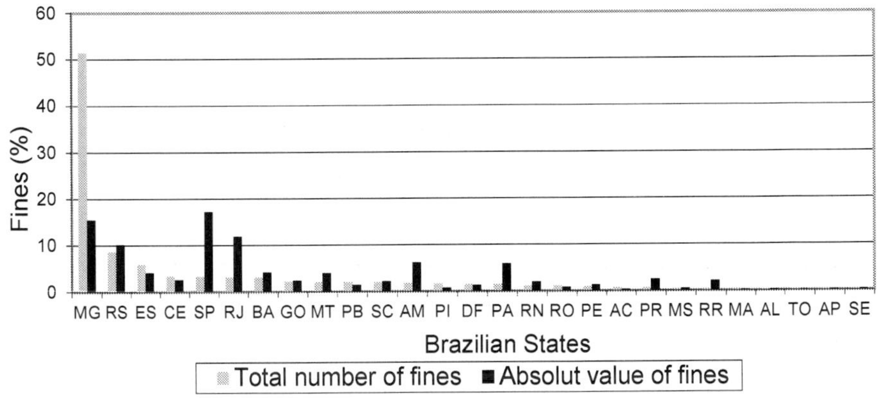

Figure 2. Percentage of Fauna fines per Brazilian State and their absolut value from 2005 to 2010.

In Figure 2, the distinction among each Brazillian State on the combat against illegal actions related to wild animals is clear. The States of Rio Grande do Sul (RS), Minas Gerais (MG), Espírito Santo (ES), São Paulo (SP) and Rio de Janeiro (RJ) were the ones with the highest numbers of fines applied between 2005 and 2010. The last four are located in the southeastern region, where is the demand from the majority of animals from traffic. The States of Sergipe (SE) and Tocantins (TO) emerged with the lowest numbers. The States with the highest absolute values applied in fines were São Paulo (SP), Minas Gerais (MG), Rio de Janeiro (RJ), Rio Grande do Sul (RS), Amazonas (AM) and Pará (PA) (both last ones are located in the rainforrest region and represent one of the main sites where some taxon are captured), unlike the States of Maranhão (MA) and Tocantins (TO), which had the lowest absolute number.

We emphasize that the animals confiscated by Brazilian environmental agencies represent only a portion of the damage [8]. The task of estimating the amount of animals withdrawn from nature per year becomes even more difficult if we consider that the possession of a wild animal captured from nature in Brazil is a common practice, despite being prohibited by law. However, IBAMA's Department of Fish and Wildlife points out that the CETAS alone received in 2008 more than 60,000 animals and were destined more than 40,000 (Figure 4). We noticed, yet, that this number is still small. It happens because most of the animals confiscated in actions of inspection are released into the wild, due to the fact that they are still in savage condition.

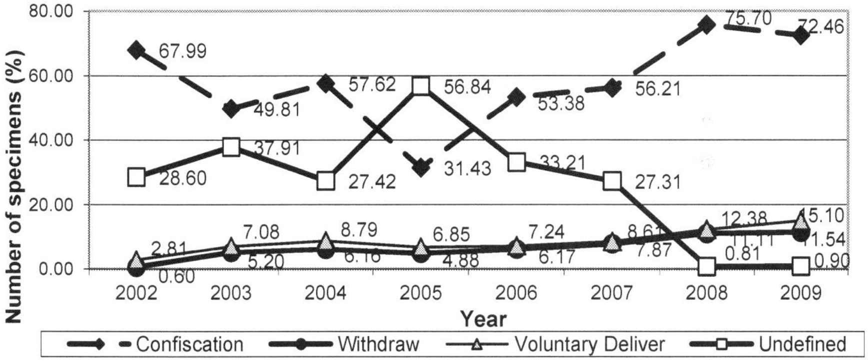

Figure 3. Number ofwild animals received by CETAS and their different forms of admission.

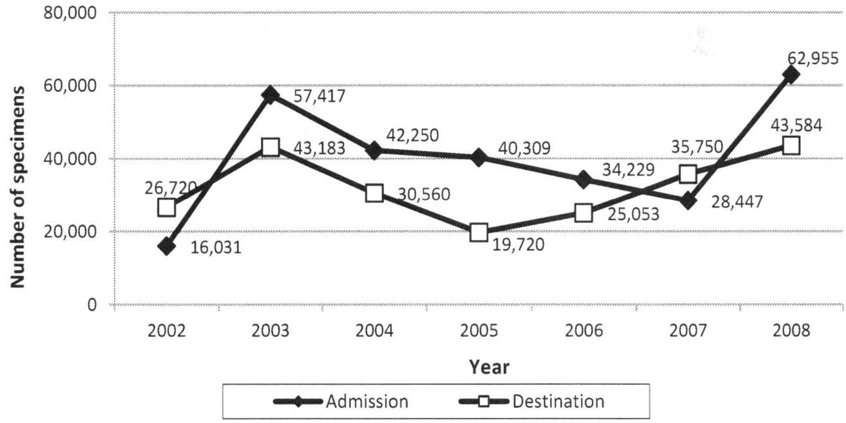

Figure 4. Relationship between admission and destination of specimens in CETAS between 2002 and 2008, in absolute numbers.

The Figure 5 and the Table 2 gather the destinations given to animals from CETAS, between 2002 and 2009. We observed that the releases, after the declining trend observed between 2004 and 2007, re-emerged as the main destination given to the confiscated animals in Brazil, reaching almost 23,000 specimens released into the wild in 2008. The placement in captivity, widely used in 2006 and 2007, has a lower incidence from 2008 on, with the publication of new normative instruments, which regulated the policy to native and exotic wild animals in captivity.

The number of deaths recorded in CETAS suffered variations over the sample period, but their values remained between 16 and 26 percent. The values of escapes/evasions remained constantly low if compared with the total number of destinations.

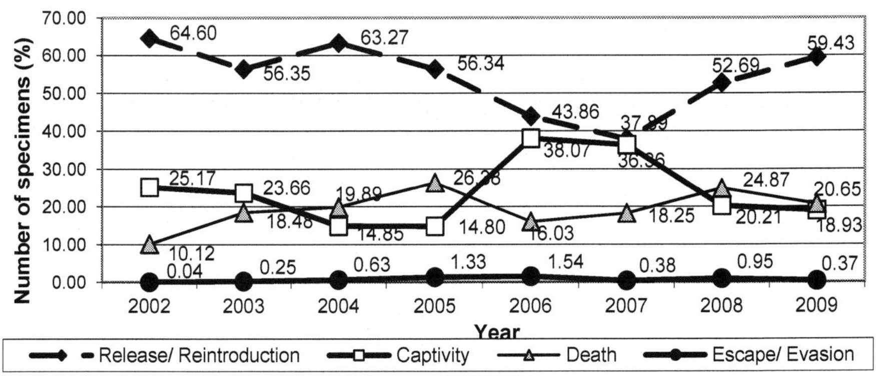

Figure 5. Destination of the animals from CETAS between 2002 and 2009.

Year	Admissions	Releases/ Reintroductions	Captivities	Deaths	Escapes/ Evasions
2002	16,031	17,260	6,725	2,705	12
2003	57,417	24,333	10,219	7,980	110
2004	42,250	19,336	4,538	6,078	191
2005	40,309	11,110	2,919	5,202	263
2006	34,229	10,988	9,537	4,015	386
2007	28,447	13,544	12,998	6,523	137
2008	62,955	22,965	8,809	10,839	413

Table 2. Number of specimens destined by CETAS between 2002 and 2008.

Figure 6. Released birds in Bahia (BA)

3. The species of confiscated animals

We found that the destination given to confiscated animals in Brazil is directly linked to the taxonomic class of animal (Figure 7). For Birds, the main form of destination was release into the wild (greater than 55%), followed by placement in captivity and death. Release into the wild was also the main destination given to reptiles (~ 60%) and mammals (~ 45%). We also noticed that reptiles obtained a lower death rate, while the exotics animals remained in captivity (~ 60%).

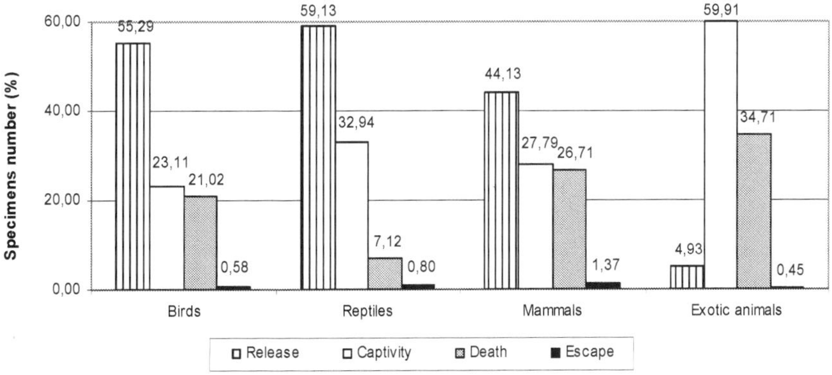

Figure 7. Destination of the animals from CETAS between 2002 and 2009, by taxonomic group.

The Table 3 presents the amount of admission and destination of animals from CETAS. Birds represented 81% of admitted specimens and 82% of released ones, between 2002 and 2009. The Birds was also the group that obtained the largest number of deaths registered (86%). In Australia and Asia, Reptilia was most targeted group of taxa for illegal trade,being also the most seized [4, 9].

Group	Entrance	Release	Captivity	Death	Escape
Birds	250,206	108,622	45,395	41,294	1,135
Reptiles	34,835	17,198	9,581	2,072	233
Mammals	17,936	7,233	4,554	4,377	225
Exotics	4,577	44	535	310	4

Table 3. Amount of specimens that entered and left the CETAS between 2002 and 2009.

We listed, in the Table 4, the 30 species most confiscated by IBAMA and accredited institutions between 2005 and 2009, according to SICAFI. The class Aves was the most representative (80%), followed by Reptilia (16.67%). The most significant families were Emberizidae (30%), Thraupidae (13.33%) and Podocnemididae (10%). The most commonly confiscated species was *Sicalis flaveola* (Saffron Finch), followed by *Saltator similis* (Green-winged Saltator) and *Sporophila caerulescens* (Double-collared Seedeater), (Figure 8).

Classif	Type	Class	Family	Specie[3]	Common name
1º	Wild animal	Aves	Emberizidae	*Sicalis flaveola* (Linnaeus, 1766)	Saffron Finch
2º	Wild animal	Aves	Thraupidae	*Saltator similis* d'Orbigny & Lafresnaye, 1837	Green-winged Saltator
3º	Wild animal	Aves	Emberizidae	*Sporophila caerulescens* (Vieillot, 1823)	Double-collared Seedeater
4º	Wild animal	Aves	Cardinalidae	*Cyanoloxia brissonii* (Lichtenstein, 1823)	Ultramarine Grosbeak
5º	Wild animal	Aves	Emberizidae	*Sporophila angolensis* (Linnaeus, 1766)	Chestnut-bellied Seed-Finch
6º	Wild animal	Reptilia	Podocnemididae	*Podocnemis expansa* Schweigger, 1812	Giant South American River Turtle
7º	Wild animal	Aves	Icteridae	*Gnorimopsar chopi* (Vieillot, 1819)	Chopi Blackbird
8º	Domestic	Aves	Phasianidae	*Gallus gallus*	Domestic

[3] Nomenclature according to the updated lists CBRO, 2011 (www.cbro.org.br) and SBH, 2011 (http://www.sbherpetologia.org.br)

Classif	Type	Class	Family	Specie[3]	Common name
	animal			(Linnaeus, 1758)	Chicken
9º	Wild animal	Aves	Thraupidae	*Paroaria dominicana* (Linnaeus, 1758)	Red-cowled Cardinal
10º	Wild animal	Aves	Emberizidae	*Sporophila lineola* (Linnaeus, 1758)	Lined Seedeater
11º	Wild animal	Reptilia	Podocnemididae	*Podocnemis sextuberculata* Cornalia, 1849	Six-tubercled Amazon River Turtle
12º	Wild animal	Aves	Emberizidae	*Zonotrichia capensis* (Statius Muller, 1776)	Rufous-collared Sparrow
13º	Wild animal	Aves	Emberizidae	*Sporophila nigricollis* (Vieillot, 1823)	Yellow-bellied Seedeater
14º	Wild animal	Aves	Emberizidae	*Sporophila collaris* (Boddaert, 1783)	Rusty-collared Seedeater
15º	Wild animal	Aves	Psittacidae	*Amazona aestiva* (Linnaeus, 1758)	Blue-fronted Parrot
16º	Wild animal	Reptilia	Alligatoridae	*Caiman crocodilus* (Linnaeus, 1758 [originally Lacerta])	Common Caiman
17º	Wild animal	Aves	Turdidae	*Turdus rufiventris* Vieillot, 1818	Rufous-bellied Thrush
18º	Wild animal	Aves	Thraupidae	*Paroaria* sp. Bonaparte, 1832	Cardinal
19º	Wild animal	Aves	---	Not specified	Bird
20º	Wild animal	Aves	Columbidae	*Zenaida auriculata* (Des Murs, 1847)	Eared Dove
21º	Wild animal	Aves	Emberizidae	*Sporophila albogularis* (Spix, 1825)	White-throated Seedeater
22º	Domestic animal	Mammalia	Bovidae	*Bos* taurus Linnaeus, 1758	Domestic Cattle
23º	Wild animal	Aves	Psittacidae	Many species	Parrot
24º	Wild animal	Aves	Fringillidae	*Sporagra magellanica* (Vieillot, 1805)	Hooded Siskin
25º	Wild animal	Reptilia	Podocnemididae	*Podocnemis unifilis* (Troschel, 1848)	Yellow-spotted Amazon River Turtle
26º	Wild	Aves	Icteridae	*Icterus jamacaii*	Campo Troupial

Classif	Type	Class	Family	Specie[3]	Common name
	animal			(Gmelin, 1788)	
27º	Wild animal	Aves	Emberizidae	*Sporophila maximiliani* (Cabanis, 1851)	Great-billed Seed-Finch
28º	Wild animal	Reptilia	Testudinidae	*Chelonoidis* sp. Fitzgerald, 1835	Tortoise
29º	Wild animal	Aves	Turdidae	*Turdus* sp. Linnaeus, 1758	Thrush
30º	Wild animal	Aves	Thraupidae	*Lanio cucullatus* (Statius Muller, 1776)	Red-crested Finch

Table 4. Most confiscated species by IBAMA and partner institutions between 2005 and 2009.

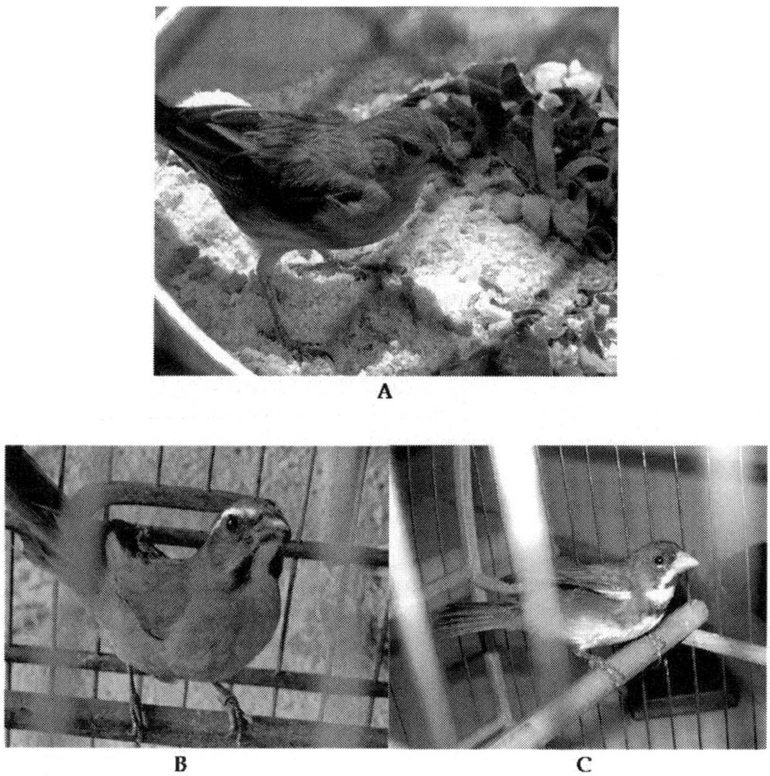

Figure 8. The three most confiscated species by environmental enforcement in Brasil: A. *Sicalis flaveola* (Saffron Finch), B. *Saltator similis* (Green-winged Saltator) and C. *Sporophila caerulescens* (Double-collared Seedeater).

According to studies conducted by [8] in south Brazil, the most commonly confiscated species by enforcement, between 1998 and 2000, was the Cardinal (*Paroaria coronata*), followed by the Saffron Finch (*Sicalis flaveola*). And the Emberizidae family presented the largest number of seized specimens, compelling evidence that the great interests of the illegal trade are the songbirds.

The Emberizidae family also excelled in seizures conducted in southeastern and northeastern Brazil [7,10,11]. According to the authors, that fact can be explained, preliminarily, because that family has many species and specimens, for being abundant in the Neotropics, for having easy occurrence in the sampled region, for the high quality of its singing, due to its low market value and for being easy to maintain. Generally, the birds most wanted for trafficking are the songbirds or those able to become pets, confering them high values of trade [8].

Some species listed in Table 4 are exclusively Amazonian, as the Giant South American river turtle, Six-tubercled Amazon River turtle and Yellow-spotted Amazon River Turtle, all very popular in regional cuisine and found in nature in large populations. The domestic chicken (*Gallus gallus*) and domestic cattle (*Bos taurus*) obtained national prominence in seizures, because the first is often used in arenas, being the target of actions against animal abuse, and the second is the subject of crime in embargoed areas due to deforestation, mainly in the Amazon region.

We also observed an intrinsic relationship between the passerines authorized breeding and the wild animals trafficking: the five species more seized are also the taxa of greatest interest for commercial and amateur breeders of passerines (Table 5). All other passerines listed in Table 4 are species authorized for commercial and amateur activity.

Classif.	*Species*[4]	Common name	Total of breeders	Total of specimens
1º	*Saltator similis* d'Orbigny & Lafresnaye, 1837	Green-winged Saltator	133.699	528.621
2º	*Sporophila angolensis* (Linnaeus, 1766)	Chestnut-bellied Seed-Finch	89.083	535.195
3º	*Sporophila caerulescens* (Vieillot, 1823)	Double-collared Seedeater	86.666	279.888
4º	*Sicalis flaveola* (Linnaeus, 1766)	Saffron Finch	83.281	444.160
5º	*Cyanoloxia brissonii* (Lichtenstein, 1823)	Ultramarine Grosbeak	46.364	108.703
6º	*Sporagra magellanica* (Vieillot, 1805)	Hooded Siskin	28.709	83.885
7º	*Turdus rufiventris* Vieillot, 1818	Rufous-bellied Thrush	27.250	57.960
8º	*Saltator maximus* (Statius Muller, 1776)	Buff-throated Saltator	19.129	53.203

[4] Nomenclature according to the updated lists CBRO, 2011 (www.cbro.org.br)

Classif.	*Species*[4]	Common name	Total of breeders	Total of specimens
9º	*Sporophila maximiliani* (Cabanis, 1851)	Great-billed Seed-Finch	18.142	123.832
10º	*Zonotrichia capensis* (Statius Muller, 1776)	Rufous-collared Sparrow	16.466	32.677
11º	*Sporophila lineola* (Linnaeus, 1758)	Lined Seedeater	13.868	25.317
12º	*Gnorimopsar chopi* *(Vieillot, 1819)*	Chopi Blackbird	12.540	21.716
13º	*Cyanoloxia cyanoides* (Lafresnaye, 1847)	Blue-black Grosbeak	11.435	23.435
14º	*Paroaria coronata* (Miller, 1776)	Red-crested Cardinal	11.310	33.110
15º	Sporophila frontalis *(Verreaux, 1869)*	Buffy-fronted Seedeater	9.301	22.073
16º	*Sporophila nigricollis* (Vieillot, 1823)	Yellow-bellied Seedeater	9.264	22.135
17º	*Molothrus oryzivorus* (Gmelin, 1788)	Giant Cowbird	8.878	18.858
18º	*Lanio cucullatus* (Statius Muller, 1776)	Red-crested Finch	6.922	13.635
19º	*Saltator fuliginosus* (Daudin, 1800)	Black-throated Grosbeak	6.756	14.533
20º	*Paroaria dominicana* *(Linnaeus, 1758)*	Red-cowled Cardinal	6.123	11.675

Table 5. Species of greatest interest to the passerine breeders in Brazil.

For [1], one of the ways to reduce the pressure on the populations for trafficking would be the encouragement of captive breeding programs to meet commercial demand. However, this strategy can be of great concern, since those animals cannot achieve the low prices offered by the trafficking [7].

4. The trafficking routes

In Figure 9 we grouped the main trafficking routes of wild animals in Brazil, including major airports, trade and source areas. We observed that, in general, the Brazilian fauna has been removed from the North, Northeast and Midwest of the Country and it is being sent to the Southeast, South and other regions of Northeast, by land or river, fuelling the national trade. In relation to the international illegal trade, we emphasize cities located in border regions in the North, Midwest and South of Brazil, as well as in ports and airports located in the Northern, Northeastern, Southern and Southeastern Brazilian regions.

For [1], beyond the States of Para (PA) and Amazonas (AM), which had national prominence in the amount of fines, other Amazonian frontiers must be of particular concern, such as the borders with the Guianas, Venezuela and Colombia, and the route of the Madeira River.

Figure 9. Main routes for the wild animals trafficking in Brazil.

The situation at the tri-border area (Brazil, Paraguay and Argentina) is also a matter of worry. According to [2], many animals are taken from the Iguaçu National Park and illegally sold during daylight or taken by peddlers to other Brazilian regions. Also in southern Brazil, the authors highlight as important areas for capturing and trading wild animals the towns of Laranjeiras do Sul (PR) and Santana do Livramento (RS), close to the border with Uruguay.

In [10], also emphasized the trafficking in the Southwest Bahia (BA) region, and they say that it is a socio-environmental problem with serious consequences to the local avifauna. According to them, the main trade in this region occurs along the BR-116 road, as well as in fairs and small shops roadside.

Specialists point the absence of alternative income for people who use the trafficking as a means of livelihood. The report elaborated by the Brazilian National Congress in 2001 [12] recommends that the Union, States and Municipalities, in an articulated manner, must develop and implement programs to generate alternative income for poor communities involved in the illegal trade of wild animals.

However, the impact of trafficking in society needs to be further studied and its actors mapped. The capture of animals in nature is part of the culture and popular tradition, being one of the main livelihoods of the poor in some regions of Brazil [10]. However, in [11] found that in many regions people are using the illegal trade of animals only as an additional source of income. Thus, mechanism for control of wildlife use and trade should be formulated that take into consideration the special ethnic conditions of each region [13].

In global scale, it is recommended a multi-pronged approach including community-scale education and empowering local people to value wildlife, coordinated international regulation, and a greater allocation of national resources to on-the-ground enforcement for effective control of trafficking and illegal trade [9]. In Brazil, we noticed that the actions against illicit related to wildlife, although increasingly more organized and efficient, still require specific structural measures, among which we may highlight:

- Improving the number and the practice of IBAMA's agents and of Environmental Military Policeman through public competition and specific and continuous training;
- Increase the volume of public resources towards the activities of control and environmental monitoring;
- Increasing the incentive for the creation, implementation and maintenance of CETAS (Wild Animals Rehabilitation Centers);
- Reviewing the penal types of Law number 9.605/1998 due to provide harsher penalties for those who engage in wild animals trafficking such as large-scale commercial activity or international and interstate trafficking;
- Increasing responsibilities and sharing information among different agencies responsible for controlling and monitoring, through formal terms and shared systems;
- Maintening permanent negotiation between the federal government and neighboring countries through bilateral agreements, so that policies or environmental standards more flexible than the Brazilian ones are not used to support the illegal activities;
- Increasing the control over the sale of wild animals by internet and their exit to abroad through joint action among different government agencies such as IBAMA, the Federal Revenue Secretariat, Ministry of Health, Federal Police, etc..;
- Promoting Specific Environmental Education Campaigns aimed at minimizing the wild animals trafficking, as well as joint efforts among the various ministries involved, including the ones of Transport, Environment, Health and Tourism.

Lastly, we hope that this paper provides important and necessary subsidies for the decision-making to combat the animal trafficking in Brazil and abroad, helping the effective protection and conservation of the nature.

5. Conclusion

We conclude that the Minas Gerais State was the largest contributor to the large volume of specimens seized in Brazil in the analized term, being *Sicalis flaveola* (Saffron Finch), *Saltator similis* (Green-winged Saltator) and *Sporophila caerulescens* (Double-collared Seedeater) the species most confiscated by environmental enforcement.

Furthermore, we noticed that releasing into the wild was the most common destination for mammals, birds and reptiles seized. The Wild Animals Rehabilitation Centers are essential support structures for the environmental enforcement actions related to fauna in Brazil.

Author details

Guilherme Fernando Gomes Destro, Tatiana Lucena Pimentel, Raquel Monti Sabaini, Roberto Cabral Borges and Raquel Barreto
Coordination of Enforcement Operations, Brazilian Institute of Environment and Renewable Natural Resources – IBAMA. SCEN, Trecho II, Ed. Sede, Brasília/DF, Brazil

6. References

[1] RENCTAS (2011) Rede Nacional de Combate ao Tráfico de Animais Silvestres. 1º Relatório Nacional sobre o Tráfico de Fauna Silvestre. Available: http://www.renctas.org.br/. Acessed 2010 jul 15.
[2] Hernandez, E.F.T.; Carvalho, M.S. de. (2006) O tráfico de animais silvestres no Estado do Paraná. Acta Scientiarum: Human and Social Sciences, j. 28, n. 2: 257-266.
[3] Wasser, S.K.; Clark, W.J.; Drori, O.; Kisamo, E.S.; Mailand, C.; Mutayoba, B.; Stephens, M. (2008) Combating the Illegal Trade in African Elephant Ivory with DNA Forensics. Conservation Biology, j. 22, n. 4: 1065–1071.
[4] Alacs, E.; Georges, A. (2008) Wildlife across our borders: a review of the illegal trade in Australia. Australian Journal of Forensic Sciences. j. 40, n. 2: 147–160.
[5] Johnson, R.N. (2010). The use of DNA identification in prosecuting wildlife-traffickers in Australia: do the penalties fit the crimes? Forensic Scienci, Medicine and Pathology, j. 6: 211–216.
[6] Pavlin, B.I.; Schloegel, L.M.; Daszak, P. (2009) Risk of Importing Zoonotic Diseases through Wildlife Trade, United States. Emerging Infectious Diseases, j. 15, n. 11: 1721-1726.
[7] Borges, R.C.; Oliveira, A. de; Bernardo, N.; Martoni, R.; Costa, M.C. da. (2006) Diagnóstico da fauna silvestre apreendida e recolhida pela Polícia Militar de Meio Ambiente de Juiz de Fora, MG (1998 e 1999). Revista Brasileira de Zoociências, j. 8, n. 1: 23-33.
[8] Ferreira, C.M.; Glock, L. (2004) Diagnóstico preliminar sobre a avifauna traficada no Rio Grande do Sul, Brasil. Biociências, j. 12, n. 1: 21-30.
[9] Rosen, G.E.; Smith, K.F. (2010) Summarizing the evidence on the international trade in illegal wildlife. EcoHealth, j. 7: 24–32.

[10] Souza, G.M. de; Soares Filho, A. de O. (2005) O comércio ilegal de aves silvestres na região do Paraguaçu e Sudoeste da Bahia. Enciclopédia Biosfera, j. 1: 1-10.

[11] Rocha, M. da S.P.; Cavalcanti, P.C. de M.; Souza, R. de L.; Alves, R.R. da N. (2006) Aspectos de comercialização ilegal de aves nas feiras livres de Campina Grande, Paraíba, Brasil. Revista de Biologia e Ciências da Terra, j. 6, n. 2: 204-221.

[12] BRASIL (2001) Comissão Parlamentar de Inquérito destinada a investigar o tráfico ilegal de animais e plantas silvestres da fauna e da flora brasileiras – cpitrafi. Relatório Final. Available: <http://www2.camara.gov.br/atividade-legislativa/comissoes/comissoes - temporarias/parlamentar-de-inquerito/51-legislatura/cpitrafi/relatorio/relatoriofinal.pdf>. Acessed 2010 sep 20.

[13] Yi-Ming L., Zenxiang, G.; Xinhai, L.; Sung, W.; Niemela, J. (2000) Illegal wildlife trade in the Himalayan region of China. Biodiversity and Conservation, j. 9: 901–918.

Section 5

Genetics and Hereditary

The Influence of Geochemistry on Biological Diversity in Fennoscandia and Estonia

Ylo Joann Systra

Additional information is available at the end of the chapter

http://dx.doi.org/10.5772/48386

1. Introduction

The Earth's crust is predominantly composed of a relatively small number of chemical elements. Only eight of them: oxygen, silicon, aluminum, iron, calcium, magnesium, sodium and potassium are present in amounts exceeding one weight percent and together they comprise almost 99% of the entire crust [1, 2]. Some elements are exceedingly rare in the Earth, or have short-lived radioactive isotopes. For example, promethium (Pm) occurs in the crust in only very small concentrations in certain uranium ores, being produced as result of nuclear fission, with the longest-lived isotope having a half-life of only 17.7 years. Similarly, technetium is a relatively light radioactive metal with atomic number 43, having no stable isotopes, while the longest-lived radioactive isotope (Te-98) has a half-life of 4.2 million years. Technetium only occurs naturally, in trace amounts, in uranium ores produced by nuclear fission [3]. When such elements are excluded, there remain 90 naturally occurring chemical elements [4] to form the geochemical basis of the life of the Earth.

2. Distribution of chemical elements in the biosphere of the Earth

The Earth's biosphere is consists of crust, hydrosphere and atmosphere. The Earth's crust is predominantly composed of 9 chemical elements, each having abundances, expressed in atomic weight, of more than one percent: oxygen (53.39%), hydrogen (17.25%), silicon (16.11%), aluminum (4.80%), sodium (1.82%), magnesium (1.72%), calcium (1.41%), iron (1.31%) and potassium (1.05%). The next most abundant elements are: carbon (0.51%), titanium (0.22%), chlorine (0.10%), fluorine (0.07%) and sulfur (0.05%). Together these 14 elements comprise 99.81% of the crust in terms of atomic weight, the remaining 76 elements representing only 0.19% or on average 0.0025 atomic weight percent each [5]. Some of these elements are therefore referred to as rare earth elements (REE). Zinc, copper, nickel, chromium and manganese, which are common in everyday use, also represent trace elements. The number of minerals recognized is also in accordance with crustal element abundances. A total of 2909 containing oxygen minerals are known, 1921 minerals contain hydrogen, 906 contain silicon, 714 contain aluminum, 560 minerals contain Na, 555 contain, Mg, 272 contain K, 272 also contain C and 172 minerals contain Ti. The chemically active elements Cl and F occur in 220 and 221 elements respectively. However, due to their greater chemical activity, relatively more minerals have been described containing Ca (867), Fe (883) and S (761) [5].

There are some differences between mean cosmic abundances and those of the Earth's crust, but the higher abundances of light elements are shared by both environments, implying that the process of nuclide synthesis obeys the same rules. The Oddo-Harkins rule states that for any two neighboring elements, the abundance of the element with an even atomic number is higher than that of the odd one. Of the 28 first elements of Earth's crust the even elements in the periodic table constitute 86.36 weight percents, while odd elements comprise only 13.64%. There are some exceptions from the rule. The noble gases (atomic numbers in brackets): helium (2), neon (10), argon (18), krypton (36), xenon (54) abundances are much lower than precited by the Oddo-Harkins rule, because they have stable nuclei and are chemically inert, and hence do not participate in chemical reactions or form compounds. The noble gases are predominantly present in the atmosphere, with argon abundance being nearly 1%, while the others are present in small amounts, because they can easily escape into space [2].

It must also be noted that especially high abundances are associated with elements for which numbers differ by 6 or multiple 6 and plus 2, such as O (8), Si (14), Ca (20), Fe (26), Sr (38), Sn (50), Ba (60). Of the most abundant elements in the Earth's crust elements, 13 have atomic numbers between 1 and 20, with only iron having a higher number, of 26 [6]. On the diagram of cosmic nuclear abundances plotted against atomic mass numbers, a significant peak indeed appears in the region of iron. This implies that that Earth's core may also contain an abundance of elements of lower atomic number, including iron. The dominance of the lighter elements is easily understood, because less energy is required for the formation. Heavy nuclei have higher electric charges than light nuclei, while for reactions between them also tend to require higher temperatures [7, 8].

Chemical elements as a rule are represented by several different isotopes, for which atoms have the same atomic number, but differ in terms of atomic mass. Tin (Sn) has the largest number (10) of naturally occurring stable isotopes, while cadmium and tellurium both have 8 isotopes. Many other elements have from 2-6 isotopes and only 22 elements, including F, Na, P, V, Mn, Au, have one single stable isotope. The highest isotopic abundances also tend to correlate with those isotopes for, which the atomic mass can be divided by 4. Oxygen isotopes have the following order of abundance: ^{16}O – 99.76%, odd ^{17}O – 0.04%, even ^{18}O – 0.2%; for silicon, isotope abundances are: ^{28}Si – 92%, ^{29}Si – 5% and ^{30}Si – 3%. In the case of carbon isotopes, ^{12}C represents 99% and ^{13}C about 1%; both of these are non-radioactive, the third isotope ^{14}C is radioactive and occurs in trace amounts in the atmosphere [3].

The 90 naturally occurring elements are found throughout the Earth's crust, in bedrock, groundwater, unconsolidated sediments and soils. More than 70% of Earth's surface is covered by oceans and seas, which provide habitats and diverse environments for numerous species of very different size. One ton of seawater contains the following elements by mass: Cl (19.4 kg), Na (10.8 kg), Mg (1,29 kg), S (0.9 kg), Ca (411 g), K (392 g), Br (67 g) and 10 trace elements (in order of abundance from 8.1-0.003 g): Sr, B, F, Li, Ru, Ba, I, As and Cs [2]. The composition of atmosphere by volume is: nitrogen (78.09%), oxygen (20.95%), argon (0.95%), CO_2 (0.03%), and neon (0.0018%); trace gases are in order of abundance (from 0.0052 to 0.000 008%): He, methane, Kr, nitrous oxide, O, O_3 and Xe. The abundance of ozone is notable in that it increases with altitude [2].

The biosphere contains innumerable species of plants, animals, fish, birds, insects and other forms of biological life. None of them can exist without water and numerous chemical elements as nutrients.

3. Chemical elements as nutrients for all forms of biological life

All plants and animals need a range of different chemical elements as nutrients for normal and healthy growth and development [9]. However, not all of the most widely distributed elements are needed in large amounts. Of the 90 naturally occurring elements on Earth, there are 11 elements found throughout the atmosphere, hydrosphere and within bedrock and soils, that are essential for all plants and animals, namely oxygen, hydrogen, carbon, nitrogen, calcium, magnesium, sodium, potassium, sulfur, phosphorus and chlorine. These elements are also known as macronutrients, of which the first four can be easily obtained from air (C, N, O) or water (H, O), while the remaining 7 elements are more common in bedrock or are present at required concentrations in groundwater. All these elements are essential to life and for example, an adult human requires a daily intake of 100 mg or more of each of these elements [10].

The sources of mineral elements are soils, drinking water and food – each kind of food has its own distinct composition of macro- and micronutrients. For example, phosphorus, calcium, sodium and magnesium are present in every kind of food, but sulfur is usually absent from dried produce [11]. The abundances of the main macronutrients in marine and terrestrial plants, marine and terrestrial animals, and in bacteria (in dry matter) are remarkably similar to one another, which indicate that all forms of living life require the same macro-elements (Table 1).

All macro-biogenic elements are essential for normal life functions and are vital for manufacturing cells and tissues in plants and animals. There is some variation in the actual element abundances according to specific environmental conditions and plant and animal requirements. Terrestrial animals need stronger skeletal support than marine organisms and their proportional abundances of phosphorus and fluorine are consequently 3 and 7 times higher. Other elements, such as potassium, sodium, sulfur and magnesium are more readily available in marine environments than on land and their concentrations in marine plants and animals are accordingly higher. Bacteria do not differ very markedly from other groups, indicating that all life forms on the Earth are essentially composed from the same chemical elements. The human being is not exception, a typical 70 kg human body consisting of the following major elements: oxygen (61%), carbon (23%), hydrogen (10%), nitrogen (2.6%), calcium (1.4%), phosphorus (1.1%), potassium (0.2%), sulfur (0,2%), sodium (0.14%), chlorine (0.12%) and magnesium (0.027%). However, not all of these widely distributed chemical elements are needed in large amounts by plants and animals. Four elements in particular: silicon, aluminum, iron and titanium, are only required in very small amounts as micro-nutrients for the regulation of specific processes [3].

Macronutrients, grams in 100 g	Marine plants	Terrestrial plants	Marine animals	Terrestrial animals	Bacteria	Average of all groups
Carbon – C	34.5	45.4	40.0	46.6	54.0	44.1
Oxygen – O	47.0	41.0	40.0	18.6	23.0	33.7
Hydrogen – H	1.5	3.0	7.5	10.0	9.6	6.32
Nitrogen – N	4.1	5.5	5.2	7.0	7.4	5.06
Potassium – K	5.2	1.4	0.5-3.0	0.74	11.5	4.12
Calcium – Ca	1.0	1.8	0.15-2.0	0.02-8.5	0.51	1.44
Sodium – Na	3.3	0.12	0.4-4.8	0.4	0.46	1.38
Phosphorus – P	0.35	0.23	0.4-1.8	1.7-4.4	3.0	1.29
Sulphur – S	1.2	0.34	0.5-1.9	0.5	0.53	0.75
Magnesium – Mg	0.52	0.32	0.5	0.1	0.7	0.43
Chlorine - Cl	0.47	0.2	0.5-9.0	0.28	0.23	0.29

Table 1. Macro-elements content in marine and terrestrial plants and animals, and bacteria (in grams to 100 g dry matter), modified after Barabanov [12]

The distribution of macro- and micronutrients in different types of bedrock, soils, plants and animals have been studies intensively for more than 50 years, but due to the large ranges in composition, no consensus has been reached on the precise abundances of essential trace element concentrations in living organisms [9-10, 13-22]. The reason for this is that micronutrients are present and needed in very small amounts. For example, selenium is an important microelement for humans but daily intake must not exceed 28-55 μg [22]. The most complete list of micronutrients is given by Emsley [15], which was published on the Internet by Uthman [20]. For an average 70 kg human body mass the abundances of 59 elements have been determined, including 11 major elements. Of 48 trace elements 43 are considered essential, while the role of thorium, uranium, samarium, beryllium and tungsten is not known. Vanadium is the least abundant biologically necessary element in the human body (0.11 mg) whereas rubidium is the most abundant element in the body (0.68 g) that lacks any biological role. Silicon (1.0 g) may or may not have a metabolic function. Only four elements exceed the 1.0 g level: Fe (4.2 g), F (2.6 g), Zn (2.3 g) and Si (1.0 g), and another four have levels greater than 0.1 g: Rb (0.68 g), Sr (0.32 g), Br (0.26 g) and Pb (0.12 g). The abundances of the remaining elements fall within the interval between 72–0.11 mg [15, 20].

Table 2 shows 29 trace element concentrations in marine and terrestrial plants and animals, and also in bacteria for some elements. The situation with respect to microelements is nearly the same as for major elements. The contents of silicon, iron, zinc, aluminum, rubidium, nickel, cobalt, molybdenum in each group are quite comparable, although some other elements show greater differences in concentration than in the case of macro-nutrients. The abundances of bromine, strontium, iodine, boron, titanium lithium and chromium are higher in marine plants and animals, which is to be expected given that sea water contains more of these elements [2] and that uptake from solution is relatively easy. Excess concentrations of many of the trace elements present in soils, including: As, Cd, Co, Cu, Cr, Hg, Mo, Ni, Pb, Se, Zn, U, Th and Zn, may be toxic to plants and animals or may affect the quality of foodstuff for human consumption [14]. They are potentially toxic to plants and animals, but can have adverse effects at relatively low, insufficient level [21]. In many countries, including Estonia, there are governmental regulations concerning maximum permitted concentrations for toxic and dangerous elements in agricultural lands, industrial and urban environments, and in drinking water.

Microelements, mg in 100 g dry matter	Marine plants	Terrestrial plants	Marine animals	Terrestrial animals	Bacteria	Average of all groups
Silicon – Si	150-2000	20-500	7-100	12-600	18	342.5
Iron – Fe	70	14	40	16	25	33.0
Bromine – Br	74	1.5	6-100	0.6	–	32.3
Zinc – Zn	15	10	0.6-150	16	–	29.1
Strontium – Sr	26-140	2.6	2-50	1.4	–	28.3
Iodine – I	3-150	0.042	0.1-15	0.043	–	21.0
Manganese – Mn	5.3	63	0.1-6	0.02	–	17.8
Aluminum – Al	6	50 (0.05-400)	1.5	0.4-10	–	15.7
Fluorine – F	0.45	0.005-4	0.2	15-50	–	8.79
Boron – B	12	5	2-5	0.05	–	5.13
Copper – Cu	1	1.4	0.4-5	0.24	4.2	1.91
Rubidium – Rb	0.74	2	2	1.7	–	1.61
Titanium – Ti	1.2-8	0.1	0.02-2	0.02	–	1.43
Barium – Ba	3	1.4	0.02-0.3	0.075	–	1.16
Arsenic – As	3	0.02	0.0005-0.03	≤ 0.02	–	0.76
Nickel – Ni	0.3	0.3	0.04-2.5	0.08	–	0.49
Lead – Pb	0.84	0.27	0.05	0.2	–	0.34
Tin – Sn	0.1	< 0.03	0.02-2	< 0.015	–	0.28
Vanadium – V	0.2	0.16	0.014-0.2	0.015	–	0.21
Silver – Ag	0.025	0.006	0.3-1.1	0.0006 (?)	–	0.18
Lithium – Li	0.5	0.01	0.1	<0.002	–	0.15
Cobalt – Co	0.07	0.05	0.05-0.5	0.003	–	0.099
Selenium – Se	0.08	0.02	–	0.17	–	0.090
Molybdenum – Mo	0,045	0.09	0.06-0.25	0.02	–	0.078
Cadmium – Cd	0.04	0.06	0.015-0.3	≤ 0.05	–	0.070
Chromium – Cr	0.1	0.023	0.02-0.1	0.0075	–	0.048
Antimony – Sb	–	0.006	0.02	0.0006	—	0.009
Mercury – Hg	0.003	0.0015	–	0.0046	–	0.003
Tungsten – W	0.0035	0.007	0.00005-	?	–	0.0016

Table 2. The content of chemical microelements in marine and terrestrial plants and animals, and bacteria, modified after [12]

The role of chemical elements in the natural environment first became the subject of study during the early decades of the twentieth century, with the first significant findings emerging in the middle of the century. In some countries, as in Great Britain, the importance of geochemical investigations has been recognized and has been supported by government-funded research. The discovery of irregularities in the distribution of diseases in the early 1960's, particularly with respect to the incidence of cancer in southwest England, initiated a detailed geochemical survey of Britain by a multi-disciplinary research team [23]. Some 20 years later, following numerous international conferences and workshops in environmental geochemistry and health, the main results of investigations were published in a special book [14], which presented for the first time, information concerning all aspects of environmental geochemistry and the distribution of harmful elements in the environment in which people live [23].

Such geochemical studies represent an important component of the environmental and geological research programs in USA, Canada, Australia, Russia, Japan, China, Germany, Poland, Sweden, Norway, Finland and many other countries.

4. Methods of studying the influence of geochemistry on biodiversity in Fennoscandia and Estonia

Specialized studies within areas covered by old-growth forests were initiated during the late 1980's in the Republic of Karelia. Much of the forest in this part of the former Soviet Union was intensively logged during the fifty years or so following the 2nd World War. As demand by the forestry and paper industry increased, there was increasing recognition of the need to preserve areas of taiga forest within a network of National Parks. As a result, the bilateral Russian-Finnish project "Inventory and studies of biological diversity in Republic of Karelia" was initiated in 1997-2000 by the Karelian Research Centre of the Russian Academy of Sciences. This project included investigations into bedrock, Quaternary sediments and soil geochemistry, vegetation, forests, mires, fish populations, algal flora and zooplankton, bird fauna, mammals and insect populations in different environments of the Republic of Karelia, with particular emphases on areas of high nature conservation value [24]. The author of the current paper was involved in the project as a geologist and as a result of 30 years experience, is familiar with bedrock composition and geochemistry in all kinds of terrain. Fieldwork during the years 1962-1984 in North Karelia and another parts of Karelia principally comprised many thousand kilometers of geological traverses on foot, from the White Sea to Russian-Finnish border, where significant areas of old growth forest remained [Figure 1].

The research group at the Karelian Research Center of Russian Academy of Sciences was characterized by close collaboration between specialists in different fields, including forestry, vascular plants, mosses, lichens, mammal species, bird fauna, insects, algal flora, periphyton-, zooplankton-, macrozoobenthos-, and fish communities. The interactions between these specialists during the field studies resulted in much new information being obtained and published at first in Russian, as progress reports of field studies, and finally in

Russian and English. The final report [24] provides extensive material for researchers in widest wide range of ecological and biological fields. Geological reviews of the Karelian Region and more detailed local studies, with numerous chemical analysis of bedrock have been published in Systra [27-36] and in [37]. The general geochemical characteristics for soils in the Republic of Karelia were published by Fedorets et al. [38], while a geochemical study of the bedrock of Valaam Island, Lake Ladoga, was done by Sviridenko and Svetov [39], forest litter and underlying mineral material were studied by Shiltsova et al. [40], and the distribution of vascular plants in Republic of Karelia was reported by Kravchenko et al. [41].

Detailed fieldwork was done in the Onega Synclinorium and the central part of the Zaonezhje Peninsula, the Paanajärvi National Park, and along the western and northwestern shore of Lake Ladoga. The results of previous research also facilitated characterization of the Western White Sea area and Central Karelia. Later, in 2002-2003 and 2008-2009, LAPBIAT and LAPBIAT-2 financial support made it possible to visit key areas for studying the influence of bedrock geochemistry on vegetation in subarctic Lapland in Finland, in the surroundings of Kilpisjärvi and Kevo, and in the Oulanka National Park, Kuusamo. The influence of carbonate rocks on vegetation in the Kilpisjärvi Region in Lapland, at latitude 69º N, was already noted by Pesola [42] at the beginning of the last century, leading to the creation in 1916 of the first protected area in Finland (now Malla Strict Reserve). Subsequently, in 1956, the Oulanka River valley near the Arctic Circle, the Oulanka National Park was declared. On the Russian side of the border the Paanajärvi National Park is larger still (1040 sq. km) and was created in 1992. The geochemical control of vegetation type is clearly marked, especially in the old forests, which are in near pristine condition and the chemical composition of bedrock is well studied. A comprehensive description of the bedrock of Finland is given in Lehtinen et al. [43, 44] and bedrock chemical characteristics in Rasilainen et al. [45].

The composition of bedrock and soils in Estonia are also well studied [46-49]. Environmental protection and the study of vegetation, birds and mammals has has a history extending back more than 100 years. Based on data from about 1550 analyzed samples, the geochemical atlas of the humus horizon of Estonian soils was prepared by Petersell et al. [49]. The 33 geochemical maps in this study, and the "Atlas of Estonian Flora" with 1353 species maps of vascular plant distribution [50] was used for accessing the influence of bedrock on vegetation.

The area considered here embraces the eastern part of Fennoscandian Shield from Kilpisjärvi and Halti Mountain and Kevo in Finnish Lapland (69º45′ N) to northern and northwestern Ladoga Lake (61°15′ N) and Estonia (59°40′- 57°30′ N). The ancient shield is represented geologically by crystalline metamorphic and igneous bedrock of Archean and Early Proterozoic ages. In Estonia these crystalline rocks are completely covered by Late Proterozoic and Paleozoic sedimentary rocks. Crystalline and sedimentary rocks differ in their respective physical and mechanical properties, and chemical composition [2]. The geology of Finland is well known and covered by comprehensive geological and geochemical maps [25-26, 45].

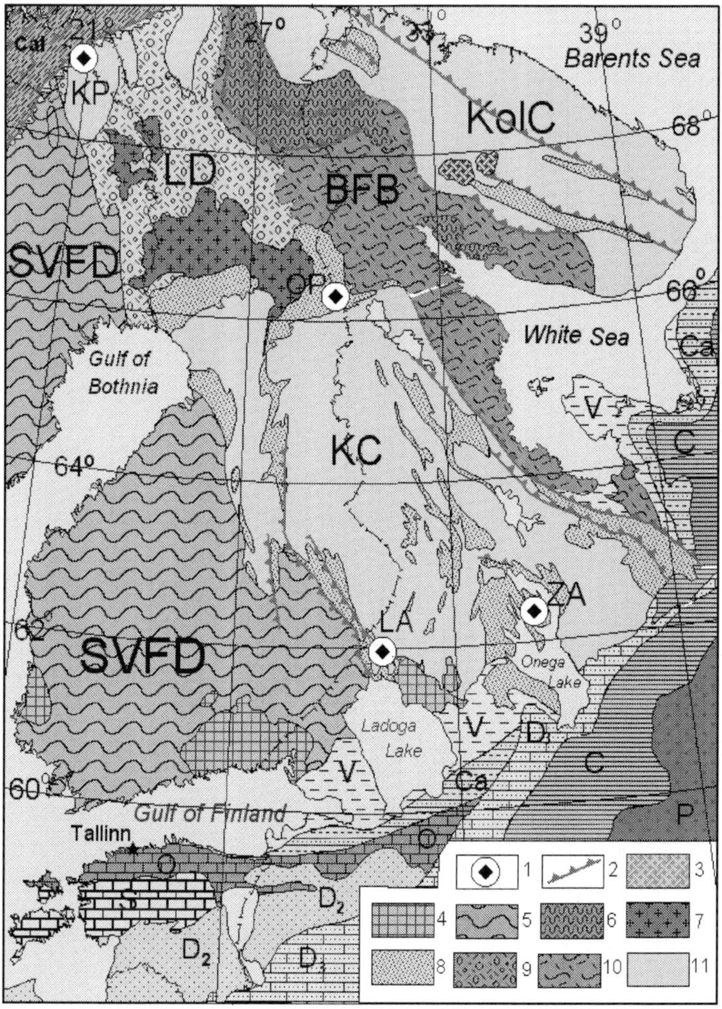

Legend: 1 – Places of detailed study: KP – Kilpisjärvi, OP – Oulanka-Paanajärvi area, ZA – Zaonezhje peninsula, Onega synclinorium, LA – northwest Ladoga Lake shore. 2 – thrusts; 3 – Devonian nepheline syenite; 4 – rabakivi granites; 5 – Svecofennian folded domain (SVFD); 6 – Lapland-Belomorian Granulite Belt; 7 – Kemijärvi granites.
Early Proterozoic volcanic-sedimentary rocks: 8 – Lapland Domain (LD); 9 – cover on the Archean basement. Archean: 10 – Belomorian Fold Belt (BFB); 11 - Archean cratons: Kola Craton (KolC) and Karelian Craton (KC).
Neoproterozoic-Paleozoic sedimentary cover (from youngest): P – Permian and younger rocks:
C – Carbon (carbonate rock, coal, clay, sandstone); D_3 – Upper Devonian (carbonate rocks, sandstone, clay); D_2 – Middle Devonian (sandstones); Cal – thrusted Scandinavian Caledonides; S and O – Silurian and Ordovician (carbonate rock, oil shale); Ca – Cambrian (sandstone, claystone); Ediacaran, Vendian complex (arkose, sandstone, claystone).

Figure 1. Geological map of the eastern part of Fennoscandian Shield and northwest marginal zone of the Russian Platform. Compiled by Y.J.Systra with using geological map of Finland [25] Fennoscandia [26] and unpublished author's map „Tectonics", Karelia (1996).

For all of the areas studied, sufficient data were available concerning bedrock chemical composition. In the Onega region some 20 complete and partial soil profiles were analyzed for major and trace element chemistry, including sampling from both bedrock and soil horizons, in order to quantitatively assess the influence of bedrock on soil composition. Geochemical data were carefully compared with the distribution of each group of plant and animal distribution; for rare and endangered species the Red Data books of Eastern Fennoscandia [51] and the Republic of Karelia [52] were useful.

Comparisons between geochemistry in different environments and the diversity of vegetation and, especially with the distribution of rare and protected plants, showed that biodiversity is influenced by different geological and geographic features, such as latitude and its solar aspect, relief and orientation of landforms, the presence of fault zones as conduits for mineralized groundwater and springs, the color of exposed bedrock and the presence of migration corridors. Nevertheless, the main control on biodiversity is usually geochemical, where the bedrock and overlying regolith contains an abundance of essential macro- and micronutrients. The Fennoscandian Shield and the surrounding Russian Platform marginal zone with its 150-700 m thick sedimentary cover sequence is an excellent area for studying the interdependence of geochemistry and biological diversity. Extensive areas have never been used for agriculture and many are currently protected from commercial exploitation. The biodiversity of the region is also relatively young, in that the continental ice sheet last withdrew from Southern Estonia between 13 500–11 700 years ago, and from northern Fennoscandia between 9000–6500 years ago. Indeed, natural exchange and colonization plants and animals, especially birds, between Oulanka National Park, (Finland) on the western side of the Maanselkä topographic divide and Paanajärvi National Park in Russian Karelia to the east, is continuing to the present day.

5. Results

The results of the study are presented separately for each region, because their geological and geochemical conditions are very different. The Republic of Karelia, in the Russian Federation, has a total area of 172 400 km^2 [53], and occupies the southeastern part of the Fennoscandian Shield. The Karelian part of the Shield comprises 3 major northwesterly trending geological domains, the Karelian Craton in the centre, the Archean Belomorian fold belt to the northeast and the Svecofennian fold terrane to southwest [35]. The exposed part of the Karelian Craton is about 600 km long and 300 km wide, with the southeastern edge being covered by Late Proterozoic Ediacaran and Paleozoic sedimentary rocks, which are contiguous with those in Estonia. The Karelian Craton formed during two distinct orogenic events, the first of which is represented by the Archean (3.8-2.5 Ga) basement, composed of Paleoarchean gneiss, gneissose diorite and migmatites, and narrow Meso- and Neoarchean greenstone belts, consisting of volcanic-sedimentary rocks. The younger phase of evolution of the craton is recorded by Paleoproterozoic (2.5-1.8 Ga) volcanic-sedimentary sequences which are preserved in synclinorial structures. The Archean and Proterozoic rocks differ greatly in geochemical characteristics: basement granites and migmatites are rich in SiO_2, but contain only very small amounts of micronutrients, while Paleoproterozoic rocks are more diverse in composition, including many types of volcanic rocks, sandstone, quartzite, carbonate rocks, and shungite-bearing black schists (Figure 2). The areas selected for detailed study in the Republic of Karelia were in the Paanajärvi national park near the Arctic Circle, supplemented by brief studies in the adjoining Oulanka National Park in Finland the northwestern shoreline of Lake Ladoga, focusing on the influence of the Mesoproterozoic Valaam gabbro-dolerite sill on soil composition and vegetation, and the Onega Synclinorium, each of which represent Paleoproterozoic bedrock having favorable compositions for supporting diverse vegetation.

Estonia is geologically situated on the northwest margin of the East-European Platform, where the Precambrian crystalline basement is covered by Ediacaran and Paleozoic sedimentary cover, varying in thickness from 120-150 m in northern Estonia to 500-780 m in southern parts of the country. The basement surface and all sedimentary cover layers are tilted gently southwards at a gradient of nearly 2.5-3 m per kilometre. Cambrian, Ordovician and Silurian sediments are exposed in Northern Estonia as east-west trending belts, whereas to the south these older sequences are covered by younger Middle and Late Devonian deposits. As everywhere in Fennoscandia the bedrock is usually covered by a thin veneer of Quaternary soft sediments, which contain material derived from both local bedrock as well as clasts transported from the crystalline bedrock of Finland and Sweden. Accordingly, since the southern marginal zone of the Fennoscandian shield is represented mostly by migmatites and gneisses, more than 90% of the transported glacial deposits consist of granitic material. This in turn has an influence on soil composition and vegetation in Estonia.

6. Paanajärvi-Oulanka protected reserves

Conservation areas near the Arctic Circle (66°10′- 66º30′ N) include two significant and contiguous national parks straddling the national borderline between Finland and Russia. The older and smaller Oulanka National Park (277.2 km^2) was established in Finland in 1956, while the larger Paanajärvi National Park (1045 km^2) in Russian Karelia was declared in 1992. Both parks share similar geological features and vegetation types. The Paanajärvi–Oulanka area has a significantly higher level of biodiversity compared to surrounding areas, due to a combination of mountainous terrain, the east-west orientation of deep valleys, microclimate and distinctive bedrock composition (Table 3).

Paleoarchean basement gneisses and granites (Table 3, No 2906, No 2969) are rich by SiO_2, but contain limited amounts of Mg and Ca and biologically needed micronutrients and (Figure 2). Small metamorphosed Meso- and Neoarchean ultramafic bodies (Table 3, No 2957-2) and thin komatiite layers between basalts contain appropriate microelements, but have very restricted distribution and do not significantly influence soil composition. The nutrient richness of soils is instead due to the Paleoproterozoic intermediate, acid, mafic, ultramafic volcanic rocks and different types of sedimentary rocks, including dolomite marbles and carbon-rich black schist. Effusive rock types such as felsic quartz porphyry (Table 3, No 3299) also contains higher concentrations of the trace metals Ti, Fe, K, Ba, Mn and V than granites. Intermediate basaltic andesite (Table 3, No 3319) has still higher abundances of Fe, Mg, Ca, Co, Cr, Cu and Ni. The mafic volcanic diabase (Table 3, No 2932-3) and leucogabbro (Table 3, No 3337) have similar compositions, although there are some differences in macro-element abundances; both rock types nevertheless contain adequate amounts of necessary microelements. Because mafic volcanic rocks are widely distributed in both national parks, there is no deficiency in microelements in the region. Although conglomerates (Table 3, No 2942) and quartzites (Table 3, No 3167-10) are SiO_2-rich, they nevertheless contain some Co, Cr, Mn, Ni and V. Quartzite sampled from a fault zone is brecciated and mineralized with fuchsite (Cr), Ca, Ba and carbonates, as indicated by the higher weight losses on ignition (LOI) and Ca abundances. The dominant source of Mg and Ca in soil is dolomitic marble (Table 3 No 2944), which is common throughout the bedrock of the area and consists of about 95% dolomite $(Ca,Mg)CO_3$: CaO 32.22%, MgO 19.51% and CO_2 43.12% (LOI).

All of the above-mentioned rock types are well represented in Quaternary glacial and glaciofluvial deposits, as glacial erosion for some 50-60 km along the deep Oulankajoki River-Paanajärvi Lake-Olanga River valley was entirely within bedrock of the Paleoproterozoic volcanic and sedimentary cover sequence. Fragments of marble are common in till and fluvioglacial sediments everywhere in the studied area, their influence on soil composition can be observed some km to east, outside of the Paanajärvi syncline, in areas underlying by Archean basement (Figure 2).

This favorable soil composition is the main reason for the unusual biological diversity of the Paanajärvi-Oulanka national parks. About 600 species of vascular plants have been identified, 67 of which have never have been reported from other parts of Karelia [24, 30]. There are 298 species of, 42 of which are listed in the Red Book of Karelia [52], while 443 species of lichen species have been recorded [56], of which 10% are rare or endangered. This number of taxa represents twice that of the corresponding figure for surrounding terrain. The Oulanka and the Paanajärvi national parks are underlain by Paleoproterozoic (2.5-1.9 Ga) volcanic and sedimentary rocks – including carbonates, while the bedrock in the surrounding region consists of diorites, granites and gneisses of the Karelian Craton [27-30, 35].

The mountainous nature of the terrain, including the higher peaks, which are over 450 m high and lie above the tree line, means that the area lies within the mountain tundra belt. Alpine-Arctic vegetation, including *Loiseleuria procumbens* L., *Phyllodoce caerulea* (L.) Bab., *Diphasiastrum alpinum* L. and other species is present on the highest mountains Nuorunen (576 m), Mäntytunturi (550 m) and Kivakka (499.5 m). Along the Kiutaköngäs rapids in the Oulanka River valley, a relict community of *Dryas octapetala* L. has survived since the Ice Age, together with *Calipso bulbosa* L. In contrast, there are no more than 30-35 species of more northerly affinity. On the Paleoproterozoic volcanic and sedimentary rocks, soli acidity is near-neutral, and soils are enriched in all essential trace elements. Bedrock geochemistry is the main reason for the high biological diversity here, principally due to the presence of carbonate rocks; the dolomitic marbles are responsible for buffering the neutral soil composition. Most occurrences of *Cypripedium calceolus* L. (Lady's Slipper) are confined to carbonate rock exposures, where up to 2000 individuals may be found an area of some hundred square meters. Slope aspect and exposure also has a marked effect on microclimate. Thus, southerly species inhabit south-facing slopes and deep valleys which show a greenhouse effect, whereas northern species persist under favorable conditions on cold north-facing slopes and in deep shady valleys.

Component	2906	2969	2957-2	3299	3319	2932-3	3337	2942	3167-10	2944
SiO_2	69.66	74.64	44.74	69.40	55.00	50.46	51.10	84.64	85.36	3.54
TiO_2	0.27	0.10	0.29	0.77	0.65	0.73	0.43	0.10	0.12	0.04
Al_2O_3	15.19	14.67	6.90	12.86	13.00	14.84	16.26	8.40	2.80	0.26
Fe_2O_3	0.60	0.18	4.22	2.29	2.40	3.51	1.47	0.77	0.23	0.17
FeO	2.15	1.01	7.90	3.23	7.54	9.05	5.60	0.21	0.73	0.93
MgO	1.31	0.45	25.60	0.50	7.10	8.80	7.60	1.35	0.30	19.51
CaO	1.82	1.96	4.28	1.96	7.43	6.09	11.43	0.14	4.76	32.22
K_2O	2.66	0.93	0.03	4.19	2.13	0.85	1.52	2.86	0.79	0.05
Na_2O	5.00	5.45	0.13	3.50	2.50	2.53	2.15	0.03	0.02	0,03
P_2O_5	–	–	0.06	–	–	0.124	–	0.072	–	–
S	0.01	0.04	–	–	–	0.02	–	0.01	–	0.04
LOI	0.61	0.26	5.08	0.67	1.61	2.54	2.00	1.10	3.83	43.12
Ba	536	626	–	1584	806	–	985	–	4475	–
Co	20	20	–	–	63	63	55	16	–	–
Cr	109	146	796	78	260	135	114	104	598	–
Cu	16	8	–	–	64	32	24	8	–	–
Mn	349	202	1550	891	1550	837	1085	39	202	1465
Ni	20	20	632	–	95	150	142	39	–	102
V	31	10	78	62	47	125	292	31	61	–
Li	–	–	–	35	39	–	10	–	–	–
Rb	–	–	–	71	66	–	22	–	–	–
Zn	–	–	–	146	–	–	–	–	–	—

Table 3. Chemical composition of bedrock in the North-Karelian synclinal zone in Russia (represents the Oulanka–Paanajärvi parks area). Analyses from Systra [27, 28]. Macro-element oxides in weight % (sulphur as element), microelements in ppm, analyses made in Laboratory of Institute Geology, Karelian Research Centre, Russian Academy of Sciences. (–) component not analyzed, Archean bedrock: No 2906 – biotite gneiss, No 2969 – gneissose plagiogranite, No 2957-2 – metapyroxenite. Paleoproterozoic bed-rock: No 3299 – quartz porphyry, No 3319 – andesitic basalt, No 2932-3 – diabase, No 3337 – leucogabbro, No 3167-10 – quartzite with fuchsite from fault zone, No 2942 – Proterozoic basal quartz rich conglomerate, No 2944 – dolomite marble

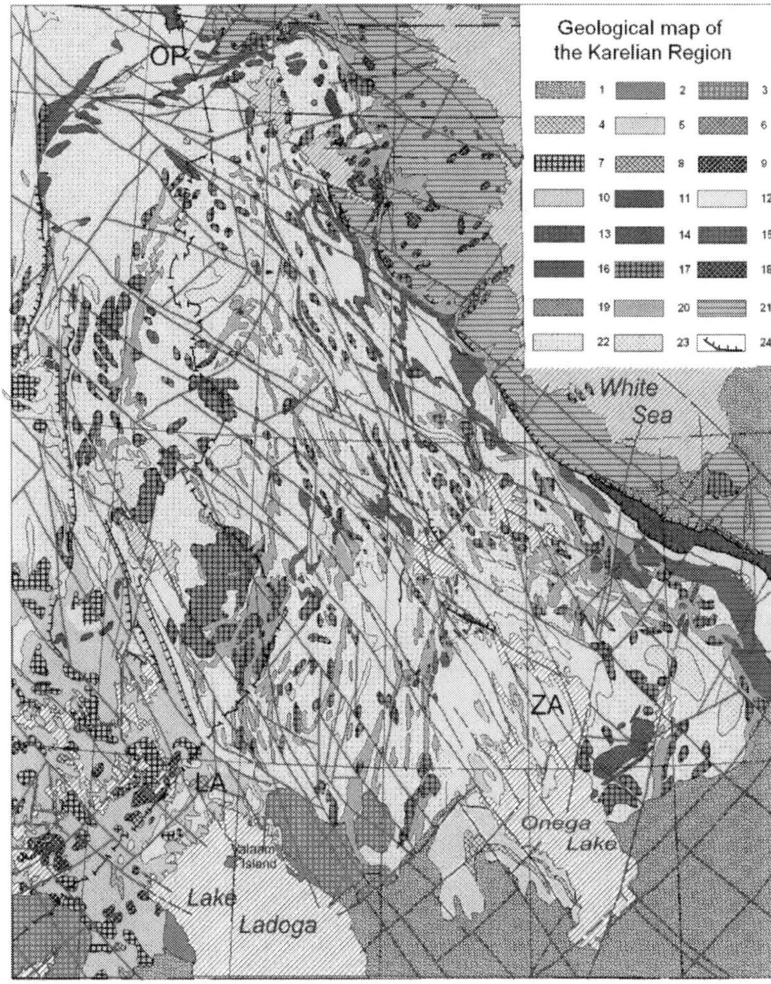

Legend: Post-Svecofennian bedrocks: 1 – Ediacara-Paleozoic platform cover; 2 – Ladoga aulacogene; 3 – post-orogenic rabakivi granites; 4 – Vepsian dolerite sills; 5 – Vepsian sedimentary rocks. Rocks of the Svecofennian (1.92-1.77 Ga): alkaline diorites and gabbro; 7 – granites; 8 – gabbro; 9 – diorites; 10 – Kalevian volcanic-sedimentary rocks; 11 – 1.98-1.95 Ga ultramafic rocks; 12 – Ludicovian and Jatulian sedimentary and volcanic rocks, including carbonates. Sumian and Sariolian (2.5-2.3 Ga) bedrocks: 13 – komatiitic basalts; 14 – layered mafic-ultramafic rocks; 15 – paligenetic granites; 16 – sedimentary-volcanic bedrocks. Neo- and Mesoarchean (2.6-3.0 Ga) bedrocks: 17 – granites; 18 – diorites; 19 – gabbro; 20 – sedimentary-volcanic rocks of greenstone belts; 21 – Belomorian Foded Belt, undivided. The oldest rocks in the region: 22 – basement granitic gneisses in the Karelian Craton; 23 – relics of the oldest sedimentary and volcanic sequences; 24 – thrust zones. Red lineaments: the main fault zones of Karelian Region. The places of study the bedrock geochemistry influence to vegetation: OP – Oulanka-Paanajärvi national parks are with surrounding territory; ZA – Zaonezhje Peninsula in the Lake Onega; LA – northeastern shore area of the Lake Ladoga and Valaam Island.

Figure 2. Geological map of the Karelian Region. Compiled by Y.J.Systra with using geological maps of Finland [25], Fennoscandia [26] and author's unpublished map "Tectonics, Karelia"(1996).

Figure 3. On left Kivakka Hill (499.5 m) in the Paanajärvi NP presents one the best layered peridotite-gabbronorite massifs in Europe. Strong bedrock from magma is not eroded so quickly as surrounding gneisses and diorites. On the top is developed mountain tundra zone with typical vegetation. On the photo below is flowering one of protected northern species - *Loiseleuria procumbens* L. On the southern slope of hill found places for nesting some southern species of birds

Numerous springs occur on the deep valley slopes, with discharge rates usually about 0.2-3 L/s. The chemical composition of groundwater in the upper vadose zone above the water-table is determined principally by topographic relief. In the eastern part of Paanajärvi, the summits of the hills are at 305-342 m above sea level, which is from 100-185 m above the level of lakes and rivers, which lie at elevations between 136-208 m [55]. Groundwater infiltrates bedrock to some extent, yielding ultra-fresh hydro-carbonate-Ca waters, with very low mineral contents, less than 100 mg/L. The temperature of spring water is 3.5-4.2° C, and pH is within the range 7.39-6.76. Although the chemical components occur at very low concentrations in spring waters, the cumulative annual flux can be considerable, even at relatively low discharge rates. This is illustrated by calculations for spring no. 6 (discharge rate = 0,5 l/s and (spring no. 9 (discharge rate = 0.3 l/s) respectively: macro-elements Ca (216 and192 kg), SO_4^{2-} (86.7 and 54.6 kg), MgO (34.7 and 14.2 kg), Na (22 and 18 kg), K (17.3 and 10.4 kg), PO_4^{3-} (0.47 and 0.19 kg), NO_3^- (0.63 and 0,095 kg), Cl (11 and 6.6 kg) and microelements: F (2 and 1.8 kg), Fe (0.93 and 1.2 kg), Zn (1.2 and 0.32 kg), Al (140 and 643 g), Ba (227 and 457 g), Sr (219 and 304 g), Mn (14.7 and 45.5 g), I (22 and 12.3 g), Cr (8.4 and 8.5 g), V (9.6 and 2.9 g), Ni (3.9 and 2.4 g), Cu (8.5 and (6.0 g), Co (0.6 and 0.6 g), Se (14.9 and 0.95 g), Sb (10.1 and 4.7 g), Cd (0.16 and 7.8 g), Pb (2 and 15.3 g), Hg (1.9 and 0.28 g), As (0.47 and 0.47 g) [57]. Small mires often form near springs, typically covering an area of several hundred square meters. Some of the water discharge is through capillary flow and in vapour phase, further enriching the soil in nutrients required by vegetation. It is only in such mires, with cold water, that cold-resistant calciphilous species are found, such as: *Saxifraga hirculus* L., *Epilobium davuricum* Hornem., *E. alsinifolium* Vill., *Angelica archangelica ssp. norvegica* (Rupr.) Nordh., *Juncus triglumis* L.

One prominent cliff section – Ruskeakallio (= Brown Cliffs) on the northern shore of lake Paanajärvi, composed by albitite dyke with numerous carbonate veins, is about 60m high and more than 300 m long. This vertical sunny wall with hanging gardens is unique, with its botanical rarities including *Gypsophila fastigiata* L.,*Aspenium ruta-muraria* L., *Draba cinerea* Adams, *D.hirta* DC, *Potentilla nivea* L., *Androsace serpentronalis* L., *Hackelia deflexa* (Wahlenb.) Opiz and more than 20 additional rare plants. During the last 150 years many generations of famous Scandinavian, and after the last war, Russian botanists have visited the Ruskeakallio cliffs. Now it is one of the most picturesque destinations in the Paanajärvi National Park, although it is not permitted to set foot on the shore, in order to protect the unique vegetation. The Paanajärvi National Park is also the only locality in Eastern Fennoscandia from which the lichen *Usnea longissima* Ach. has been recorded during the last 50 years [51,56]. A total of 97 vascular plants present at Paanajärvi have been listed in Red Book of Karelia [52]. One of the indications that the soils at Paanajärvi are compositionally favorable is the abundance of old spruce forest, which covers more than 60% of the terrain. Pine forests prevail on the coarse-grained glaciofluvial Quaternary gravels in the Oulankajoki River and Olanga River valleys, which formed through the action of very powerful melt-water-streams [32] during melting of the last ice sheet, about 10 000 years ago.

Figure 4. Ruskeakallio (Brown Cliff) is 60 m high vertical wall with unique vegetation (above left). One of very rare plant *Gypsophila fastigiata* L. (above right). Lady's slipper (*Cypripedium calceolus* L.) is common in both Oulanka and Paanajärvi National Parks (below left). *Bartsia alpina* L. - common plant for Paanajärvi and Oulanka (below right).

The deep valley of the containing Lake Paanajärvi and through which the Oulankajoki River flows has been incised into the Maanselkä Uplands, the highest points of which are between 400-600 m above sea level. This rugged topography has given rise to a special migration corridor at 109 m above sea level on Lake Pääjärvi to elevations of 200 m above sea level along the western border of the Oulanka National Park. Over the last 20 years a number of new species of birds and plants have been found on both sides of the Maanselkä topographic divide. Lake Paanajärvi has a maximum depth of 128 m, and its deepest part is thus only 8 m above the mean level of the White Sea. Nevertheless, at no stage during its history has Lake Paanajärvi been in contact with the White Sea, and the lake is inhabited by relict populations of smelt, sea trout, arctic char, whitefish, grayling and other typical Karelian lacustrine fishes. The small crustaceans *Mysis*, *Monoporeia* and *Pallacea* provide an important food source for valuable fish species.

Favorable geochemistry, hilly relief, and soils and floral diversity also contribute to the unusual diversity of birds, animals, fishes and insect species for this the latitude near the Arctic Circle.

7. Onega Synclinorium and Zaonezhje peninsula

Onega Synclinorium in one of the largest Paleoproterozoic structural features preserved in the Karelian Craton and covers about 10 thousand square km surrounding Lake Onega and Zaonezhje peninsula (Figure 1, 2). The Paleoproterozoic succession in the Onega Synclinorium begins usually with Jatulian conglomerate, sandstone and dolomite marble, and includes a remarkable salt and gypsum horizon, discovered during drilling in 2008, at depth of about 2 km, and finishes with Ludicovian black shungite-bearing schist, mafic lava and numerous sills, of age 2.2-1.8 Ga. The metamorphic grade these rocks are relatively low.

The synclinorium represents two separate folding episodes intersected by numerous NW-trending fault zones, parallel to the axial plane of the open later generation folds. The Zaonezhje peninsula sequence contains sandstones, carbonate and mafic volcanic rocks, shungite-bearing black schists, shungite-bearing aleurolites etc. The main types of bedrock are given in Table 4 and some more in [37]. Archean granite-gneisses are rich by SiO_2 as well lydit, carbon-bearing silicate bedrock, but gneisses content verysmall amounts of microelements, while shungite-bearing black schists are more rich by Cr, Cu, V, Zn. As usual carbonate rocks have high concentrations of MgO and CaO, higher Ba, Sr and sometimes Mn content. More rich by metallic microelements are mafic and ultramafic volcanites, picrite-basalts, pyroxene diabase and also tuff-aleurite and shungite bearing rocks.

The synclinorium represents two separate folding episodes intersected by numerous NW-trending fault zones, parallel to the axial plane of the open later generation folds. The Zaonezhje peninsula sequence contains sandstones, carbonate and mafic volcanic rocks, shungite-bearing black schists, shungite-bearing aleurolites etc. The main types of bedrock are given in Table 4 and some more in [37]. Archean granite-gneiss (Table 4, No 1) is rich by SiO_2, but contains small amounts of CaO and MgO, and all trace elements. Typical andesite basalt (Table 4, No 2) contains some more FeO, F_2O_3, CaO and MgO, Mn, V and Zn, but other trace elements stay in the low level. Gabbro-dolerite (Table 4, No 3), picrite-basalt (Table 4, No 4) and pyroxene diabase (Table 4, No 5) contain more FeO, CaO, MgO and the most of needed for vegetation trace elements. Dolomite marble (Table 4, No 6 and No 7) are the main sources of the Ca and Mg for soils. In Onega Synclinorium are widely developed carbon (shungite)-bearing tuff-aleurolites (Table 4, No 8, shungite-bearing black schist (Table 4, No 9), shungite-bearing aleurolite (Table 4, No 10) and lydit (Table 4, No 11), which usually contain notable amounts of microelements: Ba, Co, Cr, Cu, Mn, Ni, V, Sr and Zn. In the places, where shungite–bearing rocks are near the surface, the soils are black color and sunlight keeps soils warm earlier than on the other soils. Especially it works on agricultural territories for getting crops of vegetables and potatoes.

Component	1	2	3	4	5	6	7	8	9	10	11
SiO_2	71,38	57.20	48.08	43.54	47.49	11.78	17.90	40.28	29,11	53.92	93.89
TiO_2	0.32	0.98	2.03	1.42	1.51	0.03	0.07	6.60	0.60	1.95	0.08
Al_2O_3	14.50	14.07	13.77	8.97	13.21	0.26	2.08	12.31	7.56	14.20	1.18
Fe_2O_3	1.33	1.21	3,38	1.74	2.46	0.12	1.52	2.58	6.23	1.93	1.08
FeO	0.86	8.62	10.94	10.58	12.73	0	1.09	20.18	0	11.26	1.05
MgO	1.47	4.09	5.97	18.22	7.74	20.33	15.80	4.11	1.36	3.58	0.03
CaO	0.87	5.89	8.59	8.56	5,65	25.95	23.10	3.00	0.56	1.58	0.17
Na_2O	3.98	4.92	2.05	0.49	2.23	0.02	0.98	1.11	4.78	3.15	0.10
K_2O	2.85	0.63	1.19	0.15	0.40	0.04	0.16	0.10	0.25	1.53	0.52
P_2O_5	0.12	0.19	–	0.18	0.17	0.12	–	0.20	0.02	0.30	–
H_2O	0.17	0.15	0.78	0.47	0.69	0.04	–	0.29	0.20	0.23	0.10
LOI	1.78	1.54	3.83	5.43	5.89	41.06	–	8.54	51.44	5.83	1.36
Ba	–	-	52	82	140	280	127	384	306	169	108
Co	–	47	43	76	51	10	43	55	60	68	18
Cr	–	34	230	710	130	8	62	314	142	274	299
Cu	–	32	77	86	120	12	54	102	42	88	80
Mn	225	1178	1472	1627	1395	124	1318	2325	620	1163	155
Ni	–	63	70	290	75	12	96	201	184	182	49
V	–	117	240	300	240	32	75	107	260	252	221
Sr	–	–	170	98	97	–	319	103	120	187	84
S	–	–	–	500	500	200	–	–	1700	–	10100
Zn	24	80	65	100	–	48	56	122	184	103	140

Table 4. Chemical analyses of main bedrock types of Onega synclinorium (macro-elements in weight %, microelements – in ppm). Sample bedrock and place of sampling: 1 – graniit, 5 km to north from Kumsa River; 2 – andesite basalt, Kumsa River, Central Karelia; 3 – gabbro-dolerite, Unitsa Bay, Lake Onega; 4 – picrite-basalt, Rovkozero; 5 – pyroxene diabase, Radkola neck; 6 –dolomite marble, Pyalozero village; 7 – carbonate bedrock, Zaonezhje; 8 – shungite-bearing tuff-aleurolite, Nigrozero; 9 – shungite-bearing schist, Mednyje Jamy village, Zaonezhje peninsula; 10 – shungite-bearing aleurolite, Nigrozero; 11 – lydit, Pustoshi

In western limb of the Onega Synclinorium were studied soils in reduced and full profiles directly overlying on different bedrock, which shows that soils chemical composition is influenced by bedrock and Quaternary sediments composition (Table 5). In soil overlying shungite schist (carbon rich black schist)) the concentration of most elements is higher (Table 5, No 15, 16) than in bedrock (Table 5, No 17). This indicates that the soil has been from the surrounding environment. About 70% of the Republic of Karelia is covered by granites, granite gneisses and migmatites of the Karelian Craton, which are rich by Si, Al, Ca, Na, K, but contain very low concentrations of vital micronutrients. For example chromium concentrations in soil is some less than in bedrock, Ni, Cu and S concentrations are reduced in soil lower horizons. Shungite, consisting of native carbon, is burned on ignition resulting in high weight losses (LOI). The direct influence of bedrock geochemistry on soils is less pronounced, when soil thickness over bedrock exceeds 0.5 m. Most of soils are enriched in SiO_2, Fe_2O_3, Na_2O, K_2O and Cr from Quaternary soft sediments (Table 5).

Soil and bedrock	SiO_2	Al_2O_3	Fe_2O_3+FeO	MgO	CaO	Na_2O	K_2O	LOI	Ni	Co	Cr	Cu	V	Zn
1.5-20cm	77.23	10.74	2.33	0.94	1.44	2.77	1.95	2.17	24	8	27	8	39	16
2.20–45cm	77.20	10.05	2.44	1.06	1.56	2.97	1.86	1.38	32	8	41	8	61	56
3. Granite	**71.38**	**14.50**	**2.19**	**1.47**	**8.87**	**3.98**	**2.85**	**1.78**	**–**	**–**	**–**	**–**	**–**	**24**
4. 0–5cm	16.97	3.19	1.50	0.99	1.22	0.70	0.57	74.05	24	0	0	6	28	112
5. 5–16 cm	44.91	8.68	3.91	2.25	3.12	2.23	1.24	31.82	71	16	34	72	34	104
6. 6–32 cm	50.46	10.09	6.59	3.28	3.09	2.42	1.48	21.08	55	16	137	144	224	72
7.Peridotite	**41.26**	**5.28**	**11.64**	**26.8**	**3.79**	**0.07**	**0.02**	**8.46**	**1493**	**95**	**1680**	**32**	**168**	**803**
8. 0–4 cm	22.92	5.51	2.23	1.01	2.79	0.82	0.61	62.26	32	24	41	7	73	473
9. 4–16 cm	59.29	11.99	5.62	1.80	2.50	2.79	1.23	12.71	79	24	89	24	162	264
10.16-45cm	59.31	11.58	5.66	1,97	2.10	2.74	1.32	13.56	24	16	75	24	246	225
11.Diabase	**49.20**	**12.94**	**10.45**	**9.75**	**5.39**	**3.42**	**0.80**	**4.20**	**134**	**47**	**226**	**32**	**221**	**104**
12. 0-5cm	14.22	2.10	4.16	11.7	14.5	0.43	0.42	51.35	24	16	27	16	34	136
13. 5-15cm	17.50	2.39	2.55	16.3	16.4	0.51	0.45	42.40	16	16	7	24	67	72
14. Marble	**1.04**	**0.04**	**1.59**	**21.3**	**28.7**	**0.05**	**0.01**	**46.16**	**10**	**8**	**5**	**10**	**6**	**62**
15. 0-15cm	69.29	9.05	6.35	1.56	1.15	0.90	1.99	8.58	71	8	96	88	302	145
16.15-71cm	78.38	7.59	2.91	1.06	0.59	0.39	1.94	6.21	39	8	89	32	263	104
17.Shungite	**59.80**	**3.8**	**1.39**	**0.60**	**0.21**	**0.05**	**1.55**	**31.86**	**40**	**8**	**102**	**40**	**225**	**16**

Table 5. Composition of reduced soil profiles on bedrocks in Onega Synclinorium (macro-elements and LOI in wt%, micro-elements in ppm). Analyses made in laboratories Institute of Geology and Forest Research Institute, the Karelian Research Centre, Russian Academy of Sciences, Petrozavodsk, Russia. Soil sampling intervals in cm above bedrock (in bold): 1-3 – soils and Archean basement granite, 5 km to north from Kumsa River; 4-7 – soils and peridotite, Lake Konchozero: 8-11 – soil and gabbro-diabase, Hirvas village; 12-14 – soil and dolomite marble, Pyalozero village; 15-17 – soil and shungite schist, Zazhogino quarry, Zaonezhje peninsula, Lake Onega '

The Archean basement granites with soils were studied at northwestern limb of the Onega structure (Figure 2). Soils on Archean plagio-microcline granite (Table 5, No 3) are enriched by SiO_2, Fe_2O_3, Cr, V and Zn (Table 5, No 1-2). On the diabase, peridotite and marble the organic matter content in soil humus horizon is higher, up to 58-65%, if counting differences in LOI, and SiO_2 content is reduced to 59-14%. Soils on peridotite (Table 5, No 7) have tendency of diminishing of elements in soils characterizes Fe, MgO, CaO, SiO_2, Al_2O_3, Na_2O, K_2O, Ni, Cr, Cu, V, Co, but S and Zn show opposite tendency (Table 5, No 4-6). Soils on gabbro-diabase in Hirvas Village are more acid, contain SiO_2 59% in horizons 4-16 and 16-45 cm, and less macro-element (Table 5, No 8-10) than bedrock (Table 5, No 11). Usually the element concentration diminishes gradually from bedrock to humus horizon: A_2O_3, FeO, MgO, K_2O, V, Cu, S, but for CaO, Co and Zn trend is opposite. Zn content grows from 104 ppm in bedrock to 473 ppm in humus horizon (Table 5, No 8-11). Dolomite marble theoretical composition is CaO 30,4%, MgO 21,7%, CO_2 47.9%, quite close to that is content of these elements in dolomite – $CaMg(CO_3)_2$ from Pyalozero Village (Table 5, No 14), in soil intervals only MgO and CaO are originated from bedrock, all other nutrients are added from Quaternary sediments (Table 5, No 12-13). It is likely that the abundance of CaO, MgO and necessary micronutrients has resulted in the formation of a rich humus horizon, which is the basis for biodiversity. Phosphorus and sulphur are enriched in the humus horizon, due to the activity of animal and microbial life in the soil [58, 59], even if they are absent from the underlying bedrock, as for example, sulphur in the case of ultramafic magmatic rock peridotite (Table 5, No 4-7).

Quaternary sediments are not usually transported by ice over distance greater than 50 km, and typically less than 5 km. For the Onega Synclinorium this means, that favorable volcanic and carbonate rock influences are present everywhere, including fluvioglacial eskers and deltas. It must be noted that the most of bedrock contains notable amounts of Mg, Ca, Na, Fe, Mn; mafic and ultramafic volcanic and intrusive rocks, shungite (carbon)-bearing aleurolites, in addition to essential metallic micronutrients such as Co, Cu, Cr, Ni, V, Zr etc. About 800 vascular plants species have been recorded here as well as many rare plants from the Red Data Book for Fennoscandia [51] and Republic of Karelia [52], which is twice the number of vascular plant species (500) in the very large (5206 square km) Vodlozero National Park, which is located at the same latitude and shares the climatic conditions, but which is situated 50 km to east, where thick Quaternary till and glaciofluvial deposits overlie Archean granites. To the east of the Vodlozero NP, in the Archangelsk Region of Russia is the Kenozero NP, which has similar physiographic conditions, but differs geologically due to the presence of Paleozoic carbonate rock and till derived from mixed sources, including granite from the Karelian Craton and local limestone. The number of vascular plant species present here is 534, of which 61 is listed into Red Book.

8. Western and Northwestern Lake Ladoga area and the Valaam Archipelago

The northeastern marginal zone of the Svecofennian Domain near Lake Ladoga is characterized by mantled granite gneiss domes. Narrow synclinal zones between basement domes comprise equivalents of the same Paleoproterozoic volcanic and sedimentary rocks, including dolomite marbles that occur in the Paanajärvi – Oulanka area and in the Onega

Synclinorium. In the atlas of soils of the Republic of Karelia pH-values in the upper parts of soil horizons in the northwestern part of Lake Ladoga and in the central part of Onega Synclinorium are much higher (pH = 3.5-5.0) than in Central Karelia, where bedrock consists mostly of granites and diorites, associated with pH values of 3.5 or less [38]. Due the small number of samples from the Paanajärvi National Park the influence of carbonate bedrock is not apparent in the Soil Atlas. Early Proterozoic Jatulian and Ludicovian rock units in the Svecofennian Domain in the Ladoga Zone are overlain by younger Kalevian sediments. The Precambrian basement is divided into separate blocks by numerous fault zones, which are reflected topographically as deep narrow valleys in which some northern species of mosses and lichens are found. The total number of vascular plant species in the area reaches 750, of which 550 are also present in the Valaam archipelago. The number of lichens and lichenicolous fungi exceeds 800 and 269 mosses have also been identified here [24, 52]. The main reason for such biodiversity is the extensive occurrence of carbonate and mafic volcanic rocks with their high concentrations of essential micro-nutrients.

The central part of Lake Ladoga coincides with the Mesoproterozoic sedimentary-volcanic bedrock of Ladoga Graben (aulacogene), which includes the thick dolerite sill outcropping on the islands of the Valaam Archipelago. Lake Ladoga is the largest freshwater body in Europe and shows a significant effect on microclimate. The coastline is much more favourable for agriculture and natural vegetation compared to terrain some 100-150 m above lake level, where snow melts some weeks in springtime and where autumn frosts occur several weeks earlier. The bedrock of the entire Valaam Archipelago is composed of a gabbro-dolerite sill, 220 m in thickness, associated with monzonite-quartz syenite, granite porphyry and granophyre [39]. A study of heavy metals and sulfur distribution in the soils of Valaam Archipelago has also recently been completed [40].

Comparison of the distribution of heavy metals (Table 6) in the main types of bedrock, namely gabbro-dolerite No 3211-6 and an average composite from monzonite – quartz syenites No 1488-3-12 [39] and soil horizons, shows good correlation between bedrock and soil composition (Table 6). The whole archipelago is formed from gabbro-dolerite and monzonite-syenite sill, while the surrounding lake is more than 100 m deep [61]. Therefore only small amounts of material would have been transported from the western and northwestern shores of the lake to the Valaam Island during glacial time. The local doleritic bedrock might therefore be expected to be enriched in microelements. Zinc is often concentrated in the upper parts of soil profiles (Table 5), while Cr, Cu, Ni and V abundances are close to those in bedrock. Only Pb shows some higher concentrations in soil horizons, but this may be due the small number of studied analyses in bedrock.

V. Koval'skij [13] gave next concentration intervals for normal life and development: Co – 7-30, Cu – 15-60, Zn – 30-70, Mn – 400-3000, Mo – 1,5-4, I – 2-40, B – 3-30, Sr – 0-10 [16]. At levels in excess of 500 ppm, zinc in soil interferes with the uptake of essential metals: Fe, Mn and B [3]. Anyway, so rare bedrock had significant influence to plant diversity. The Valaam Archipelago Nature Park has at least 590 vascular plant species in addition to introduced plant species, which grow here in great abundance. The finding of 61 Red Data Book species indicates that Valaam Archipelago Nature Park is of considerable floristic value. Some species, as *Plantanthera chlorantha* (Cust.) Reichenb., *Potentilla neumanniania* Reichenb., *Corydalis intermedia* (L.) Merát, *Cotoneaster integerrimus* Medik and *Myosotis ramosissima* Rockel ex Schult do not occur in other protected areas of Karelia [60]. The content of Ca in Valaam bedrock usually does not exceed 6-7% and is connected with other elements in hard mineral lattice. At weathering Na and Ca ions are carried out the first order, so in soil forms deficit of Ca and in archipelago does not grow beautiful plant *Cypripedium calceolus* L.

Bedrock or soil	Cd	Co	Cr	Cu	Mn	Ni	Pb	V	S	Zn	FeO+ Fe_2O_3	Fe
Gabbro-dolerite	–	24	82	8	953	16	–	67	–	137	12.2%	–
Monzonite-syenite	–	20	34	12	973	9	16	54	–	166	10.9%	–
Forest litter	0.74	2.6	10	14	746	7.7	26	–	1240	170	–	11900
Soil beneath litter	0.76	6.4	23	15	612	4.9	25	–	612	129	–	38000

Table 6. Heavy metals contents in bedrock [39] and average abundances in forest litter and the uppermost mineral soil horizon beneath the litter [40] of Valaam Island. Sum Fe oxides in weight%, microelements in ppm.

9. Kilpisjärvi area, Finnish Lapland, NW Finland

This area belongs geologically to the eastern marginal zone of the Scandinavian Caledonide nappes, far to the north of Arctic Circle (69º 03′). The relief on the eastern slope of Scandinavian Mountain Ridge lies between 472 m (Lake Kilpisjärvi) and 1328 m above sea level, vegetation is typical for tundra zone, trees are represented with tundra birch, willows and some aspen trees on the southern slope of Saana Mountain (1029 m) near the Lake Kilpisjärvi, from coniferous grows only juniper, which may reach very high age, preliminary study shows, that more than 1000 years.

The oldest gneissose granodiorites, felsic, mafic and ultramafic volcanics and mica gneisses belong to the 2.7-2.8 Ga Neoarchean basement, which represents the autochthonous foreland to the Caledonian Orogen. osition, cratonised ago. This deformed and metamorphosed basement is unconformably overlain by Early Cambrian basal conglomerate, with silty and quartz-rich intercalations. Dolomitic marble is typical of the Jerta Nappe, which has been thrust somewhat to the southeast and forms a layer from 1-40 m thick , in some places strongly folded, and which may cover extensive parts of the hill slopes. Although there are 3-4 more nappes, with a total thickness of 450 m near the Kilpisjärvi Lake and up to 1250 m near the Halti Mountain [62-64], the most interesting rock unit with respect to the diversity vegetation is the dolomitic marble of the Jerta Nappe and the numerous small springs on the southern slope of Saana Mountain. The important influence of carbonate rocks on vegetation was first noted here by A. Pesola [42]. There are now three botanical protected areas that include all large dolomite outcrops in the Kilpisjärvi area: the Malla Strict Nature Reserve, 3000 ha, has been under protection since 1916, and was declared as a strict reserve in 1938, the Saana Protected Territory and the Annjalonji Protected Area were declared in 1988. Although there are many other hills with the same elevation and slope aspect, most of the rare and protected species are absent, where the bedrock is other than dolomite.

Figure 5. *Rhododendron lapponicum* L.flowering on the Paleozoic marbles of the Malla Strict Protected Territory (above). Slope of Pikku-Malla Hill with dolomite marble layer (below, light layer).

On the upper part of the Pikku-Malla fell these dolomite outcrops coincide with most of the rare plant finds: *Erigeron acer* L., *E. uniflorus* , *Rhododendron lapponicum* L., *Polystichum lonchitis* L., *Pseudorchis albida* (Fernald), *Silene uralensis* Rupr., *Veronica fruticans* Jacq. These and other protected rare and endangered vascular plant species are only found in the presence of carbonate rocks. During the flowering of *Dryas octapetala* L. the folded dolomite layers on the hill-slopes resemble natural flowerbeds. The influence of dolomite continues down slope and also in the spring waters. Numerous small springs discharge in the lower part of slopes and are associated with great floral diversity. In winter time the springs may be frozen but during the growing season, discharge flow rates of 0.2 l/s were recorded for Spring 1 and 0.1 l/s for Spring 2. Respective mineral components for Spring 1 and Spring 2 (where data are available)are: Ca(45 and 47 kg), Mg (18 and22.2 kg), K (3.7 and 1.8 kg), Na (7.4 and 2.4 kg), Si (9.5 and 2.7 kg), S (26.6 and 23.8 kg), SO_4^{2-} (71.9 and 68.6 kg), NO_3^- (1.3 and 0.3 kg), Cl (4.3 and 1.4 kg), Sr (262 and 180 g), Al (55.2 and 5.5 g), Ba (23.2 and 30.6 g), B (7,3 and 4.8 g), Li (2.4 and 3.4 g), Zn (6.1 and 3.1 g), Rb (3.3 and 2.1 g), Mo (0.95 and 2.1 g), V

(4.3 and 0.36 g), Cu (3.1 and 0.86 g), U (0.72 and 2.4 g), Co (190 mg), Bi (63 and 110 mg), Mn (63 and 47 mg), As (252 mg, Spring 1 only), Sb (31.5 mg, Spring 1 only) and Th (95 and 16 mg). Elements such as Ag, Be, Cd, Cr, Ni, P, Pb, Se, Tl, Fe, Br, F have concentrations below detection limits. Water samples were analyzed at the Geological Survey of Finland. Because discharge from such springs is low, much of the water evaporates, leading to enrichment of essential elements in soils. On the soils developed over Precambrian granodiorite near these springs, some rare species were found, such as *Saxifraga aizoides* L., *Pseudorchis albida* (Fernald).

Figure 6. *Saxifraga aizoides* L. is flowering near the spring on the southern slope of Saana Mountain (above). *Polystrichum lonchitis* (L.) Roth is very rare plant in the Northern Finland, in Kilpisjärvi area is possible to meet it only on carbonate bedrock (below left). Another protected species *Pseudorchis albida* L. also likes Ca-rich soils (below right).

About 470 vascular plants species have been recorded in the surroundings of Kilpisjärvi, most of them occurring on exposures of dolomitic marble. Pelitic Cambrian schists contains more microelements than quartzite and may be a more favored substrate for some species, including *Dryas octapetala* L.

10. Areas surrounding Kevo Lake and Kevo Subarctic Research Station

The bedrock of this area consists of Archean migmatites and gneisses, mostly hornblede-bearing gneisses, which are divided into blocks with fault zones. Surficial deposits are well-sorted glaciofluvial gravels with clasts of granite and gneiss, which are thus rather poor sources of nutrients. Scots pine (P. sylvestris) is common in valleys and on hillslopes, but the Kevo area (69 °45′ N) is some 150-200km beyond the northern limit for spruce. The region is characterized by typical northern tundra species, and lacks rare species, which need more abundant macro- and micronutrients. Common species are *Loiseleuria procumbens* L., *Phyllodoce caerulea* (L.) Bab., *Diphasiastrum alpinum* L., *Pinguicula vulgaris* L. etc.

11. Influence of bedrock and Quaternary sedimentary geochemistry on biodiversity in Estonia

Estonia is located along the southern shore of the Gulf Finland and the Precambrian crystalline basement is everywhere covered by Ediacaran and Paleozoic sedimentary cover. The sedimentary cover together with basement is tilted gently southwards at a gradient of nearly 3 m per kilometre and its thickness increases from 125-140 m in the north to 600 m and more in southern and southwestern Estonia. Cambrian, Ordovician and Silurian bedrock are exposed in northern Estonia as east-west trending belts, whereas to the south these older sequences are covered by younger rocks. The resistant Ordovician limestone forms cliffs up to 56 m high, known as the North Estonian Klint (Figure 7).

During the last 400 000 years, the region was repeatedly covered by ice during several glacial events, which advanced towards the south or southeast from the Scandinavian Mountains, transporting metamorphic and igneous rock material to Estonia. The Svecofennian domain is covered mostly by granites, migmatites and gneisses, with mafic rocks comprising only 3.5% of till and boulder material [65]. Northern Estonia belongs to the zone of glacial erosion, where the thickness of Quaternary cover seldom exceeds 5-10 m, being in many places on alvars less than 1 m thick; eskers and glaciofluvial deltas may however, exceed 20 m in thickness. Southern Estonia is in contrast characterized by moderate sedimentary accumulation, with till cover in the Otepää and Haanja Uplands commonly exceeding 100 m and in the ancient buried valley of Abja a local maximum thickness of 207 m is attained. Local carbonate rock cobbles and pebbles predominate in the thick till sequences in central Estonia, but near the southern border the crystalline Fennoscandian bedrock becomes prevalent. These have a strong influence on soil composition, the Ca content in the humus horizon falling to 0.2% or less, while in carbonate bedrock Ca contents are between 1-8%. The Mg content of sediments derived from carbonate rocks is usually 0.5-0.8%, whereas in southern Estonia it is commonly less than

0.09%. The minimum content of Mn in soils for healthy growth and development of plants is 400 ppm [16], but much of Estonia has Mn concentrations less than 230 ppm [49].

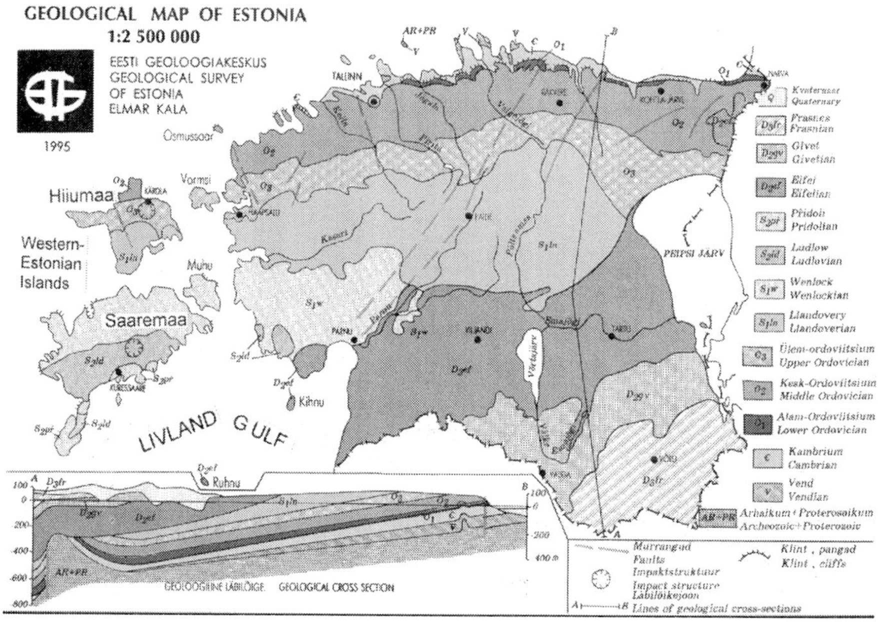

Figure 7. Geological map of Estonia. Compiled by E.Kala, 1995. Is published with permission of the Geological Survey of Estonia

Cambrian claystones and sandstones outcrop in the narrow zone between the klint and Baltic Sea shore. A number of micro- and macronutrients: K, Mg, P, B, Co, Cr, Cu, Ni, V and Zn have higher concentrations in these clays (Table 7, Es-1) than in carbonate rocks and sandstones. The influence of the clays is however restricted to this narrow zone. The Ordovician sequence begins with the Pakerordi Stage sandstone and the kerogenous dark-brown argillite horizon, which has a thickness of 7.7 m in northwestern Estonia, diminishing to about 2.0 m in northeastern Estonia. This argillite is notable for high concentrations of U (up to 400 ppm), K, As, B, Co, Cr, Cu, Hg, Mo, Ni, Pb, V and Zn (Table 7, Es-2). Soils are enriched in F, Mn. Mo, P, U and Y, but only within a narrow zone to the south, about 10 km wide, parallel to the klint [49].

The central and northern part of the Estonian mainland and the Western Islands are composed of Ordovician and Silurian carbonate rocks: marl, limestone and dolostone. Clayey limestone (Table 7, Es-17) contents some more SiO_2, Al_2O_3, MgO, K_2O, Ba, F, Rb and Sr. Limestone (Table 7, Es-3 and Es-9) contain maximum amount of CaO (50.51-52.67%), some SiO_2, Al_2O_3, MgO, Mn, Sr, sometimes also F. Dolostone of Estonia (Table 7, Es-16, Es-18 and Es-4) usually contain some per cents impurities due the short distance from Fennoscandian Shield, as source of sandy material. MgO content in dolostone often exceeds 20%, CaO content 28,87-30.87%.

Limestone and dolostone as bedrock and as pebbles and cobbles in gravel enrich all soils of Central Estonia and Western Estonia Islands with needed Ca and Mg.

Components	Es-1	Es-2	Es-3	Es-16	Es-9	Es-17	Es-18	Es-4	Es-15	Es-5
SiO_2	59.24	52.14	4.84	4.17	1,21	9.83	4.69	2.84	63.00	95.1
TiO_2	0.88	0.76	0.077	0.084	0.022	0.089	0.062	0.041	0.20	0.23
Al_2O_3	17.38	13.15	1.10	1.20	1.25	2.45	0.88	0.75	18.04	1.81
Fe_2O_3	4.29	0.85	0.06	0.49	0.5	0.38	0.17	0.1	0.59	0.11
FeO	2.60	3.02	0.44	2.58	–	0.38	0.30	0.35	0.14	1.33
MgO	2.58	1.11	0.85	16.10	1.25	5.36	20.16	20.42	0.76	0.05
CaO	0.84	0.22	50.51	30.87	52.67	40.79	28.87	29.35	0.29	0.047
K_2O	5.84	7.95	0.51	0.59	0.16	1.72	0.413	0.07	14.91	0.07
Na_2O	0.13	0.10	0.08	0.04	<0.03	0.043	0.04	0.26	0.05	1.03
P_2O_5	0.31	0.13	0.42	0.232	0.043	0.015	0.026	0.012	0.077	0.022
Cl	0.13	0.02	0.03	0.05	0.04	0.03	0.06	0.07	0.04	0.01
S	0.05	2.19	0.12	0.132	0.19	0.21	0.02	0.11	0.04	0.01
LOI	4.95	19.91	40.39	42.49	42.61	37.70	44.02	45.33	1.37	0.29
As	1.3	37	<1	<1	0.80	<1	<1	<1	2.6	<3
B	150	53	21	–	8.9	–	–	12	40	23
Ba	420	379	29	41.6	20.09	46.88	31.92	14	119	147
Cd	<1	<1	<1	0.04	<5	0.01	0.065	<1	<1	<1
Co	20	12	1.8	1.43	<3	0.98	1.28	1.5	4.0	2.6
Cr	78	80	9	9.68	8.25	7.28	10.67	9	9	31
Cu	25	105	3	3.95	4.0	3.64	1.95	<4	6	10
F	1195	570	545	350	<100	210	180	<100	1200	<100
Hg	<0.01	0.162	<0.01	<10	–	<10	<10	<0.01	0.023	<0.01
Mn	320	158	460	2060	371.4	229	399	214	16	98
Mo	<2	56	<2	0.41	<1	0.57	0.31	<2	1.0	<2
Ni	40	98	4	5.11	3.99	5.37	2.91	4	6	9
Pb	11	77	5	2.84	13.94	3.54	1.63	4	9	7
Rb	176	118	10	10.53	2.77	15.16	8.92	7	63	20
Se	<4	2.3	<6	<1	0.50	<1	<1	<3	<3	<4
Sn	3,9	3.2	<1	0.13	<1	0.25	0.14	<1	7.2	0.73
Sr	91	53	178	59.64	113.1	137.4	51.43	32	10	19
V	109	509	9	14.00	3.40	9.60	20.8	11	14	11
Zn	176	133	4.0	14.88	8.46	8.88	7.94	5	38	4

Table 7. Macro- and micronutrients content in the main bedrock groups of Estonia (compiled after reference analyses [47, 48] macro-elements and LOI in weight %, micro-components in ppm. Samples: Es-1 – claystone (clay), Ca_2, Kunda; Es-2 – kerogenic argillite, O_1, Tallinn; Es-3 – limestone,O_2, Tallinn; Es-16 – dolostone O_2, Maardu, Tallinn; Es-9 – limestone, O_3, Vasalemma, Harjumaa; Es-17 – clayey limestone, S_1, Valgu, Läänemaa; Es-18 – dolostone, S_1, Anelema quarry, Pärnumaa; Es-4 – dolostone, S_1, Mündi quarry; Järvamaa; Es-15 – metabentonite, Kinnekulle bed, O_3, Pääsküla, Tallinn; Es-5 – sandstone, Middle Devonian, Suur-Taevaskoda outcrop, Ahja River.

The entire region, except in areas of thick Quaternary cover, is enriched in Ca (1.25-6.06%), Mg (0.44-2.12%), and sporadically in Mn and F. High fluorine contents are typical for felsic volcanic ash (metabentonite) layers (Table 7, Es-15). All other micronutrients in carbonate rocks are present in small concentrations (Table 7), which is also reflected in the compositions of soils developed on such bedrock. The central and northern part of the Estonian mainland and the Western Islands are composed of Ordovician and Silurian carbonate rocks: marl (clayey limestone), limestone and dolostone. Clayey limestone (Table 7, Es-17) contents some more SiO_2, Al_2O_3, MgO, K_2O, Ba, F, Rb and Sr. Limestone (Tabel Es-3 and Es-9) contain maximum amount of CaO (50.51-52.67%), some SiO_2, Al_2O_3, MgO, Mn, Sr, in some places also F. Dolostones in Estonia (Table 7, Es-16, Es-18, Es-4) are usually secondary origin, so they contain some silicate minerals impurities due the short distance to Fennoscandian Shield. MgO content stays between 16.10 and 20.42%, CaO – 28.87-30.87%, some higher is Ba, Sr, sometimes Mn and F concentrations. The entire region, except in areas of thick Quaternary cover, is enriched in Ca (1.25-6.06%), Mg (0.44-2.12%), and sporadically in Mn and F. High fluorine content is typical for felsic volcanic ash (metabentonite) layers (Table 7, Es-15). All other micronutrients in carbonate rocks are present in small concentrations (Table 7) as they are also in soils developed on such bedrock [49].

Southern Estonia is covered mostly by sandstones, which usually contain more than 95% quartz (Table 7, Es-5). As might be expected, concentrations of macro- and micronutrients in bedrock and Quaternary cover are low. The Sakala and Otepää Uplands contain material transported by ice from northern Estonia – mostly carbonate rocks - and from the Fennoscandian Shield – crystalline rock, mostly granite and migmatite. Many elements, such as B, Cd, Co, Cr, Cu, F, Hg, K, Mo, Na, Ni, Sn, Sr, U, V and Zn, occur at very low concentrations in Fennoscandian bedrock and on the geochemical maps of the humus horizon of soil of Estonia they are distributed more evenly than elements derived from local bedrock. Because Estonian soils thus represent mixing of two components of differing composition, the influence of soils on vegetation is generally not so marked when compared with soils in Fennoscandia, where differences in vegetation are much greater. Nevertheless, *Asplenium ruta-muraria* L., A. *septentrionale* L., *Equisetum hyemale* L., E. *x moorei* Newman, *etc* grow only on the carbonate rocks of northwestern Estonia, near the Baltic Sea shore.

The most comprehensive listing of Estonian vascular plants and their distribution [50], has made it possible to assess how bedrock composition influences vegetation. The list records a total of 1353 plant species, about 50 of have uncertain occurrence, while some 700 are distributed more or less evenly, with no preference for bedrock type. A further 160 rare plant species likewise show no particular correlation with rock type, while 67 species are only found in proximity to the Baltic coast. There are 55 species that are endemic to carbonate bedrock and 35 species that grow exclusively on the sandstones of southern Estonia and 76 species that are completely absent from sandstone terrain. There are 137 species that occur predominantly on carbonate rocks, and compared to only 58 species that grow preferentially on sandstone.

The Western Estonian Islands emerged from the Baltic Sea later than the continental part of Estonia, which is also reflected in vegetation diversity: 46 vascular plant species are absent from Saaremaa, but another 50 species occur only on Saaremaa and have not been found on the mainland. The soil in Estonia has formed from two main sources, local sedimentary bedrock and crystalline material transported by glaciers from Fennoscandia. Geochemical maps of the humus horizon show that the concentrations of some elements - B, Ba, Cd, Co, Cr, K, Na, Ni, Sn, U, Th, V, Zn - show little or no influence from local bedrock. Conversely, abundances of Ca, Mg and F in soil are correlated with bedrock composition, Mo and P concentrations in particular being closely associated with the kerogenic argillite near the North Estonian Klint; bedrock controls on Fe and Hg distribution is not so evident. In summary, material derived from the Fennoscandian Shield provided most of the following microelements in Estonian soils: B, Ba, Cd, Co, Cr, K, Na, Ni, Sn, U, Th, V and Zn. Recommended maximum permissible concentrations have been defined and legislated for most elements, but natural minimum concentrations should also be considered, below which there may be adverse influences on plant and animal health and life. For Mo this lower limit is 1.5 ppm and for Mn it is 400 ppm, but normal ecological functions require a concentration of 3000 ppm [16]. In Estonia the average Mo content is 2.5 ppm, but almost a third of the country has concentrations less than 1.2 ppm, Concentrations of Mn are between 75-2400 ppm, but half of the country records concentrations less than 400 ppm.

Southern Estonia is covered mostly by sandstones, which usually contain more than 95% quartz (Table 7, Es-5). As Might be expected, concentrations of macro- and micronutrients in bedrock and Quaternary cover are low. The Sakala and Otepää Uplands contain material transported by ice from northern Estonia – mostly carbonate rocks - and from the Fennoscandian Shield – crystalline rock, mostly granite and migmatite. Many elements, such as B, Cd, Co, Cr, Cu, F, Hg, K, Mo, Na, Ni, Sn, Sr, U, V and Zn, occur at very low concentration in Fennoscandian bedrock and on the geochemical maps of the humus horizon of soil of Estonia they are distributed more evenly than elements derived from local bedrock.

Comparison of average microelement concentrations in bedrock and humus horizons in Karelia, Russia (Precambrian Fennoscandian Shield) and Estonia (mainly Paleozoic sedimentary bedrock) clearly shows that soils on Precambrian bedrocks in Karelia contain much more microelements than sedimentary bedrock and soils on sedimentary bedrocks of Estonia (Table 8). It means that magmatic processes are bringing out from crust deeper levels enriched by many trace metals as Cd, Co, Cr, Cu, Ni, V, Zn and others. At studying geochemistry of soil humus horizon in Zaonežje peninsula in 2000 only in single probes Cu exceeded 2-3 times the highest permitted concentrations.

Bedrock/ Elements	Cd	Co	Cr	Cu	Mn	Ni	V	Zn
Karelian bedrock	0.74	41	168	50	942	99	124	85
Estonian bedrock	0.04	2.8	11.7	4.6	481	5.0	12.3	15.9

Table 8. Comparison of average micronutrients content in bedrocks of Karelia Tables 3-5, Karelia, Russia and in bedrocks of Estonia (Table 7), excluded are bedrock analyses with extreemly high concentrations as peridotite in Karelia, clay and kerogenous argillite in Estonia

The comparison shows clearly that soils on metamorphic or magmatic rocks of Karelia content much more trace elements, than sedimentary bedrocks in Estonia. Only Mn content in limestone may be reach the same level than in volcanic or intrusive bedrocks, all other elements have in magma concentrations 10-20 times higher.

12. Discussion

The geochemistry of bedrock and Quaternary sediments strongly influences soil nutrient content and at the same time biodiversity. Where beneficial nutrients are abundant in soils, southern species may spread far beyond their normal range, as in the case of *Fragaria vesca* L. and *Rebes rubrum* L. which are found near the Arctic Circle in the Paanajärvi National Park. The importance of micronutrients for wild animals and birds is also well known. In Northern Karelia in autumn, when blueberries are ripe, forest birds such as the capercaillie (*Tetrao urgallus*), black grouse (*Tetrao tetrix* L.) and even ducks tend to be selective in feeding, preferentially choosing plants growing on gabbroic and ultramafic rock, where berries are larger and contain more micronutrients. Larger mammals, including moose (*Alces alces* L.), brown bear (*Ursus arctos* L.), and reindeer (*Rangifer tarandus* L.) feed on mushrooms in summer and autumn time, because of the need for microelements. Animals that have access to all necessary nutrients are healthier and stronger, and hence better equipped to defend territory and offspring. Areas with abundant flowering plants are favored by insects, and therefore also by birds which then make nests in these places.

Geochemistry is therefore a valuable tool in assessing regional biodiversity in national parks, and in protection of species or areas of natural significance, or in designating areas for construction and urban development and agricultural use. A relationship between bedrock geochemistry and endemic diseases is established in many places [14, 66, 67, 68]. We must not forget that all life is built from the same building blocks, known as elements [3].

Author details

Ylo Joann Systra
Department of Mining, Chair of Applied Geology, Tallinn University of Technology, Tallinn, Estonia

Acknowledgement

Field work in Karelia was done in part with financial support from Ministry of the Environment of Finland. We acknowledge the support of the European Community - Research Infrastructure Action under the FP6 "Structuring the European Research Area" Programme, LAPBIAT (RITA-CT-2006-025969) for studies in the Kilpisjärvi, Kevo and Oulanka areas in 2008-2009 and project No SF0140093s08 from the Estonian Ministry of Education and Research. This study is also part of the Estonian Science Foundation Grants No 7499 and No 8123. The author would like to thank Peter Sorjonen-Ward for correcting the English of the manuscript and numerous colleagues from different institutes of Karelian Research Center, who took part in the Finnish-Russian biodiversity project 1997-2000.

13. References

[1] Mason BH (1966) Principles of Geochemistry. 3rd edit., New York, John Wiley and Sons, Inc.329 p.

[2] Mason BH (1995) Geochemical Distribution of the Elements. In: New Encyclopedia Britannica, 15, Macropedia. pp. 939-950.

[3] Emsley, J (2003) Nature's Building Blocks, An A-Z. Guide to the Elements. Oxford University Press. 539 p.

[4] Albarède F (2009) Geochemistry. An Introduction. 2nd ed. Cambridge University Press. 342 p.

[5] Bulakh AG (2002) General mineralogy. Sankt-Petersburg University Publisher. 356 p.(in Russian).

[6] Saukov AA (1975) Geochemistry. Nauka, Moscow. 480 p. (in Russian).

[7] Tayler RJ (1972) Origin of the Chemical Elements. 3rd edition, London & Winchester. 170 p.

[8] Tayler RJ (1995) Origin of the elements. In: The New Encyclopedia Britannica, 15, Macropedia. pp. 934-938.

[9] Thornton I (1983a) (ed) Applied Environmental Geochemistry. Academic Press, London). 501 p.

[10] Crounse RG, Pories WJ, Bray TB, Mauger RL (1983). Geochemistry and Man: Healt and Disease. 1. Essential Elements. In: Thornton I (ed.) Applied Environmental Geochemistry. Academic Press, London. pp. 267-308.

[11] Food and Nutrition III (1997) Tables of the chemical composition of food. Tallinn Technial University. 87 p.

[12] Barabanov, VF 1985 Geochemistry. Nedra, Leningrad. 423 p. (in Russian).

[13] Koval'skij, VV 1982 Geochemical environment and life. Moscow. 282 p. (in Russian).

[14] Thornton I (1983b) Geochemistry Applied to Agriculture. In: Thornton I (ed.) Applied Environmental Geochemistry, Academic Press. London. pp. 231-266.

[15] Emsley, J (1998) The Elements, 3rd ed. Oxford, Clarendon Press. 251 p.

[16] Trofimov VT, Ziling DG, Baraboshkina TA et al. (2000) Ecologic functions of the lithosphere. Moscow University Press. 432 p (in Russian).

[17] Kabata-Pendias A, Pendias H (2001) Trace Elements in Soils and Plants. 3rd Edition. CRC Press, Boca Raton. 413 p.

[18] Merian E, Anke M, Ihnat M, Stoeppler M, (eds) (2004) Elements and their Compounds it the Environment. 2nd Edition. Wiley-VCH. V. 1–3. 1773 p.

[19] Nies D.H. 2004. Essential and Toxic Effects of Elements on Microorganisms. In: *Elements and their Compounds in the Environment*. V.1, 2nd edit. Merian E. et al. (eds), pp 257-304.Wiley-VCH Verlag CmbH & Co.KGaA.

[20] Uthman E (2000) Elemental Composition of the Human body: http://www.abti.net/ABEBackup/Elemental_Composition_of_ the Body.doc 27.02.2012.

[21] Plant JA, Raiswell R (1983) Principles of Environmental Geochemistry. In: Thornton I. (ed) Applied Environmental Geochemistry. Academic Press, London. pp.1-39.

[22] Himeno S, Imura N (2002) Selenuim in Nutrion and Toxicology. In: Sankar B. (ed.). Heavy Metals in the Environment. New York, NY, USA: Marcel Dekker Inc.pp.587-629
[23] Webb JS (1983) Foreword. In: Thornton I (ed) Applied Environmental Geochemistry. Academic Press, London. pp. vii-viii.
[24] Gromtsev, A.N., Kitaev, S.P., Krutov, V.I. et al. (eds) (2003) Biotic diversity of Karelia: conditions of formation, communities and species. Russian Academy of Sciences, Karelian Research Centre, Petrozavodsk. 244 p.
[25] Korsman K, Koistinen T, Kohonen J, Wennerström M, Ekdahl E, Honkamo M, Idman H, Pekkala Y (eds.) (1997) Bedrock map of Finland 1:1 000 000. Geol. Survey of Finland, Espoo.
[26] Koistinen T, Stephens MB, Bogatshev V, Nordgulen Ø, Wennerström M, Korhonen J. (comps.) (2001) Geological map of Fennoscandian Shield 1:2 000 000. Espoo, Trondheim, Uppsala, Moscow. Geol surveys of Finland, Norway, Sweden, Ministry of Natural Resources, Russia, 2 sheets.
[27] Systra YJ (1987) Geological formations and tectonics of the Kukasozero-Hankusjärvi-Kuzhjärvi area. In: Early Precambrian of Karelia (geology, petrology, tectonics). Petrozavodsk, Karelian Branch Academy of Sciences USSR. pp.35-54 (in Russian).
[28] Systra YJ (1990) Structural evolution of the eastern part of Paanajärvi Synecline. In: Precambrian of Northern Karelia. Petrozavodsk, Karelian Research Center, Academy of Sciences USSR. pp. 40-61 (in Russian).
[29] Systra, YJ (1991) Tectonics of the Karelian Region. St. Petersburg, Nauka. 176 p. (in Russian).
[30] Systra YJ (1998a) Geological diversity: the main reason for the biodiversity of the Paanajärvi National Park. Oulanka Reports 19, 23-26.
[31] Systra Y (1998b) Vegetation map of the Oulanka-Paanajärvi national parks. In: Hautala H, Rautiainen L (eds) Oulanka-Paanajärvi. Acticmedia, Kuusamo.1 sheet.
[32] Systra, YJ (1998c) Role of geological-geomorphological factors in the formation of biodiversity in the Paanajärvi National Park. In: Biodiversity inventories and studies in the areas of Republic of Karelia bordering on Finland (express inform. materials). RAS, Karelian Research Center, Forest Research Institute. Krutov VI, Gromtsev AN (eds). Petrozavodsk. pp.27-32 (in Russian).
[33] Systra YJ (2000) Northern shore of Lake Ladoga. Geological characteristics of the territory. In: Biodiversity inventories and studies in Zaonezhskij peninsula and on the northern shore of Lake Ladoga (express inform. materials) Gromtsev AN, Krutov VI (eds) Petrozavodsk. pp. 194-197 (in Russian).
[34] Systra YJ (2003) Geological characteristics. In: Gromtsev, A.N., Kitaev, S.P., Krutov, V.I. et al. (eds.). 2003. Biotic diversity of Karelia: conditions of formation, communities and species. Petrozavodsk, Karelian Research Center of RAS. pp. 7-12
[35] Systra YJ (2004a) Main features of the Karelian region structures. In: Sharov, V.N. (ed.). Deep structure and seismicity of the Karelian region and its margins. pp.14-29. Karelian Research Centre, Russian Academy of Sciences, Petrozavodsk.

[36] Systra YJ (2004b) Geological background for biodiversity in the eastern Fennoscandia, Estonia and Latvia. In: Parkes, M.A. Natural and Cultural Landscapes - The Geological Foundation, Royal Irish Academy, Dublin. pp. 73-76.

[37] Golubev AI, Systra YJ (2000) Zaonezhskij peninsula. Geological characteristics of the territory. In: Biodiversity inventories and studies in Zaonezhskij peninsula and on the northern shore of Lake Ladoga (express inform. materials) Gromtsev AN, Krutov VI (eds) Petrozavodsk. pp. 9-15 (in Russian).

[38] Fedorets NG, Bahmet, ON, Solodovnikov AN, Morozov AK (2008) Soils of Karelia: geochemistry atlas. Moscow, Nauka. 47 p (in Russian).

[39] Sviridenko LP, Svetov AP (2008) Valaam gabbro-dolerite sill and geodynamics of the Ladoga Lake trough. Petrozavodsk, Institute of Geology, Karelian Research Centre, Russian Academy of Sciences. 123 p. (in Russian).

[40] Shiltsova GV, Morozova RM, Litinski PJu (2008) Heavy metals and sulphur in Valaam archipelago soils. Petrozavodsk, Karelian Research Center, RAS, Forest Research Institute. 109 p. (in Russian with English conclusions and abstract).

[41] Kravchenko AV, Gnatiuk EP, Kuznetsov OL (2000) Distribution and Occurrence of Vascular Plants in Floristic Districts of Karelia. Petrozavodsk, Karelian Research Centre of RAS, Forest Research Institute, Institute of Biology. 76 pp.

[42] Pesola VA (1928) Kaltsiumkarbonaatti kasvimaantieteellisenä tekijänä Suomessa. Ann.Soc. "Vanamo" 9 (1), 246 pp. (in Finnish with summary: Calcium carbonate as a factor in the distribution of plants in Finland).

[43] Lehtinen M, Nurmi P, Rämö T (eds) (1998) Suomen kallioperä – 3000 vuosimiljoonaa. Helsinki, Geological Society of Finland. 373 p (in Finnish).

[44] Lehtinen M, Nurmi P, Rämö T (eds) (2005) Precambrian Geology of Finland: Key to the Evolution of the Fennoscandian Shield. Elsevier BV, Amsterdam. 736 p.

[45] Rasilainen K, Lahtinen R, Bornhorst TJ (2008) Chemical characteristics of Finnish Bedrock – 1:1 000 000 scale bedrock map units. Geological Survey of Finland. Report of Investigations 171. Espoo. 94 p.

[46] Raukas A, Teedumäe A (eds.) (1997) Geology and mineral resources of Estonia. Tallinn, Estonian Academy Publishers. 436 p.

[47] Kiipli T, Batchelor RA, Bernal JP, Cowing Ch, Hagel-Brunnström M. et al. (2000) Seven sedimentary rock reference samples from Estonia. Oil Shale, 17, 3, 215-223.

[48] Kiipli T (2005) Database of the Chemical Quality of Carbonate Rocks. In: Yearbook of the Geological Survey of Estonia 2004, Tallinn, pp. 34-37 (in Estonian).

[49] Petersell V, Ressar H, Carlsson M, Mõttus V, Enel M, Mardla A, Täht K (1997). The Geochemical Atlas of The Humus Horizon of Estonian Soil. Tallinn-Uppsala: Geol. Survey of Estonia & Geol.Survey of Sweden. 37 maps and Explanatory text. 75 p.

[50] Kukk T, Kull T (eds.) (2005) Atlas of the Estonian Flora. Tartu: Estonian University of Life Sciences. 527 pp.

[51] Kotiranta H, Uotila P, Sulkava S & Peltonen S-L (eds) (1998) Red Data Book of East Fennoscandia. Ministry of Environment, Finnish Environment Institute & Botanical Museum, Finnish Museum of Natural History, Helsinki. 351 p.

[52] Red Book: Republic of Karelia (2007) Ministry of agriculture, fish husbandry and ecology the Republic of Karelia. Petrozavodsk, Karelia. 368 p. (in Russian)
[53] Atlas Karel'skoi ASSR (1989) Peihvasser VN (ed) Moscow. 40 p (in Russian).
[54] Söyrinki N, Saari V (1980) Die Flora im Nationalpark Oulanka, Nord-Finnland (The flora of Oulanka National Park, northern Finland). Acta Bot. Fennica 114. 150 p.
[55] Hautala A, Rautiainen L (1998) Oulanka – Paanajärvi. Photos: Hautala H, Rautiainen, L, texts: Systra Y, Viramo, J. Kuusamo, Articmedia. 144 p.
[56] Halonen P (1993) The lichen flora of the Paanajärvi National Park. Oulanka Reports 12, 45-54.
[57] Borodulina GS, Systra YJ (2001) Springs along the nature path in Paanajärvi National Park. Oulanka Reports 25, 5-8.
[58] Ponge J-F (2003) Humus forms in terrestrial ecosystems: a framework to biodiversity. Soil Biol.& Biochemistry 35, 935-945.
[59] Lukina NV, Poljanskaja LM & Orlova MA (2008) Nourishing regime of soils of the northern taiga forests. Moscow, Nauka. 342p. (in Russian).
[60] Kravchenko AV, Kuznetsov OL (2003) The role of protected areas in Karelia's border zone in the conservation of floristic biodiversity. In: Gromtsev, A.N., Kitaev, S.P., Krutov, V.I. et al. (eds.). 2003. Biotic diversity of Karelia: conditions of formation, communities and species. pp.69-76. Russian Academy of Sciences, Karelian Research Centre, Petrozavodsk.
[61] Ladoga Lake 2002) Atlas. Sankt-Peterburg. 129 p. (in Russian).
[62] Lehtovaara JJ (1994a) Geological map of Finland. Sheet 1823, Kilpisjärvi., Pre-Quaternary rocks, Geol.Survey of Finland.
[63] Lehtovaara JJ (1994b) Geological map of Finland, Sheet 1842 Halti, Pre-Quaternary rocks.Geolo. Survey of Finland.
[64] Lehtovaara JJ (1995) Kilpisjärven ja Haltin kartta-alueiden kallioperä. Summary: Pre-Quaternary rocks of the Kilpisjärvi and Halti map-sheet areas. Geol. Survey of Finland, Espoo 1995. 64 p.
[65] Pirrus E (2009) Big stones of Estonia. Story of large erratic boulders. Tallinn, Teaduste Akadeemia kirjastus (in Estonian, summary in English).
[66] Golovin A, Krinochkin L (2004) Geochemical specialization of bedrock and soils as indicator of regional geochemical endemicity. Geologija 48, 22-28, Vilnius.
[67] Gilder SSB (1964) The London Letter. Canad.Med.Ass.J. v. 91, 1081-1082.
[68] Thornton I, Farago ME, Thims CR et al (2008) Urban geochemistry: research strategies to assist risk assessment and remediation of brownfield sites in urban areas. Environ.Geochem. Health 30, 565-576.

Chapter 18

Tree Species Diversity and Forest Stand Structure of Pahang National Park, Malaysia

Mohd Nazip Suratman

Additional information is available at the end of the chapter

http://dx.doi.org/10.5772/50339

1. Introduction

Information on composition, diversity of tree species and species-rich communities is of primary importance in the planning and implementation of biodiversity conservation efforts. In addition, the diversity of trees is fundamental to the total tropical rainforest diversity as trees provide resources and habitat structure for almost other forest species (Cannon *et al.*, 1998). According to Singh (2002), biodiversity is essential for human survival and economic well being and ecosystem function and stability. UNEP (2001) reported that habitat destruction, over exploitation, pollution and species introduction are identified as major causes of biodiversity loss. Hubbel *et al.* (1999) mentioned that disturbances created by these factors determine forest dynamics and tree diversity at the local and regional scales. These disturbances have been considered as an important factor structuring communities (Sumina, 1994).

In forest management operations, inventories on biodiversity are used to determine the nature and distribution of biodiversity region at the region being managed. Quantification of tree species diversity is an important aspect as it provides resources for many species (Cannon *et al.*, 1998). Being a dominant life form, trees are easy to locate precisely and to count (Condit *et al.*, 1996) and are also relatively better known, taxonomically (Gentry, 1992).

While Pahang National Park provides both fully-protected habitats and long-term maintenance of biological diversity, the structure and composition of its flora still remain rather insufficiently known. To protect forests from declining, it is essential to examine the current status of species diversity as it will provide guidance for the management of protected areas. Therefore, using Kuala Keniam forest as an example, a study was conducted to describe quantitatively stand structure of the forests of Kuala Keniam within Pahang National Park, and to determine the level of species composition, diversity and distribution in this area. Information from this quantitative inventory will provide a valuable reference for forest assessment and improve our knowledge in identification of ecologically useful species as well as species of special concern, thus identify conservation efforts for sustainability of forest biodiversity.

2. Materials and methods

2.1. Description of study area

The data for this study were collected from Kuala Keniam forest, Pahang National Park, Malaysia (latitude 4° 31′ 07.17″ N, longitude 102° 28′ 31.26″ E) which ranges about 120 – 200 m above sea level. Kuala Keniam is located at the protected lowland dipterocarp forests within the national park in the state of Pahang. The area is administered by the Department of Wildlife and National Park (DWNP) Malaysia in collaboration with the Universiti Teknologi MARA (UiTM) which operates a research station in the area.

The weather in Pahang National Park is characterized by permanent high temperatures ranging from 20°C at night and 35°C in the day with a high relative humidity (above 80%). Periods of sunshine in the morning are usually followed by heavy thunderstorms in the afternoon, sometimes accompanied by severe gusts of wind. The highest rainfall occurs in October to November with about 312 mm of rainfall. The lowest rainfall occurs in March with only about 50 mm of rain. Sedimentary rocks account for about 83% of National Park. The last formation of sedimentary rocks belongs to the Cretaceous-Jurassic era which exists in Kuala Keniam and its vicinity. The rocks are thick cross-bedded sandstone deposits with subordinate conglomerates and mudstones. The topography consists mainly of lowland, undulating and riverine areas and gently rolling hills with slopes of between 5° to 45°.

The overall vegetation type in Pahang National Pahang is lowland dipterocarp forests in which is characterized by high proportion of species in the family of Dipterocarpaceae with Meranti (*Shorea* spp.) and Keruing (*Dipterocarpus* spp.) as the dominant species. Lowland dipterocarp forest is one of the most rich-species communities in the world, with more than 200 species ha^{-1} (Okuda *et al.*, 2003). Other vegetation communities in Pahang National Park range from the humid rainforests of the lowland, to the montane oak and ericaceous forests in the higher elevation. The highest peak is Mount Tahan 2,187 m, which also the highest point in Peninsular Malaysia. Tahan River and Tembeling River are the headstream tributaries of Pahang National Pahang with the presence riparian tree species, i.e., Gapis (*Saraca multiflora*), Keruing neram (*Dipterocarpus oblongifolius*), Merbau (*Intsia palembanica*), Kasai daun bersar (*Pometia pinnata*) and Melembu (*Pterocambium javanicum*), along river banks. The rainforest consists of tall evergreen trees which attain heights between 30 – 50 m (i.e., Tualang - *Koompassia excelsa*).

2.2. Sampling design and data collection

A topographic map was used to locate the existing forest trails and baselines in the forest area. A total of five transect lines of 100 m in length and 20 m in width (abbreviated as T1, T2, T3, T4 and T5 thereafter) were established in east-west direction using a compass (Table 1, Figure 1).

Each transect line was gridded into five plots, each 20 m × 20 m in size, as workable units. These transect lines were perpendicular to the existing baseline in the forest area and constructed 5 m after the line. The topographic position, including the gradient was measured at each plot. The slope was measured using a clinometer. A tape measure was used to mark the transect lines at the intervals of 20 m. All trees with a diameter at breast height (DBH, 1.3 m above the ground) above 10 cm were measured, tagged and identified by species. The DBH was measured using a DBH tape. If field identification was not possible, the botanical specimens were taken to the herbarium section of the Forest Research Institute Malaysia (FRIM) for identification.

Transect	No. of plots	Area (m^2)	Slope (°)	Topography
T1	5	20 × 20	5 – 35	Steep lower slope with riverine areas
T2	5	20 × 20	3 – 20	Gentle to mid-slope
T3	5	20 × 20	0 – 10	Mainly flat and gentle slope
T4	5	20 × 20	3 – 30	Mid-slope with riverine areas
T5	5	20 × 20	0 – 10	Mainly flat and gentle slope

Table 1. General features of sample plot within the five transect lines of the study area.

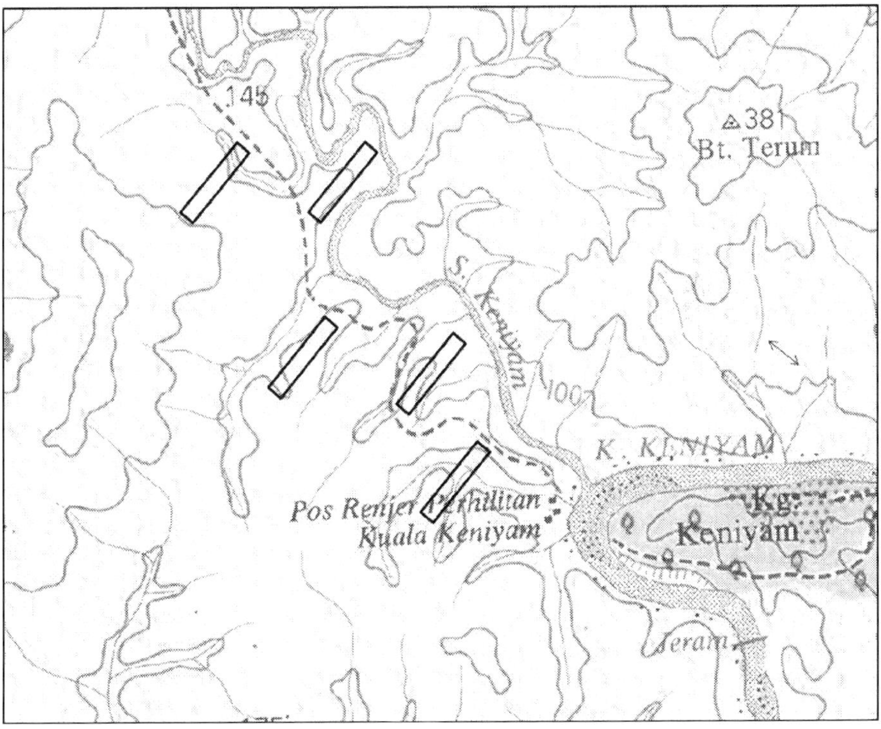

Figure 1. A map of the study area shows the location of five transect lines.

2.3. Data analysis

The means of basal area, genera, species and stem per hectare were calculated for each transect line. One-way analysis of variance (ANOVA) was used to test the differences between the means of these parameters using SAS system (SAS Institute, 2000). The relative dominance of species in each transect line was identified on the basis of relative basal area. The relative basal area of a species on transect lines was calculated as the basal area of a species divided by total basal area of the site and multiplied with 100. The dominant and co-dominant species of each site were identified based on this value. The species with the highest relative basal area was defined as dominant and that with the second highest relative basal area was defined as co-dominant.

In this study, the stand structure was described based on the distribution of species in the study sites and distribution of trees by diameter classes. Therefore, the tree data were grouped into 5 cm diameter classes e.g., the class boundaries were 10 – 14.9, 15 – 19.5 cm, etc. These gave a frequency of trees in each diameter class and were then used to draw bar char graphs.

2.4. Basal area

Basal area is a measure of tree density that defines the area of a given section of land that is occupied by the cross-section of tree. Basal area (*BA*) is calculated using the following equation that converts the DBH in cm to the basal area in m^2.

$$BA = \pi r^2$$

$$= 3.142 \times \left(\frac{\text{dbh}}{200}\right)^2$$

Where

BA = tree basal area (m2)
r = radius (cm)

2.5. Species diversity, richness and evenness indices

A variety of different diversity indices can be used as measures of some attributes of community structure because they are often seen as ecological indicators (Magurran, 1988). Diversity indices provide important information about rarity and commonness of species in a community. The indices can be used to compare diversity between habitat types (Kent and Coker, 1992). The comparison can be between different habitats or a comparison of one habitat over time. Different diversity, species richness, species evenness indices were calculated for each transect as well as pooled data for all transects.

a. Shannon-Weiner diversity index (*H'*) (Shannon and Weiner, 1949) is calculated using the following equation:

$$H' = \sum_{i=1}^{S} p_i \ln p_i$$

Where

H'= the Shannon-Wiener index
p_i= the proportion of individuals belonging to species i
ln=the natural log (i.e., 2.718)

b. The species richness (number of species per unit area) was calculated using Margalef index of species richness (Margalef, 1958) as follows:

$$\text{SR}=\frac{S-1}{\ln(N)}$$

Where

SR=the Margalef index of species richness
S =the number of species
N =the total number of individuals

c. The Whittaker's index of species evenness (Whittaker, 1972) was calculated using the following equation:

$$E_w = \frac{S}{\ln N_i - \ln N_s}$$

Where

E_w=the Whittaker's index of evenness
N_i=the abundance of most important species
N_s=the abundance of the least important species

d. α-diversity was measured based on unified indices (exponential Shannon-Weiner index and Simpson's diversity) as follows:

$$N_1 = \exp^{H'}$$

Where

N_1=the number of equally common species
H'=the Shannon-Weiner index

e. Simpson's diversity (D) (Simpson, 1949) was calculated using the following equation:

$$D = 1 - \lambda$$

Where

D=the Simpson diversity

λ= the Simpson's concentration of dominance calculated as $\sum p_i^2$.

f. The Whittaker's index of β-diversity (Whittaker, 1972) was calculated as:

$$\beta w = \frac{S_c}{\bar{S}}$$

Where

β_w= the Whittaker's index of β-diversity
S_c=the total number of species
$\bar{S}$= the average number of species per sample

g. Bray-Curtis index (C_N) (Bray and Curtis, 1947), a similarity coefficient, is used to measure similarity between transect lines.

$$C_N = \frac{2jN}{(aN + bN)}$$

Where

C_N=the Bray-Curtis index
aN=individual numbers of plot A
bN=individual numbers of plot B
jN= the sum of less individual numbers of each species common in plots A and B

3. Results and discussion

3.1. Stand structure analysis of different sites

Information on the basal area, stem, species and genera densities are efficient expression for revealing forest stand structure and spatial distribution of trees present in the landscape. These four parameters are presented in Table 2. In this study, the means of basal area ha^{-1}, stem ha^{-1}, species ha^{-1} and genera ha^{-1} were measured in every plot (20 m × 20 m) and were averaged to provide an estimate for each transect line. From the analysis of variance, it was found that the difference in the means of these parameters among transects were not statistically significant at $P \leq 0.05$.

The mean of basal area obtained in the present study ranged from 17.2 m^2 ha^{-1} (T4) to 34.3 m^2 ha^{-1} (T3) (Table 2), which is lower compared to those recorded in other tropical rainforests. Examining the structure and composition of lowland tropical rainforests in north Borneo, Burgess (1961) recorded a basal area of 73.6 m^2 ha^{-1} (≥ 10 cm DBH) over a small area (0.08 ha) at Gum Gum Sabah. In another study in an evergreen forest of Andaman Islands, basal area of 44.6 m^2 ha^{-1} has been recorded in 4.5 ha sampled area (Padalia *et al.*, 2004). A much lower basal area of 29 m^2 ha^{-1} and 5.6 m^2 ha^{-1} have been recorded in logged over forest of Sungkai, Perak (Suratman *et al.*, 2007) and secondary forests of Sungai Sator, Kelantan (Suratman *et al.*, 2009), respectively. Both are secondary forests and were put under a selection system of timber extraction in the past, and are considered to be of poor species.

Variables	T1	T2	T3	T4	T5
Mean basal area (m^2 ha^{-1})	25.2 (25.2)	24.8 (13.3)	34.3 (15.1)	17.2 (16.0)	33.3 (12.2)
Mean no. of stems ha^{-1}	510 (123.3)	430 (105.2)	505 (51.2)	315 (219.8)	480 (186.6)
Mean no. of species ha^{-1}	370 (105.2)	370 (105.2)	450 (30.6)	280 (190.7)	405 (125.5)
Mean no. of genera ha^{-1}	340 (72.0)	340 (72.0)	435 (51.8)	250 (165.8)	365 (109.8)
Total no. of species per individual	0.50	0.69	0.63	0.79	0.64

Note: The values in parentheses are standard deviation. All means for the first four parameters above are not significantly different at $P \leq 0.05$.

Table 2. The stand structure of Kuala Keniam forest.

The density and size distribution of trees contribute to the structural pattern characteristic of rainforests. In primary tropical rainforests, the density of trees varies within the limits and depends on many factors. The means number of species and stems per hectare on different transects varied from 280 (T4) – 450 (T3) and 315 (T4) – 510 (T1), respectively (Table 2), indicating a mixed nature of distribution of species and individuals in the forest at each transect, a characteristic of the tropical rainforests. The factors controlling tree density include the effects of natural and anthropogenic disturbance and soil condition (Richards, 1952). From the field observation, the reserve area of the primary forest in the study sites is generally homogenous, with no evidence of major disturbance, and appeared to be a representative example of the lowland forest of Kuala Keniam.

Information on the density-dependent status of species in the study site is important for conservation and management. Studies have classified the density of trees ha^{-1} in tropical forests ranges from low values of 245 stems ha^{-1} (Ashton, 1964; Campbell *et al.*, 1992; Richards, 1996) intermediate values of 420 – 617 stems ha^{-1} (Campbell *et al.*, 1992) in Brazilian Amazon and high values of 639 – 713 stems ha^{-1} in Central Amazon upland forests (Ferreira *et al.*, 1998). In the present study, the density of stems per hectare ranged from 315 – 510 stems ha^{-1}, reflecting spatial variability in the sampled sites. The range fell within intermediate category in the above studies. In the Neotropics, the maximum richness is found up to 300 stems ha^{-1} (Gentry, 1988). A much lower result was reported for forests in Africa where the species richness is about 60 stems ha^{-1} (Bernhard-reversat *et al.*, 1978).

Tree species composition in tropical areas varies greatly from one place to another mainly due to variation in biogeography, habitat and disturbance (Whitmore, 1998). In the tropical rainforests, the tree species per hectare ranges from about 20 to a maximum of 223 (Whitmore, 1984). Philips and Gentry (1994) reported a range of 56 – 282 species ha^{-1} (>10 cm DBH) in mature tropical forests. In the present study, a range of 280 to 450 species ha^{-1} has been recorded in the lowland rainforest of Kuala Keniam (Table 2). In the very rich rainforests, the number of species in rainforests could be as high as 400 species ha^{-1} (Nwoboshi, 1982). When compared to some rainforests around the world, the lowland rainforest of Kuala Keniam could be considered to be species rich. Tropical rainforests in

South America harbour 200 – 300 species ha^{-1} (Richards, 1996). In the tropical evergreen forest of Andaman Islands, India, Padalia *et al.* (2004) found that 58 tree species ha^{-1} were recorded belong to 176 genera and 81 families.

The mean numbers of genera per hectare varied from 340 to 435 genera ha^{-1}. These values are much higher than that obtained by Sagar *et al.* (2003) at a dry tropical forest region of India (4 – 22 genera ha^{-1}). T4 had the highest total number of species per individual when compared to the other four sites of study. The difference could be due to genetic and site difference. A study on vegetation types in Yunnan, Chiangcheng *et al.* (2007) found that slope direction had influence on the tree diversity at different altitudes. The tree diversity on the sunny slope was lower than that on shady slope. The difference in terrain, gradient and slope direction causes the difference soil, water and microclimate which may cause of differences in species adaptability.

3.2. Dominant tree species

On the basis of relative basal area, the five sites differed in the combination of dominant and co-dominant species (Appendix). *Elateriospermum tapos* was dominant in T1 and co-dominant in T4. *Koompassia malaccensis* dominated at the T2 and co-dominated at the T3. *Xanthophyllum lelacarum* was dominant in T3 while *Shorea leprosula* was dominant in T4. *Dyera costulata* and *Dipterocarpus costulatus* were dominated and co-dominated at T5, respectively. Thus, the species exhibit local dominance. These data revealed that T1 represented *Elateriospermum-Intsia* community; T2, *Koompassia-Pentaspadon* community; T3, *Xanthophyllum-Koompassia* community; T4, *Shorea-Koompassia* community; and T5, *Dyera-Dipterocarpus* community. Two tree species, i.e., *Alphonsea elliptica* and *Syzygium* sp., are common on all transects.

3.3. Species diversity

The five transect lines yielded a total of 448 stems and 198 species of trees ≥ 10 cm DBH. These species represent 116 genera and 44 families (Appendix). The number of species and individual varied from 50 to 64 species and 63 to 102 individuals per transect of 100 m × 20 m size, respectively. Table 3 shows the summary statistics for various indices of diversity, richness and evenness. It is generally recognized that the area and environmental heterogeneity have strong effects on species diversity (Rosenzweig, 1995; Whitmore, 1998; Waide *et al.*, 1999). The Shannon-Weiner index (H') was used to compare species diversity between transects. The H' for T1–T5 were 3.42, 3.91, 3.97, 3.84 and 3.91, respectively, indicating that among transects, T3 was the most complex in species diversity whereas T1 is the simplest community in terms of species composition. The Shannon-Weiner diversity index (range between 3.42 – 3.91) obtained for trees more than 10 cm DBH in this study was lower than those recorded in the tropical rainforests of Barroo Colorado Island, Panama [4.8](Knight, 1975) and Silent Valley, India [4.89](Singh *et al.*, 1981). In a more recent study in Shenzen, China, Wang *et al.* (2006) recorded a lower range of Shannon-Weiner index (i.e., 1.92 – 3.10) for trees ≥ 2 cm DBH in a subtropical forest. However, a comparison of diversity indices obtained in the present study with the ones above is difficult due to vast differences in the area sampled, plot size, and the standard diameter class taken.

Variables	T1	T2	T3	T4	T5
Shannon-Weiner index (*H′*)	3.42	3.91	3.97	3.84	3.91
Margalef index of species richness (*SR*)	10.81	13.02	13.65	11.83	13.15
Whittaker index of evenness (E_w)	16.04	36.66	35.72	44.60	31.35
The number of equally common species (N_1)	30.72	49.72	53.11	46.55	49.75
Simpson's diversity (*D*)	0.93	0.98	0.98	0.98	0.97
Whittaker index of β-diversity (β_W)	3.51	3.88	3.56	4.46	3.77

Table 3. Pattern of tree species diversity in Kuala Keniam forest.

Similar patterns were found for species richness, which was computed using Margalef index of species richness (*SR*) and the number of equally common species (N_1). The *SR* ranged from 10.81 to 3.97 and the N_1 ranged from 30.72 to 53.11. Whittaker index of evenness (E_w) ranged from 16.04 to 44.60, the highest value was recorded at T4 and the lowest at T1. In the present study, Simpson's diversity (*D*) was not a very sensitive indicator of diversity as four of five sites (T2 – T5) had somewhat similar values. Whittaker index of β-diversity (β_W) was used to compare habitat heterogeneity within a transect. The β_W value was the highest for T4 (4.46) and the lowest for T1 (3.51). Further analysis indicated that the number of species per individual had a direct positive influence on β-diversity (Figure 2). According to Condit *et al.* (1998), species richness is positively associated with species abundance. This relationship suggests that large population is less prone to extinction than small ones (Preston, 1962). Based on the relationship between abundance and diversity, habitats supporting larger numbers of individuals can support more populations and more species than habitat supporting small number of individuals.

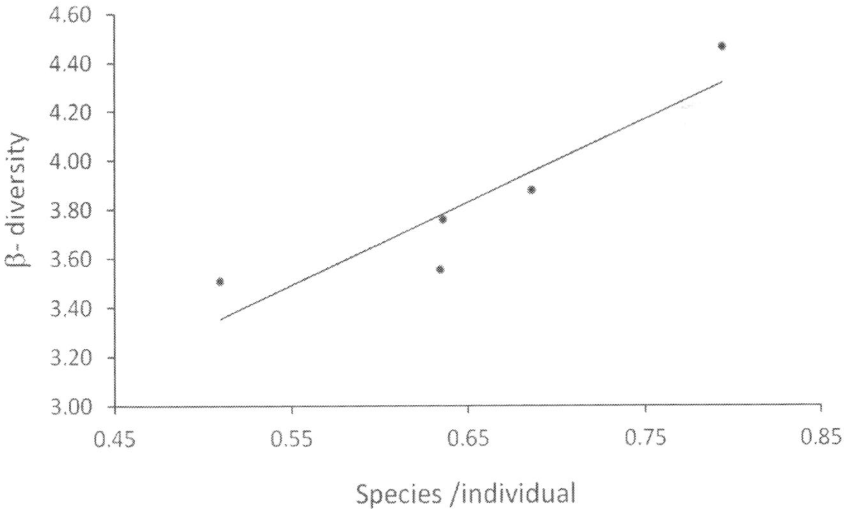

Figure 2. Relationship between β-diversity (β_w) and species/individual (S_n) according to $\beta_w = 1.624 + 3.393S_n$, r^2=83, p=0.03.

3.4. Similarity between transects

The similarity based on Bray-Curtis index (C_N) was calculated between the pair of transects, and abundance similarity matrix was constructed (Table 4). The Bray-Curtis similarity index was used because it is often a satisfactory coefficient for biological data on community structure (Clarke and Warwick, 1994). Comparison of C_N values among the five transects data indicates that the species composition of T1 was fairly different from those of the other four sites. T3 had a high species similarity to T4 and T5, and T4 had a high species similarity to T5. T2 was similar to some degree to T4 and T5.

Transect	T1	T2	T3	T4
T2	0.15			
T3	0.14	0.19		
T4	0.13	0.13	0.27	
T5	0.14	0.11	0.37	0.29

Table 4. Similarity coefficient among the five transects of Kuala Keniam forest.

3.5. Family-wise distribution

A total of 44 tree families were encountered in the forest of Kuala Keniam (Figure 3). The maximum number of tree species belongs to the family of Euphorbiaceae which accounts for 23.9% of the total individuals encountered in the study site. *Elateriospermum tapos* is the most widely occurring species from this family. Other trees from this family such as *Macaranga lowii, Mallotus leucodermis* and *Pimelodendron griffithianum* are among the important part of floristic composition in the study area. The other dominant families are Myristicaceae, Burseraceae, and Leguminosae which account for 8.3%, 5.4% and 4.5% of the total individual encountered in the study site, respectively. The fifth most dominant family is Myrtaceae with 4.2%. Earlier study also indicated that Euphorbiaceae was the dominant family in Sungkai forest with 27% of tree species belong to this family (Suratman *et al.*, 2007). Two other studies conducted in India for tree species also support the fact that Euphorbiaceae is the dominant family in Bay Islands (Dagar and Singh, 1999) and Andaman Islands (Padalia *et al.*, 2004). The dominant plant family in Neotropical lowland forests and Africa is Leguminosae (Gentry, 1988) and in Southeast Asia the dominants are Dipterocarpaceae (Richards, 1952; Whitmore, 1998).

3.6. Diameter class distribution

The stand structure of lowland rainforests of Kuala Keniam forest was studied based on the distribution of tree diameter class. The diameter distribution of trees is very variable and some forests have large numbers of trees of 40 – 60 cm DBH (Richards, 1952). In this study, the distribution of trees clearly displays the characteristic of De iocourt's factor procedure (inverse J distribution) where stems frequencies decrease with the increase in DBH (Figure 4). This generally indicates that stands are developing and regeneration in the forest is

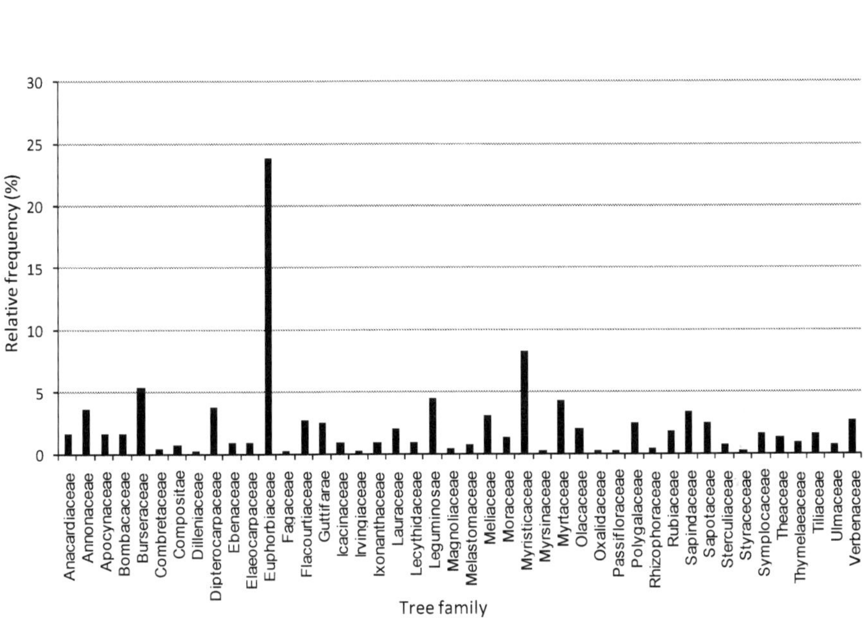

Figure 3. Family-wise distribution of tree species of Kuala Keniam forest.

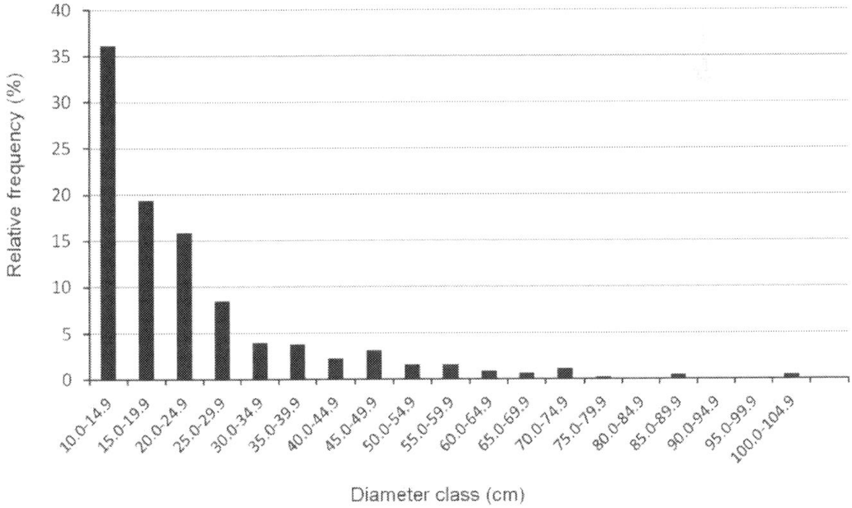

Figure 4. Diameter distribution of trees at Kuala Keniam forest.

present. Natural regeneration is dependent on the availability of mother trees, fruiting pattern and favourable conditions. As shown in the figure, the presence of growth of the forest is indicated by the movement of trees in various diameter classes. Higher number of stems for smaller diameter classes, with 36% of trees fell within the 10 – 14.9 cm, 19% fell within 15 – 19.9 cm, 16% fell within 20 – 24.9 cm, 9% fell within 25 – 29.9 cm and 4% fell within 30 – 34.9 cm. The histogram shows a less or an absent number of stems in diameter classes from 79.9 cm onwards. Under natural conditions, an old, big emergent tree may fall down and create gap. Forest regeneration via natural succession will take place if the area is not too far away from mature primary forest trees serving as source for the recalcitrant seeds.

4. Conclusion

The forests of Kuala Keniam are protected primary forests which comprises of natural vegetation and are dictated by a combination of biotic and abiotic factors like topography, altitude, geology, climatic etc. as well as historical conditions of geology and climate. The density and size distribution of trees contribute to the structural pattern characteristics of the forest. The study indicated that the forests of Kuala Keniam are characterized by a uniform distribution of individuals with mixed species composition, and the sites are represented by different combinations of the dominants and co-dominant species. The distribution of trees displays the characteristic of De iocourt's factor procedure (inverse J distribution) where stems frequencies decrease with the increase in DBH, indicating stable populations in which regeneration of forest in this area is present.

Appendix

List of species, family and the relative basal area of Kuala Keniam forest

Species	Family	T1	T2	T3	T4	T5
Aglaia sp.	Meliaceae	-	2.65	0.23	0.37	0.17
Agrostistachys longifiolia	Euphorbiaceae	-	0.26	-	-	-
Aidia densiflora	Rubiaceae	-	-	0.21	-	0.24
Alphonsea elliptica	Annonaceae	2.37	0.55	1.48	5.91	1.41
Alphonsea jengkasii	Annonaceae	-	-	-	-	0.16
Alseodaphne intermedia	Lauraceae	-	-	0.84	-	-
Anisoptera laevis	Dipterocarpaceae	-	-	4.35	-	-
Antidesma coriaceum	Euphorbiaceae	-	-	0.47	-	-
Antisdesma sp.	Euphorbiaceae	-	0.28	-	-	-
Aporosa arborea	Euphorbiaceae	0.51	-	-	-	-
Aporosa aurea	Euphorbiaceae	-	0.28	-	-	-

Species	Family	T1	T2	T3	T4	T5
Aporosa falcifera	Euphorbiaceae	-	-	-	1.63	0.21
Aporosa globifera	Euphorbiaceae	-	0.25	-	-	-
Aporosa microstachya	Euphorbiaceae	-	-	-	-	0.22
Aporosa nigricans	Euphorbiaceae	-	0.29	-	-	-
Aporosa prainiana	Euphorbiaceae	-	-	-	0.24	0.12
Aporosa symplocoides	Euphorbiaceae	-	-	-	-	0.17
Archidendron ellipticum	Leguminosae	-	-	0.32	-	-
Ardisia sp.	Myrsinaceae	-	0.21	-	-	-
Aromadendron elegans	Myristicaceae	-	1.42	-	-	-
Artocarpus griffithii	Moraceae	-	-	0.48	-	-
Artocarpus lowii	Moraceae	-	-	-	-	0.73
Austrobuxus nitidus	Euphorbiaceae	-	-	-	-	5.10
Baccaurea brevipes	Euphorbiaceae	-	-	-	0.24	-
Baccaurea kunstleri	Euphorbiaceae	-	-	-	0.42	-
Baccaurea minor	Euphorbiaceae	-	0.72	2.77	-	-
Baccaurea reticulata	Euphorbiaceae	-	-	-	0.38	3.46
Barringtonia macrostachya	Lecythidaceae	-	-	0.28	0.37	0.20
Beilschmiedia lucidula	Lauraceae	-	-	0.99	-	-
Blumeodendron kurzii	Euphorbiaceae	-	-	0.82	-	-
Buchanania sessifolia	Anacardiaceae	-	-	0.79	-	0.28
Callicarpa maingayi	Verbenaceae	-	-	-	-	0.39
Callophyllum sp.	Guttiferae	-	0.40	-	-	-
Canarium littorale	Burseraceae	1.20	-	2.00	-	-
Carallia brachiata	Rhizophoraceae	-	-	-	-	0.75
Casearia clarkei	Flacourtiaceae	-	-	0.61	0.94	0.13
Casearia sp.1	Flacourtiaceae	-	0.91	-	-	-
Cheilosa malayana	Euphorbiaceae	-	0.21	-	0.35	-
Chisocheton sp.	Meliaceae	0.25	-	-	-	-
Cinnamomum iners	Lauraceae	-	1.55	-	-	-
Croton levifolium	Euphorbiaceae	-	-	-	-	0.29
Cryptocarya densiflora	Lauraceae	-	-	0.61	-	-
Cryptocarya infectoria	Lauraceae	-	-	-	-	0.45
Cryptocarya kurzii	Lauraceae	-	0.63	0.50	-	-
Dacryodes costata	Burseraceae	-	-	0.24	-	-
Dacryodes rostrata	Burseraceae	0.59	-	0.13	0.31	4.79

Species	Family	T1	T2	T3	T4	T5
Dacryodes rugosa	Burseraceae	0.48	-	0.19	1.08	0.18
Dialium indum L. var. *indum*	Leguminosae	0.74	-	-	-	-
Dialium platysepalum	Leguminosae	-	0.82	-	-	-
Dillenia reticulata	Dilleniaceae	-	6.69	-	-	-
Diospyros buxifolia	Ebenaceae	-	-	0.18	-	-
Diospyros maingayi	Ebenaceae	-	-	-	-	1.87
Diospyros sumatrana	Ebenaceae	-	-	-	-	0.28
Diplospora malaccensis	Rubiaceae	-	0.28	-	-	0.57
Dipterocarpus costulatus	Dipterocarpaceae	-	-	-	-	12.63
Durio griffithii	Bombacaceae	0.28	-	1.67	0.33	0.68
Durio lowianus	Bombacaceae	-	-	-	5.03	-
Dyera costulata	Apocynaceae	-	-	0.84	2.97	13.38
Dysoxylum flavescens	Meliaceae	-	-	1.71	-	-
Dysoxylum sp.	Meliaceae	1.19	-	-	-	-
Dysoxylum sp1.	Meliaceae	-	-	2.91	-	-
Elaeocarpus nitidus	Elaeocarpaceae	-	-	0.30	1.13	-
Elaeocarpus palembanicus	Elaeocarpaceae	-	-	-	-	0.13
Elaeocarpus petiolatus	Elaeocarpaceae	-	-	-	0.68	-
Elateriospermum tapos	Euphorbiaceae	16.11	-	4.48	7.73	3.40
Erythrospermum candidum	Flacourtiaceae	-	0.34	-	-	-
Flacourtia rukam	Flacourtiaceae	0.33	-	-	-	-
Garcinia bancana	Guttiferae	-	-	-	-	3.96
Garcinia griffithii	Guttiferae	-	-	-	0.75	3.18
Garcinia nervosa	Guttiferae	-	-	-	-	1.79
Garcinia parvifolia	Guttiferae	-	-	1.03	-	-
Garcinia pyrifera	Guttiferae	-	-	0.40	-	-
Gironniera nervosa	Ulmaceae	0.26	-	-	-	-
Gironniera subaequalis	Ulmaceae	-	-	0.77	-	-
Gonocaryum gracile	Icacinaceae	-	-	-	0.27	-
Gonystylus maingayi	Thymelaeaceae	-	4.24	0.24	2.89	-
Gordonia penangensis	Theaceae	-	0.69	0.98	-	1.77
Guioa sp.	Sapindaceae	-	-	-	-	-
Homalium longifolium	Flacourtiaceae	-	-	1.95	-	-
Hopea sulcata	Dipterocarpaceae	-	-	0.84	-	-
Horsfieldia fulva	Myristicaceae		-	1.12	-	-

Species	Family	T1	T2	T3	T4	T5
Horsfieldia sucosa	Myristicaceae	0.74	-	-	-	-
Horsfieldia tomentosa	Myristicaceae	0.19	-	-	-	-
Hosfieldia polyspherula var. sumatrana	Myristicaceae	-	-	-	-	0.57
Hunteria zeylanica	Apocynaceae	-	-	3.41	-	-
Hydnocarpus woodii	Flacourtiaceae	1.72	-	-	-	-
Intsia palembanica	Leguminosae	12.80	-	-	-	-
Irvingia malayana	Irvingiaceae	-	-	-	0.58	-
Ixonanthes icosandra	Ixonanthaceae	-	3.28	0.61	-	0.81
Kibatalia maingayi	Apocynaceae	0.31	-	-	-	-
Knema furfuracea	Myristicaceae	-	-	-	1.46	0.96
Knema hookeriana	Myristicaceae	1.70	0.27	-	-	-
Knema intermedia	Myristicaceae	1.62	0.60	-	-	-
Knema laurina	Myristicaceae	0.38	-	-	-	-
Knema patentinervia	Myristicaceae	-	1.53	1.35	2.51	0.30
Knema scortechinii	Myristicaceae	1.78	0.43	-	-	-
Knema stenophylla	Myristicaceae	-	-	-	-	0.57
Koompassia excelsa	Leguminosae	-	2.08	-	2.79	-
Koompassia malaccensis	Leguminosae	0.97	14.87	7.90	-	-
Lasianthus sp.	Rubiaceae	-	-	-	0.34	-
Lithocarpus curtisii	Fagaceae	-	-	-	3.48	-
Litsea machilifolia	Lauraceae	-	-	-	0.26	-
Macaranga hypoleuca	Euphorbiaceae	-	-	-	0.94	0.33
Macaranga lowii	Euphorbiaceae	-	-	1.06	0.78	2.57
Magnolia liliifera	Magnoliaceae	-	-	1.07	-	-
Mallotus leucodermis	Euphorbiaceae	6.22	1.05	-	-	-
Mallotus oblongifolius	Euphorbiaceae	0.24	-	0.24	-	-
Mallotus sp.	Euphorbiaceae	1.70	-	-	-	-
Mangifera griffithii	Anacardiaceae	-	-	-	1.33	-
Medusanthera gracilis	Icacinaceae	-	-	0.21	-	0.45
Meiogyne monosperma	Annonaceae	-	0.27	-	-	-
Memecylon minutiflorum	Melastomaceae	-	0.26	-	-	-
Memecylon pubescens	Melastomaceae	-	0.87	-	-	-
Mesua ferrea	Guttiferae	-	-	0.21	-	-
Mesua lepidota	Guttiferae	1.00	-	-	-	-

Species	Family	T1	T2	T3	T4	T5
Mesua racemosa	Guttiferae	-	-	-	-	0.47
Mezzettia elliptica	Annonaceae	-	-	-	-	2.54
Microcos antidesmifolia	Tiliaceae	-	0.72	-	-	-
Microcos fibrocarpa	Tiliaceae	0.17	-	-	-	-
Microcos laurifolia	Tiliaceae	-	1.26	-	-	-
Microcos tomentosa	Tiliaceae	-	0.16	-	-	-
Monocarpia marginalis	Annonaceae	-	-	0.35	-	-
Myristica gigantea	Myristicaceae	-	-	-	-	1.81
Nauclea officinalis	Rubiaceae	-	-	-	1.89	-
Neoscortechinia kingii	Euphorbiaceae	-	-	0.23	-	-
Nephelium costatum sub-species oppoides	Sapindaceae	-	0.41	-	-	-
Nephelium cuspidatum	Sapindaceae	4.77	-	-	1.05	-
Nephelium maingayi	Sapindaceae	-	0.16	-	1.62	-
Nothaphoebe umbelliflora	Lauraceae	0.90	-	-	-	-
Ochanostachys amentacea	Olacaceae	1.10	-	0.12	-	0.47
Palaquium clarkeanum	Sapotaceae	-	-	-	0.75	-
Palaquium gutta	Sapotaceae	-	-	2.01	4.02	1.22
Palaquium hexandrum	Sapotaceae	-	-	0.64	-	-
Palaquium hispidum	Sapotaceae	-	2.71	-	-	-
Palaquium maingayi	Sapotaceae	-	-	2.83	-	-
Palaquium microcarpum	Sapotaceae	-	0.57	-	-	-
Parartocarpus venenosus	Moraceae	-	-	0.39	-	-
Paratocarpus bracteatus	Moraceae	-	-	-	1.72	-
Paropsia vareciformis	Passifloraceae	-	0.23	-	-	-
Payena dasyphylla	Sapotaceae	-	-	0.94	-	-
Payena lanceolata var. lanceolata	Sapotaceae	-	-	-	5.07	-
Payena maingayi	Sapotaceae	-	1.88	-	-	-
Pellacalyx saccardianus	Rhizophoraceae	-	-	-	0.67	-
Pentace curtisii	Tiliaceae	0.17	-	-	-	-
Pentace strychnoidea	Tiliaceae	-	-	-	-	0.98
Pentaspadon velutinus	Anacardiaceae	7.40	7.76	-	-	-
Pimelodendron griffithianum	Euphorbiaceae	0.62	-	3.51	0.72	0.14
Planchonia grandis	Lecythidaceae	-	-	-	2.71	-
Polyalthia jenkensii	Annonaceae	-	0.56	-	-	0.13

Species	Family	T1	T2	T3	T4	T5
Polyalthia rumphii	Annonaceae	-	1.08	-	-	-
Pometia ridleyi	Sapindaceae	2.65	-	-	-	-
Pseudoclausena chrysogyne	Meliaceae	-	0.29	-	-	-
Pternandra echinata	Melastomaceae	-	-	-	0.38	-
Pterocymbium javanicum	Sterculiaceae	1.44	-	-	-	-
Ptychopyxis caput-medusae	Euphorbiaceae	3.29	-	-	-	-
Pyrenaria acuminata	Theaceae	-	-	-	0.75	-
Santiria griffithii	Burseraceae	-	-	0.18	-	-
Santiria laevigata	Burseraceae	1.03	0.96	10.47	-	-
Santiria sp.	Burseraceae	0.17	-	-	-	-
Santiria tomentosa	Burseraceae	-	-	-	-	1.64
Sapium baccatum	Euphorbiaceae	-	6.52	-	-	-
Saraca declinata	Leguminosae	-	0.46	0.25	-	-
Sarcotheca griffithii	Oxalidaceae	-	-	-	-	0.24
Scaphium linearicarpum	Sterculiaceae	1.13	-	-	-	-
Scaphium macropodum	Sterculiaceae	-	-	-	-	0.26
Schoutenia accrescens	Tiliaceae	-	0.20	-	-	-
Shorea leprosula	Dipterocarpaceae	5.18	-	-	12.67	6.73
Shorea multiflora	Dipterocarpaceae	-	-	-	3.26	-
Shorea ovalis	Dipterocarpaceae	-	-	0.28	-	6.19
Shorea parvifolia	Dipterocarpaceae	3.15	-	2.86	-	0.28
Species A	Meliaceae	0.73	-	-	-	-
Streblus elongatus	Moraceae	0.88	-	-	-	-
Strombosia ceylanica	Olacaceae	-	-	-	-	0.62
Strombosia javanica	Olacaceae	-	1.33	-	-	-
Strombosia sp.	Olacaceae	-	-	-	8.59	-
Styrax benzoin	Styraceceae	-	0.50	-	-	-
Symplocos fasciculata	Symplocaceae	2.37	1.75	-	-	-
Symplocos sp.	Symplocaceae	-	-	-	1.00	-
Syzygium chloranthus	Myrtaceae	0.63	-	-	-	-
Syzygium densiflora	Myrtaceae	-	-	-	0.25	-
Syzygium duthieanum	Myrtaceae	0.32	-	-	-	-
Syzygium griffithii	Myrtaceae	-	-	0.13	3.46	0.33
Syzygium lineatum	Myrtaceae	0.90	-	-	-	-
Syzygium protulata	Myrtaceae	0.69	-	-	-	-

Species	Family	T1	T2	T3	T4	T5
Syzygium pustulatum	Myrtaceae	-	-	0.24	-	-
Syzygium sp.	Myrtaceae	0.79	1.38	9.23	0.92	1.32
Tarenna mollis	Rubiaceae	1.13	-	-	-	-
Teijsmanniodendron coriaceum	Verbenaceae	-	-	1.11	-	2.35
Terminalia citrina	Combretaceae	-	7.03	-	-	-
Timonius wallichianus	Rubiaceae	1.35	-	-	-	-
Vernonia arborea	Compositae	-	0.59	-	-	0.64
Vitex pinnata	Verbenaceae	1.35	-	-	-	-
Xanthophyllum griffithii	Polygalaceae	-	-	0.35	-	1.53
Xanthophyllum lelacarum	Polygalaceae	-	-	10.08	-	-
Xerospermum laevigatum	Sapindaceae	-	0.50	-	-	-
Xerospermum noronhianum	Sapindaceae	-	3.94	-	-	-
Xylopia magna	Annonaceae	-	7.76	-	-	-
Xylopia malayana	Annonaceae	-	0.16	-	-	0.43

Table 5. List of species, family and the relative basal area of Kuala Keniam forest.

Author details

Mohd Nazip Suratman
Faculty of Applied Sciences, University of Technology MARA, Shah Alam, Malaysia

5. References

Ashton, P. S. 1964. A quantitative phytosociological technique applied to tropical mixed rainforest vegetation. Malays. For., 27, 304–307.

Bernhard-reversat, F, Huttel, C., and G. Lemee. 1978. Structure and functioning of evergreen rain forest ecosystem of the Ivory Coast. In: Tropical Forest Ecosystems: A State-of-Knowledge Report, UNESCO, Paris, pp. 557 – 574.

Bray, J. R. and J. T. Curtis. 1957. An ordination of upland forest communities of southern Wisconsin. Ecological Monographs 27:325-349.

Burgess, P. F. 1961. The structure and composition of lowland tropical rain forest in north Borneo. Malays. For., 24: 66–80.

Changcheng, T., Xuelong, J., Hua, P., Pengfei, F. and Z. Shoubiao. 2007. Tree species diversity of black-crested gibbons (*Nomascus concolor*). Acta Ecologica Sinica. 27(10): 4002 – 4010.

Campbell, D.G., J.L. Stone and A. Rosas Jr. 1992. A comparison of the phytosociology and dynamics of three floodplain (Varzea) forest of known ages, Rio Jurua, western Brazilian Amazon. Botanical Journal of the Linnean Society 108: 231-237.

Cannon, C.H., Peart, D.R., and M. Leighton. 1998. Tree species diversity in commercially logged Bornean rain forest. Science 28, 1366–1368.

Condit, R., Hubbell, S.P., La Frankie, J.V., Sukumar, R., Manokaran, N., Foster, R.B. and P.S. Ashton. 1996. Species–area and species–individual relationships for tropical trees a comparison of three 50 ha plots. J. Ecol. 84, 549–562.

Dagar, R. J. C. and N.T. Singh. 1999. Plant Resources of the Andaman and Nicobar Islands, Vol. 81, 211–0165.

Ferreira, L. V. and G.T. Prance. 1998. Species richness and floristic composition in four hectares in the Jaú National Park in upland forests in Central Amazonia. Biodivers. Conserv , 7, 1349 - 1364.

Gentry, A.H. 1988. Changes in plant community diversity and floristic composition on environmental and geographic gradients. Annals of the Missouri Botanical Garden. 75 1-34

Gentry, A.H. 1990. Floristic similarities and differences between southern Central America and upper and central Amazonia. In: Gentry, A.H. (Ed,) Four Neotropical Rainforests. Yale University Press, New Haven, CT, pp. 141 - 157.

Gentry, A.H., 1992. Tropical forest biodiversity: distributional patterns and their conservational significance. Oikos 63, 19–28.

Hubbell, S.P., Foster, R.B., O'Brien, S.T., Harms, K.E., Condit, R., Wechsler, B., Wright, S.J. and S. Loode Lao. 1999. Light-gap disturbance, recruitment limitation, and tree diversity in a Neotropical forest. Science 283, 554–557.

Kent, M., and P. Coker. 1992. Vegetation description and analysis: a practical approach. Belhaven Press, London.

Knight, D. H. 1975. A phytosociological analysis of species rich tropical forest on Barro Colorada Islands, Panama. Ecol. Monogra., 45, 259–284.

Magurran, A.E. 1988. Ecological Diversity and Its Measurement. Princeton University Press, Princeton, N J.

Margalef, F. R. 1958. Information theory in ecology. Gen. Syst., 3:36–71.

Okuda, T., Suzuki, M., Adachi, N., Quah, E. S., Hussein, N. A. And N. Manokaran. 2003. Effect of selective logging on canopy and stand structure and tree species composition in a lowland dipterocarp forest in Peninsular Malaysia. Forest Ecology and Management 175: 297–320.

Philips, O.L. and A.H. Gentry. 1994. Increasing turnover through time in tropical forests. Science. 263, 954 - 958.

Preston, F.W. 1962. The canonical distribution of commonness and rarity, Part I. Ecology, 43:185 - 215.

Richards, P. W. 1952. The Tropical Rain Forest, Cambridge University Press.

Richards, P. W. 1996. The Tropical Rain Forest: An Ecological Study, Cambridge University Press, Cambridge.

SAS Institute Inc. 2000. SAS/STAT®User's guide, version 8, vol. 1, Cary, NC, USA.

Shannon, C. E. and W. Weiner, 1949. The Mathematical Theory of Communication, University of Illinois Press, Urbana, USA.

Simpson, E. H. 1949. Measurement of diversity. Nature, 163: 688.

Singh, J. S., Singh, S. P., Saxena, A. K. and Y.S. Rawat. 1981. Report on the Silent Valley Study, Ecology Research Circle, Kumaun University, Nainital, 86.

Singh, J.S., 2002. The biodiversity crisis: a multifaceted review. Curr. Sci. 82, 638–647.

Sumina, O.I., 1994. Plant communities on anthropogenically disturbed sites on Chukotka Peninsula, Russia. J. Veg. Sci. 5, 885–896.

Suratman, M.N., M. Kusin, and S.A.K. Yamani. 2007. Study of tree species composition in Sungkai Forest. Paper presented at the National Biodiversity Seminar, Department of Wildlife and National Parks. 20 – 21 Nov. 2007, Seremban.

Suratman, M.N., M.S. Daim, S.A.K. Yamani, M. Kusin and R. Embong. 2009. Report of Flora and Fauna Surveys for Sungai Sator Forest, Jeli, Kelantan. Environmental Impact Assessment (EIA) Report for the Development of UMK campus. 26pp.

UNEP, 2001. India: State of the Environment. United Nations Environment Programme.

Waide, R.B., Willig, M.R., Steiner, C.F., Mittelbach, G., Gough, L., Dodson, S.I., Juday, G.P. and R. Parmenter. 1999. The relationship between productivity and species richness. Annu. Rev. Ecol. Systematics, 30, 257 – 300.

Wang, D.P., Ji, S.Y., Chen, F.P., Xing, F.W. and S. L. Peng. 2006. Diversity and relationship with succession of naturally regenerated southern subtropical forests in Shenzhen, China and its comparison with the zonal climax of Hong Kong. Forest Ecology and Management 222: 384-390.

Whitmore, T. C. 1998. An Introduction to Tropical Forests, Clarendon Press, Oxford and University of Illinois Press, Urbana, 2nd, Ed. pp. 117.

Whitmore, T.C. 1984. Tropical Rain Forests of the Far East. Clarendon Press, Oxford.

Whittaker, R. H. 1972. Evolution and measurements of species diversity. Taxon, 21: 213–251.

Chapter 19

Image Processing for Pollen Classification

Marcos del Pozo-Baños, Jaime R. Ticay-Rivas, Jousé Cabrera-Falcón, Jorge Arroyo, Carlos M. Travieso-González, Luis Sánchez-Chavez, Santiago T. Pérez, Jesús B. Alonso and Melvín Ramírez-Bogantes

Additional information is available at the end of the chapter

http://dx.doi.org/10.5772/47527

1. Introduction

Palynology - *"The study of pollen grains and other spores, especially as found in archaeological or geological deposits. Pollen extracted from such deposits may be used for radiocarbon dating and for studying past climates and environments by identifying plants then growing."* [1]

Over 20% of all the world's plants are already at the edge of becoming extinct [2]. Saving earth's biodiversity for future generations is an important global task [3] and as many methods as available must be combined to achieve this goal. This involves mapping plants distribution by collecting pollen and identifying them in a laboratory environment.

Pollen grain classification has been an expensive qualitative process, involving observation and discrimination of features by a highly qualified palynologist. It is still the most accurate and effective method. But it certainly limits research progress, taking considerable amounts of time and resources [4].

Automatic recognition of pollen grains can overcome these problems, producing purely objective results faster. Such a tool would provide invaluable in the studies of flora. This advantages were obvious for Flenley [5] [6], who proposed the implementation of an automatic pollen grain classification system in 1968. However, the idea was intractable at that time. Mainly, because of technology restrictions. Nowadays, technology is not a barrier any more, and the discussed system is a reality thanks to computer vision.

This chapter presents the latest results obtained by the authors in the field of automatic pollen grain classification. This will be done by introducing a developed system, paying special attention to the phases of *preprocessing* (section 3.1) and *feature extraction* (section 4). Results for a 17 pollen species database obtained with the commented system will also be shown (section 6).

2. Related work

The begins of automatic pollen identification were based on scanning electron microscope (SEM) images. Langford applied statistical classifiers on texture parameters on 1988, reporting a 94.30% of accuracy on a six pollen class database [7]. Later, artificial neural networks (ANN) were used on the classification task, achieving a success rate of 100% with 3 classes [8].

However, SEM images are expensive and difficult to produce and the use of light microscope (LM) images were explored in 1998 [9]. Again, first attempts were not fruitful due to the low quality images provided by the technology of the time. But recent works has demonstrated that the use of LMs images is, in fact, possible.

For example, [10] reported a 100% of success with a small database containing 4 classes. Moreover, it was one of the first works using artificial neural networks for the classification phase, along with texture parameters. Again, [11] used artificial neural networks for classification. This time, brightness and shape descriptors were extracted as pollen features. A 90% of accuracy with a 3 class database was reported.

[12] and [13] presented a more complex work, combining shape and ornamentation of the grains; using simple geometric measures, and concurrence matrices applied for the measurement of texture. Again, artificial neural networks were used for classification. These works reported a 87.7% recognition rate for a 5 classes database and a 97.7% for a three class database respectively.

[14] describes an automatic optical recognition and classification of pollen grains system. This is able to locate pollen grains on slides, focus and photograph them before identify the species applying a trained neural network. The system achieved a 90% of recognition rate with a 3 class database.

Other works use more sophisticate capture methods, achieving 3 dimensional representations of the pollen grains. [15] presented a combination of statistical reasoning, feature learning and expertise knowledge. A feature extraction algorithm was applied alternating 2D and 3D representations. Iterative refinement of hypotheses was used during the classification process. This work reported a 77% of accurate rate in a database with 30 classes and 97% when only 4 classes were used. An other example, [16], which used a confocal laser microscope to create the 3D models, achieved a 90% recognition rate with 3 classes database.

3. Pollen extraction

At the actual development stage of the system, the detection of pollen grain is highly but not fully automatic. This should not be of any surprise, as the task of pollen location inside sample images is itself a different problem, which is as much complicated as the problem studied in this chapter.

Thus, users should first select and area with a pollen grain inside. Preferably, an area, as small as possible, where an isolated pollen grain is located. This user selected region of interest (ROI) is then automatically preprocessed to detect the contour of the grain (see figure 1).

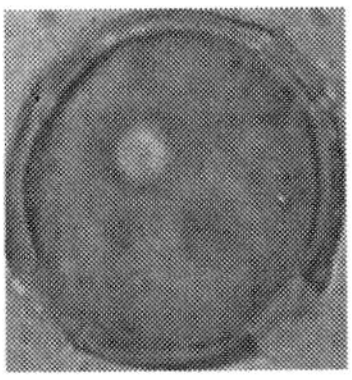

Figure 1. An example of a pollen grain manually selected by the user.

3.1. Preprocessing

This section introduces the automatic preprocessing algorithm used for pollen extraction and preparation. It is important to remind that this process is applied to the image area manually selected by the user, like that showed on figure 1. The preprocessing steps are (see figure 2):

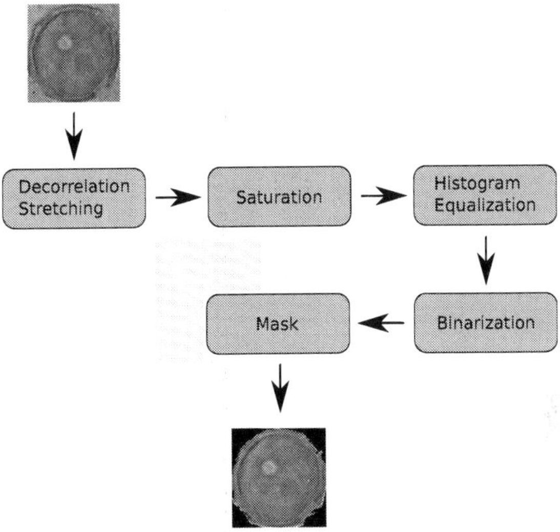

Figure 2. Automatic preprocessing steps for pollen extraction.

1. *Decorrelation stretching*: This process aims to reduce the autocorrelation of the information contained in the image [17]. This is done as a three steps process:
   (a) The original bands are transformed to their principal components.
   (b) The principal components are then stretched separately.
   (c) The resulting data is transformed back to the original space applying the inverse of the principal component transformation.

   The results is a linear transformation of the spectral bands, resulting in uncorrelated variables with unit variance, and enhancing displays. The result can be seen in figure 3.

2. *Saturation*:The saturation channel of the image represents the amount of colour used at each pixel, i.e. the lower the saturation is the greyer the pixel is. This channel is actually extracted from the HSV image representation [18].

   In particular, the saturation channel is computed as:

$$S = \begin{cases} 0, & \text{if } MAX = 0 \\ 1 - \frac{MIN}{MAX} & \text{otherwise} \end{cases} \tag{1}$$

   The result of computing the saturation channel of the docorrelation stretched image is shown on figure 3. The simplification of the task of differentiating pollen and background is obvious.

3. *Histogram equalization*: Equalizing the histogram of an image aims to obtain a uniform distribution of the pixel values. This maximizes the contrast without loosing structural information, i.e., conserving the entropy [19].
4. *Binarization*: The binarization of an image consist of transform each pixel's value to '0' or '1' depending on whether it has a value lower or higher (respectively) than a set threshold. This results on a simple image containing pure geometric information.
5. *Mask*: Finally, in a bid to obtain a clear mask of the pollen grain, several image processing functions are applied such as "imfill" and "bwareaopen" provided by the Image Processing Toolbox of Matlab [20].

The resulting mask can be either used for feature extraction or to remove the background of the pollen grain image. The result of applying each preprocessing step can be seen on figure 3

4. Feature extraction

Pollen images by their own does not prove to be a high quality information for the task of automatic pollen grain classification. Although they contain the necessary information, this information is hidden and diffused around the image and behind other unimportant data. In order to extract the relevant information from raw samples, they need to be further processed by the *feature extractor*.

A total of 50 features are extracted from the pollen images. I.e. the output of the *feature extraction* block is a vector with length 50. These 50 features corresponds to 24 geometric parameters carrying information regarding size and basic shape, and 26 texture parameters with information about how pixel intensities are distributed on the image. A detailed view of each of these features will be given here.

Certainly, colour may be an attractive source of information. However, since the preparation of pollen grain samples imply the use of a stain, it is not recommended to use it. Moreover, the stain effects is not constant along time and the colour of the same sample may change.

4.1. Geometric parameters

Geometric parameters contain information about the size and the basic shape of the pollen grains. The 24 geometric parameters extracted in the systems presented in this chapter are:

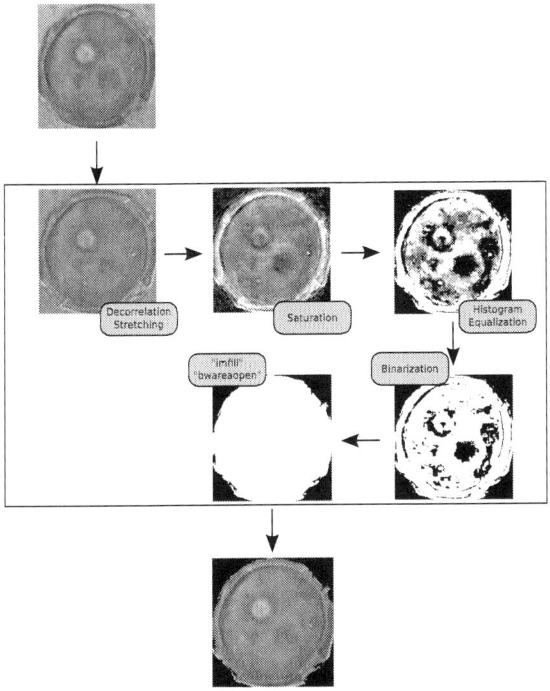

Figure 3. The result of applying each preprocessing step to a pollen grain image. Note that the sequence followed is the same as in figure 2.

- *Area*: Refers to the amount of pixels with level '1' in the pollen mask.
- *BoundingBox*: Smallest rectangle enclosing the pollen. In particular, parameters width and hight are used as:

$$\begin{aligned} BoundingBox(1) &= \text{width} \\ BoundingBox(2) &= \text{hight} \end{aligned} \tag{2}$$

- *Centroide*: Refers to the mass centre of the pollen grain. Coordinates (x, y).
- *MajorAxisLength*: Length of the major axis of the ellipse with the same second order normalized central moment of the object.
- *MinorAxisLength*: Length of the minor axis of the ellipse with the same second order normalized central moment of the object.
- *ConvexArea*: Area of the smallest convex shape enclosing the object.
- *EquivDiameter*: Diameter of the circle with the same area as the object.

$$EquivDiameter = \sqrt{\frac{4 \times Area}{\pi}} \tag{3}$$

- *Solidity*: Portion of the area of the convex region contained in the pollen.

$$Solidity = \frac{Area}{ConvexArea} \tag{4}$$

- *Perimeter*: Length of the perimeter of the mask image.
- *Extent*: Portion of the area of the bounding box contained in the pollen.

$$Extent = \frac{Area}{Area_{BoundingBox}} \tag{5}$$

- *Eccentricity*: Relation between the distance of the focus of the ellipse and the length of the principal axis.
- *WeightedCentroid*: This is a centroid computing weighted by the pixel values of the grey-scale image.
- *Shape*: Measures how circular is the pollen. Its values are in the range [0,1], where 1 corresponds to a perfect circle.

$$Shape = \frac{4 \times \pi \times Area}{Perimeter^2} \tag{6}$$

- *Thickness*: This is the number of times that the mask has to be eroded with a 3x3 square filter, until it disappears, e.i. the image gets black.
- *Box*: These are the coordinates of an inner rectangle area computed from the *BoundingBox* parameters as:

$$\begin{aligned}
Box(1) &= \frac{BoundingBox(1)}{4} \\
Box(2) &= \frac{BoundingBox(2)}{4} \\
Box(3) &= \frac{BoundingBox(1)}{2} \\
Box(4) &= \frac{BoundingBox(2)}{2}
\end{aligned} \tag{7}$$

- *Hight*: Length of the largest line enclosed in the pollen.
- *Width*: Length of the largest line enclosed in the pollen and perpendicular to *Hight*.

4.2. Texture parameters

Texture parameters provide information regarding how pixels are distributed on the image, such as contour changes or objects inside the pollen grain.

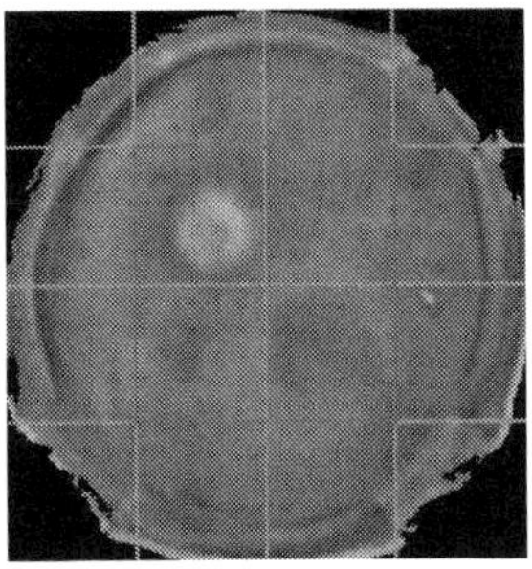

Figure 4. Example of the inner rectangle area computing from the *BoudingBox*.

The first 4 of the 26 texture parameters introduced in this section are computed using the grey level co-occurrence matrix (GLCM). This matrix gives information about the frequency of pixel value pairs combinations. In particular, the value of GLCM(i,j) is the number of times that a pixel with value 'j' sits next and at the left of a pixel with value 'i'. Figure 5 shows and example of this.

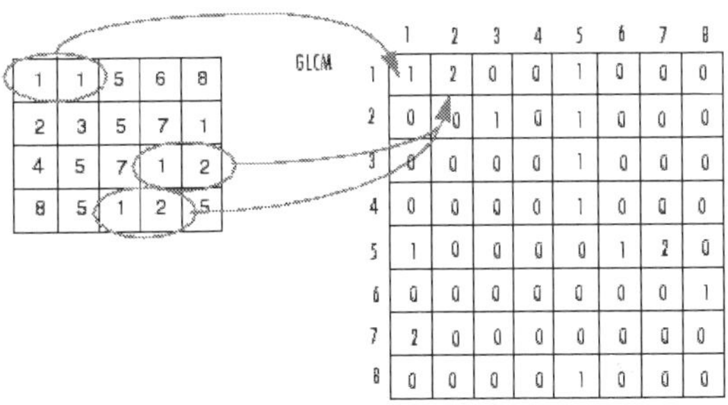

Figure 5. Example of a grey level co-occurrence matrix.

- *Contrast*: Mean intensity difference between a pixel and its neighbours. This value is computed as:

$$Contrast = \sum_{i,j} |i-j|^2 p(i,j) \tag{8}$$

- *Correlation*: Measures how must correlated it a pixel with respect to its neighbours. This value is computed as:

$$Correlation = \sum_{i,j} \frac{(i-\mu_i)(j-\mu_j)p(i,j)}{\sigma_i \sigma_j} \tag{9}$$

- *Energy*: Sum of the squared elements of the GLCM. This is:

$$Energy = \sum_{i,j} p(i,j) \tag{10}$$

- *Homogeneity*: Measures how close the distribution of objects of the GLCM are to the diagonal of the GLCM. This is:

$$Homogeneity = \sum_{i,j} \frac{p(i,j)}{1 + |i - j|} \tag{11}$$

- *Entropy*: This measure is applied to six different images derived from the original pollen grain image. These images are the the outer and inner bounding box (*BoundingBox* and *Box*) of the blue channel of the RGB representation, the saturation and the value channels of the HSV representation. A representation can be seen in figure 6.

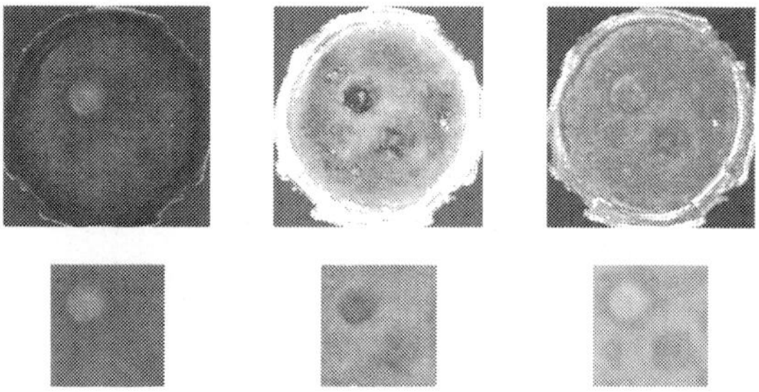

Figure 6. Images used to compute the *entropy* measures. They correspond to channels blue, saturation and value (left-right) and outer and inner bounding box (up-down).

Entropies are are scalar values representing a statistical measure of the randomness of the pixel values. Each value is computed as:

$$Entropy = \sum p \log_2 p, \tag{12}$$

where p is the histogram count of the corresponding image.

- *Fourier Descriptors*: These measures are based on the analysis of the pollen contour points, and it provides information about the pollen shape. It is worth it to mention that a major property of the *fourier descriptors* is its invariance to geometric transformations, such as rotation, scale and sift.

To compute these parameters, the complex representation of the contour $z_i = x_i + jy_i$ is used, where $i = 0, 1, 2..., N_c - 1$ with N_c the number of points of the contour. Moreover, the contour is sampled every 2 degrees. Now, the discrete Fourier transform (DFT) of z is:

$$a(u) = \frac{1}{N_c} \sum_{i=0}^{N_c - 1} z_i e^{-j2\pi u / N_b} \quad u = 0, 1, 2, ..., N_b - 1 \tag{13}$$

The resultant complex coefficients $a(u)$ are transformer in a power spectrum $|a(u)|^2$. Finally, the discrete cosine transform (DCT) is applied to reduce the dimensionality of the vector, ending up with a vector of length 5.

- *Relative areas*: This is a 5 elements vector which values correspond to the number of active pixels (pixels with value '1') after binarizing the pollen image with different thresholds. In particular, the thresholds used are 0.3, 0.4, 0.5, 0.6 and 0.7. Figure 7 shows an example of this.

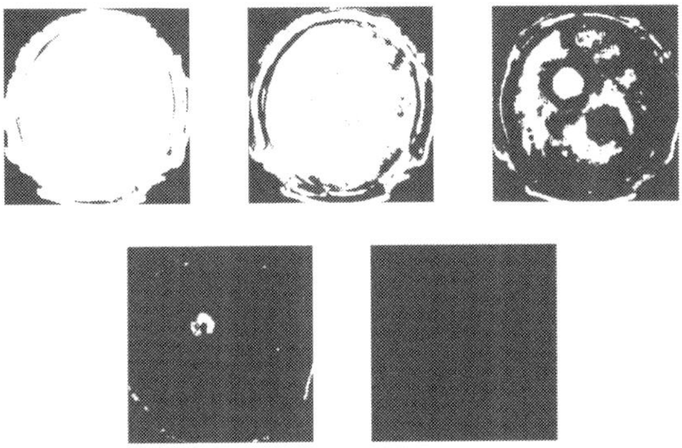

Figure 7. Results of applying thresholds 0.3, 0.4, 0.5, 0.6 and 0.7 respectively to a pollen image.

- *Relative objects*: In this case the number of objects (group of connected pixels with value '1' and surrounded of pixels with value '0') contained inside the pollen grain are counted, using an inverted and masked version of the binarized images computed the *relative areas*. See figure 8 for an example.

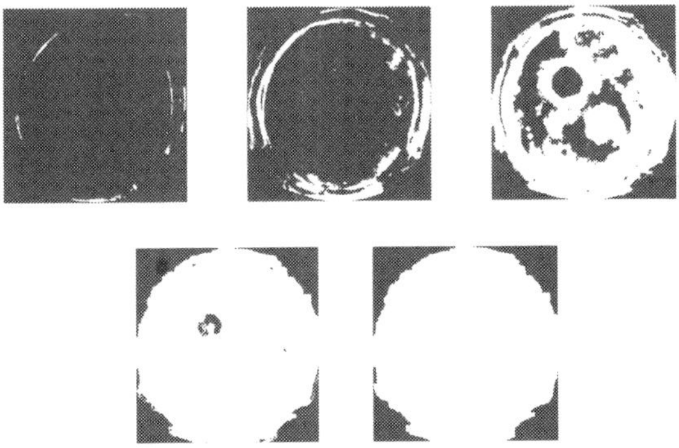

Figure 8. Images used to compute the *relative objects*.

5. Classification

Several works such as [10], [22], [11] and [13] used artificial neural networks (ANNs) as classifiers. These algorithms works as follow:

- Parameters are computed from a set of training samples.

- The computed parameters are passed to the ANN so that it gets trained. This means that the ANN automatically adjusts its parameters to solve the problem of classify the parameters in different classes.
- After the training process, a new testing parameters vector can be passed to the ANN and it will produce an output regarding the sample class.

An ANN is a mathematical model inspired in the structure and functional aspects of the biological neural networks. It could be defined as a set of simple computational elements massively interconnected following a hierarchical organization [21].

In this case, a multilayer perceptron architecture trained by a back propagation algorithm (MLP-BP) is proposed. The principal characteristic of this algorithm is its ability to solve non-lineal problems. Its architecture is composed of several layers. Each layer corresponds to a set of neurons receiving data from the previous layer and transmit data to the next layer. This layer can be divided in "input layer", "hidden layer" and "output layer" as shown in figure 9. In this case, the number of hidden layers is set to one.

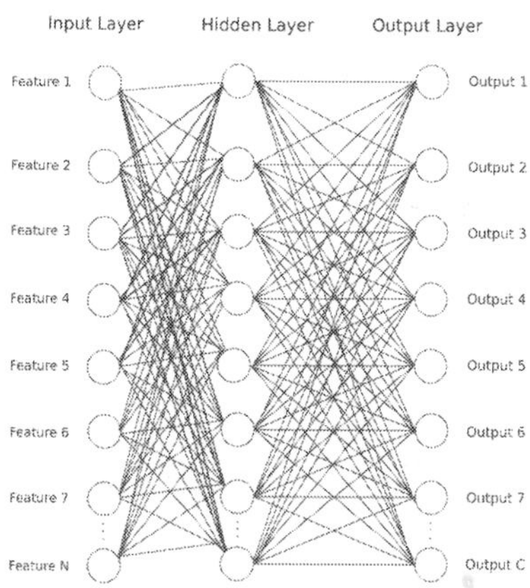

Figure 9. Architecture of the multilayer perceptron.

It is important note that the training process of the ANN contain an aleatory factor which determines the solution found. In other words, the training process does not avoid local minimums. To overcome this limitation, the proposed classifier implements 11 individual ANNs and sum their resulting scores to obtain a final response. The idea behind this fusion is that the set of computed solutions complement each other, i.e. some solutions correct the errors produced by others.

6. Experimentation methodology, results and discussion

A system were implemented in order to test the quality of the proposed approach. This system uses all the techniques introduced in previous sections (preprocessing, feature extraction and classification). This section gives the details about the database used and the experimental procedure, along with a detailed explanation of the obtained results.

6.1. Database

The database used for the experimentation contains 345 images of 17 different pollen grain classes. Images has been captured with a 2 mega-pixels digital camera connected to a microscope set to apply a 40 times zoom.

More precisely, these images correspond to 17 sub-genders and species of 11 different families of tropical honey plants situated in Costa Rica (Central America). Table 1 shows the exact information about family, gender and specie.

Class	Family	Gender	Specie	Samples
1	Asteraceae	Baltimora	Recta	24
2	Asteraceae	Tridats	Procumbels	47
3	Asteraceae	Critonia	Morifolia	21
4	Asteraceae	Elephentopus	Mollis	17
5	Bombacaceae	Bombacptis	Quinata	18
6	Caesalpinaceae	Cassea	Gradis	35
7	Combretaceae	Combretum	Fructicosum	25
8	Comvulvulaceae	Ipomea	Batatas	15
9	Fabaceae	Aeschynomene	Sensitiva	24
10	Fabaceae	Cassia	Fistula	36
11	Fabaceae	Miroespermyn	Frutesens	18
12	Fabaceae	Enterolobium	Cyclocarpun	18
13	Myrsinaceae	Ardisia	Revoluta	18
14	Malpighiaceae	Bunchosin	Cornifolia	36
15	Saphindaceae	Cardioesperman	Grandiflorus	20
16	Saphindaceae	Melicocca	Bijuga	26
17	Verbenaceae	Lantana	Camara	25

Table 1. The exact information about family, gender and specie of the 17 classes included in the DDBB used. The last column expresses the number of samples of pollen grains extracted from the database.

Applying the pollen grain extraction algorithm introduced in section 3, a total of 423 pollen images distributed on all species were obtained. The number of samples extracted for each sample was greater than one. This was possible thanks to images such as that shown in figure 10 where more than one pollen grain could be extracted. Figure 11 shows a sample of each pollen specie included in the DDBB.

6.2. Experiments

First, remember from section 5 that the design of the classifier include 30 ANNs fused at the score level. Thus, the number of hidden units on the ANNs had to be specified. To do so, a set of experiments with different configurations were executed to find the optimal value. To obtain a valid measure of the performance of the system, 30 iterations of a hold 50% out cross-validation procedure was executed. Results will be shown and discussed in sections 6.3

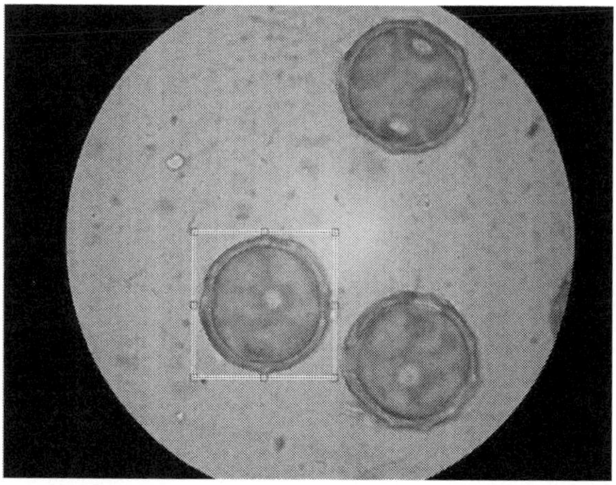

Figure 10. Database image sample. Note that more than one pollen grain can be extracted from this image.

Class	Sample	Nº	Class	Sample	Nº	Class	Sample	Nº
1	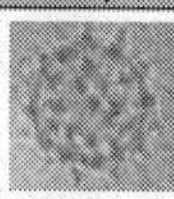	24	7		25	13		18
2	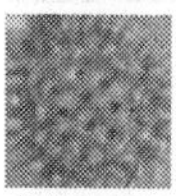	47	8		15	14		36
3		21	9		24	15		20
4		17	10	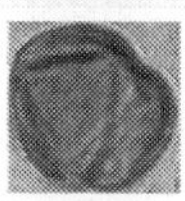	36	16		26
5		18	11	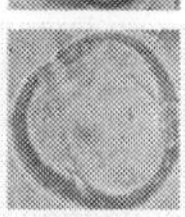	18	17		25
6		35	12		18			

Figure 11. Samples of the 17 different pollen grain species.

and 6.4 respectively. For now, it is enough to note that the optimal value were found with a 30 neurons hidden layer.

Thus, using this optimal configuration of the ANNs, further experiments were executed to evaluate the performance of the designed system. In this case, 30 iterations of a K-folds cross-validation procedure were applied with values of 'K' equal 3, 5, 7 and 10.

Note that the set of all experimental procedures (hold-50%-out and 3, 5, 7 and 10 folds) are based on divisions of the database in disjoint training and test sets. Moreover, this experiments can be seen as using different proportions of the database of training, i.e. using a different number of samples for training. In particular, the proportions of samples used for training are 1/2, 2/3, 4/5, 6/7 and 9/10 respectively.

6.3. Results

It is important to note that every experiment was repeated 30 times in order to obtain a valid measure of the system's performance. Therefore, results are given in terms of mean percentage and standard deviation (mean % and std).

The first experiment tested different configurations of the ANNs. Figure 12 shows the progress of the success rate when the number of units in the hidden layer increased from 10 to 150. A highest rate of 90.54% of success rate were obtained with 30 units (see table 2).

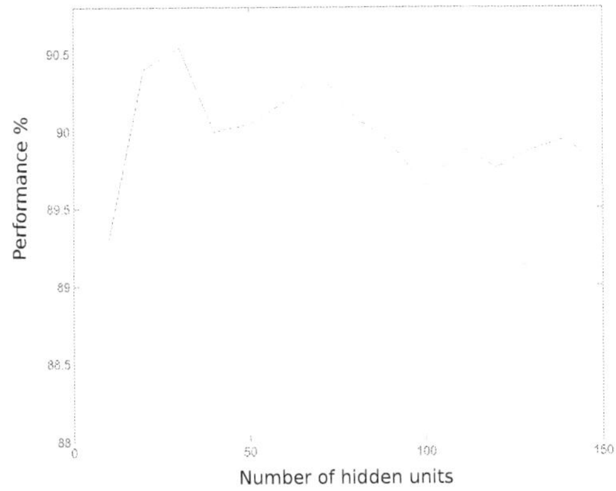

Figure 12. Performance progress for different number of units in the hidden layer of the ANN.

A second group of experiments aimed to measure the system's performance with different number of samples for training. Table 3 shows the results obtained for 3, 5, 7 and 10 folds. Note that the success rate increased with the number of training samples (from 90.54% to 92.81%), while the std decreased (from 1.29 to 0.74).

6.4. Discussion

It can be argued that the number of hidden units of the ANNs could be further optimized executing a finer search around the point found. However, based on the similar accuracy measures obtained between 10 and 80 units and stds higher than the range of accuracy

Neurons in the hidden unit	Mean % ± std
10	89.29% ± 2.11
20	90.40% ± 1.69
30	**90.54% ± 1.29**
40	90.00% ± 1.66
50	90.05% ± 1.78
60	90.19% ± 1.33
70	90.38% ± 1.52
80	90.09% ± 1.37
90	89.92% ± 1.42
100	89.64% ± 1.49
110	89.90% ± 1.34
120	89.76% ± 1.55
130	89.87% ± 1.37
140	89.95% ± 1.42
150	89.72% ± 1.64

Table 2. Performance progress for different number of units in the hidden layer of the ANN.

Experiment	Mean % ± std
Hold-50%-Out	90.54% ± 1.29
3 *k*-folds	91.40% ± 1.05
5 *k*-folds	92.38% ± 0.75
7 *k*-folds	92.43% ± 0.82
10 *k*-folds	92.81% ± 0.74

Table 3. Results for 30 iterations of different experiments.

percentages, paying the cost of running a finer search for a minimal increment of performance was not worth it. Therefore, 30 units were chosen as the optimal point.

On the other hand, the results obtained for the second round of experiments show an increasing in both system's performance and stability. This seems to indicate that the performance of the system may increase with a bigger training database.

7. Conclusions

This chapter has introduced the problem of automatic pollen grain classification, which is vital for biologists and flora researches among others. As pointed out in section 3, the task of automatically detecting the pollen grains from samples is a complex problem itself and fall beyond the scope of this chapter. Thus, a semi-automatic algorithm for pollen extraction was explored instead,

The chapter mainly focused its attention in giving a fair amount of both geometric and texture parameters. Moreover, the extraction of this parameters relied on the good work performed by the preprocessing block during the pollen's perimeter definition.

Finally, these parameters were tested implementing a completed system. In particular, the system used the semi-automatic pollen detection and preprocessing algorithms introduced, along with the mentioned feature extraction techniques and a classifier based on the fusion of

11 ANNs at the score level. The system was tested executing a number of experiments using different hold-out and k-folds cross-validation procedures. The results showed success rates between 90.54% and 92,81%, pointing out the quality of the presented parameters for pollen grain classification. Moreover, these results improve those achieved by other authors such as [10], [22], [11] and [13], even though the number of classified species was significantly larger.

Acknowledgements

This work has been supported by Spanish Government, in particular by "Agencia Española de Cooperación Internacional para el Desarrollo" under funds from D/027406/09 for 2010, D/033858/10 for 2011, and A1/039531/11 for 2012. To M.Sc. Luis A. Sánchez Chaves, Tropical Bee Research Center (CINAT) at Universidad Nacional de Costa Rica, for provide the database images and palynology knowledge.

Author details

del Pozo-Baños Marcos, Ticay-Rivas Jaime R., Cabrera-Falcón Jousé, Travieso Carlos M., Pérez Santiago T. and Alonso Jesús B.
Signal and Communication Department Institute for Technological Development and Innovation in Communications University of Las Palmas de Gran Canaria IdeTIC-ULPGC. Las Palmas de Gran Canaria, Spain

Arroyo Jorge, Sánchez-Chavez Luis and Ramírez-Bogantes Melvin
Escuela de Matemáticas, Universidad Nacional, Costa Rica

8. References

[1] Definition of "palynology" by the Oxford Dictionaies site (http://oxford dictionaries.com/definition/palynology). Visited last time on April 2011.

[2] Plants under pressure: a global assessment. The first report of the IUCN Sampled Red List Index for Plants. Royal Botanic Gardens, Kew, UK (2010)

[3] Sytnik KM (2010) Preservation of biological diversity: Top-priority tasks of society and state. Ukrainian Journal of Physical Optics 11(suppl. 1), S2-S10

[4] Stillman EC, Flenley JR (1996) The Needs and Prospects for Automation in Palynology. Quaternary Science Reviews 15, 1-5.

[5] Flenley JR (1968) The problem of pollen recognition. In: Problems in Picture Interpretation, (ed. M. B. Clowes and J. P. Penny), pp. 141-145. CSIRO, Canberra.

[6] Flenley JR (1990) Some prospects for palynology in the South- West Pacific Region. Massey University Faculty of Social Sciences Occasional Papers, No. 1.

[7] Langford M, Taylor GE, Flenley JR (1990) Computerized identification of pollen grains by texture analysis. Review of Palaeobotany and Palynology, Volume 64, Issues 1-4, 23 October 1990, pp.197-203.

[8] Treloar WJ (1992) Digital image processing techniques and their application to the automation of palynology. Ph.D. Thes., University of Hull, Hull UK.

[9] Treloar WJ and Flenley JR (1996) An investigation into the potential of light microscopy for the automatic identification of pollen grains by the analysis of their surface texture. Poster Presentation, 9th International Palynological Congress, Houston TX.

[10] Li P and Flenley J (199) Pollen texture identification using neural networks. Grana 38(1), pp.59-64.

[11] Rodriguez-Damian M, Cernadas E, Formella A, Sa-Otero R (2004) Pollen classification using brightness-based and shape-based descriptors. Proceedings of the 17th International Conference on Pattern Recognition, ICPR 2004, August 23-26, vol.2, pp.23-26.

[12] Zhang Y, Fountain DW, Hodgson RM, Flenley JR and Gunetileke S (2004) Towards automation of palynology 3: pollen pattern recognition using Gabor transforms and digital moments. Journals of quaternary science, vol.19, 2004, pp.763-768.

[13] Rodriguez-Damian M, Cernadas E, Formella A, Fernandez-Delgado M and De Sa-Otero P (2005) Automatic detection and classification of grains of pollen based on shape and texture. IEEE Trans. on Systems, Man, and Cybernetics, Part C: Applications and Reviews 36(4), 531-542.

[14] Allen GP, Hodgson RM, Marsland SR and Flenley JR (2008) Machine vision for automated optical recognition and classification of pollen grains or other singulated microscopic objects. Mechatronics and Machine Vision in Practice, 15th International Conference on , 2008, pp.221-226.

[15] Boucher A and Thonnat M (2002) Object recognition from 3D blurred images, 16th International Conference on Pattern Recognition, vol.1, pp.800-803.

[16] Ronneberger, Burkhardt O, Schultz H, General E (2002) Purpose object recognition in 3D volume data sets using gray-scale invariants-classification of airborne pollen-grains recorded with a confocal laser scanning microscope. Proceedings on Pattern Recognition, 2002. 16th International Conference, vol.2, pp.290- 295.

[17] Campbell NA (1996) The decorrelation stretch transformation. INT. J. Remote Sensing, vol.17, no.10, pp.1939-1949.

[18] Agoston MK (2005) Computer graphics and geometric modeling: implementation and algorithms. Springer Verlag.

[19] Baxes GA (1994) Digital Image Processing: Principles and Applications. John Wiley & Sons.

[20] Matlab documentation at Mathworks. http://www.mathworks.com. Visited last time on April 2011.

[21] Hopfield JJ (1982) Neural networks and physical systems with emergent collective computational abilities. Proc. NatL Acad. Sci. USA Biophysics, vol.79, pp.2554-2558.

[22] France I, Duller A, Duller G and Lamb H (2000) A new approach to automated pollen analysis. Quaternary Science Reviews 18, pp.536-537.